ADVANCES IN IMAGING AND ELECTRON PHYSICS

VOLUME 137

DOGMA OF THE CONTINUUM AND THE CALCULUS OF FINITE DIFFERENCES IN QUANTUM PHYSICS

Advances in
Imaging and Electron Physics

Dogma of the Continuum and the Calculus of Finite Differences in Quantum Physics

HENNING F. HARMUTH
Retired, The Catholic University of America
Washington, DC, USA

BEATE MEFFERT
Humboldt-Universität
Berlin, Germany

VOLUME 137

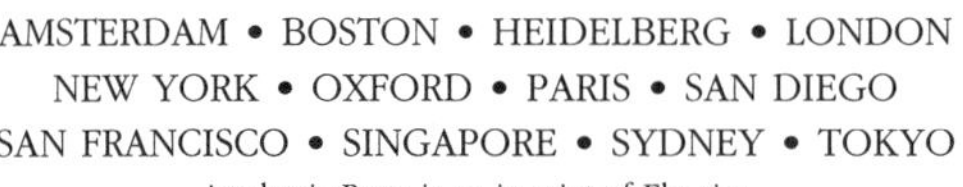
AMSTERDAM • BOSTON • HEIDELBERG • LONDON
NEW YORK • OXFORD • PARIS • SAN DIEGO
SAN FRANCISCO • SINGAPORE • SYDNEY • TOKYO
Academic Press is an imprint of Elsevier

Elsevier Academic Press
525 B Street, Suite 1900, San Diego, California 92101-4495, USA
84 Theobald's Road, London WC1X 8RR, UK

This book is printed on acid-free paper. 

For all information on all Elsevier Academic Press publications visit our Web site at www.books.elsevier.com

ISBN-13: 978-0-12-014779-3
ISBN-10: 0-12-014779-3

PRINTED IN THE UNITED STATES OF AMERICA
05 06 07 08 09 9 8 7 6 5 4 3 2 1

CONTENTS

Equations are numbered consecutively within each of Sections 1.1 to 6.10. Reference to an equation in a different section is made by writing the number of the section in front of the number of the equation, e.g., Eq. (1.1-45) for Eq. (45) in Section 1.1.

Illustrations are numbered consecutively within each section, with the number of the section given first, e.g., Figure 1.2-1.

References are listed by the name of the author(s), the year of publication, and a lowercase Latin letter if more than one reference by the same author(s) is listed for that year.

3 Inhomogeneous Difference Equation

4 Klein-Gordon Difference Equation for Small Distances

5 Difference Equation in Spherical Coordinates

6 Appendix

FUTURE CONTRIBUTIONS

G. Abbate
New developments in liquid-crystal-based photonic devices

S. Ando
Gradient operators and edge and corner detection

A. Asif
Applications of noncausal Gauss-Markov random processes in multidimensional image processing

C. Beeli
Structure and microscopy of quasicrystals

M. Bianchini, F. Scarselli, and L. Sarti
Recursive neural networks and object detection in images

G. Borgefors
Distance transforms

A. Bottino
Retrieval of shape from silhouette

A. Buchau
Boundary element or integral equation methods for static and time-dependent problems

B. Buchberger
Gröbner bases

J. Caulfield
Optics and information sciences

C. Cervellera and M. Muselli
The discrepancy-based approach to neural network learning

T. Cremer
Neutron microscopy

H. Delingette
Surface reconstruction based on simplex meshes

A. R. Faruqi
Direct detection devices for electron microscopy

R. G. Forbes
Liquid metal ion sources

J. Y.-l. Forrest
Grey systems and grey information

E. Förster and F. N. Chukhovsky
X-ray optics

A. Fox
The critical-voltage effect

L. Godo & V. Torra
Aggregation operators

A. Gölzhäuser
Recent advances in electron holography with point sources

K. Hayashi
X-ray holography

M. I. Herrera
The development of electron microscopy in Spain

D. Hitz
Recent progress on HF ECR ion sources

D. P. Huijsmans and N. Sebe
Ranking metrics and evaluation measures

K. Ishizuka
Contrast transfer and crystal images

J. Isenberg
Imaging IR-techniques for the characterization of solar cells

K. Jensen
Field-emission source mechanisms

L. Kipp
Photon sieves

G. Kögel
Positron microscopy

T. Kohashi
Spin-polarized scanning electron microscopy

W. Krakow
Sideband imaging

R. Leitgeb
Fourier domain and time domain optical coherence tomography

B. Lencová
Modern developments in electron optical calculations

R. Lenz (vol. 138)
Aspects of colour image processing

W. Lodwick
Interval analysis and fuzzy possibility theory

R. Lukac
Weighted directional filters and colour imaging

L. Macaire, N. Vandenbroucke, and J.-G. Postaire
Color spaces and segmentation

M. Matsuya
Calculation of aberration coefficients using Lie algebra

S. McVitie
Microscopy of magnetic specimens

L. Mugnier, A. Blanc, and J. Idier
Phase diversity

K. Nagayama (vol. 138)
Electron phase microscopy

M. A. O’Kccfc
Electron image simulation

J. Orloff and X. Liu (vol. 138)
Optics of a gas field-ionization source

D. Oulton and H. Owens
Colorimetric imaging

N. Papamarkos and A. Kesidis
The inverse Hough transform

K. S. Pedersen, A. Lee, and M. Nielsen
The scale-space properties of natural images

E. Rau
Energy analysers for electron microscopes

H. Rauch
The wave-particle dualism

E. Recami
Superluminal solutions to wave equations

J. Řehácek, Z. Hradil, J. Peřina, S. Pascazio, P. Facchi, and M. Zawisky
Neutron imaging and sensing of physical fields

G. Ritter
Lattice-based artifical neural networks

J.-F. Rivest
Complex morphology

G. Schmahl
X-ray microscopy

G. Schönhense, C. M. Schneider, and S. A. Nepijko
Time-resolved photoemission electron microscopy

F. Shih
General sweep mathematical morphology

R. Shimizu, T. Ikuta, and Y. Takai
Defocus image modulation processing in real time

S. Shirai
CRT gun design methods

N. Silvis-Cividjian and C. W. Hagen
Electron-beam-induced deposition

T. Soma
Focus-deflection systems and their applications

Q. F. Sugon
Geometrical optics in terms of Clifford algebra

W. Szmaja
Recent developments in the imaging of magnetic domains

I. Talmon
Study of complex fluids by transmission electron microscopy

I. J. Taneja (vol. 138)
Divergence measures and their applications

M. E. Testorf and M. Fiddy
Imaging from scattered electromagnetic fields, investigations into an unsolved problem

M. Tonouchi
Terahertz radiation imaging

N. M. Towghi
I_p norm optimal filters

Y. Uchikawa
Electron gun optics

K. Vaeth and G. Rajeswaran
Organic light-emitting arrays

J. Valdés (vol. 138)
Units and measures, the future of the SI

D. Walsh (vol. 138)
The importance-sampling Hough transform

G. G. Walter
Recent studies on prolate spheroidal wave functions

B. Yazici
Stochastic deconvolution over groups

FOREWORD

The ancient Greeks did not distinguish between mathematics as a science of the thinkable and physics as a science of the observable. This cast long shadows over the development of physics. As a first example we cite the dogma of the circle. Ptolemy expressed it as follows:

> ... we believe it is the necessary purpose and aim of the mathematician to show forth all the appearances of the heavens as products of regular and circular motion. (Ptolemy, 1952, Almagest, Book III, 1; p. 83, §2)

It is generally assumed that Kepler ended the dogma of the circle, but this is true only for astronomy. The superposition of deferents and epicycles of Ptolemy and Copernicus developed into the Fourier series in complex notation. We meet the old circle under the new name *exponential function* $e^{i\omega t}$ *in the complex plane*. Another circle in disguise is *the character group* $\{e^{iyx}\}$ *of the topologic group of real numbers*. The word *the* in the title provides the connection with Greek thinking. Mathematics justifies only the name *a character group*. ...

A second long shadow was cast by Euclid's geometry. Navigators had been using spherical trigonometry since about 1500 to chart their course across the oceans. But the greatest mathematicians struggled three centuries later with the question of whether Euclid's geometry was the only possible one.

Here we are concerned with a third long shadow, the dogma of the continuum of physical space and time. It can be traced back to the Eleatic school of the Greeks in southern Italy. Zeno of Elea (c. 490–c. 430 BCE) advanced the paradox of the race between Achilles and the turtle as well as that of the arrow that does not fly, to refute the continuum or the infinite divisibility of distances in space and time. Zeno's paradoxes were in turn refuted by Aristotle in his *Physica*. Aristotle's arguments in favor of a mathematical continuum for the physical space and time were so convincing that they were questioned rarely since. The physics of space and time became a branch of mathematics.

Newton demonstrated the perception of physics as a branch of mathematics when he wrote

> Absolute, true and mathematical time, of itself, and from its nature, flows equably without connection with anything external, ... (Newton, 1971, p. 6)

Newton and Leibniz carried the concept of infinite divisibility from the *denumerable infinite* of the Greeks to the *nondenumerable infinite* of differential calculus.

The development of non-Euclidean geometries and the experimental verification of the acoustic Doppler effect changed our thinking about time and space to the concepts used in the special and the general theory of relativity, and beyond. A quotation of Einstein from his later years shows this development:

> But to connect every instant of time with a number, by the use of a clock, to regard time as an one-dimensional continuum, is already an invention. So also are the concepts of Euclidean and non-Euclidean geometry, and our space understood as a three-dimensional continuum. (Einstein and Infeld, 1938, p. 311)

The straightforward proof of a continuum of physical space and time would be the observation of events at two spatial points x and $x + dx$ or two times t and $t + dt$. What is physically possible are observations at x and $x + \Delta x$ or t and $t + \Delta t$, where Δx and Δt may be very small but must be finite. Any finite interval Δx, Δt can be divided into nondenumerably many subintervals dx, dt, which means we are a long way from a mathematical continuum.

If we want to use finite differences Δx, Δt instead of differentials dx, dt we must use the calculus of finite differences instead of the differential calculus. This is a true generalization since no fixed values for Δx, Δt are specified at the beginning of the calculation. When solving for the eigenfunctions of a difference equation in relativistic quantum physics we typically get well-behaved functions if the spatial resolution Δx is large enough, but sequences of random numbers for too small values of Δx. This is how the calculation represents the Compton effect. The theory goes beyond Heisenberg's uncertainty relation since it puts a lower limit on Δx rather than on the product $\Delta x \Delta p$.

Consider elementary particles within the framework of differential calculus. We must match the physical situation to the mathematical method and we do so by defining elementary particles to be "point-like" to avoid giving them any spatial features. Using the calculus of finite differences we must demand only that an elementary particle is smaller than an arbitrarily small but finite distance Δx to avoid any observable spatial feature. The difference theory clearly offers the better choice.

A particle with mass m_0 can become an antiparticle with mass $-m_0$ without a quantum jump in the difference theory. Under certain conditions a finite spatial resolution Δx permits such a transition without violating any physical laws.

Generally, the theorem of Hölder stated in 1887 that the gamma function can be defined by a simple difference equation $\Gamma(x + 1) = x\Gamma(x)$ but by no algebraic differential equation. This implies that differential and difference equations define different classes of functions.

The calculus of finite differences predates the differential calculus since differentials are obtained as limits of finite differences. The success of differential calculus in science and engineering stimulated its enormous development. There were no comparable applications for the calculus of finite differences before its usefulness for relativistic quantum physics was discovered and there was thus little development. Our Bibliography lists only 10 mathematical books on the calculus of finite differences published in the twentieth century[1].

We want to thank Humboldt-Universität of Berlin for help with computer and library services.

Henning F. Harmuth

[1]Our search was limited to books in English, French, German, Russian, and Spanish. We would be grateful for information about books in Chinese or Japanese.

List of Frequently Used Symbols

$\mathbf{A}_e$	As/m	electric vector potential
$\mathbf{A}_m$	Vs/m	magnetic vector potential
$\mathbf{B}$	Vs/m^2	magnetic flux density
b_κ, b_κ^*	–	Eq. (2.5-28)
b_κ^-, b_κ^+	–	Eq. (2.5-29)
c	m/s	299 792 458; velocity of light (definition)
$D_{\kappa i}(\theta)$	–	Eqs. (6.3-17)–(6.3-21)
$\hat{D}_{\kappa i}(\theta)$	–	Eqs. (6.3-25)–(6.3-28)
$\tilde{D}_{\kappa i}(\theta)$	–	Eqs. (6.4-6)–(6.4-9)
$\mathbf{D}$	As/m^2	electric flux density
$d(\kappa)$	–	Eq. (2.5-18)
$d_{\kappa i}$, $d_{\kappa i}(\theta)$	–	Eqs. (6.3-47)–(6.3-50)
$\mathbf{E}$, E	V/m	electric field strength
E	VAs	energy
e	As	electric charge
$F(\zeta)$	–	Eqs. (2.3-7), (2.3-15)
$\mathbf{g}_e$	A/m^2	electric current density
$\mathbf{g}_m$	V/m^2	magnetic current density
$G_1(\zeta, \theta)$	–	Eq. (3.2-4)
$G_2(\zeta, \theta)$	–	Eq. (4.2-3)
$G_{c\kappa}(\theta)$	–	Eq. (3.2-32)
$G_{s\kappa}(\theta)$	–	Eq. (3.2-33)
$\mathbf{H}$, H	A/m	magnetic field strength
$H_{c\kappa}(\theta)$	–	Eq. (6.3-29)
$H_{s\kappa}(\theta)$	–	Eq. (6.3-1)
$\mathcal{H}$	–	Hamilton function
h	Js	$6.626\ 075\ 5 \times 10^{-34}$, Planck's constant
$\hbar = h/2\pi$	Js	$1.054\ 572\ 7 \times 10^{-34}$
I_T	–	Eq. (2.4-29)
m_0	kg	rest mass
N	–	$T/\Delta t$, Eq. (2.2-6)
$p_1(\zeta, \theta)$	–	Eq. (4.2-7)
p_C	–	Eq. (4.4-10)
$p_N(\zeta, \theta)$	–	Eq. (4.2-5)
Q	–	Eq. (1.1-45)
$\tilde{r}$	–	$\Delta\tilde{r}$, Eq. (5.5-2)
$S_\kappa(\theta)$	–	Eqs. (3.2-20), (3.2-54)

(Continued)

s	V/Am	magnetic conductivity
T	s	arbitrarily large but finite time interval
$T_\kappa(\theta)$	–	Eqs. (3.2-20), (3.2-55)
t	s	time variable
Δt	s	arbitrarily small but finite time interval
U	VAs	Eq. (2.5-2)
$\hat{u}(\zeta, \theta)$	–	Eqs. (3.2-8), (3.2-10)–(3.2-13)
V	m/s	velocity
$\hat{v}(\zeta, \theta)$	–	Eqs. (3.2-20), (3.2-21)
$Z = \mu/c$	V/A	376.730 314; wave impedance of empty space
Z	–	1, 2, ...; charge number
α	–	$Ze^2/2h7.297\ 535 \times 10^{-3}$, Eq. (1.1-45)
α_e	–	$ZecA_e/m_0c^2$, Eq. (1.1-45)
$\tilde{\alpha}$	–	Eq. (5.5-2)
β_κ	–	Eqs. (2.4-11), (2.4-13), (2.4-14)
$\tilde{\gamma}$	–	$4\pi Z\alpha$, Eq. (5.5-2)
$\tilde{\delta}$	–	Eq. (5.5-2)
$\epsilon = 1/Zc$	As/Vm	$1/\mu c^2$; permittivity
$\tilde{\Delta}$	–	symbol for difference quotient: $\tilde{\Delta}F/\tilde{\Delta}x$, Eq. (1.2-1)
$\tilde{\Delta}_l$	–	left difference quotient, Eq. (1.2-5)
$\tilde{\Delta}_r$	–	right difference quotient, Eq. (1.2-4)
Δ	–	symbol for finite difference: $x + \Delta x$
ζ	–	$x_j/c\Delta t$, normalized distance; Eqs. (2.2-6), (2.3-1)
θ	–	$t/\Delta t$, normalized time; Eq. (2.2-6)
ι	–	Eq. (2.3-18)
κ_0	–	Eq. (2.4-32)
λ_1	–	$ec\Delta t A_{m0x}/\hbar$; Eqs. (2.3-2), (3.2-36)
λ_2	–	Eqs. (2.3-2), (3.2-36)
λ_3	–	ϕ_{e0}/cA_{m0x}; Eqs. (2.3-2), (3.2-37)
λ_C	m	$h/m_0c = 8.89 \times 10^{-15}$ for π^+ and π^-, Eq. (1.1-45)
$\tilde{\lambda}$	–	Eq. (5.5-2)
$\mu = Z/c$	Vs/Am	$4\pi \times 10^{-7}$; permeability
ρ_e	As/m^3	electric charge density
ρ_m	Vs/m^3	magnetic charge density
ρ_r	–	$\tilde{\alpha}r$, Eq. (5.5-1)
ρ_κ	–	constant, Eq. (2.3-25)
σ	A/Vm	electric conductivity, Eq. (1.1-7)

ϕ_{e}	V	electric scalar potential
ϕ_{m}	A	magnetic scalar potential
φ_{κ}	–	Eq. (2.3-38)
Ψ_0, Ψ_1	–	Eq. (2.1-38)
Ψ_{x}	–	Eq. (2.1-8)
$\Psi_{\mathrm{x}0}$, $\Psi_{\mathrm{x}1}$	–	Eq. (2.1-8)
$\Psi_{\mathrm{x}0x_j}$	–	Eq. (2.1-25), (2.2-26), (2.1-37)
$\Psi_{\mathrm{x}1x_j}$	–	Eq. (2.1-37)

1 Introduction

1.1 Modified Maxwell Equations and Basic Relations

During the last century an enormous number of books and journal articles has been published on solutions of Maxwell's equations. Practically all of them were about sinusoidal waves. These solutions were outside the conservation law of energy, since periodic sinusoidal waves, infinitely extended, must have infinite energy unless their power is zero. We have found only two papers that used Gaussian pulses rather than periodic sinusoidal functions to keep the energy finite (King and Harrison 1968, King 1993). The Gaussian pulse still starts at minus infinity, which prevents its use as a signal. A signal has to be zero before a certain finite time. This need was recognized by mathematicians who coined the term *causal functions* for functions that are zero before a finite time. Since signals—like any producible or observable electromagnetic wave—must have a finite energy, they are mathematically represented by quadratically integrable causal functions or *signal solutions*.

The lack of signal solutions of Maxwell's equations must always have been a problem for its serious students. The best equations of the theory of electromagnetism did generally not provide the only solutions that could be produced or observed experimentally. Stratton (1941) tried to overcome this problem but his mathematical derivations were not well received and we do not find them in the textbooks published later.

From 1986 on it was recognized that Maxwell's equations could generally not have solutions that satisfied the causality law, which explained the absence of signal solutions (Harmuth 1986a,b,c; Hillion 1991, 1992a,b; 1993). The addition of a magnetic (dipole) current density term corrected this shortcoming[1] (Harmuth 1986a,b,c; Anastasovski et al. 2001). Rotating magnetic dipoles produce magnetic dipole currents just as rotating electric dipoles—e.g. in a material like barium-titanate—produce electric dipole currents.

Maxwell's equations modified by a magnetic dipole current density can be written in the following form with international units in a coordinate system at rest:

[1]Publications of Lehnert (1995, 1996) as well as Lehnert and Roy (1998) call for an "extended electromagnetic theory" that keeps Maxwell's $\mathbf{g}_\mathrm{m} = 0$ in the following Eq.(2) but replaces ρ_e [As/m^3] by σ/c [(A/m^2)(s/m)] in Eq.(3). For an overview see Barrett (1993).

ISSN 1076-5670/05
DOI: 10.1016/S1076-5670(05)37001-7

$$\operatorname{curl} \mathbf{H} = \frac{\partial \mathbf{D}}{\partial t} + \mathbf{g}_{\mathrm{e}} \tag{1}$$

$$-\operatorname{curl} \mathbf{E} = \frac{\partial \mathbf{B}}{\partial t} + \mathbf{g}_{\mathrm{m}} \tag{2}$$

$$\operatorname{div} \mathbf{D} = \rho_{\mathrm{e}} \tag{3}$$

$$\operatorname{div} \mathbf{B} = 0 \quad \text{or} \ \operatorname{div} \mathbf{B} = \rho_{\mathrm{m}} \tag{4}$$

We use here an old-fashioned notation but the problem of Maxwell's equations with the causality law was found with this notation. It may well be that the physical meaning of equations is easier to grasp with this old notation. The operators ∇ and $\Box$ will be used when mathematical compactness is more important than physical meaning. The symbols $\mathbf{E}$ and $\mathbf{H}$ stand for the electric and magnetic field strength, $\mathbf{D}$ and $\mathbf{B}$ for the electric and magnetic flux density, $\mathbf{g}_{\mathrm{e}}$ and $\mathbf{g}_{\mathrm{m}}$ for the electric and magnetic current density, ρ_{e} and ρ_{m} for the electric and a possible magnetic charge density. The existence of magnetic charges is not accepted as confirmed experimentally, but there are serious theoretical arguments for their existence. The magnetic dipole current density $\mathbf{g}_{\mathrm{m}}$ does not depend on the existence of magnetic charges. The existence of magnetic dipoles is not disputed and rotating dipoles create dipole currents. The electric current density $\mathbf{g}_{\mathrm{e}}$ has always stood for electric monopole current densities requiring charges, but also for electric dipole and higher order multipole current densities whose total charge is zero. Without dipole currents no electric current could flow through a capacitor whose dielectric is an insulator for electric monopole currents.

The use of the term signal solution should not mislead one to believe that the modified Maxwell equations are only of interest in low-energy effects typically associated with information transmission. The electromagnetic pulse of an atomic bomb explosion or the electromagnetic radiation produced by a supernova explosion fit the definition of a signal solution too. The electric and magnetic field strengths produced in these examples are generally not defined by Maxwell's equation but they are defined by the modified Maxwell equations.

Equations (1) to (4) are augmented by *constitutive equations* that connect $\mathbf{D}$ with $\mathbf{E}$, $\mathbf{B}$ with $\mathbf{H}$, $\mathbf{g}_{\mathrm{e}}$ with $\mathbf{E}$, and $\mathbf{g}_{\mathrm{m}}$ with $\mathbf{H}$. In the simplest case this connection is provided by scalar constants called permittivity ϵ, permeability μ, electric conductivity σ, and magnetic conductivity s. The electric and magnetic conductivities may be *monopole current conductivities* as well as *dipole* or higher order *multipole current conductivities*:

$$\mathbf{D} = \epsilon \mathbf{E} \tag{5}$$

$$\mathbf{B} = \mu \mathbf{H} \tag{6}$$

$$\mathbf{g}_{\mathrm{e}} = \sigma \mathbf{E} \tag{7}$$

$$\mathbf{g}_{\mathrm{m}} = s \mathbf{H} \tag{8}$$

In more complicated cases ϵ, μ, σ, and s may vary with location, time, and direction, which requires time-variable tensors for their representation. In still more complicated cases Eqs.(5) to (8) may be replaced by partial differential equations. For sinusoidal time variation of $\mathbf{E}$, $\mathbf{H}$, $\mathbf{D}$, $\mathbf{B}$, $\mathbf{g}_\mathrm{e}$, and $\mathbf{g}_\mathrm{m}$ one may use functions of frequency $\epsilon(\omega)$, $\mu(\omega)$, $\sigma(\omega)$, and $s(\omega)$ but this takes one beyond the conservation law of energy and the causality law. The widespread use of $\epsilon(\omega)$, $\mu(\omega)$, and $\sigma(\omega)$ demonstrates our ability to derive useful results from wrong theories.

A number of basic relations derived from the modified Maxwell equations will be needed. They are listed here without derivation. References for their derivation are given.

The electric and magnetic field strength in Maxwell's equations are related to a vector potential $\mathbf{A}_\mathrm{m}$ and a scalar potential ϕ_e:

$$\mathbf{E} = -\frac{\partial \mathbf{A}_\mathrm{m}}{\partial t} - \operatorname{grad} \phi_\mathrm{e} \tag{9}$$

$$\mathbf{H} = \frac{c}{Z} \operatorname{curl} \mathbf{A}_\mathrm{m} \tag{10}$$

For the modified Maxwell equations we have to add a vector potential $\mathbf{A}_\mathrm{e}$ and a scalar potential ϕ_m. Equations (9) and (10) are replaced by the following relations[2]:

$$\mathbf{E} = -Zc \operatorname{curl} \mathbf{A}_\mathrm{e} - \frac{\partial \mathbf{A}_\mathrm{m}}{\partial t} - \operatorname{grad} \phi_\mathrm{e} \tag{11}$$

$$\mathbf{H} = \frac{c}{Z} \operatorname{curl} \mathbf{A}_\mathrm{m} - \frac{\partial \mathbf{A}_\mathrm{e}}{\partial t} - \operatorname{grad} \phi_\mathrm{m} \tag{12}$$

The vector potentials are not completely specified since Eqs.(11) and (12) define only $\operatorname{curl} \mathbf{A}_\mathrm{e}$ and $\operatorname{curl} \mathbf{A}_\mathrm{m}$. Two additional conditions can be chosen that we call the *extended Lorentz convention*:

$$\operatorname{div} \mathbf{A}_\mathrm{m} + \frac{1}{c^2} \frac{\partial \phi_\mathrm{e}}{\partial t} = 0 \tag{13}$$

$$\operatorname{div} \mathbf{A}_\mathrm{e} + \frac{1}{c^2} \frac{\partial \phi_\mathrm{m}}{\partial t} = 0 \tag{14}$$

The potentials of Eq.(11) and (12) then satisfy the following inhomogeneous partial differential equations:

[2]Harmuth and Husain 1994, Sec. 1.7; Harmuth et. al. 2001, Sec. 1.6.

$$\nabla^2 \mathbf{A}_{\mathrm{e}} - \frac{1}{c^2}\frac{\partial^2 \mathbf{A}_{\mathrm{e}}}{\partial t^2} \equiv \Box \mathbf{A}_{\mathrm{e}} = -\frac{1}{Zc}\mathbf{g}_{\mathrm{m}} \tag{15}$$

$$\nabla^2 \mathbf{A}_{\mathrm{m}} - \frac{1}{c^2}\frac{\partial^2 \mathbf{A}_{\mathrm{m}}}{\partial t^2} \equiv \Box \mathbf{A}_{\mathrm{m}} = -\frac{Z}{c}\mathbf{g}_{\mathrm{e}} \tag{16}$$

$$\nabla^2 \phi_{\mathrm{e}} - \frac{1}{c^2}\frac{\partial^2 \phi_{\mathrm{e}}}{\partial t^2} \equiv \Box \phi_{\mathrm{e}} = -Zc\rho_{\mathrm{e}} \tag{17}$$

$$\nabla^2 \phi_{\mathrm{m}} - \frac{1}{c^2}\frac{\partial^2 \phi_{\mathrm{m}}}{\partial t^2} \equiv \Box \phi_{\mathrm{m}} = -\frac{c}{Z}\rho_{\mathrm{m}} \tag{18}$$

Particular solutions of these partial differential equations may be represented by integrals taken over the whole space. We note that the magnetic charge density ρ_{m} may be always zero, which implies $\phi_{\mathrm{m}} \equiv 0$:

$$\mathbf{A}_{\mathrm{e}}(x,y,z,t) = \frac{1}{4\pi Zc}\iiint \frac{\mathbf{g}_{\mathrm{m}}(\xi,\eta,\zeta,t-r/c)}{r}\,d\xi\,d\eta\,d\zeta \tag{19}$$

$$\mathbf{A}_{\mathrm{m}}(x,y,z,t) = \frac{Z}{4\pi c}\iiint \frac{\mathbf{g}_{\mathrm{e}}(\xi,\eta,\zeta,t-r/c)}{r}\,d\xi\,d\eta\,d\zeta \tag{20}$$

$$\phi_{\mathrm{e}}(x,y,z,t) = \frac{Zc}{4\pi}\iiint \frac{\rho_{\mathrm{e}}(\xi,\eta,\zeta,t-r/c)}{r}\,d\xi\,d\eta\,d\zeta \tag{21}$$

$$\phi_{\mathrm{m}}(x,y,z,t) = \frac{c}{4\pi Z}\iiint \frac{\rho_{\mathrm{m}}(\xi,\eta,\zeta,t-r/c)}{r}\,d\xi\,d\eta\,d\zeta \tag{22}$$

Here r is the distance between the coordinates ξ, η, ζ of the current and charge densities and the coordinates x, y, z of the potentials:

$$r = \left[(x-\xi)^2 + (y-\eta)^2 + (z-\zeta)^2\right]^{1/2} \tag{23}$$

If there is no magnetic charge ρ_{m}, the scalar potential ϕ_{m} drops out; if in addition there are no magnetic dipole current densities $\mathbf{g}_{\mathrm{m}}$, the vector potential $\mathbf{A}_{\mathrm{e}}$ drops out too. Equations (11) and (12) are then reduced to the conventional Eqs.(9) and (10). The field strength $\mathbf{H}(\zeta,\theta)$ does not have defined values but $\mathbf{E}(\zeta,\theta)$ has. This implies that $\mathbf{A}_{\mathrm{m}}$ in Eq.(10) is undefined but $\partial\mathbf{A}_{\mathrm{m}}/\partial t$ in Eq.(9) must be defined to yield defined values for $\mathbf{E}(\zeta,\theta)$. Hence, Eqs.(9) and (10) contain a contradiction and cannot be used[3].

We note that only Eqs.(1)–(4) are needed to derive Eqs.(11)–(22), the constitutive equations (5)–(8) are not used.

The Lagrange function and the Hamilton function shall be needed for a particle with mass m, charge e, and velocity $\mathbf{v}$ in an electromagnetic field. From the Lorentz equation of motion

[3]For a more detailed discussion of this contradiction see Harmuth et. al. 2001, Sec. 3.1.

$$\frac{\partial}{\partial t}(m\mathbf{v}) = e\mathbf{E} + \frac{Ze}{c}\mathbf{v} \times \mathbf{H} \tag{24}$$

one can derive for $v \ll c$ from the original Maxwell equations the Lagrange function[4] $\mathfrak{L}_\mathrm{M}$:

$$\mathfrak{L}_\mathrm{M} = \frac{1}{2}m(\dot{x}^2 + \dot{y}^2 + \dot{z}^2) + e(-\phi_\mathrm{e} + A_{\mathrm{m}x}\dot{x} + A_{\mathrm{m}y}\dot{y} + A_{\mathrm{m}z}\dot{z}) \tag{25}$$

The modified Maxwell equations yield a Lagrange function represented by a matrix[5] $\boldsymbol{\mathfrak{L}}$ that is derived in some detail in Section 1.3 and is shown in Eq.(1.3-24). A more compact representation is by means of unit vectors $\mathbf{e}_x$, $\mathbf{e}_y$, $\mathbf{e}_z$:

$$\boldsymbol{\mathfrak{L}} = \boldsymbol{\mathfrak{L}}_\mathrm{M} + \boldsymbol{\mathfrak{L}}_\mathrm{c} = (\mathfrak{L}_\mathrm{M} + \mathfrak{L}_{\mathrm{c}x})\mathbf{e}_x + (\mathfrak{L}_\mathrm{M} + \mathfrak{L}_{\mathrm{c}y})\mathbf{e}_y + (\mathfrak{L}_\mathrm{M} + \mathfrak{L}_{\mathrm{c}z})\mathbf{e}_z \tag{26}$$

The term $\mathfrak{L}_\mathrm{M}$ is the same as in Eq.(25) while $\mathfrak{L}_{\mathrm{c}x}$ is defined by

$$\begin{aligned}\mathfrak{L}_{\mathrm{c}x} = \frac{Ze}{c}\dot{x}(\dot{y}A_{\mathrm{e}z} - \dot{z}A_{\mathrm{e}y}) + \frac{Ze}{c}\int\Big[\dot{z}\frac{\partial\phi_\mathrm{m}}{\partial y} - \dot{y}\frac{\partial\phi_\mathrm{m}}{\partial z} + A_{\mathrm{e}z}\ddot{y} - A_{\mathrm{e}y}\ddot{z} \\ - c^2\left(\frac{\partial A_{\mathrm{e}z}}{\partial y} - \frac{\partial A_{\mathrm{e}y}}{\partial z}\right) + \left(\dot{y}\frac{\partial}{\partial y} + \dot{z}\frac{\partial}{\partial z}\right)(\dot{y}A_{\mathrm{e}z} - \dot{z}A_{\mathrm{e}y})\Big]dx\end{aligned} \tag{27}$$

The terms $\mathfrak{L}_{\mathrm{c}y}$ and $\mathfrak{L}_{\mathrm{c}z}$ are obtained from $\mathfrak{L}_{\mathrm{c}x}$ by the cyclical replacements $x \to y \to z \to x$ and $x \to z \to y \to x$.

The derivatives $\dot{x}$, $\dot{y}$, $\dot{z}$ in Eq.(27) can and should be replaced by the components of the moment $\mathbf{p}$:

$$\mathbf{p} = p_x\mathbf{e}_x + p_y\mathbf{e}_y + p_z\mathbf{e}_z \tag{28}$$

$$p_x = \frac{\partial\mathfrak{L}_x}{\partial\dot{x}} = \frac{\partial(\mathfrak{L}_\mathrm{m} + \mathfrak{L}_{\mathrm{c}x})}{\partial\dot{x}} = m\dot{x} + eA_{\mathrm{m}x} + \frac{Ze}{c}(A_{\mathrm{e}z}\dot{y} - A_{\mathrm{e}y}\dot{z}) \tag{29}$$

$$p_y = \frac{\partial\mathfrak{L}_y}{\partial\dot{y}} = m\dot{y} + eA_{\mathrm{m}y} + \frac{Ze}{c}(A_{\mathrm{e}x}\dot{z} - A_{\mathrm{e}z}\dot{x}) \tag{30}$$

$$p_z = \frac{\partial\mathfrak{L}_z}{\partial\dot{z}} = m\dot{z} + eA_{\mathrm{m}z} + \frac{Ze}{c}(A_{\mathrm{e}y}\dot{x} - A_{\mathrm{e}x}\dot{y}) \tag{31}$$

This is a major effort and we rewrite only the first component of $\mathfrak{L}_{\mathrm{c}x}$, denoted $\mathfrak{L}_{\mathrm{c}x1}$, in this form[6]:

---

[4]The subscript M refers to 'Maxwell'.

[5]The subscript c refers to 'correction'.

[6]For the other components see Harmuth et. al. 2001, Sec. 3.2.

$$\begin{aligned}
\mathcal{L}_{cx1} = \frac{Ze}{c}(A_{ez}\dot{y} - A_{ey}\dot{z})\dot{x} = \frac{Ze}{m^2c^2}\Big(& A_{ex}(\mathbf{p} - e\mathbf{A}_{\mathrm{m}})_y - A_{ey}(\mathbf{p} - e\mathbf{A}_{\mathrm{m}})_z \\
& + \frac{Ze}{mc}\Big\{ A_{ez}[\mathbf{A}_{\mathrm{e}} \times (\mathbf{p} - e\mathbf{A}_{\mathrm{m}})]_y - A_{ey}[\mathbf{A}_{\mathrm{e}} \times (\mathbf{p} - e\mathbf{A}_{\mathrm{m}})]_z \Big\}\Big) \\
& \times \Big[(\mathbf{p} - e\mathbf{A}_{\mathrm{m}})_x + \left(\frac{Ze}{mc}\right)^2 A_{ex}\mathbf{A}_{\mathrm{e}} \cdot (\mathbf{p} - e\mathbf{A}_{\mathrm{m}}) \\
& + \frac{Ze}{mc}[\mathbf{A}_{\mathrm{e}} \times (\mathbf{p} - e\mathbf{A}_{\mathrm{m}})]_x \Big] \left[1 + \left(\frac{Ze}{mc}\right)^2 \mathbf{A}_{\mathrm{e}}^2 \right]^{-2} \qquad (32)
\end{aligned}$$

The Hamilton function $\mathbf{\mathcal{H}}$ derived from the Lagrange function $\mathbf{\mathcal{L}}$ of Eq.(26) can be represented as a vector too. If either the energy mc^2 is large compared with the energy due to the potential $\mathbf{A}_{\mathrm{e}}$ or the magnitude of the potential $\mathbf{A}_{\mathrm{m}}$ is large compared with the magnitude of $\mathbf{A}_{\mathrm{e}}$ we obtain the following simple equations:

$$\mathbf{\mathcal{H}} = \mathcal{H}_x\mathbf{e}_x + \mathcal{H}_y\mathbf{e}_y + \mathcal{H}_z\mathbf{e}_z \qquad (33)$$

$$\mathcal{H}_x = \frac{1}{2m}(\mathbf{p} - e\mathbf{A}_{\mathrm{m}})^2 + e\phi_{\mathrm{e}} - \mathcal{L}_{cx} \qquad (34)$$

$$\mathcal{H}_y = \frac{1}{2m}(\mathbf{p} - e\mathbf{A}_{\mathrm{m}})^2 + e\phi_{\mathrm{e}} - \mathcal{L}_{cy} \qquad (35)$$

$$\mathcal{H}_z = \frac{1}{2m}(\mathbf{p} - e\mathbf{A}_{\mathrm{m}})^2 + e\phi_{\mathrm{e}} - \mathcal{L}_{cz} \qquad (36)$$

The terms $(1/2m)(\mathbf{p} - e\mathbf{A}_{\mathrm{m}})^2 + e\phi_{\mathrm{e}}$ equal the conventional one derived from Maxwell's original equations.

If the simplifying assumptions made for the derivation of Eqs.(34)–(36) are not satisfied one obtains the following exact but much more complicated Hamilton function:

$$\begin{aligned}
\mathbf{\mathcal{H}} = \frac{1}{2m}\Big[& (\mathbf{p} - e\mathbf{A}_{\mathrm{m}})^2 + \left(\frac{Ze}{mc}\right)^2 \Big\{ 2[\mathbf{A}_{\mathrm{e}} \cdot (\mathbf{p} - e\mathbf{A}_{\mathrm{m}})]^2 + [\mathbf{A}_{\mathrm{e}} \times (\mathbf{p} - e\mathbf{A}_{\mathrm{m}})]^2 \Big\} \\
& + \left(\frac{Ze}{mc}\right)^4 \mathbf{A}_{\mathrm{e}}^2 [\mathbf{A}_{\mathrm{e}} \cdot (\mathbf{p} - e\mathbf{A}_{\mathrm{m}})]^2 \Big] \left[1 + \left(\frac{Ze}{mc}\right)^2 \mathbf{A}_{\mathrm{e}}^2 \right]^{-2} + e\phi_{\mathrm{e}} - \mathbf{\mathcal{L}}_{\mathrm{c}} \qquad (37)
\end{aligned}$$

Terms multiplied by $(ZecA_{\mathrm{e}}/mc^2)^2$ or $(ZecA_{\mathrm{e}}/mc^2)^4$ have been added to the simplified terms of Eqs.(34)–(36).

Dropping the simplifying restriction $v \ll c$ we obtain more complicated expressions. In particular, the Hamilton function can be written with the help of series expansions only[7]. The relativistic generalization of the Lagrange function of Eq.(26) is:

$$\mathcal{L} = -m_0c^2(1-v^2/c^2)^{1/2} + e(-\phi_{\mathrm{e}} + \mathbf{A}_{\mathrm{m}}\cdot\mathbf{v}) + \mathcal{L}_{\mathrm{c}} \tag{38}$$

In analogy to Eqs.(34)–(36) we first write an approximation for the three components of the Hamilton function that holds if the energy due to the potential $\mathbf{A}_{\mathrm{e}}$ is small compared with the energy $m_0c^2/(1-v^2/c^2)^{1/2}$ and the magnitude of $\mathbf{A}_{\mathrm{e}}$ is small compared with the magnitude of $\mathbf{A}_{\mathrm{m}}$:

$$\mathcal{H}_x = c\big[(\mathbf{p}-e\mathbf{A}_{\mathrm{m}})^2 + m_0^2c^2\big]^{1/2} + e\phi_{\mathrm{e}} - \mathcal{L}_{\mathrm{c}x} \tag{39}$$

$$\mathcal{H}_y = c\big[(\mathbf{p}-e\mathbf{A}_{\mathrm{m}})^2 + m_0^2c^2\big]^{1/2} + e\phi_{\mathrm{e}} - \mathcal{L}_{\mathrm{c}y} \tag{40}$$

$$\mathcal{H}_z = c\big[(\mathbf{p}-e\mathbf{A}_{\mathrm{m}})^2 + m_0^2c^2\big]^{1/2} + e\phi_{\mathrm{e}} - \mathcal{L}_{\mathrm{c}z} \tag{41}$$

If we leave out the correcting terms $\mathcal{L}_{\mathrm{c}x}$, $\mathcal{L}_{\mathrm{c}y}$, $\mathcal{L}_{\mathrm{c}z}$ we have the conventional relativistic Hamilton function for a charged particle in an electromagnetic field, written with three components rather than one. We call these equations the zero order approximation in $\alpha_{\mathrm{e}} = \alpha_{\mathrm{e}}(\mathbf{r},t) = ZecA_{\mathrm{e}}/m_0c^2$. Let us note that α_{e} is a dimension-free normalization of the magnitude of the potential $\mathbf{A}_{\mathrm{e}}(\mathbf{r},t)$. A first-order approximation in α_{e} is provided by the following equations:

$$\mathcal{H}_x = c\big[(\mathbf{p}-e\mathbf{A}_{\mathrm{m}})^2 + m_0^2c^2\big]^{1/2}(1+\alpha_{\mathrm{e}}Q) + e\phi_{\mathrm{e}} - \mathcal{L}_{\mathrm{c}x} \tag{42}$$

$$\mathcal{H}_y = c\big[(\mathbf{p}-e\mathbf{A}_{\mathrm{m}})^2 + m_0^2c^2\big]^{1/2}(1+\alpha_{\mathrm{e}}Q) + e\phi_{\mathrm{e}} - \mathcal{L}_{\mathrm{c}y} \tag{43}$$

$$\mathcal{H}_z = c\big[(\mathbf{p}-e\mathbf{A}_{\mathrm{m}})^2 + m_0^2c^2\big]^{1/2}(1+\alpha_{\mathrm{e}}Q) + e\phi_{\mathrm{e}} - \mathcal{L}_{\mathrm{c}z} \tag{44}$$

$$Q = \frac{1}{m_0^2c^2}\frac{(\mathbf{p}-e\mathbf{A}_{\mathrm{m}})^2[\mathbf{A}_{\mathrm{e}}\cdot(\mathbf{p}-e\mathbf{A}_{\mathrm{m}})]^2}{[1+(\mathbf{p}-e\mathbf{A}_{\mathrm{m}})^2/m_0^2c^2]^{3/2}\,\mathbf{A}_{\mathrm{e}}^2\,(\mathbf{p}-e\mathbf{A}_{\mathrm{m}})^2}$$

$$\alpha_{\mathrm{e}} = \alpha_{\mathrm{e}}(\mathbf{r},t) = \frac{ZecA_{\mathrm{e}}(\mathbf{r},t)}{m_0c^2} = 2\frac{Ze^2}{2h}\frac{h}{m_0c}\frac{A_{\mathrm{e}}(\mathbf{r},t)}{e}$$
$$= 2\alpha\frac{h}{m_0c}\frac{A_{\mathrm{e}}(\mathbf{r},t)}{e} = 2\alpha\frac{\lambda_{\mathrm{C}}A_{\mathrm{e}}(\mathbf{r},t)}{e}$$

$\alpha_{\mathrm{e}} = 2.210\times10^5A_{\mathrm{e}}(\mathbf{r},t)$ for electron, $\alpha_{\mathrm{e}} = 1.204\times10^2A_{\mathrm{e}}(\mathbf{r},t)$ for proton

$$\alpha = \frac{Ze^2}{2h} \approxeq 7.297\,535\times10^{-3}\ \text{fine structure constant},\ \lambda_{\mathrm{C}} = \frac{h}{m_0c} \tag{45}$$

The fine structure constant α is a universal constant of quantum physics. The factor $(h/c)A_{\mathrm{e}}/m_0e$ normalizes the magnitude A_{e} of the potential $\mathbf{A}_{\mathrm{e}}$ by the

[7]Harmuth et al. 2001, Sec. 3.3.

$$
\begin{aligned}
\frac{dA(\theta)}{d\theta} \rightarrow & \lim_{\Delta\theta\rightarrow d\theta} \frac{1}{2\Delta\theta}[A(\theta+\Delta\theta) - A(\theta-\Delta\theta)] \\
& \lim_{\Delta\theta\rightarrow d\theta} \frac{1}{\Delta\theta}[A(\theta+\Delta\theta) - A(\theta)] \\
& \lim_{\Delta\theta\rightarrow d\theta} \frac{1}{\Delta\theta}[A(\theta) - A(\theta-\Delta\theta)]
\end{aligned} \tag{6}
$$

The second-order difference quotient is practically always used in the symmetric form:

$$
\begin{aligned}
\frac{d^2A(\theta)}{d\theta^2} \rightarrow \frac{\tilde{\Delta}^2 A(\theta)}{(\tilde{\Delta}\theta)^2} = \frac{\tilde{\Delta}^2 A(\theta)}{\tilde{\Delta}\theta^2} &= \frac{A(\theta+\Delta\theta) - 2A(\theta) + A(\theta-\Delta\theta)}{(\Delta\theta)^2} \\
&= A(\theta+1) - 2A(\theta) + A(\theta-1) \quad \text{for } \Delta\theta = 1
\end{aligned} \tag{7}
$$

The second-order difference quotient does not formally follow from using twice the first-order symmetric difference quotient. However, one may use first the left and then the right first-order difference quotient or vice versa:

$$
\begin{aligned}
\frac{\tilde{\Delta}_{\mathrm{r}}}{\tilde{\Delta}\theta}\left(\frac{\tilde{\Delta}_{\mathrm{l}} A(\theta)}{\tilde{\Delta}\theta}\right) &= \frac{\tilde{\Delta}_{\mathrm{r}}}{\tilde{\Delta}\theta}[A(\theta) - A(\theta-1)] = [A(\theta+1) - A(\theta)] - [A(\theta) - A(\theta-1)] \\
&= A(\theta+1) - 2A(\theta) + A(\theta-1) \\
\frac{\tilde{\Delta}_{\mathrm{l}}}{\tilde{\Delta}\theta}\left(\frac{\tilde{\Delta}_{\mathrm{r}} A(\theta)}{\tilde{\Delta}\theta}\right) &= \frac{\tilde{\Delta}_{\mathrm{l}}}{\tilde{\Delta}\theta}[A(\theta+1) - A(\theta)] = [A(\theta+1) - A(\theta)] - [A(\theta) - A(\theta-1)] \\
&= A(\theta+1) - 2A(\theta) + A(\theta-1)
\end{aligned} \tag{8}
$$

Mathematicians are usually satisfied with the right difference quotient. But its use in physics introduces an asymmetry that is strictly due to mathematics and that may lead to divergencies that are avoided by the symmetric difference quotient[1]. Hence, we use whenever possible Eqs.(1), (3), and (7) for the first- and second-order difference quotient.

As a further example of the importance of symmetry consider the difference quotient of the product $u(\theta)v(\theta)$ in various notations:

$$
\begin{aligned}
\frac{\tilde{\Delta}[u(\theta)v(\theta)]}{\tilde{\Delta}\theta} &= \frac{1}{2\Delta\theta}[u(\theta+\Delta\theta)v(\theta+\Delta\theta) - u(\theta-\Delta\theta)v(\theta-\Delta\theta)] \\
&= \frac{1}{2\Delta\theta}\{u(\theta+\Delta\theta)[v(\theta+\Delta\theta) - v(\theta-\Delta\theta)] \\
&\qquad + v(\theta-\Delta\theta)[u(\theta+\Delta\theta) - u(\theta-\Delta\theta)]\} \\
&= u(\theta+\Delta\theta)\frac{\tilde{\Delta}v(\theta)}{\tilde{\Delta}\theta} + v(\theta-\Delta\theta)\frac{\tilde{\Delta}u(\theta)}{\tilde{\Delta}\theta}
\end{aligned} \tag{9}
$$

[1]Harmuth 1989, Sec. 8.2.

Since $u(\theta)$ and $v(\theta)$ are not treated equally on the right side of Eq.(9) even though they are treated equally on the left side, we rewrite the second line of Eq.(9):

$$\begin{aligned}\frac{\tilde{\Delta} u(\theta)v(\theta)}{\tilde{\Delta}\theta} &= \frac{1}{2\Delta\theta}\{v(\theta+\Delta\theta)[u(\theta+\Delta\theta)-u(\theta-\Delta\theta)] \\ &\qquad + u(\theta-\Delta\theta)[v(\theta+\Delta\theta)-v(\theta-\Delta\theta)]\} \\ &= u(\theta-\Delta\theta)\frac{\tilde{\Delta} v(\theta)}{\tilde{\Delta}\theta} + v(\theta+\Delta\theta)\frac{\tilde{\Delta} u(\theta)}{\tilde{\Delta}\theta} \end{aligned} \tag{10}$$

The sum of Eqs.(9) and (10) divided by 2 is symmetric in $u(\theta)$ and $v(\theta)$:

$$\begin{aligned}\frac{\tilde{\Delta} u(\theta)v(\theta)}{\tilde{\Delta}\theta} &= \frac{1}{2}[u(\theta+\Delta\theta)+u(\theta-\Delta\theta)]\frac{\tilde{\Delta} v(\theta)}{\tilde{\Delta}\theta} \\ &\qquad + \frac{1}{2}[v(\theta+\Delta\theta)+v(\theta-\Delta\theta)]\frac{\tilde{\Delta} u(\theta)}{\tilde{\Delta}\theta} \end{aligned} \tag{11}$$

According to Eq.(7) we have the relations

$$\begin{aligned} u(\theta+\Delta\theta)+u(\theta-\Delta\theta) &= 2u(\theta) + \frac{\tilde{\Delta}^2 u(\theta)}{\tilde{\Delta}\theta^2}(\Delta\theta)^2 \\ v(\theta+\Delta\theta)+v(\theta-\Delta\theta) &= 2v(\theta) + \frac{\tilde{\Delta}^2 v(\theta)}{\tilde{\Delta}\theta^2}(\Delta\theta)^2 \end{aligned}$$

and Eq.(11) becomes:

$$\begin{aligned}\frac{\tilde{\Delta} u(\theta)v(\theta)}{\tilde{\Delta}\theta} &= \left(u(\theta) + \frac{(\Delta\theta)^2}{2}\frac{\tilde{\Delta}^2 u(\theta)}{\tilde{\Delta}\theta^2}\right)\frac{\tilde{\Delta} v(\theta)}{\tilde{\Delta}\theta} \\ &\qquad + \left(v(\theta) + \frac{(\Delta\theta)^2}{2}\frac{\tilde{\Delta}^2 v(\theta)}{\tilde{\Delta}\theta^2}\right)\frac{\tilde{\Delta} u(\theta)}{\tilde{\Delta}\theta} \\ &= u(\theta)\frac{\tilde{\Delta} v(\theta)}{\tilde{\Delta}\theta} + v(\theta)\frac{\tilde{\Delta} u(\theta)}{\tilde{\Delta}\theta} + O(\Delta\theta)^2 \end{aligned} \tag{12}$$

The symmetrizing of Eq.(9) yields the exact equivalent of the expression for $d[u(\theta)v(\theta)]/d\theta$ of the differential calculus. If one does not like to ignore the terms $O(\Delta\theta)^2$ one can use Eq.(11) instead of Eq.(12). We do not know whether the use of Eq.(11) leads to significantly different results than the use of Eq.(12). We encounter here the first time the lack of development of the calculus of finite differences and we choose to use the simpler Eq.(12).

Differential calculus permits us to define integration by means of the differential equation

$$\frac{du(x)}{dx} = \varphi(x) \tag{13}$$

and its formal solution

$$u(x) = \int \frac{du(x)}{dx} dx = \int \varphi(x) dx \tag{14}$$

A very similar process leads in the calculus of finite differences from the difference quotient of first order to *summation* rather than integration. In order to use the results of Nörlund and Milne-Thomson[2] we follow closely their derivation. This forces us to use the notation

$$\underset{\omega}{\tilde{\Delta}} u(x) = \frac{u(x+\omega) - u(x)}{\omega} = \varphi(x) \tag{15}$$

To connect Eq.(15) with our notation in Eq.(1) we choose first $\omega = 2\Delta x$ and then $\omega = 2$:

$$\frac{1}{2\Delta x}[u(x + 2\Delta x) - u(x)] = \varphi(x) \tag{16}$$

With the substitution $x = x' - \Delta x$ we get:

$$\frac{u(x' + \Delta x) - u(x' - \Delta x)}{2\Delta x} = \varphi(x' - \Delta x), \quad x' = x + \Delta x \tag{17}$$

$$\frac{1}{2}[u(x' + 1) - u(x' - 1)] = \varphi(x' - 1), \quad x' = x + 1, \ \Delta x = 1 \tag{18}$$

Consider now a function $f(x)$

$$\begin{aligned} f(x) &= C_0 - \omega[\varphi(x) + \varphi(x + \omega) + \varphi(x + 2\omega) + \ \ldots] \\ &= C_0 - \omega \sum_{s=0}^{\infty} \varphi(x + s\omega) \end{aligned} \tag{19}$$

and the shifted function $f(x + \omega)$:

$$f(x + \omega) = C_0 - \omega \sum_{s=0}^{\infty} \varphi(x + \omega + s\omega) \tag{20}$$

[2]Nörlund 1924, Ch. 3; Milne-Thomson 1951, Ch. VIII.

A formal solution of Eq.(15) is obtained by the substitution of $f(x)$ for $u(x)$. For a reason to be seen presently we may replace the constant C_0 by a definite integral

$$C_0 = \int_0^\infty \varphi(\nu)d\nu \tag{21}$$

The *Hauptlösung* or *principal solution* of Eq.(15), which is also called the *sum of the function* $\varphi(x)$, may then be written in the following form:

$$F(x\,|\,\omega) = \int_0^\infty \varphi(\nu)d\nu - \omega\sum_{s=0}^{\infty}\varphi(x+s\omega) \tag{22}$$

The integral is used instead of the constant C_0 because a divergency of this integral may compensate a divergency of the sum, which a constant C_0 in Eq.(19) could not do. The principal solution of Eq.(15) is thus obtained by *summing* the function $\varphi(x)$. Nörlund introduced the following notation for this summation:

$$F(x\,|\,\omega) = \mathop{\mathrm{S}}_{c}^{x} \varphi(\nu)\mathop{\Delta}_{\omega}\nu = \int_0^\infty \varphi(\nu)d\nu - \omega\sum_{s=0}^{\infty}\varphi(x+s\omega) \tag{23}$$

The function $F(x\,|\,\omega)$ is said to be obtained by *summing* $\varphi(x)$ *from* c *to* x. The integral in Eq.(23) represents an 'integration' or 'summing' constant like the c in an integral $\int_c^x f(x')dx'$ if it and the sum converge. In this case one may write C_0 for the integral rather than evaluate Eq.(21).

Nörlund generalized the definition of $F(x\,|\,\omega)$ beyond what is shown in Eq.(22) to functions that can be made summable by means of an exponential function:

$$F(x\,|\,\omega) = \lim_{\mu\to 0}\left(\int_0^\infty \varphi(\nu)e^{-\mu\lambda(\nu)}d\nu - \omega\sum_{s=0}^{\infty}\varphi(x+s\omega)e^{-\mu\lambda(x+s\omega)}\right) \tag{24}$$

This limit process makes some functions $\varphi(x)$ summable. It is needed to obtain the sum of the constant a in Table 1.2-1 (Milne-Thomson 1951, p. 203).

We write $F(x\,|\,\omega)$ for the symmetric difference quotient on the left side of Eq.(17), substituting first $\omega = 2\Delta x$, $x = x' - \Delta x$ and then $\omega = 2$, $x = x' - 1$:

$$\text{for } \omega = 2\Delta x,\ x = x' - \Delta x$$

$$F(x'-\Delta x\,|\,2\Delta x) = \mathop{\mathfrak{S}}_{c}^{x'-\Delta x} \varphi(\nu)\Delta\nu = \int_0^\infty \varphi(\nu)d\nu - 2\Delta x \sum_{s=0}^{\infty} \varphi(x' - \Delta x + 2s\Delta x)$$

$$G(x) = F(x\,|\,2\Delta x) = \mathop{\mathfrak{S}}_{c}^{x} \varphi(\nu)\Delta\nu = \int_c^\infty \varphi(\nu)d\nu - 2\Delta x \sum_{s=0}^{\infty} \varphi(x + 2s\Delta x)$$

$$\mathop{\Delta}_{\omega}\nu = \Delta\nu \text{ for } \omega = 2\Delta x \tag{25}$$

For the choice $\omega = 2$, $x = x' - 1$ we get the following result that will be used from here on unless Eq.(25) is specifically indicated:

$$\text{for } \omega = 2,\ x = x' - 1$$

$$F(x' - 1\,|\,2) = \mathop{\mathfrak{S}}_{c}^{x'-1} \varphi(\nu)\Delta\nu = \int_0^\infty \varphi(\nu)d\nu - 2\sum_{s=0}^{\infty} \varphi(x' - 1 + 2s)$$

$$G(x) = F(x\,|\,2) = \mathop{\mathfrak{S}}_{c}^{x} \varphi(\nu)\Delta\nu = \int_c^\infty \varphi(\nu)d\nu - 2\sum_{s=0}^{\infty} \varphi(x + 2s)$$

$$\mathop{\Delta}_{\omega}\nu = \Delta\nu \quad \text{for } \omega = 2 \tag{26}$$

Let us derive the summation directly in detail from Eq.(17) by the substitutions

$$\begin{aligned} f(x' - \Delta x) &= C_0 - 2\Delta x[\varphi(x' - \Delta x) + \varphi(x' - \Delta x + 2\Delta x) + \varphi(x' - \Delta x + 4\Delta x) + \ldots] \\ &= C_0 - 2\Delta x \sum_{s=0}^{\infty} \varphi(x' - \Delta x + 2s\Delta x) \end{aligned}$$

$$\begin{aligned} f(x' + \Delta x) &= C_0 - 2\Delta x[\varphi(x' + \Delta x) + \varphi(x' + \Delta x + 2\Delta x) + \varphi(x' + \Delta x + 4\Delta x) + \ldots] \\ &= C_0 - 2\Delta x \sum_{s=0}^{\infty} \varphi(x' + \Delta x + 2s\Delta x) \end{aligned}$$

Substitution into Eq.(17) yields:

$$\frac{1}{2\Delta x}[f(x' + \Delta x) - f(x' - \Delta x)] = \varphi(x' - \Delta x)$$

Hence, we may define the principal solution in analogy to Eq.(22):

$$F(x'\,|\,2\Delta x) = \int\limits_0^\infty \varphi(\nu)d\nu - 2\Delta x \sum_{s=0}^\infty \varphi(x' + 2s\Delta x)$$

$$F(x' - \Delta x\,|\,2\Delta x) = \int\limits_0^\infty \varphi(\nu)d\nu - 2\Delta x \sum_{s=0}^\infty \varphi(x' - \Delta x + 2s\Delta x)$$

$$F(x\,|\,2\Delta x) = \int\limits_0^\infty \varphi(\nu)d\nu - 2\Delta x \sum_{s=0}^\infty \varphi(x + 2s\Delta x), \quad x = x' - \Delta x$$

This is again Eq.(25) for $\omega = 2\Delta x$.

Let $\varphi(x)$ in Eq.(15) be the exponential function e^{-x} and let ω as well as x be real and positive. We obtain:

$$\begin{aligned} u(x) = F(x\,|\,\omega) &= \mathop{\mathbb{S}}_c^x e^{-\nu} \mathop{\Delta\nu}_\omega = \int\limits_c^\infty e^{-\nu}d\nu - \omega \sum_{s=0}^\infty e^{-(x+s\omega)} \\ &= e^{-c} - \frac{\omega e^{-x}}{1 - e^{-\omega}} \end{aligned} \tag{27}$$

Consider now the following difference equation to see what the symmetric difference quotient and the choice $\omega = 2$ do to Eq.(27):

$$\begin{aligned} \tilde{\Delta} v(x) &= \frac{1}{2}[v(x' + 1) - v(x' - 1)] = e^{-x'} = \varphi(x'), \quad u(x') = ev(x') \\ &= \frac{1}{2}[u(x' + 1) - u(x' - 1)] = e^{-(x'-1)} = \varphi(x' - 1) \end{aligned} \tag{28}$$

This equation corresponds to Eq.(13). We get with the help of Eq.(26) and $x' = x + 1$:

$$\begin{aligned} v(x') &= e^{-1}u(x') \\ &= e^{-1}G(x') = \mathop{\mathbb{S}}_c^{x'} e^{-(\nu+1)}\Delta\nu = e^{-1}\Bigg(\int\limits_c^\infty e^{-\nu}d\nu - 2\sum_{s=0}^\infty e^{-(x'+2s)}\Bigg) \\ &= e^{-1}\left(e^{-c} - \frac{2e^{-x'}}{1-e^{-2}}\right) = e^{-(c+1)} - \frac{2e^{x'}}{e-e^{-1}} = \frac{2}{e-e^{-1}}\left(C - e^{-x'}\right) \end{aligned} \tag{29}$$

The integral $\int \exp(-x')dx'$ yields $C - \exp(-x')$. Equation (29) differs by a factor $2/(e - e^{-1}) \doteq 0.85092$ from this result.

It has been shown[3] that Eq.(27) not only holds for the exponential function $\varphi(x) = e^{-x}$ in Eq.(15) but generally for the exponential function $e^{\gamma x}$ in the complex plane as long as the condition $|\omega| < 2\pi/|\gamma|$ is satisfied:

$$\mathop{\mathfrak{S}}_{c}^{x} e^{\gamma\nu} \mathop{\Delta\nu}_{\omega} = -\frac{e^{\gamma c}}{\gamma} - \frac{\omega e^{\gamma x}}{1 - e^{\gamma\omega}}, \quad |\omega| < \frac{2\pi}{|\gamma|}$$

$$\mathop{\mathfrak{S}}_{c}^{x'} e^{\gamma(\nu+1)} \Delta\nu = -\frac{e^{\gamma(c+1)}}{\gamma} - \frac{2e^{\gamma x'}}{e^{-\gamma} - e^{\gamma}}, \quad |\gamma| < \pi \tag{30}$$

One may use this relationship to solve Eq.(15) for $\varphi(x) = \sin\gamma x$:

$$\frac{1}{\omega}[u(x+\omega) - u(x)] = \sin\gamma x = \frac{1}{2i}\left(e^{i\gamma x} - e^{-i\gamma x}\right)$$

$$u(x) = \frac{1}{2i} \mathop{\mathfrak{S}}_{c}^{x} \left(e^{i\gamma\nu} - e^{-i\gamma\nu}\right) \mathop{\Delta\nu}_{\omega}$$

$$= \frac{1}{2i}\left(-\frac{e^{i\gamma c}}{i\gamma} - \frac{\omega e^{i\gamma x}}{1 - e^{i\gamma\omega}} + \frac{e^{-i\gamma c}}{-i\gamma} + \frac{\omega e^{-i\gamma x}}{1 - e^{-i\gamma\omega}}\right)$$

$$= \frac{\cos\gamma c}{\gamma} - \frac{\omega}{2}\frac{\cos\gamma(x - \omega/2)}{\sin(\gamma\omega/2)}, \quad |\omega| < \frac{2\pi}{|\gamma|} \tag{31}$$

We need the corresponding result for Eq.(28). Intermediate steps are given to help with verification:

$$\tilde{\Delta}v(x') = \frac{1}{2}[v(x'+1) - v(x'-1)] = \sin\gamma x' = \frac{1}{2i}\left(e^{i\gamma x'} - e^{-i\gamma x'}\right) \tag{32}$$

$$\frac{1}{2}[v_1(x'+1) - v_1(x'-1)] = \frac{1}{2i}e^{i\gamma x'}, \quad u_1(x') = 2ie^{-i\gamma}v_1(x')$$

$$\frac{1}{2}[u_1(x'+1) - u_1(x'-1)] = e^{i\gamma(x'-1)} = \varphi(x'-1) \tag{33}$$

$$\frac{1}{2}[v_2(x'+1) - v_2(x'-1)] = -\frac{1}{2i}e^{-i\gamma x'}, \quad u_2(x') = -2ie^{i\gamma}v_2(x')$$

$$\frac{1}{2}[u_2(x'+1) - u_2(x'-1)] = e^{-i\gamma(x'-1)} = \varphi(x'-1) \tag{34}$$

[3] Nörlund 1924, p. 81; Milne-Thomson 1951, p. 231.

TABLE 1.2-1

SUMS $v(x') = v(x)$ OF CERTAIN FUNCTIONS $\varphi(x') = \varphi(x)$ ACCORDING TO EQ.(26) FOR $\omega = 2$. THE INTEGRALS OF $\varphi(x)$ ARE SHOWN FOR COMPARISON.

$\varphi(x)$	$v(x) = \mathop{S}\limits_{c}^{x} \varphi(\nu+1)\Delta\nu$	$\int \varphi(x)dx$
a	$a(x-c-1)$	$ax + C$
e^{-x}	$-\frac{e^{-x}}{\operatorname{sh} 1} + e^{-(c+1)}$	$-e^{-x} + C$
$e^{\gamma x}$	$\frac{e^{\gamma x}}{\operatorname{sh}\gamma} - \frac{e^{\gamma(c+1)}}{\gamma}$ γ complex, $\lvert\gamma\rvert < \pi$	$+\frac{e^{\gamma x}}{\gamma} + C$
$\sin\gamma x$	$-\frac{\cos\gamma x}{\sin\gamma} + \frac{\cos[\gamma(c+1)]}{\gamma}, \; \lvert\gamma\rvert < \pi$	$-\frac{\cos\gamma x}{\gamma} + C$
$\cos\gamma x$	$\frac{\sin\gamma x}{\sin\gamma} + \frac{\sin[\gamma(c+1)]}{\gamma}, \; \lvert\gamma\rvert < \pi$	$+\frac{\sin\gamma x}{\gamma} + C$
$\operatorname{sh}\gamma x$	$\frac{\operatorname{ch}\gamma x}{\operatorname{sh}\gamma} - \frac{\operatorname{ch}\gamma(c+1)}{\gamma}, \; \lvert\gamma\rvert < \pi$	$\frac{\operatorname{ch}\gamma x}{\gamma} + C$
$\operatorname{ch}\gamma x$	$\frac{\operatorname{sh}\gamma x}{\operatorname{sh}\gamma} - \frac{\operatorname{sh}\gamma(c+1)}{\gamma}, \; \lvert\gamma\rvert < \pi$	$\frac{\operatorname{sh}\gamma x}{\gamma} + C$
$e^{\lambda x}\sin\gamma x$	$\frac{e^{\lambda x}}{\lambda^2+\gamma^2}(\lambda_0\sin\gamma x - \gamma_0\cos\gamma x) + C_0$ $\lambda_0 = -\frac{2(\lambda^2+\gamma^2)\operatorname{sh}\lambda\cos\gamma}{\cos 2\gamma - \operatorname{ch} 2\lambda}$ $\gamma_0 = \frac{2(\lambda^2+\gamma^2)\operatorname{ch}\lambda\sin\gamma}{\cos 2\gamma - \operatorname{ch} 2\lambda}$ $C_0 = -\frac{e^{\lambda(c+1)}}{\lambda^2+\gamma^2}\{\lambda\sin[\gamma(c+1)]$ $-\gamma\cos[\gamma(c+1)]\}$	$\frac{e^{\lambda x}}{\lambda^2+\gamma^2}(\lambda\sin\gamma x - \gamma\cos\gamma x) + C$
$e^{\lambda x}\cos\gamma x$	$\frac{e^{\lambda x}}{\lambda^2+\gamma^2}(\lambda_0\cos\gamma x + \gamma_0\sin\gamma x) + C_1$ λ_0 and γ_0 are shown above $C_1 = -\frac{e^{\lambda(c+1)}}{\lambda^2+\gamma^2}\{\lambda\cos[\gamma(c+1)]$ $-\gamma\sin[\gamma(c+1)]\}$	$\frac{e^{\lambda x}}{\lambda^2+\gamma^2}(\lambda\cos\gamma x + \gamma\sin\gamma x) + C$

$$
\begin{aligned}
v(x') &= v_1(x') + v_2(x') = \frac{1}{2i}e^{i\gamma}u_1(x') - \frac{1}{2i}e^{-i\gamma}u_2(x') \\
&= \mathop{\mathrm{S}}_{c}^{x'} \sin\gamma(\nu+1)\Delta\nu = \frac{1}{2i}e^{i\gamma}\mathop{\mathrm{S}}_{c}^{x'} e^{i\gamma\nu}\Delta\nu - \frac{1}{2i}e^{-i\gamma}\mathop{\mathrm{S}}_{c}^{x'} e^{-i\gamma\nu}\Delta\nu \\
&= \frac{1}{2i}\left(-\frac{e^{i\gamma(c+1)}}{i\gamma} - \frac{2e^{i\gamma(x'+1)}}{1-e^{2i\gamma}} + \frac{e^{-i\gamma(c+1)}}{-i\gamma} + \frac{2e^{-i\gamma(x'+1)}}{1-e^{-2i\gamma}}\right) \\
&= \frac{\cos[\gamma(c+1)]}{\gamma} - \frac{\cos\gamma x'}{\sin\gamma}, \quad |\gamma| < \pi
\end{aligned}
\tag{35}
$$

The integral $\int \sin\gamma x'\,dx'$ yields $-(\cos\gamma x')/\gamma + C$. Equation (35) has the factor $1/\sin\gamma$ rather than $1/\gamma$.

Table 1.2-1 shows a collection of sums. Except for the last two they are all due to Nörlund. There is a long way to go to turn Table 1.2-1 into something comparable to our available tables of integrals.

The integral $\int \varphi(x)dx$ in the first row of Table 1.2-1 is a simplified form of $\int_c^x \varphi(\nu)d\nu$. A corresponding simplification will sometimes be used here for the summation in the first row of Table 1.2-1:

$$
u(x) = \mathop{\mathrm{S}}_{c}^{x} \varphi(\nu+1)\Delta\nu = \mathop{\mathrm{S}}_{c}^{x} \varphi(x+1)\Delta x = \mathrm{S}\, \varphi(x+1)\Delta x \tag{36}
$$

Summation and *differenciation* are inverse operations in the sense that integration and differentiation are inverse operations[4]:

$$
\underset{\omega}{\tilde{\Delta}} \mathop{\mathrm{S}}_{c}^{x} \varphi(\nu)\underset{\omega}{\Delta}\nu = \varphi(x), \qquad \mathop{\mathrm{S}}_{c}^{x} \left[\underset{\omega}{\tilde{\Delta}}\varphi(\nu)\right] \underset{\omega}{\Delta}\nu = \varphi(x) + C \tag{37}
$$

For the symmetric first-order difference quotient of Eq.(1) and the second-order difference quotient of Eq.(7) we must adopt the convention:

$$
\mathop{\mathrm{S}}_{c}^{\theta} \frac{\tilde{\Delta}^2 A(\nu)}{\tilde{\Delta}\nu^2}\Delta\nu = \frac{\tilde{\Delta}A(\theta)}{\tilde{\Delta}\theta} + C \tag{38}
$$

1.3 Lagrange Function and Modified Maxwell Equations

We have previously derived a Lagrange function for the modified Maxwell equations[1]. We give here a more detailed and hopefully more understandable derivation. The starting point is the Lorentz equation in MKSA notation:

---

[4]Nörlund 1924, Ch. 3, § 3/22; Milne-Thomson 1951, Sec. 8.1.

[1]Harmuth et al. 2001, Sec. 3.2.

$$\frac{d}{dt}(m\mathbf{v}) = e\mathbf{E} + \frac{Ze}{c}\mathbf{v} \times \mathbf{H} \tag{1}$$

The field strengths **E** and **H** of the modified Maxwell equations are written in the potential form. We use the old-fashioned notation curl and grad instead of $\nabla\times$ and ∇ since it seems to make the physical processes more lucid:

$$\mathbf{E} = -Zc\,\mathrm{curl}\,\mathbf{A}_\mathrm{e} - \frac{\partial \mathbf{A}_\mathrm{m}}{\partial t} - \mathrm{grad}\,\phi_\mathrm{e} \tag{2}$$

$$\mathbf{H} = \frac{c}{Z}\,\mathrm{curl}\,\mathbf{A}_\mathrm{m} - \frac{\partial \mathbf{A}_\mathrm{e}}{\partial t} - \mathrm{grad}\,\phi_\mathrm{m} \tag{3}$$

The substitution of **E** and **H** into Eq.(1) yields

$$\begin{aligned}\frac{d}{dt}(m\mathbf{v}) = e\Big(-\mathrm{grad}\,\phi_\mathrm{e} - \frac{\partial \mathbf{A}_\mathrm{m}}{\partial t} + \mathbf{v}\times\mathrm{curl}\,\mathbf{A}_\mathrm{m}\Big)\\ - \frac{Ze}{c}\Big(\mathbf{v}\times\mathrm{grad}\,\phi_\mathrm{m} + \mathbf{v}\times\frac{\partial \mathbf{A}_\mathrm{e}}{\partial t} + c^2\,\mathrm{curl}\,\mathbf{A}_\mathrm{e}\Big)\end{aligned} \tag{4}$$

The first line is the conventional equation of motion of an electrically charged particle in an electromagnetic field. The second line contains additional terms due to the modification of Maxwell's equations. Two of them with the potential $\mathbf{A}_\mathrm{e}$ apply to the undisputed magnetic dipole currents produced by the rotation of magnetic dipoles as well as to hypothetical magnetic monopole currents produced by so far not reliably observed magnetic charges. The term with the potential ϕ_m is strictly hypothetical until the observations of magnetic charges are widely accepted.

Equation (4) is to be rewritten with the help of a Lagrange function L, spatial variables x_j and velocity variables $\partial x_j/\partial t = \dot{x}_j$. The Lagrange function must satisfy the Euler equation

$$\frac{d}{dt}\left(\frac{\partial L}{\partial \dot{x}_j}\right) = \frac{\partial L}{\partial x_j}, \quad j = 1,\ 2,\ 3 \tag{5}$$

which derives the equation of motion from the principle of least action by Maupertuis and Hamilton.

In order to transform Eq.(4) we use a relation found in the literature (Morse and Feshbach 1953, part I, p. 295)

$$\frac{d\mathbf{A}_\mathrm{m}}{dt} = \frac{\partial \mathbf{A}_\mathrm{m}}{\partial t} + \mathbf{v}\cdot(\nabla\mathbf{A}_\mathrm{m}) \tag{6}$$

which assumes for Cartesian coordinates the previously used form

$$\frac{d\mathbf{A}_{\mathrm{m}}}{dt} = \frac{\partial \mathbf{A}_{\mathrm{m}}}{\partial t} + \frac{\partial \mathbf{A}_{\mathrm{m}}}{\partial x}\dot{x} + \frac{\partial \mathbf{A}_{\mathrm{m}}}{\partial y}\dot{y} + \frac{\partial \mathbf{A}_{\mathrm{m}}}{\partial z}\dot{z} \tag{7}$$

In analogy to Eq.(6) we get the further equation

$$\begin{aligned}\frac{d(\mathbf{v}\times\mathbf{A}_{\mathrm{e}})}{dt} &= \frac{\partial(\mathbf{v}\times\mathbf{A}_{\mathrm{e}})}{\partial t} + \mathbf{v}\cdot[\nabla(\mathbf{v}\times\mathbf{A}_{\mathrm{e}})]\\ &= \frac{\partial \mathbf{v}}{\partial t}\times\mathbf{A}_{\mathrm{e}} + \mathbf{v}\times\frac{\partial \mathbf{A}_{\mathrm{e}}}{\partial t} + \mathbf{v}\cdot[\nabla(\mathbf{v}\times\mathbf{A}_{\mathrm{e}})]\end{aligned} \tag{8}$$

For Cartesian coordinates we again get the previously used form

$$\frac{d}{dt}(\mathbf{v}\times\mathbf{A}_{\mathrm{e}}) = \frac{\partial \mathbf{v}}{\partial t}\times\mathbf{A}_{\mathrm{e}} + \mathbf{v}\times\frac{\partial \mathbf{A}_{\mathrm{e}}}{\partial t} + \left(\dot{x}\frac{\partial}{\partial x} + \dot{y}\frac{\partial}{\partial y} + \dot{z}\frac{\partial}{\partial z}\right)(\mathbf{v}\times\mathbf{A}_{\mathrm{e}}) \tag{9}$$

Substitution of Eqs.(6) and (8) into Eq.(4) brings:

$$\begin{aligned}&\frac{d}{dt}(m\mathbf{v} + e\mathbf{A}_{\mathrm{m}}) + \frac{Ze}{c}\frac{d}{dt}(\mathbf{v}\times\mathbf{A}_{\mathrm{e}})\\ &\qquad = e[-\operatorname{grad}\phi_{\mathrm{m}} + \mathbf{v}\cdot(\nabla\mathbf{A}_{\mathrm{m}}) + \mathbf{v}\times\operatorname{curl}\mathbf{A}_{\mathrm{m}}]\\ &\quad + \frac{Ze}{c}\left(-\mathbf{v}\times\operatorname{grad}\phi_{\mathrm{m}} + \frac{\partial \mathbf{v}}{\partial t}\times\mathbf{A}_{\mathrm{e}} + \mathbf{v}\cdot[\nabla(\mathbf{v}\times\mathbf{A}_{\mathrm{e}})] - c^2\operatorname{curl}\mathbf{A}_{\mathrm{e}}\right)\end{aligned} \tag{10}$$

For Cartesian coordinates we had obtained:

$$\begin{aligned}&\frac{d}{dt}(m\mathbf{v} + e\mathbf{A}_{\mathrm{m}}) + \frac{Ze}{c}\frac{d}{dt}(\mathbf{v}\times\mathbf{A}_{\mathrm{e}})\\ &\qquad = e\left(-\operatorname{grad}\phi_{\mathrm{e}} + \frac{\partial \mathbf{A}_{\mathrm{m}}}{\partial x}\dot{x} + \frac{\partial \mathbf{A}_{\mathrm{m}}}{\partial y}\dot{y} + \frac{\partial \mathbf{A}_{\mathrm{m}}}{\partial z}\dot{z} + \mathbf{v}\times\operatorname{curl}\mathbf{A}_{\mathrm{m}}\right)\\ &\quad + \frac{Ze}{c}\Bigg[-\mathbf{v}\times\operatorname{grad}\phi_{\mathrm{m}} + \frac{\partial \mathbf{v}}{\partial t}\times\mathbf{A}_{\mathrm{e}} + \left(\dot{x}\frac{\partial}{\partial x} + \dot{y}\frac{\partial}{\partial y} + \dot{z}\frac{\partial}{\partial z}\right)(\mathbf{v}\times\mathbf{A}_{\mathrm{e}})\\ &\qquad\qquad - c^2\operatorname{curl}\mathbf{A}_{\mathrm{e}}\Bigg]\end{aligned} \tag{11}$$

We may break Eq.(10) into two equations. The first equals the one obtained from the original Maxwell equations and the second contains the terms due to the modification of Maxwell's equations:

$$\frac{d}{dt}(m\mathbf{v} + e\mathbf{A}_{\mathrm{m}}) = e[-\operatorname{grad}\phi_{\mathrm{e}} + \mathbf{v}\cdot(\nabla\mathbf{A}_{\mathrm{m}}) + \mathbf{v}\times\operatorname{curl}\mathbf{A}_{\mathrm{m}}] \tag{12}$$

$$\frac{Ze}{c}\frac{d}{dt}(\mathbf{v}\times\mathbf{A}_{\mathrm{e}}) = \frac{Ze}{c}\left(-\mathbf{v}\times\operatorname{grad}\phi_{\mathrm{m}}+\frac{\partial\mathbf{v}}{\partial t}\times\mathbf{A}_{\mathrm{e}}+\mathbf{v}\cdot[\nabla(\mathbf{v}\times\mathbf{A}_{\mathrm{e}})]-c^2\operatorname{curl}\mathbf{A}_{\mathrm{e}}\right) \quad (13)$$

The use of these equations is complicated by the dyadics $\nabla\mathbf{A}_{\mathrm{m}}$ in Eq.(12) and $\nabla(\mathbf{v}\times\mathbf{A}_{\mathrm{e}})$ in Eq.(13). They are avoided for Cartesian coordinates, a fact which greatly favors them. Instead of Eqs.(12) and (13) we use the two corresponding equations derived from Eq.(11):

$$\begin{aligned}\frac{d}{dt}(m\mathbf{v}+e\mathbf{A}_{\mathrm{m}}) = e\Big(&-\operatorname{grad}\phi_{\mathrm{e}} \\ &+\frac{\partial\mathbf{A}_{\mathrm{m}}}{\partial x}\dot{x}+\frac{\partial\mathbf{A}_{\mathrm{m}}}{\partial y}\dot{y}+\frac{\partial\mathbf{A}_{\mathrm{m}}}{\partial z}\dot{z}+\mathbf{v}\times\operatorname{curl}\mathbf{A}_{\mathrm{m}}\Big)\end{aligned} \quad (14)$$

$$\begin{aligned}\frac{Ze}{c}\frac{d}{dt}(\mathbf{v}\times\mathbf{A}_{\mathrm{e}}) = \frac{Ze}{c}\Big[&-\mathbf{v}\times\operatorname{grad}\phi_{\mathrm{m}}+\frac{\partial\mathbf{v}}{\partial t}\times\mathbf{A}_{\mathrm{e}} \\ &+\left(\dot{x}\frac{\partial}{\partial x}+\dot{y}\frac{\partial}{\partial y}+\dot{z}\frac{\partial}{\partial z}\right)(\mathbf{v}\times\mathbf{A}_{\mathrm{e}})-c^2\operatorname{curl}\mathbf{A}_{\mathrm{e}}\Big]\end{aligned} \quad (15)$$

With the help of the expressions

$$\begin{aligned}\mathbf{v} &= \dot{x}\mathbf{e}_x+\dot{y}\mathbf{e}_y+\dot{z}\mathbf{e}_z \\ \mathbf{A}_{\mathrm{m}} &= A_{\mathrm{m}x}\mathbf{e}_x+A_{\mathrm{m}y}\mathbf{e}_y+A_{\mathrm{m}z}\mathbf{e}_z \\ \operatorname{grad}\phi_{\mathrm{e}} &= \frac{\partial\phi_{\mathrm{e}}}{\partial x}\mathbf{e}_x+\frac{\partial\phi_{\mathrm{e}}}{\partial y}\mathbf{e}_y+\frac{\partial\phi_{\mathrm{e}}}{\partial z}\mathbf{e}_z \\ \operatorname{curl}\mathbf{A}_{\mathrm{m}} &= \left(\frac{\partial A_{\mathrm{m}z}}{\partial y}-\frac{\partial A_{\mathrm{m}y}}{\partial z}\right)\mathbf{e}_x+\left(\frac{\partial A_{\mathrm{m}x}}{\partial z}-\frac{\partial A_{\mathrm{m}z}}{\partial x}\right)\mathbf{e}_y+\left(\frac{\partial A_{\mathrm{m}y}}{\partial x}-\frac{\partial A_{\mathrm{m}x}}{\partial y}\right)\mathbf{e}_z\end{aligned} \quad (16)$$

we rewrite Eq.(14) in component form:

$$\begin{aligned}\frac{d}{dt}(m\mathbf{v}+e\mathbf{A}_{\mathrm{m}}) &= \frac{d}{dt}[(m\dot{x}+eA_{\mathrm{m}x})\mathbf{e}_x+(m\dot{y}+eA_{\mathrm{m}y})\mathbf{e}_y+(m\dot{z}+eA_{\mathrm{m}z})\mathbf{e}_z] \\ &= e\frac{\partial}{\partial x}(-\phi_{\mathrm{e}}+\dot{x}A_{\mathrm{m}x}+\dot{y}A_{\mathrm{m}y}+\dot{z}A_{\mathrm{m}z})\mathbf{e}_x \\ &\quad+e\frac{\partial}{\partial y}(-\phi_{\mathrm{e}}+\dot{x}A_{\mathrm{m}x}+\dot{y}A_{\mathrm{m}y}+\dot{z}A_{\mathrm{m}z})\mathbf{e}_y \\ &\quad+e\frac{\partial}{\partial z}(-\phi_{\mathrm{e}}+\dot{x}A_{\mathrm{m}x}+\dot{y}A_{\mathrm{m}y}+\dot{z}A_{\mathrm{m}z})\mathbf{e}_z\end{aligned} \quad (17)$$

We note that in lines 2, 3 and 4 the factors of the unit vectors $\mathbf{e}_x$, $\mathbf{e}_y$, $\mathbf{e}_z$ have the same form except for the differential operators $\partial/\partial x$, $\partial/\partial y$, $\partial/\partial z$. These factors plus the differential operators match the right side of Eq.(5) for every j. In order to get the left side of Eq.(7) correctly we must add a term $\frac{1}{2}m(\dot{x}^2+\dot{y}^2+\dot{z}^2)$ to the common factors of lines 2, 3 and 4. We obtain the Lagrange function $\mathfrak{L}_{\rm M}$ of the conventional Maxwell equations:

$$\begin{aligned}\mathfrak{L}_{\rm M} &= \frac{1}{2}m(\dot{x}^2+\dot{y}^2+\dot{z}^2)+e(-\phi_{\rm e}+\dot{x}A_{{\rm m}x}+\dot{y}A_{{\rm m}y}+\dot{z}A_{{\rm m}z})\\ &= \frac{1}{2}mv^2+e\mathbf{v}\cdot\mathbf{A}_{\rm m}-e\phi_{\rm e}\end{aligned} \tag{18}$$

One may readily verify that $L=\mathfrak{L}_{\rm M}$ in Eq.(5) satisfies this equation for every x_j or every unit vector $\mathbf{e}_x$, $\mathbf{e}_y$, $\mathbf{e}_z$ in Eq.(17).

We turn to Eq.(15) and rewrite it in component form too. The following terms are required:

$$\begin{aligned}\mathbf{A}_{\rm e} &= A_{{\rm e}x}\mathbf{e}_x+A_{{\rm e}y}\mathbf{e}_y+A_{{\rm e}z}\mathbf{e}_z\\ \mathbf{v}\times\operatorname{grad}\phi_{\rm m} &= \left(\dot{y}\frac{\partial\phi_{\rm m}}{\partial z}-\dot{z}\frac{\partial\phi_{\rm m}}{\partial y}\right)\mathbf{e}_x+\left(\dot{z}\frac{\partial\phi_{\rm m}}{\partial x}-\dot{x}\frac{\partial\phi_{\rm m}}{\partial z}\right)\mathbf{e}_y+\left(\dot{x}\frac{\partial\phi_{\rm m}}{\partial y}-\dot{y}\frac{\partial\phi_{\rm m}}{\partial x}\right)\mathbf{e}_z\\ \frac{\partial\mathbf{v}}{\partial t}\times\mathbf{A}_{\rm e} &= (\ddot{y}A_{{\rm e}z}-\ddot{z}A_{{\rm e}y})\mathbf{e}_x+(\ddot{z}A_{{\rm e}x}-\ddot{x}A_{{\rm e}z})\mathbf{e}_y+(\ddot{x}A_{{\rm e}y}-\ddot{y}A_{{\rm e}x})\mathbf{e}_z\\ \mathbf{v}\times\mathbf{A}_{\rm e} &= (\dot{y}A_{{\rm e}z}-\dot{z}A_{{\rm e}y})\mathbf{e}_x+(\dot{z}A_{{\rm e}x}-\dot{x}A_{{\rm e}z})\mathbf{e}_y+(\dot{x}A_{{\rm e}y}-\dot{y}A_{{\rm e}x})\mathbf{e}_z\\ \operatorname{curl}\mathbf{A}_{\rm m} &\to \operatorname{curl}\mathbf{A}_{\rm e}\quad \text{for } {\rm m}\to{\rm e}\end{aligned} \tag{19}$$

We substitute Eqs.(19) into Eq.(15):

$$\begin{aligned}\frac{Ze}{c}\frac{d}{dt}(\mathbf{v}\times\mathbf{A}_{\rm e}) &= \frac{Ze}{c}\frac{d}{dt}[(\dot{y}A_{{\rm e}z}-\dot{z}A_{{\rm e}y})\mathbf{e}_x+(\dot{z}A_{{\rm e}x}-\dot{x}A_{{\rm e}z})\mathbf{e}_y+(\dot{x}A_{{\rm e}y}-\dot{y}A_{{\rm e}x})]\mathbf{e}_z\\ &= \frac{Ze}{c}\left[\dot{z}\frac{\partial\phi_{\rm m}}{\partial y}-\dot{y}\frac{\partial\phi_{\rm m}}{\partial z}+\left(\frac{\partial}{\partial t}+\dot{x}\frac{\partial}{\partial x}+\dot{y}\frac{\partial}{\partial y}+\dot{z}\frac{\partial}{\partial z}\right)(\dot{y}A_{{\rm e}z}-\dot{z}A_{{\rm e}y})\right.\\ &\qquad\left.-c^2\left(\frac{\partial A_{{\rm e}z}}{\partial y}-\frac{\partial A_{{\rm e}y}}{\partial z}\right)\right]\mathbf{e}_x\\ &+\frac{Ze}{c}\left[\dot{x}\frac{\partial\phi_{\rm m}}{\partial z}-\dot{z}\frac{\partial\phi_{\rm m}}{\partial x}+\left(\frac{\partial}{\partial t}+\dot{x}\frac{\partial}{\partial x}+\dot{y}\frac{\partial}{\partial y}+\dot{z}\frac{\partial}{\partial z}\right)(\dot{z}A_{{\rm e}x}-\dot{x}A_{{\rm e}z})\right.\\ &\qquad\left.-c^2\left(\frac{\partial A_{{\rm e}x}}{\partial z}-\frac{\partial A_{{\rm e}z}}{\partial x}\right)\right]\mathbf{e}_y\\ &+\frac{Ze}{c}\left[\dot{y}\frac{\partial\phi_{\rm m}}{\partial x}-\dot{x}\frac{\partial\phi_{\rm m}}{\partial y}+\left(\frac{\partial}{\partial t}+\dot{x}\frac{\partial}{\partial x}+\dot{y}\frac{\partial}{\partial y}+\dot{z}\frac{\partial}{\partial z}\right)(\dot{x}A_{{\rm e}y}-\dot{y}A_{{\rm e}x})\right.\\ &\qquad\left.-c^2\left(\frac{\partial A_{{\rm e}y}}{\partial x}-\frac{\partial A_{{\rm e}x}}{\partial y}\right)\right]\mathbf{e}_z\end{aligned} \tag{20}$$

The factors of $\mathbf{e}_x$, $\mathbf{e}_y$, $\mathbf{e}_z$ in lines 2–7 of this equation are not equal as the corresponding factors in Eq.(17) are. To make progress we use intuition and integrate the factor of $\mathbf{e}_x$ over x:

$$\begin{aligned}\mathfrak{L}_{cx} = \frac{Ze}{c}\int\Bigg[&\dot{z}\frac{\partial\phi_{\mathrm{m}}}{\partial y} - \dot{y}\frac{\partial\phi_{\mathrm{m}}}{\partial z} + \left(\frac{\partial}{\partial t} + \dot{x}\frac{\partial}{\partial x} + \dot{y}\frac{\partial}{\partial y} + \dot{z}\frac{\partial}{\partial z}\right)(\dot{y}A_{\mathrm{e}z} - \dot{z}A_{\mathrm{e}y})\\ &- c^2\left(\frac{\partial A_{\mathrm{e}z}}{\partial y} - \frac{\partial A_{\mathrm{e}y}}{\partial z}\right)\Bigg]dx\\ = \frac{Ze}{c}\dot{x}(\dot{y}A_{\mathrm{e}z} - \dot{z}A_{\mathrm{e}y}) &+ \frac{Ze}{c}\int\Bigg[\dot{z}\frac{\partial\phi_{\mathrm{m}}}{\partial y} - \dot{y}\frac{\partial\phi_{\mathrm{m}}}{\partial z}\\ &+ \left(\frac{\partial}{\partial t} + \dot{y}\frac{\partial}{\partial y} + \dot{z}\frac{\partial}{\partial z}\right)(\dot{y}A_{\mathrm{e}z} - \dot{z}A_{\mathrm{e}y}) - c^2\left(\frac{\partial A_{\mathrm{e}z}}{\partial y} - \frac{\partial A_{\mathrm{e}y}}{\partial z}\right)\Bigg]dx \end{aligned} \tag{21}$$

The derivative of $\mathfrak{L}_{cx}$ with respect to $\dot{x}$ yields the factor of $\mathbf{e}_x$ in the first line of Eq.(20):

$$\frac{Ze}{c}\frac{\partial\mathfrak{L}_{cx}}{\partial\dot{x}} = \frac{Ze}{c}(\dot{y}A_{\mathrm{e}z} - \dot{z}A_{\mathrm{e}y}) \tag{22}$$

This expression fits $\partial L/\partial\dot{x}_j$ on the left side of Eq.(5), while the derivative $\partial\mathfrak{L}_{cx}/\partial x$ of Eq.(21) fits the right side $\partial L/\partial x_j$. Hence, Eq.(21) yields the previously published[1] x-component $\mathfrak{L}_{cx}$ of the Lagrange function $\mathfrak{L}_{\mathrm{c}}$, where the subscript c stands for 'correction'.

The cyclical replacements $x \to y \to z \to x$ in Eq.(21) yield the components $\mathfrak{L}_{cy}$ and $\mathfrak{L}_{cz}$ of the correcting Lagrange function $\boldsymbol{\mathfrak{L}}_{\mathrm{c}}$. It can be written as a matrix:

$$\boldsymbol{\mathfrak{L}}_{\mathrm{c}} = \begin{pmatrix}\mathfrak{L}_{cx} & 0 & 0\\ 0 & \mathfrak{L}_{cy} & 0\\ 0 & 0 & \mathfrak{L}_{cz}\end{pmatrix} \tag{23}$$

The sum of the Lagrange function $\mathfrak{L}_{\mathrm{M}}$ of Eq.(18) and $\boldsymbol{\mathfrak{L}}_{\mathrm{c}}$ yields the complete Lagrange function $\boldsymbol{\mathfrak{L}}$. Using the unit matrix we can write this sum as the sum of two matrices:

$$\begin{aligned}\boldsymbol{\mathfrak{L}} &= \boldsymbol{\mathfrak{L}}_{\mathrm{M}} + \boldsymbol{\mathfrak{L}}_{\mathrm{c}}\\ &= \left(\frac{1}{2}mv^2 + e\mathbf{v}\cdot\mathbf{A}_{\mathrm{m}} - e\phi_{\mathrm{e}}\right)\begin{pmatrix}1 & 0 & 0\\ 0 & 1 & 0\\ 0 & 0 & 1\end{pmatrix} + \begin{pmatrix}\mathfrak{L}_{cx} & 0 & 0\\ 0 & \mathfrak{L}_{cy} & 0\\ 0 & 0 & \mathfrak{L}_{cz}\end{pmatrix}\end{aligned} \tag{24}$$

If one wants these matrices displayed in the vector component form of Eqs.(17) or (20) one can multiply them with the matrix

$$\mathfrak{e}_{xyz} = \begin{pmatrix} \mathbf{e}_x & 0 & 0 \\ 0 & \mathbf{e}_y & 0 \\ 0 & 0 & \mathbf{e}_z \end{pmatrix} \tag{25}$$

We observe that the term $\mathbf{g}_\mathrm{m}$ was added in Eq.(1.1-2) to overcome the problem of Maxwell's equations with causality. That this term led to the use of a matrix for the Lagrange function was an unintended consequence. The matrix notation is spread to anything that uses electromagnetic theory, e.g., the Klein-Gordon equation. We recognize here how impossible it is to foresee where a new theory will take us.

The Lagrange function $\mathfrak{L}_\mathrm{M}$ in Eq.(18) was derived for Cartesian coordinates but it could be written in the second line of Eq.(18) free of any reference to coordinates. We may check this for spherical coordinates. The three components of the velocity $\mathbf{v}$ of the point $r\mathbf{e}_r + r\vartheta\mathbf{e}_\vartheta + r\sin\vartheta\,\varphi\mathbf{e}_\varphi$ are wanted for the directions $\mathbf{e}_r$, $\mathbf{e}_\vartheta$, $\mathbf{e}_\varphi$:

$$\begin{aligned} \mathbf{v} &= \frac{\partial r}{\partial r}\frac{\partial r}{\partial t}\mathbf{e}_r + \frac{\partial(r\vartheta)}{\partial\vartheta}\frac{\partial\vartheta}{\partial t}\mathbf{e}_\vartheta + \frac{\partial(r\sin\vartheta\,\varphi)}{\partial\varphi}\frac{\partial\varphi}{\partial t}\mathbf{e}_\varphi \\ &= \dot{r}\mathbf{e}_r + r\dot{\vartheta}\mathbf{e}_\vartheta + (r\sin\vartheta)\dot{\varphi}\mathbf{e}_\varphi = \dot{x}_1\mathbf{e}_r + \dot{x}_2\mathbf{e}_\vartheta + \dot{x}_3\mathbf{e}_\varphi \qquad (26) \\ \mathbf{A}_\mathrm{m} &= A_{\mathrm{m}r}\mathbf{e}_r + A_{\mathrm{m}\vartheta}\mathbf{e}_\vartheta + A_{\mathrm{m}\varphi}\mathbf{e}_\varphi \qquad (27) \\ \mathfrak{L}_\mathrm{M} &= \frac{1}{2}m[\dot{r}^2 + r^2\dot{\vartheta}^2 + (r\sin\vartheta)^2\dot{\varphi}^2] \\ &\qquad + e[-\phi_\mathrm{e} + \dot{r}A_{\mathrm{m}r} + r\dot{\vartheta}A_{\mathrm{m}\vartheta} + (r\sin\vartheta)\dot{\varphi}A_{\mathrm{m}\varphi}] \qquad (28) \end{aligned}$$

$$\begin{aligned} &\frac{\partial\mathfrak{L}_\mathrm{M}}{\partial\dot{r}} = m\dot{r} + eA_{\mathrm{m}r}, \quad \frac{\partial\mathfrak{L}_\mathrm{M}}{\partial(r\dot{\vartheta})} = mr\dot{\vartheta} + eA_{\mathrm{m}\vartheta} \\ &\frac{\partial\mathfrak{L}_\mathrm{M}}{\partial(r\sin\vartheta\,\dot{\varphi})} = mr\sin\vartheta\,\dot{\varphi} + eA_{\mathrm{m}\varphi} \end{aligned} \tag{29}$$

$$\begin{aligned} \frac{\partial\mathfrak{L}_\mathrm{M}}{\partial r} &= e\frac{\partial}{\partial r}(-\phi_\mathrm{e} + \dot{r}A_{\mathrm{m}r} + r\dot{\vartheta}A_{\mathrm{m}\vartheta} + r\sin\vartheta\,\dot{\varphi}A_{\mathrm{m}\varphi}) \qquad (30) \\ \frac{\partial\mathfrak{L}_\mathrm{M}}{\partial(r\vartheta)} &= e\frac{\partial}{\partial(r\vartheta)}(-\phi_\mathrm{e} + \dot{r}A_{\mathrm{m}r} + r\dot{\vartheta}A_{\mathrm{m}\vartheta} + r\sin\vartheta\,\dot{\varphi}A_{\mathrm{m}\varphi}) \qquad (31) \\ \frac{\partial\mathfrak{L}_\mathrm{M}}{\partial(r\sin\vartheta\,\varphi)} &= e\frac{\partial}{\partial(r\sin\vartheta\,\varphi)}(-\phi_\mathrm{e} + \dot{r}A_{\mathrm{m}r} + r\dot{\vartheta}A_{\mathrm{m}\vartheta} + r\sin\vartheta\,\dot{\varphi}A_{\mathrm{m}\varphi}) \qquad (32) \end{aligned}$$

One recognizes the equivalence of Eqs.(29)–(32) and Eq.(17) if one chooses the variables r, $r\vartheta$, $(r\sin\vartheta)\varphi$ rather than r, ϑ, φ for the spherical coordinates.

The Lagrange function suggests using the coordinates r, $r\vartheta$, $(r\sin\vartheta)\varphi$. Both notations work. There is an evident advantage in physics of having the same dimension of length for all coordinates as in Cartesian coordinates. No such advantage exists in pure mathematics since it does not use physical dimensions.

More effort is required to write the correcting Lagrange function $\mathfrak{L}_\mathrm{c}$ of Eqs.(23) and (24) for spherical coordinates since the components $\mathfrak{L}_{\mathrm{c}x}$, $\mathfrak{L}_{\mathrm{c}y}$, $\mathfrak{L}_{\mathrm{c}z}$ apply only to Cartesian coordinates. We can obtain the correcting terms $\mathfrak{L}_{\mathrm{c}r}$, $\mathfrak{L}_{\mathrm{c}\vartheta}$, $\mathfrak{L}_{\mathrm{c}\varphi}$ for spherical coordinates from Eq.(13). The calculation is shown in Section 6.5. Here we list only $\mathfrak{L}_{\mathrm{c}r}$, $\mathfrak{L}_{\mathrm{c}\vartheta}$, $\mathfrak{L}_{\mathrm{c}\varphi}$, which have to be substituted for $\mathfrak{L}_{\mathrm{c}x}$, $\mathfrak{L}_{\mathrm{c}y}$, $\mathfrak{L}_{\mathrm{c}z}$ in Eqs.(23) and (24). The vectors $\mathbf{e}_x$, $\mathbf{e}_y$, $\mathbf{e}_z$ in Eq.(25) must be replaced by $\mathbf{e}_r$, $\mathbf{e}_\vartheta$, $\mathbf{e}_\varphi$:

$$\begin{aligned}\mathfrak{L}_{\mathrm{c}r} = {}& \frac{Ze}{c}\dot{r}(r\dot{\vartheta}A_{\mathrm{e}\varphi} - r\sin\vartheta\,\dot{\varphi}A_{\mathrm{e}\vartheta}) + \frac{Ze}{c}\int\Bigg[\sin\vartheta\dot{\varphi}\frac{\partial\phi_\mathrm{m}}{\partial\vartheta} - \frac{\dot{\vartheta}}{\sin\vartheta}\frac{\partial\phi_\mathrm{m}}{\partial\varphi} \\ &+ r\ddot{\vartheta}A_{\mathrm{e}\varphi} - r\sin\vartheta\,\ddot{\varphi}A_{\mathrm{e}\vartheta} + \left(\dot{\vartheta}\frac{\partial}{\partial\vartheta} + \dot{\varphi}\frac{\partial}{\partial\varphi}\right)(r\dot{\vartheta}A_{\mathrm{e}\varphi} - r\sin\vartheta\,\dot{\varphi}A_{\mathrm{e}\vartheta}) \\ &- \frac{c^2}{r\sin\vartheta}\left(\frac{\partial(A_{\mathrm{e}\varphi}\sin\vartheta)}{\partial\vartheta} - \frac{\partial A_{\mathrm{e}\vartheta}}{\partial\varphi}\right)\Bigg]dr\end{aligned} \tag{33}$$

$$\begin{aligned}\mathfrak{L}_{\mathrm{c}\vartheta} = {}& \frac{Ze}{c}r\dot{\vartheta}(r\sin\vartheta\,\dot{\varphi}A_{\mathrm{e}r} - \dot{r}A_{\mathrm{e}\varphi}) + \frac{Ze}{c}\int\Bigg[\frac{\dot{r}}{r\sin\vartheta}\frac{\partial\phi_\mathrm{m}}{\partial\varphi} - r\sin\vartheta\,\dot{\varphi}\frac{\partial\phi_\mathrm{m}}{\partial r} \\ &+ r\sin\vartheta\,\ddot{\varphi}A_{\mathrm{e}r} - \ddot{r}A_{\mathrm{e}\varphi} + \left(\dot{r}\frac{\partial}{\partial r} + \dot{\varphi}\frac{\partial}{\partial\varphi}\right)(r\sin\vartheta\,\dot{\varphi}A_{\mathrm{e}r} - \dot{r}A_{\mathrm{e}\varphi}) \\ &- \frac{c^2}{r}\left(\frac{1}{\sin\vartheta}\frac{\partial A_{\mathrm{e}r}}{\partial\varphi} - \frac{\partial(rA_{\mathrm{e}\varphi})}{\partial r}\right)\Bigg]r\,d\vartheta\end{aligned} \tag{34}$$

$$\begin{aligned}\mathfrak{L}_{\mathrm{c}\varphi} = {}& \frac{Ze}{c}r\sin\vartheta\,\dot{\varphi}(\dot{r}A_{\mathrm{e}\vartheta} - r\dot{\vartheta}A_{\mathrm{e}r}) + \frac{Ze}{c}\int\Bigg[r\dot{\vartheta}\frac{\partial\phi_\mathrm{m}}{\partial r} - \frac{\dot{r}}{r}\frac{\partial\phi_\mathrm{m}}{\partial\vartheta} + \ddot{r}A_{\mathrm{e}\vartheta} - r\ddot{\vartheta}A_{\mathrm{e}r} \\ &+ \left(\dot{r}\frac{\partial}{\partial r} + \dot{\vartheta}\frac{\partial}{\partial\vartheta}\right)(\dot{r}A_{\mathrm{e}\vartheta} - r\dot{\vartheta}A_{\mathrm{e}r} - \frac{c^2}{r}\left(\frac{\partial(rA_{\mathrm{e}\vartheta})}{\partial r} - \frac{\partial A_{\mathrm{e}r}}{\partial\vartheta}\right)\Bigg]r\sin\vartheta\,d\varphi\end{aligned} \tag{35}$$

For an explanation of the Lagrange function of Eq.(24) consider a spinning bullet shot upward. The kinetic energy is transformed into potential energy on the way up and back into kinetic energy on the way down. Such transformable energies are represented by the Lagrange function $\mathfrak{L}_\mathrm{M}$ of Eq.(18). The rotational energy of the spinning bullet will not be transformed into either kinetic or potential energy. A more complicated bullet that extends helicopter blades when its kinetic energy approaches zero could transform the rotational energy into potential energy—but only in the atmosphere, not above it. Which energies can be transformed into each other depends on the particular physical circumstances. An electrically charged particle traveling through an electromagnetic field permits transfer between the kinetic energy $\frac{1}{2}mv^2$ and the energies $e\mathbf{v} \cdot \mathbf{A}_{\mathrm{m}}$ or $e\phi_{\mathrm{e}}$ according to Eq.(18). The transformation between the energy $\frac{1}{2}mv^2$ and the energies $(Ze/c)v_x v_y A_{\mathrm{e}z}$ or $(Ze/c)\int v_z(\partial\phi_{\mathrm{m}}/\partial y)dx$ may be more complicated but it is not prohibited.

1.4 Dogma of the Continuum in Physics

The concept of a continuous space and time or a space–time continuum seems to have originated at the Eleatic school of the Greeks in southern Italy. Zeno of Elea (c. 490–c. 430 BCE) advanced the paradoxes of the race between Achilles and the turtle as well as of the arrow that cannot fly to show that infinite divisibility of space and time was not possible. Aristotle (384–322 BCE) made it clear that the concept of the continuum was discussed before he wrote his *Physica*:

> Now a motion is thought to be one of those things which are continuous, and it is in the continuous that the infinite first appears; and for this reason, it often happens that those who define the continuous use the formula of the infinite, that is, they say that continuous is that which is infinitely divisible. [Apostle 1969, Book III (Γ), 1, § 2]

Aristotle refuted the paradox of the race between Achilles and the turtle by pointing out that the sum of infinitely many distances $\Delta x + \Delta x/2 + \Delta x/4 + \ldots$ is not infinite. He argued the concept of the continuum for space, time, and motion so successfully, that it has been challenged very rarely ever since[1]. Isaac Newton (1642–1727) took this concept apparently so much for granted that he did not mention where it came from, even though he was very meticulous in listing and elaborating his assumptions. The Greeks thought of *denumerable infinite* when they talked about infinite divisibility. Gottfried Wilhelm Leibniz (1646–1716) and Newton generalized or abstracted this concept to *nondenumerable infinite* when they developed the differential calculus and arrived at space and time differentials dx and dt.

The concept of a continuum is clearly thinkable, which is all that is required in mathematics. When we mention space and time we are no longer in mathematics but in physics and we require that a concept like space–time continuum must be observable[2]. If we could observe something at the location x and something else at $x + dx$ or at the time t and $t + dt$ we could prove the existence of a space and time continuum by observation. One would not expect that any one proposing such a proof would be taken seriously. A spatial

[1]Book V(E) 3, § 5 defines continuous; Book VI(Z) 1, 2 elaborates the concept of the continuum further; Zeno of Elea is refuted in Book VI(Z) 2 and 9 (Apostle 1969; Aristotle 1930).

[2]We note that pure mathematics has no space or time variables. Instead it has complex variables, real variables, integer variables, random variables, rational variables, etc. No physical units like *meter* or *second* are attached to variables in pure mathematics.

resolution dx or a time resolution dt is no more observable than the number of angels that can dance on the tip of a pin.

How did we get into the situation that some of the most basic concepts of physics are inherently not observable? We treated physics as a branch of mathematics! This was quite acceptable at the time of the ancient Greeks. Today we recognize mathematics as a science of the thinkable and physics as a science of the observable. They cannot be more for each other than a tool or a source of inspiration.

To see how concepts of pure mathematics were transformed into concepts of physics consider the definition of the continuum with the help of the real numbers on the numbers axis. A one-dimensional continuum has the same topology as the real numbers. It is usual to map any real number x by means of the exponential function e^{iyx} onto the unit circle in the complex plane. The resulting map $\{e^{iyx}\}$ is called the *character group of the topologic group of real numbers*. Why do we map the real numbers on a circle and not on an ellipse or a polygon? The definition of a character group does not require the circle, it only permits it (van der Waerden 1966, Part 1, § 54). A shift along the numbers axis becomes a rotation along the unit circle. Only the circle remains unchanged under infinitesimal as well as finite rotations. If we do away with the unobservable concept of infinity and use an arbitrarily large but finite (time) interval T, and if we further do away with the unobservable infinitesimal and replace it with an arbitrarily small but finite (time) interval Δt we obtain a finite number $N = T/\Delta t$ of (time) intervals. Instead of a circle we would use a regular polygon with N corners.

It is not yet evident why the concepts of time and space should be connected either to the circle or a regular polygon. We can only assume that the use of the circle is due to the *dogma of the circle*, which has influenced European thinking at least since Plato (c. 428–c. 348 BCE). It is usually connected with astronomy but its influence goes much further. Claudius Ptolemy (c. 90–c. 168) stated it as follows:

> ... we believe it is the necessary purpose and aim of the mathematician to show forth all the appearances of the heavens as products of regular and circular motions. (Ptolemy 1952, Almagest, Book III, 1, p. 83, second paragraph)

If one believes that strongly in the circle one should have no trouble believing in the continuum derived with the help of the dogma of the circle. It is worth remembering that Nicholas Copernicus (1473–1543) could not overcome the dogma of the circle. We celebrate Johannes Kepler (1571–1630) for breaking the hold of the circle on our thinking with his book *Astronomia Nova* in 1609. Actually, this hold was never completely overcome. The deferents and epicycles of Ptolemy and Copernicus evolved into the Fourier series in complex notation. The representation of a function by a Fourier series expansion is as pervasive today as the description of the orbits of celestial objects by a superposition of circles before Kepler. It prevented electrical communications for almost a century to advance from a steady-state theory to a transient theory, even though in retrospect the use of a steady-state theory for the transmission of information is hard to believe.

This short summary of the dogma of the circle should suffice to show that deriving the dogma of the continuum from the dogma of the circle worked in its time but is no longer a good idea. What we can observe are distances between objects measured with some form of a ruler. In other words, we observe relative locations or *points in a man-made coordinate system*. Furthermore, we can observe changes relative to a standard change. This standard change was once our own aging process due to its unique importance for us, refined by the orbiting of the Earth around the Sun and the rotation of the Earth around its own axis. Progress in technology replaced this original standard of change by man-made standards, e.g., by the pendulum clock, the quartz clock, and the cesium clock. The words *space*, *time*, and *space–time* permit us to talk efficiently with few words about points in a coordinate system, relative changes, and points in moving coordinate systems, all of which are observable as long as we stay clear of the concept of the continuum (Harmuth 1989, Chapters 2–4). But these words took on a life of their own and led to the continuum of space, time, and space-time. Since these continuums cannot be observed they cannot be described in terms of physics. Mathematics was used to overcome the difficulty. This created the impression that physics is a branch of mathematics. Let us illustrate this development with the help of a reference:

> From the essence of space remains in the hands of the mathematician, using such abstraction, only one truth: that *it is a three-dimensional continuum*[3]. (Weyl 1921, 1968, vol. II, p. 213)

Einstein (1879–1955) is remembered for his curved space–time. But in his later years he had a change of mind and called it an invention:

> The psychological subjective feeling of time enables us to order our impressions, to state that one event precedes another. But to connect every instant of time with a number, by the use of a clock, to regard time as an one-dimensional continuum, is already an invention. So also are the concepts of Euclidean and non-Euclidean geometry, and our space understood as a three-dimensional continuum. (Einstein and Infeld 1938, p. 311)

A collection of quotations about the concepts of space and time from the days of Aristotle to 1976 is presented by Harmuth (1989, Sec. 1.4).

For a simple example of what finite differences Δx, Δt rather than differentials dx, dt can do for us consider the cases of infinite energy that currently are resolved by *renormalization*. It has been pointed out before that some of these cases are caused by a problem of Maxwell's equations with the causality law (Harmuth et al. 2001, Sections 4.4, 4.5). The modified Maxwell equations with a term for magnetic current densities provide one solution to the problem

[3]Vom Wesen des Raumes bleibt dem Mathematiker bei solcher Abstraktion nur eine Wahrheit in Händen: daß *er ein dreidimensionales Kontinuum ist*.

by being able to account for the magnetic dipole currents produced by rotating magnetic dipoles.

The finite differences Δx and Δt of space and time resolution provide a second mechanism to avoid infinite energies. This second mechanism is based strictly on quantum theory. Consider the interaction of an electron with itself. A photon is emitted and reabsorbed within a time Δt. The uncertainty relation $\Delta E \Delta t > \hbar$ permits a large energy of the photon if Δt is very small. For $\Delta t \rightarrow dt$ one obtains a possible infinite energy for the photon and thus an infinite *self-energy* for the electron. This infinite self-energy is currently overcome by renormalization, which leads to the subtraction $\infty - \infty$. If Δt cannot be smaller than a finite observation or resolution time we can never get infinite self-energy and we never have to use renormalization with a subtraction of infinite from infinite.

The success of a mechanism that avoids infinite energies and is based on quantum mechanics suggests applying it to charge renormalization too and then to the unsolved problem of energy renormalization in a quantum gravitation theory. We want to show here generally that all this is derived from a basic theorem of information theory that applies to observed, processed or transmitted information:

Information is always finite.

For explanation consider a measurement that yields a possible largest result Y and a possible smallest result ΔY, usually referred to as the resolution. Let Y and ΔY be chosen so that the ratio $Y/\Delta Y = N$ is an integer; the choice $N = 2^{n+1} - 1$, which yields $N = 111\ldots11$ (n digits 1) in binary notation, works best. The information obtained by one measurement is at most[4] equal to the number of binary digits required to write $N = 2^{n+1} - 1$ as a binary number. For $2^n < N < 2^{n+1}$ we need n binary digits 1 or 0 to write N and they represent the information $\log_2 2^n = n$ [bit] or $n/8$ [byte] for $N = 2^{n+1} - 1$.

We note that the units *bit* or *byte* look like physical units such as *meter* or *kilometer*, but n is a pure number without a physical dimension. A number 10011 with $n = 5$ binary digits may stand for 19 meter, Joule, Volt, ... The important fact is that Y can be arbitrarily large but finite and ΔY can be arbitrarily small but finite. Information theory excludes the infinite as well as the infinitesimal from any theory based on observation. Whenever we come across something infinitely large or infinitesimally small, that is not merely based on writing convenience like $y \rightarrow \infty$ or $y \rightarrow dy$ rather than $y \ggg 1$ or $y \lll 1$, we know that the most basic law of information theory is violated. Infinite information is no better than infinite energy or infinitesimal time intervals in a science based on observation and experimentation.

One sometimes reads that time in mathematics can run forward and back-

---

[4]This maximum of the information is obtained if the result y of a measurement has equal probability to be in any one of the intervals $0 \leq y < \Delta Y$, $\Delta Y \leq y < 2\Delta Y$, ..., $(N-1)\Delta Y \leq y < N\Delta Y = Y$.

ward equally well while this is not so in physics. Such statements ignore that the concept of time does not exist in pure mathematics since variables in mathematics do not have a physical dimension like *second*. A time variable indicates that mathematics is applied to physics. In this case we must satisfy not only mathematical axioms and rules but in addition physical laws. The causality law is such a physical law. It introduces the universally observed distinguished direction of time since an effect comes after its cause. If we can find an example in physics where a cause comes after its effect we would observe that the time runs opposite to its usual direction.

1.5 Concept of Space Based on Finite Differences

The ancient Greeks did not clearly distinguish between mathematics as a science of the thinkable and physics as a science of the observable. They thought in terms of the unobservable continuum and this had an enormous influence on our thinking about space. The continuum was not seriously challenged until the twentieth century. Two quotes from Schrödinger show the change of thinking:

> In Einstein's theory of gravitation matter and its dynamical interaction are based on the notion of an intrinsic geometric structure of the space–time continuum. The ideal aspiration, the ultimate aim, of the theory is not more and not less than this: A four-dimensional continuum endowed with a certain intrinsic geometric structure, a structure that is subject to certain inherent purely geometric laws, is to be an adequate model or picture of the 'real world around us in space and time' with all that it contains and including its total behavior, the display of all events going on in it. (Schrödinger 1950, p. 1)

Quantum theory and the general theory of relativity were the two great physical theories of the twentiethy century. Both were developed with the help of differential calculus but quantum theory was always against the concept of infinitesimally accurate observations or measurements. Six years after the quotation cited above Schrödinger wrote as follows:

> We must not admit the possibility of continuous observation .... The idea of a continuous range, so familiar to mathematicians in our days, is quite exorbitant, an enormous extrapolation of what is really accessible to us. (Schrödinger 1956)

This was the time when information theory was developed. Information could be measured in *bit* or *byte*. The thought of observing, processing, or transmitting nondenumerably infinite bit of information was preposterous.

Besides the concept of the continuum our usual geometry assumes that we can look 'from outside' on geometric structures drawn on paper or constructed in *physical space*. This is quite adequate except if we start thinking about the universe. We cannot look at the universe from the outside since we are 'in it' or part of it. To make this point more understandable consider the equilateral

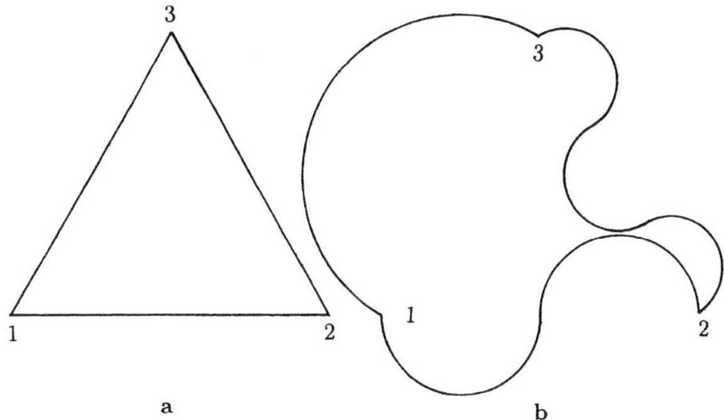

FIG.1.5-1. Two equilateral triangles.

triangle of Fig.1.5-1a. It has three vertices 1, 2, 3 and three equally long *edges*[1]. Figure 1.5-1b shows again three vertices and three equally long edges. We have no difficulty seeing a difference between the two triangles. However, if we plot the three edges in Fig.1.5-1b in planes that are perpendicular to the paper plane, the figure looks exactly like Fig.1.5-1a if we look from above at the paper plane and see the curved edges projected onto the paper plane. We can see different things if we can look from different directions. This is geometry viewed from outside.

Let the triangles of Fig.1.5-1 be implemented 'in space' by three rods or thick wires and make them very long. If we put an ant schooled in mathematics and physics on these structures it is 'in the structure' or part of the structure. The ant will observe that there are vertices where two rods come together. It can also observe that the rods are equally long by pacing, which means it counts how often it has to put its left front foot forward when walking from one vertex to the next. Finally, the ant will observe that there is a forward and a backward direction; it can crawl in either direction indefinitely. The ant will first assume it lives on an unlimited structure according to Fig.1.5-2. Giving it some more thought, it may make a scent mark at one of the vertices and run either in the forward or the backward direction. It observes that it returns to the starting point after having run along three edges. This indicates that it lives in a limited or *closed* structure according to Fig.1.5-1, but it cannot tell which of the two.

Consider a triangle connecting Earth, Alpha Centauri, and Sirius. We can measure the distances from Earth to Alpha Centauri and Sirius directly, but not the distance from Sirius to Alpha Centauri. There are various methods to measure such distances; they all use electromagnetic waves. These waves do not propagate along a straight line in the sense of Euclid since they get bended by the masses in the universe and we do not know the distribution or

[1]We use the word *edges* rather than *sides* since the word sides takes on a different meaning if we combine four triangles into a tetrahedron in space later on.

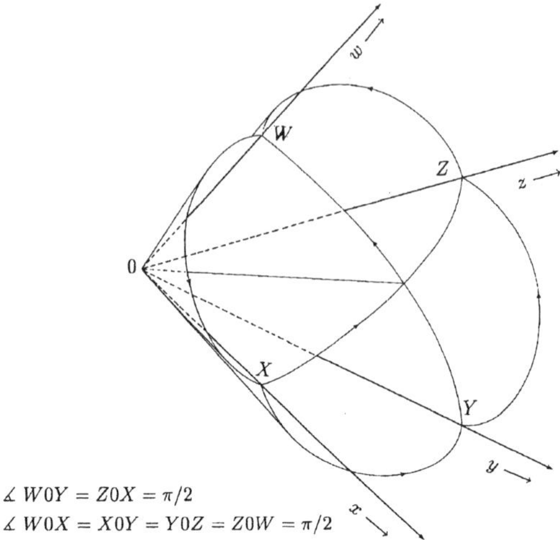

FIG.1.5-6. Four axes x, y, z, w with angles $\pi/2$ between them. Some of the angles are measured on a conical surface rather than on a Euclidean plane.

except for the increased difficulty of drafting. A discrete structure with any number of dimensions can be implemented 'in space' and drafted 'on a surface'. We can choose the surface even though it will usually be a Euclidean paper plane. But a paper plane can be bended into a cylinder or a conical surface. There is no such choice for the space in which the structures of Figs.1.5-3 to 1.5-5 can be implemented.

The claim that Figs.1.5-4 and 1.5-5 represent four- and five-dimensional structures that can be implemented in space deviates from the usual thinking. One frequent objection is that we can only have three mutually perpendicular vectors, axes, or directions in space. Consider Fig.1.5-6 to see where this belief comes from. It shows four axes x, y, z, w with angles of 90° between any two of them. This is possible because four of the angles are measured on conical surfaces rather than planar ones. It is evident that the principle of Fig.1.5-6 can be generalized to more than four dimensions. Since a Euclidean plane can be bent into a conical surface without changing the Gaussian curvature it would be difficult for inhabitants of the surfaces of Fig.1.5-6 to distinguish between the planar and the conical surfaces. An outside observer has no such difficulty.

A second problem of Fig.1.5-6 is the existence of surfaces on which one can define angles. Our Figs.1.5-1 to 1.5-5 require the paper surface for drafting but we always pointed out that the structures could be implemented in space with rods and vertices. Since at least some of the rods in Figs.1.5-4 and 1.5-5 must be bent one cannot use the structures to define planar surfaces as one does with the tetrahedron of Fig.1.5-3 in Euclidean geometry. We shall show

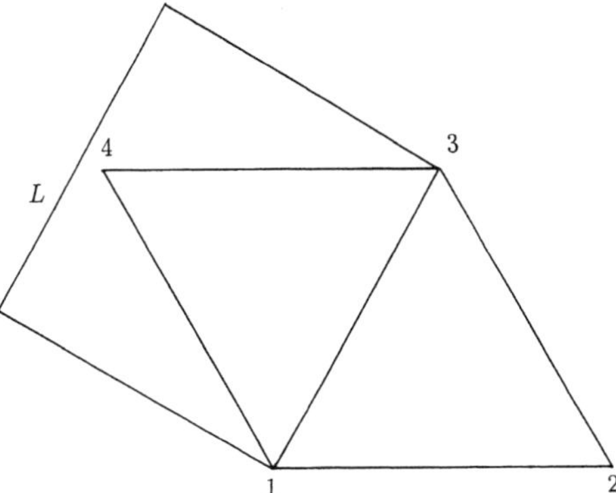

FIG.1.5-7. Making a third dimension invisible by wrapping the plane L with the vertex 4 around the edge between vertices 1 and 3.

later on that the lack of surfaces is an important feature of a geometry based on finite differences.

The concept of hiding dimensions has attracted attention lately. To see how it works in a geometry of finite differences consider Fig.1.5-7. We again have the triangle $\overline{123}$ of Fig.1.5-1 that shall be implemented in space with three rods. In addition, we have a triangle $\overline{134}$ that shall be drafted on a sheet of paper L. Since three edges meet at the vertices 1 and 3 we must call this structure a three-dimensional structure. Let the sheet L of paper be wrapped around the edge connecting the vertices 1 and 3. The third dimension disappears from view but it is, of course, still there.

For a generalization of this concept refer to Fig.1.5-8. We have the three-dimensional triangle of Fig.1.5-3. Three planes L_1, L_0, L_{-1} are shown in which the vertices 3, 1, and 4, as well as 2 are located. We compress these three planes into one plane L_0. Then we replace the plane L in Fig.1.5-7 by the plane L_0. The compressed structure of Fig.1.5-8 may be wrapped around the edge $\overline{13}$ of Fig.1.5-7. Eureka, the three dimensions of Fig.1.5-8 have disappeared from view. Without compression and wrapping the combination of the structure of Fig.1.5-8 and the triangle $\overline{123}$ of Fig.1.5-7 would have yielded a five-dimensional structure.

The use of ever higher unobservable spatial dimensions reminds one of the use of ever more epicycles for the representation of planetary orbits before Kepler. In a differential theory a particle moves from x to $x + \Delta x$ in infinitely many steps of length dx. The theory can describe the particle at any one of these steps. The assumption that the particle does not move in one dimension but in many provides one with degrees of freedom that make the theory reconcilable with observation. This is quite different in a geometry based on finite differences Δx. We may observe a particle at the location x that disappears. Shortly later an equal or related particle appears at the location $x + \Delta x$. Since

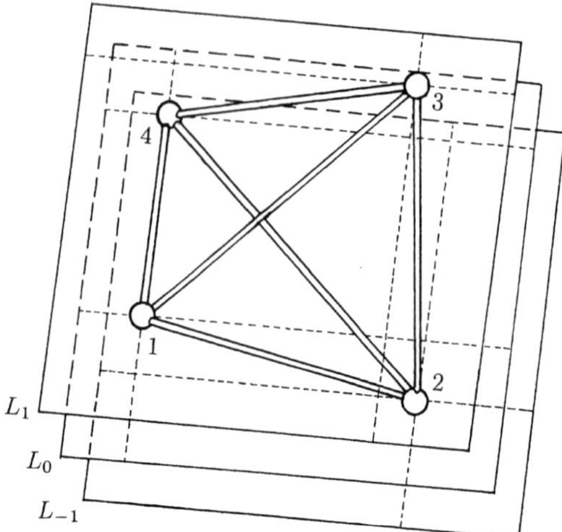

FIG.1.5-8. Compression of three dimensions into one paper plane L_0 that is then wrapped around the edge from vertex 1 to 3 in Fig.1.5-7 instead of the plane L shown there.

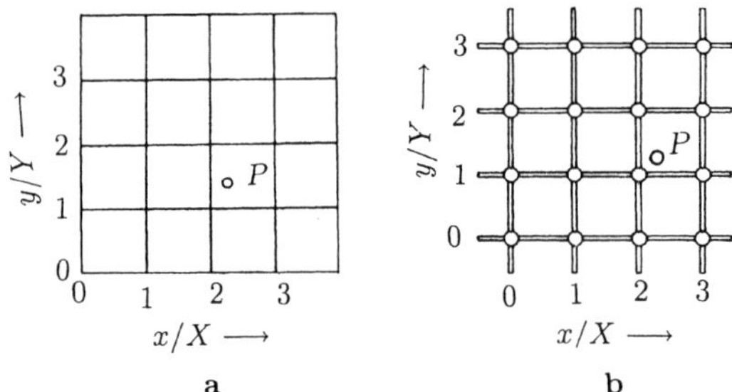

FIG.1.5-9. A two-dimensional continuous coordinate system (a) and a corresponding discrete coordinate system (b).

we can neither observe the particle continuously nor hang a license plate on it we are never sure that the same particle disappeared at x and reappeared at $x + \Delta x$. Hence, a first assumption is required to decide whether the particles are the same or not. Since we cannot observe what happened to the particle during the transition from x to $x + \Delta x$ we can make as many assumptions as needed to match the theory to the observation. We do not need unobservable spatial dimensions since the transition from x to $x + \Delta x$ is unobservable.

We turn to coordinate systems. The continuous coordinate system of Fig.1.5-9a defines a point for any real variable x/X and y/Y, which means it defines a surface. This is not evident since the paper plane used for the drafting and the defined surface are not readily discriminated. We will return

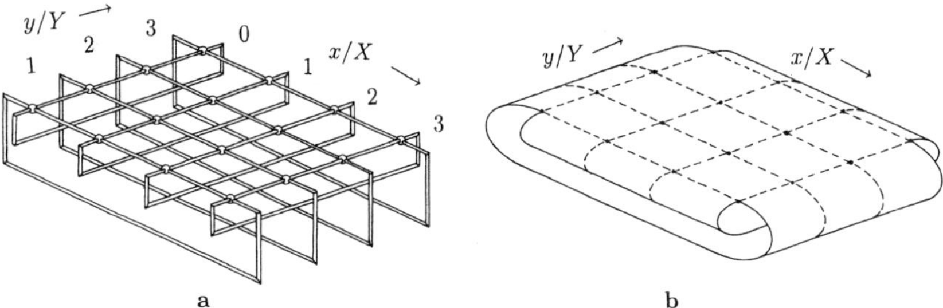

FIG.1.5-10. A closed two-dimensional discrete coordinate system with 16 points (a) and the surface obtained by replacing the 16 discrete points by nondenumerable many points (b).

to this point presently. In Fig.1.5-9b we have a discrete coordinate system that defines points for integer values x/X and y/Y. There is no suggestion that anything more than a finite or a denumerable number of points is defined. We use rods and small spheres to show the discrete coordinate system. Such a coordinate system will be called n-dimensional if no more than n coordinate 'rods' intersect at any coordinate point.

The coordinate systems in Fig.1.5-9 have a limited extension. We can either extend them arbitrarily far—but not infinitely far—beyond their limits or we can turn them into *closed* systems. The closed discrete coordinate system is shown in Fig.1.5-10a and the closed continuous systems in Fig.1.5-10b. There is no surprise with Fig.1.5-10a; it clearly shows a man-made structure. This is not so for Fig.1.5-10b. We see a surface with coordinate lines painted on. The difficulty we had in Fig.1.5-9a with the distinction between the paper plane and a surface defined by the coordinate system has been resolved. The coordinate lines in Fig.1.5-10b are clearly man-made but the surface appears to be mathematics-made. We note that most points in Fig.1.5-10b are 'on' the surface, but there are edge points and four corner points. No such edge or corner points exist in Fig.1.5-10a; all 16 points are equal.

Let us advance to three dimensions. Figure 1.5-11 shows the extension of Fig.1.5-9b to the three dimensions x, y, z. We recognize 8 interior points where three coordinate lines or rods intersect, 24 surface points where two rods intersect and one terminates, 24 edge points where one rod goes through and two terminate, and finally 8 corner points where three rods terminate.

The extension of Fig.1.5-11 to a continuous coordinate system would produce a cube with coordinate lines painted on the surface and through its interior. We shall use the word *space*—without the modifier three-dimensional—for this extension of the term surface[2]. There is no reason why the physical con-

[2]The connection between space and surface is not so close in everyday language. We say, e.g., "I have space" but not "I have surface".

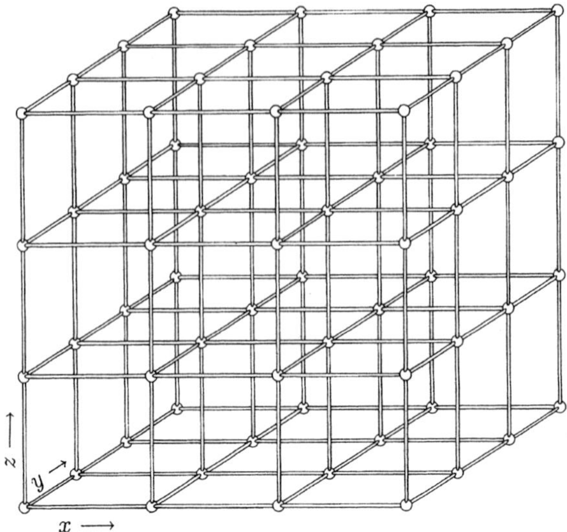

FIG.1.5-11. Three-dimensional discrete coordinate system with 64 points of which 8 are interior points, 24 surface points, 24 edge points, and 8 corner points.

cept of space should not be distinguished from the mathematical concept. The problem of distinguishing between the physical concept of space and the mathematical concept of a three-dimensional continuum does not exist in a discrete geometry. Neither Fig.1.5-9b nor 1.5-10a suggests the concept of 'surface'.

The coordinate system of Fig.1.5-11 can be transformed into a closed one but the drafting problem becomes overwhelming. We show Fig.1.5-11 again in Fig.1.5-12 but there are now three additional lines shown: dotted, dashed, and dashed-dotted. These lines connect the points 0,1,1 and 3,2,2 of the open coordinate system. Furthermore, the points 1,0,1 and 2,3,2 as well as 1,1,0 and 2,2,3 are connected. There are a total of $24 + 24 + 8 = 56$ non-interior or outside points in Fig.1.5-12, which would require 28 connections rather than the three shown. If this seems complicated one might consider the equivalent of Fig.1.5-12 for a continuous, three-dimensional coordinate system.

The advancement from the three-dimensional coordinate systems of Figs. 1.5-11 and 1.5-12 to four dimensions is a challenge to one's drafting skills. First we reduce the coordinate points or spheres from $4^3 = 64$ in Fig.1.5-11 to the absolute minimum $3^3 = 27$, which becomes $3^4 = 81$ in four dimensions. This is a large number but much better than $4^4 = 256$. Figure 1.5-13 shows such a four-dimensional coordinate system with the variables x, y, z, w. To help one understand it we show in Fig.1.5-14 the corresponding three-dimensional coordinate system with $3^3 = 27$ coordinate points. Figure 1.5-13 consists of three structures according to Fig.1.5-14 shifted in the direction w from $w = 0$ to $w = 1$ and $w = 2$. We can draft such a structure on a surface and we can

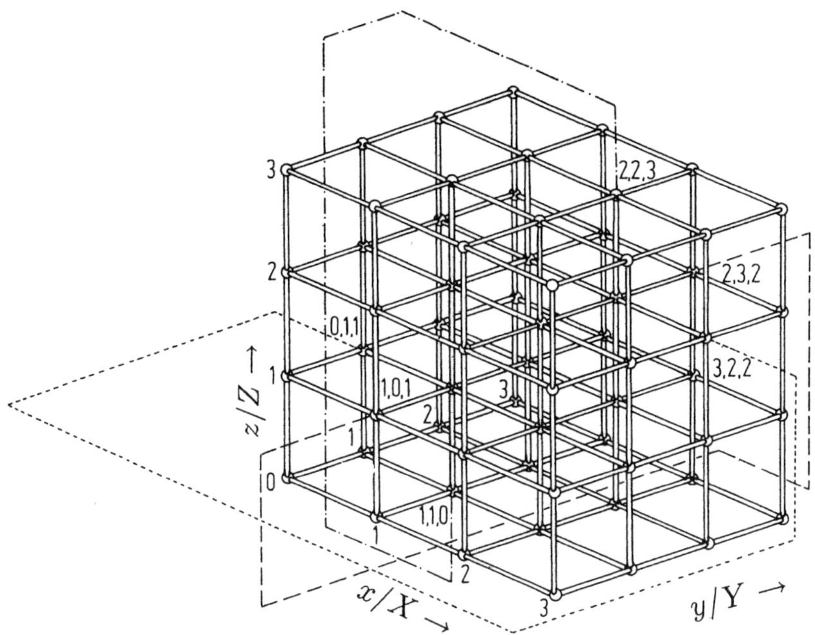

FIG.1.5-12. Three-dimensional closed or bounded coordinate system. Three typical connections of surface points that change the open to a closed coordinate system are shown by dotted, dashed, and dashed-dotted lines.

construct it with rods and small spheres in space.

To make Fig.1.5-13 more understandable consider an inspector for hidden things who is restricted to the three dimensions x, y, z. He/she can visit only the $3^3 = 27$ points with $w = 0$ but not the $2 \times 3^3 = 54$ points with $w = 1$ or $w = 2$. The fourth dimension appears to be the ideal hiding place. But it is easier to restrain mathematical inspectors to the dimensions x, y, z than physical inspectors. Furthermore, even though our inspectors cannot walk to the points with coordinate $w = 1$ or $w = 2$ they can still see them, since the propagation of light is independent of the coordinate system.

We have shown that a geometry based on finite differences permits one to implement structures with more than three dimensions in physical space. An equally surprising result is that a discrete coordinate system can always be reduced to a one-dimensional coordinate system. Figure 1.5-15a shows the two points A and B in the usual two-dimensional coordinate system. The coordinate distances $x = 5 - 2 = 3$ and $y = 5 - 1 = 4$ yield the Pythagorean distance $\sqrt{3^2 + 4^2} = 5$.

In Fig.1.5-15b the 64 coordinate points of Fig.1.5-15a are shown again but they are now numbered consecutively from 0 to 63, which eliminates y

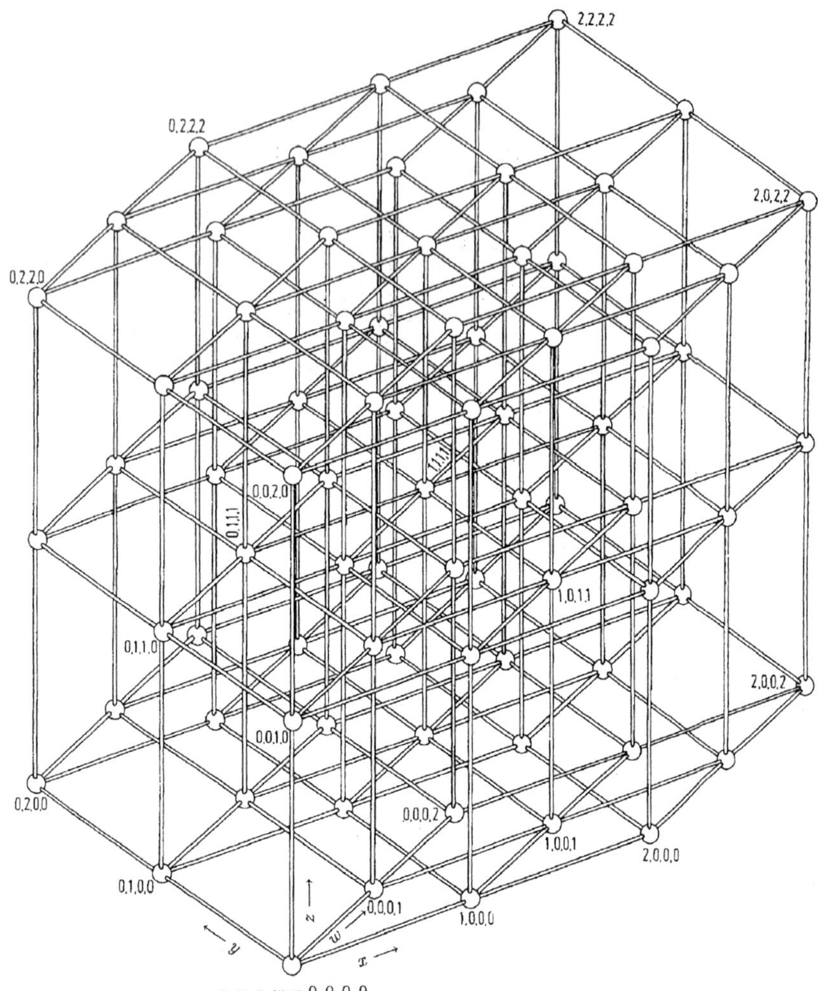

FIG.1.5-13. Four-dimensional, discrete Cartesian coordinate system with $n^r = 3^4 = 81$ grid points and $rn^{r-1} = 4 \times 3^3 = 108$ rods.

and leaves us with only one dimension x. It becomes difficult to calculate the Pythagorean distance, but it may no longer be of much interest. An ant running along the spatial axis x will state the distance from A to B as $45 - 10 = 35$, and this will be the distance of interest. As a subtle aside we note that no square root is needed.

Figure 1.5-15c shows a different way to connect the coordinate points of Fig.1.5-15b. The distance between A and B along the coordinate axis is now

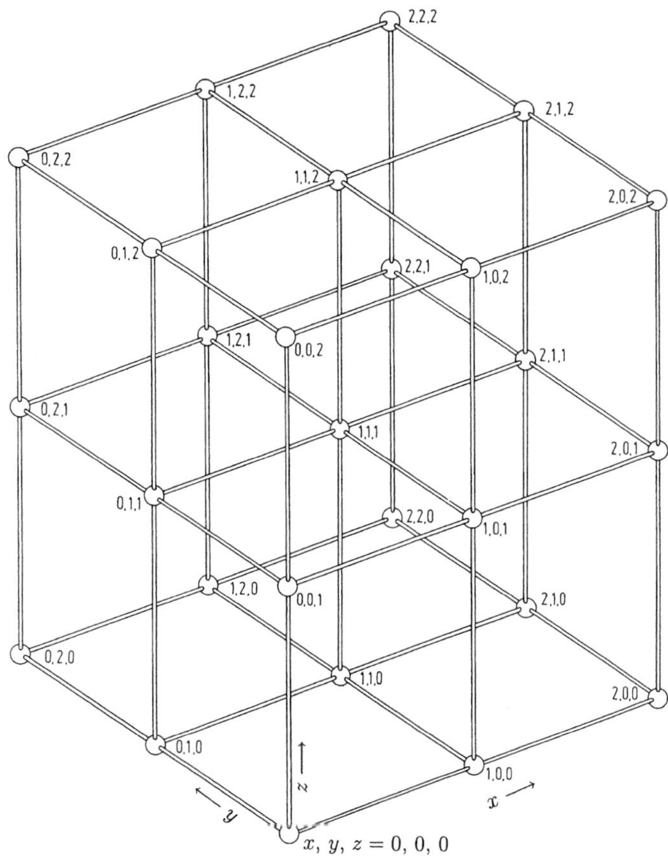

FIG.1.5-14. Three-dimensional version of Fig.1.5-13. Figure 1.5-13 shows three such structures shifted in the direction w from $w = 0$ to $w = 1$ and $w = 2$.

$33 - 26 = 7$, again without a square root. Measuring the distance according to Fig.1.5-15a makes good sense to an outsider looking at the 64 coordinate points. An insider having to run along the coordinate axis will not find the Pythagorean distance very helpful.

A possible extension of Fig.1.5-15c for 'three dimensions' is shown in Fig.1.5-16. We need only one dimension x to connect all discrete coordinate points in three dimensions.

Figures 1.5-15 and 1.5-16 make it clear that the concept of 'dimension' is flexible in a geometry of finite differences. But the illustrations also show the importance of the concept of topology. If we consider Fig.1.5-15a to represent an area of the surface of Earth we will consider it plausible that the temperature

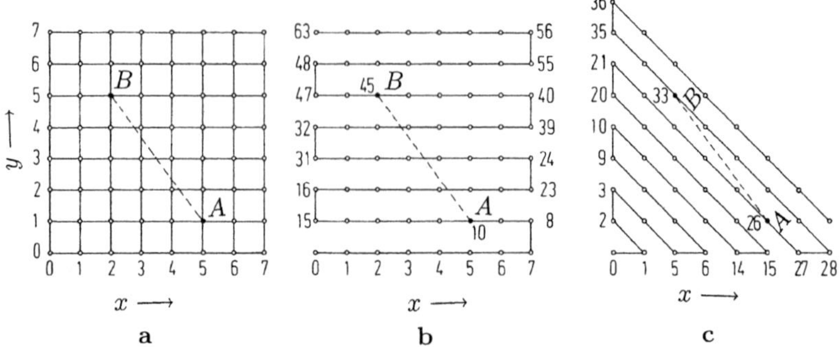

FIG.1.5-15. Two-dimensional, discrete Cartesian coordinate system (a), its replacement by a one-dimensional coordinate system (b), and the generalization of the principle to coordinate systems with denumerable many points.

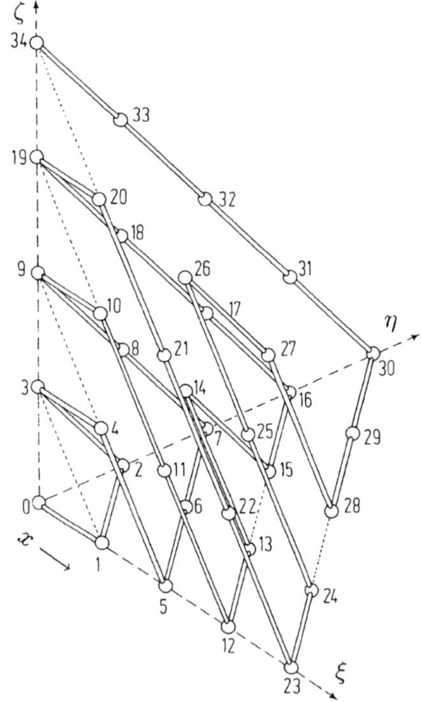

FIG.1.5-16. Replacement of a three-dimensional, discrete Cartesian coordinate system with denumerable many points according to Fig.1.5-14 by a one-dimensional coordinate system.

in point B at the time $t + \Delta t$ will depend on the temperature at the time t in point B as well as in the neighboring points x, $y = 1$, 5; 2, 4; 2, 6; and 3, 5. The topology of Fig.1.5-15b does not appear to be useful in this case. However, the situation changes if we consider Fig.1.5-15b to represent a rope or string folded up in the way shown. The topologies of Figs.1.5-15b, c and 1.5-16 appear more useful for a folded string than the topology of Fig.1.5-15a.

We hope this short discussion of discrete geometries with inside or outside observers demonstrates the difference between mathematical geometries based on the continuum and geometries restrained by the physical requirement of finite resolution of any observation. Physics is not a branch of mathematics, we can never demand that mathematical assumptions and their results must apply to physics. We must always demand that the mathematical assumptions match the physical requirements.

2 Modified Klein-Gordon Equation

2.1 Differential Equation with Magnetic Current Density

The modified Klein-Gordon differential equation is obtained by adding a magnetic (dipole) current density term to Maxwell's equations. The modified Klein-Gordon difference equation is produced by replacing differential operators with finite difference operators.

We start with the Hamilton function of the usual Klein-Gordon equation without magnetic dipole current density

$$\mathcal{H} = c[(\mathbf{p} - e\mathbf{A}_{\mathrm{m}})^2 + m_0^2 c^2]^{1/2} + e\phi_{\mathrm{e}} \tag{1}$$

which may be rewritten as follows:

$$\begin{gathered}(\mathbf{p} - e\mathbf{A}_{\mathrm{m}})^2 - \frac{1}{c^2}(\mathcal{H} - e\phi_{\mathrm{e}})^2 = -m_0^2 c^2 \\ (p_x - eA_{\mathrm{m}x})^2 + (p_y - eA_{\mathrm{m}y})^2 + (p_z - eA_{\mathrm{m}z})^2 - \frac{1}{c^2}(\mathcal{H} - e\phi_{\mathrm{e}})^2 = -m_0^2 c^2\end{gathered} \tag{2}$$

Using the substitutions

$$p_x \rightarrow \frac{\hbar}{i}\frac{\partial}{\partial x},\ p_y \rightarrow \frac{\hbar}{i}\frac{\partial}{\partial y},\ p_z \rightarrow \frac{\hbar}{i}\frac{\partial}{\partial z},\ \mathcal{H} \rightarrow -\frac{\hbar}{i}\frac{\partial}{\partial t} \tag{3}$$

and applying the operators to a function Ψ yields the conventional Klein-Gordon equation:

$$\begin{gathered}\left[\sum_{j=1}^{3}\left(\frac{\hbar}{i}\frac{\partial}{\partial x_j} - eA_{\mathrm{m}x_j}\right)^2 - \frac{1}{c^2}\left(\frac{\hbar}{i}\frac{\partial}{\partial t} + e\phi_{\mathrm{e}}\right)^2\right]\Psi = -m_0^2 c^2 \Psi \\ x_1 = x,\ x_2 = y,\ x_3 = z\end{gathered} \tag{4}$$

The addition of a magnetic current density term to Maxwell's equations calls for a Hamilton function with three components $\mathcal{H}_x$, $\mathcal{H}_y$, $\mathcal{H}_z$ in Eq.(1). Equations (1.1-42) to (1.1-44) show a first-order approximation in α of these three components. We copy $\mathcal{H}_x$:

ISSN 1076-5670/05
DOI: 10.1016/S1076-5670(05)37002-9

$$\mathcal{H}_x = c[(\mathbf{p} - e\mathbf{A}_\mathrm{m})^2 + m_0^2c^2]^{1/2}\left(1 + \alpha\frac{2\lambda_\mathrm{C}A_\mathrm{e}}{e}Q\right) + e\phi_\mathrm{e} - \mathcal{L}_{cx} \tag{5}$$

$$(\mathcal{H}_x - e\phi_\mathrm{e} + \mathcal{L}_{cx})^2 = c^2[(\mathbf{p} - e\mathbf{A}_\mathrm{m})^2 + m_0^2c^2]\left(1 + \alpha\frac{2\lambda_\mathrm{C}A_\mathrm{e}}{e}Q\right)^2 \tag{6}$$

$$2\alpha\lambda_\mathrm{C}A_\mathrm{e}/e = 2\alpha\lambda_\mathrm{C}A_\mathrm{e}(\mathbf{r},t)/e = \alpha_\mathrm{e}(\mathbf{r},t),\ \lambda_\mathrm{C} = h/m_0c,\ \alpha = Ze^2/2h$$

The terms Q and $\mathcal{L}_{cx}$ are defined in Eqs.(1.1-45) and (1.1-27). They require the potentials $\mathbf{A}_\mathrm{e}$, $\mathbf{A}_\mathrm{m}$, ϕ_e, and ϕ_m, which are defined by Eqs.(1.1-15) to (1.1-22).

We may replace $[1+\alpha(2\lambda_\mathrm{C}A_\mathrm{e}/e)Q]^2$ in Eq.(6) by $1+2\alpha(2\lambda_\mathrm{C}A_\mathrm{e}/e)Q$ since Eqs.(5) and (6) hold only in first order of α. Equation (6) can be written as follows in first order of α, using Eq.(1.1-47) for factors that may not commute:

$$(\mathbf{p} - e\mathbf{A}_\mathrm{m})^2 - \frac{1}{c^2}(\mathcal{H}_x - e\phi_\mathrm{e})^2 + \alpha\left[\frac{4\lambda_\mathrm{C}A_\mathrm{e}}{e}[(\mathbf{p} - e\mathbf{A}_\mathrm{m})^2 + m_0^2c^2]Q\right.$$
$$\left.- \frac{1}{c^2}\left((\mathcal{H}_x - e\phi_\mathrm{e})\frac{\mathcal{L}_{cx}}{\alpha} + \frac{\mathcal{L}_{cx}}{\alpha}(\mathcal{H}_x - e\phi_\mathrm{e}) + \frac{\mathcal{L}_{cx}^2}{\alpha}\right)\right] = -m_0^2c^2 \tag{7}$$

A solution as an expansion in powers of α is obtained if we replace the function Ψ in Eq.(4) by the function Ψ_x:

$$\Psi_x = \Psi_{x0} + \alpha\Psi_{x1} \tag{8}$$

The following equation in first order of α is obtained if we apply Eq.(7) to the function Ψ_x:

$$\left\{(\mathbf{p} - e\mathbf{A}_\mathrm{m})^2 - \frac{1}{c^2}(\mathcal{H}_x - e\phi_\mathrm{e})^2 + \alpha\left[\frac{4\lambda_\mathrm{C}A_\mathrm{e}}{e}[(\mathbf{p} - e\mathbf{A}_\mathrm{m})^2 + m_0^2c^2]Q\right.\right.$$
$$\left.\left.- \frac{1}{c^2}\left((\mathcal{H}_x - e\phi_\mathrm{e})\frac{\mathcal{L}_{cx}}{\alpha} + \frac{\mathcal{L}_{cx}}{\alpha}(\mathcal{H}_x - e\phi_\mathrm{e}) + \frac{\mathcal{L}_{cx}^2}{\alpha}\right)\right]\right\}(\Psi_{x0} + \alpha\Psi_{x1})$$
$$= -m_0^2c^2(\Psi_{x0} + \alpha\Psi_{x1}) \tag{9}$$

We assume that $\mathcal{L}_{cx}/\alpha$ and $\mathcal{L}_{cx}^2/\alpha$ are of order $O(1)$ or less. This assumption will have to be checked later on. Equation (9) may then be separated into one equation of order $O(1)$ and a second one of order $O(\alpha)$:

$$\left((\mathbf{p} - e\mathbf{A}_\mathrm{m})^2 - \frac{1}{c^2}(\mathcal{H}_x - e\phi_\mathrm{e})^2 + m_0^2c^2\right)\Psi_{x0} = 0 \tag{10}$$

$$\left((\mathbf{p}-e\mathbf{A}_\mathrm{m})^2-\frac{1}{c^2}(\mathcal{H}_x-e\phi_\mathrm{e})^2+m_0^2c^2\right)\Psi_{x1} = -\left[\frac{4\lambda_\mathrm{C}A_\mathrm{e}}{e}[(\mathbf{p}-e\mathbf{A}_\mathrm{m})^2+m_0^2c^2]Q\right.$$
$$\left.- \frac{1}{c^2}\left((\mathcal{H}_x - e\phi_\mathrm{e})\frac{\mathcal{L}_{cx}}{\alpha} + \frac{\mathcal{L}_{cx}}{\alpha}(\mathcal{H}_x - e\phi_\mathrm{e}) + \frac{\mathcal{L}_{cx}^2}{\alpha}\right)\right]\Psi_{x0} \tag{11}$$

We recognize in Eq.(10) the usual Klein-Gordon equation of Eq.(4). Equation (11) is the same equation with an inhomogeneous term added.

The factor Q in Eq.(11) has a term $[1+(\mathbf{p}-e\mathbf{A}_\mathrm{m})^2/m_0^2c^2]^{-3/2}$ according to Eq.(1.1-45). We want to replace the moment $\mathbf{p}$ by the operators of Eq.(3). This requires an explanation of what to do with Q. A straightforward approach would be to multiply Eq.(11) with $[1+(\mathbf{p}-e\mathbf{A}_\mathrm{m})^2/m_0^2c^2]^{3/2}$, shift all terms multiplied with a square root to one side, and square both sides of the equation. The terms $\mathfrak{L}_{cx}$ cause a problem. A significant simplification is possible if one restricts the calculation to small values of $(\mathbf{p}-e\mathbf{A}_\mathrm{m})^2/m_0^2c^2$:

$$(\mathbf{p}-e\mathbf{A}_\mathrm{m})^2/m_0^2c^2 \ll 1 \tag{12}$$

The following series expansion may then be used; the factor k rather than $-3/2$ is shown since the same expansion will be needed for $k=-1/2$ later on:

$$[1+(\mathbf{p}-e\mathbf{A}_\mathrm{m})^2/m_0^2c^2]^k \doteq 1+k(\mathbf{p}-e\mathbf{A}_\mathrm{m})^2/m_0^2c^2 \tag{13}$$

This takes care of part of Q in Eq.(1.1-45), but we still have a factor

$$Q_{02}=\frac{[\mathbf{A}_\mathrm{e}\cdot(\mathbf{p}-e\mathbf{A}_\mathrm{m})]^2}{\mathbf{A}_\mathrm{e}^2(\mathbf{p}-e\mathbf{A}_\mathrm{m})^2} \tag{14}$$

to deal with. Only the factor $(\mathbf{p}-e\mathbf{A}_\mathrm{m})^2$ of Q in Eq.(1.1-45) can be rewritten with the substitutions of Eq.(3) without any explanation. The factor Q_{02} of Eq.(14) can be eliminated if we replace the vectors $\mathbf{A}_\mathrm{e}$ and $(\mathbf{p}-e\mathbf{A}_\mathrm{m})$ by matrices of rank 3 whose components are vectors:

$$\mathbf{A}_\mathrm{e}=\begin{pmatrix} A_{\mathrm{e}x}\mathbf{e}_x & 0 & 0 \\ 0 & A_{\mathrm{e}y}\mathbf{e}_y & 0 \\ 0 & 0 & A_{\mathrm{e}z}\mathbf{e}_z \end{pmatrix} \tag{15}$$

$$\mathbf{p}-e\mathbf{A}_\mathrm{m}=\begin{pmatrix} (p_x-eA_{\mathrm{m}x})\mathbf{e}_x & 0 & 0 \\ 0 & (p_y-eA_{\mathrm{m}y})\mathbf{e}_y & 0 \\ 0 & 0 & (p_z-eA_{\mathrm{m}z})\mathbf{e}_z \end{pmatrix} \tag{16}$$

The substitution of Eqs.(15) and (16) into Eq.(14) yields

$$Q_{02}=1 \tag{17}$$

This is a satisfying simple result but it replaces the usual Klein-Gordon equation by a matrix equation of rank 3. The form of Q in Eq.(1.1-45) is reduced to

$$Q=\frac{1}{m_0^2c^2}\frac{(\mathbf{p}-e\mathbf{A}_\mathrm{m})^2}{[1+(\mathbf{p}-e\mathbf{A}_\mathrm{m})^2/m_0^2c^2]^{3/2}} \tag{18}$$

The denominator of Q can be eliminated with Eq.(13) and Q is brought into a form that permits the substitution of the operators of Eq.(3):

$$Q \doteq \frac{(\mathbf{p}-e\mathbf{A}_{\mathrm{m}})^2}{m_0^2c^2}\left(1-\frac{3}{2}\frac{(\mathbf{p}-e\mathbf{A}_{\mathrm{m}})^2}{m_0^2c^2}\right) \tag{19}$$

We note that there is no difference whether we multiply $(\mathbf{p}-e\mathbf{A}_{\mathrm{m}})^2$ on the right with $m_0^2c^2-(3/2)(\mathbf{p}-e\mathbf{A}_{\mathrm{m}})^2$ as done in Eq.(19) or on the left, since $(\mathbf{p}-e\mathbf{A}_{\mathrm{m}})^2$ has only to commute with a constant or itself to transform a multiplication on the right to one on the left.

The term $\mathfrak{L}_{cx}$ in Eq.(11) must be written explicitly in a form that permits the substitutions of Eq.(3). This term is excessively long and we must break it into the five components shown by Eqs.(1.1-27) and (1.1-32). We refer to the literature for their derivation[1]:

$$\begin{aligned}
\frac{1}{\alpha}\mathfrak{L}_{cx1} &= \frac{Ze}{\alpha c}(A_{ex}\dot{y}-A_{ey}\dot{z})\dot{x}\\
&\doteq \frac{Ze}{\alpha m_0^2c}\frac{[A_{ez}(\mathbf{p}-e\mathbf{A}_{\mathrm{m}})_y-A_{ey}(\mathbf{p}-e\mathbf{A}_{\mathrm{m}})_z](\mathbf{p}-e\mathbf{A}_{\mathrm{m}})_x}{1+(\mathbf{p}-e\mathbf{A}_{\mathrm{m}})^2/m_0^2c^2}\\
&\doteq \frac{Ze}{2\alpha m_0^2c}\Bigg[[A_{ez}(\mathbf{p}-e\mathbf{A}_{\mathrm{m}})_y-A_{ey}(\mathbf{p}-e\mathbf{A}_{\mathrm{m}})_z](\mathbf{p}-e\mathbf{A}_{\mathrm{m}})_x\\
&\qquad\times\left(1-\frac{(\mathbf{p}-e\mathbf{A}_{\mathrm{m}})^2}{m_0^2c^2}\right)+\left(1-\frac{(\mathbf{p}-e\mathbf{A}_{\mathrm{m}})^2}{m_0^2c^2}\right)\\
&\qquad\qquad\times[A_{ez}(\mathbf{p}-e\mathbf{A}_{\mathrm{m}})_y-A_{ey}(\mathbf{p}-e\mathbf{A}_{\mathrm{m}})_z](\mathbf{p}-e\mathbf{A}_{\mathrm{m}})_x\Bigg]
\end{aligned} \tag{20}$$

$$\begin{aligned}
\frac{1}{\alpha}\mathfrak{L}_{cx2} &= \frac{Ze}{\alpha c}\int\left(\frac{\partial\phi_{\mathrm{m}}}{\partial y}\dot{z}-\frac{\partial\phi_{\mathrm{m}}}{\partial z}\dot{y}\right)dx\\
&\doteq \frac{Ze}{\alpha m_0c}\int\left(\frac{\partial\phi_{\mathrm{m}}}{\partial y}(\mathbf{p}-e\mathbf{A}_{\mathrm{m}})_z-\frac{\partial\phi_{\mathrm{m}}}{\partial z}(\mathbf{p}-e\mathbf{A}_{\mathrm{m}})_y\right)\\
&\qquad\qquad\times\left(1+\frac{(\mathbf{p}-e\mathbf{A}_{\mathrm{m}})^2}{m_0^2c^2}\right)^{-1/2}dx\\
&\doteq \frac{Ze}{2\alpha m_0c}\int\Bigg[\left(\frac{\partial\phi_{\mathrm{m}}}{\partial y}(\mathbf{p}-e\mathbf{A}_{\mathrm{m}})_z-\frac{\partial\phi_{\mathrm{m}}}{\partial z}(\mathbf{p}-e\mathbf{A}_{\mathrm{m}})_y\right)\\
&\qquad\times\left(1-\frac{(\mathbf{p}-e\mathbf{A}_{\mathrm{m}})^2}{2m_0^2c^2}\right)+\left(1-\frac{(\mathbf{p}-e\mathbf{A}_{\mathrm{m}})^2}{2m_0^2c^2}\right)\\
&\qquad\qquad\times\left(\frac{\partial\phi_{\mathrm{m}}}{\partial y}(\mathbf{p}-e\mathbf{A}_{\mathrm{m}})_z-\frac{\partial\phi_{\mathrm{m}}}{\partial z}(\mathbf{p}-e\mathbf{A}_{\mathrm{m}})_y\right)\Bigg]dx
\end{aligned} \tag{21}$$

[1]Harmuth et al. (2001), Eqs.(3.3-53) to (3.3-57).

$$
\begin{aligned}
\frac{1}{\alpha}\mathfrak{L}_{cx3} &= \frac{Ze}{\alpha c}\int (A_{ez}\ddot{y} - A_{ey}\ddot{z})dx \\
&\doteq \frac{Ze}{\alpha m_0 c}\int \Bigg[A_{ez}\frac{\partial}{\partial t}\left(\frac{(\mathbf{p}-e\mathbf{A}_\mathrm{m})_y}{[1+(\mathbf{p}-e\mathbf{A}_\mathrm{m})^2/m_0^2c^2]^{1/2}}\right) \\
&\qquad - A_{ey}\frac{\partial}{\partial t}\left(\frac{(\mathbf{p}-e\mathbf{A}_\mathrm{m})_z}{[1+(\mathbf{p}-e\mathbf{A}_\mathrm{m})^2/m_0^2c^2]^{1/2}}\right)\Bigg]dx \\
&\doteq \frac{Ze}{2\alpha m_0 c}\int \Bigg\{A_{ez}\frac{\partial}{\partial t}\Bigg[\left(1-\frac{(\mathbf{p}-e\mathbf{A}_\mathrm{m})^2}{2m_0^2c^2}\right)(\mathbf{p}-e\mathbf{A}_\mathrm{m})_y \\
&\qquad + (\mathbf{p}-e\mathbf{A}_\mathrm{m})_y\left(1-\frac{(\mathbf{p}-e\mathbf{A}_\mathrm{m})^2}{2m_0^2c^2}\right)\Bigg] \\
&\qquad - A_{ez}\frac{\partial}{\partial t}\Bigg[\left(1-\frac{(\mathbf{p}-e\mathbf{A}_\mathrm{m})^2}{2m_0^2c^2}\right)(\mathbf{p}-e\mathbf{A}_\mathrm{m})_z \\
&\qquad + (\mathbf{p}-e\mathbf{A}_\mathrm{m})_z\left(1-\frac{(\mathbf{p}-e\mathbf{A}_\mathrm{m})^2}{2m_0^2c^2}\right)\Bigg]\Bigg\}dx
\end{aligned}
\tag{22}
$$

$$
\frac{1}{\alpha}\mathfrak{L}_{cx4} = \frac{Zec}{\alpha}\int\left(\frac{\partial A_\mathrm{ey}}{\partial z}-\frac{\partial A_\mathrm{ez}}{\partial y}\right)dx
\tag{23}
$$

$$
\begin{aligned}
\frac{1}{\alpha}\mathfrak{L}_{cx5} &= \frac{Ze}{\alpha c}\int\left(\dot{y}\frac{\partial}{\partial y}+\dot{z}\frac{\partial}{\partial z}\right)(A_{ez}\dot{y}-A_{ey}\dot{z})dx \\
&\doteq \frac{Ze}{4\alpha m_0^2 c}\int\Bigg\{\Bigg[\left(1-\frac{(\mathbf{p}-e\mathbf{A}_\mathrm{m})^2}{2m_0^2c^2}\right)(\mathbf{p}-e\mathbf{A}_\mathrm{m})_y \\
&\qquad + (\mathbf{p}-e\mathbf{A}_\mathrm{m})_y\left(1-\frac{(\mathbf{p}-e\mathbf{A}_\mathrm{m})^2}{2m_0^2c^2}\right)\Bigg]\frac{\partial}{\partial y} \\
&\qquad + \Bigg[\left(1-\frac{(\mathbf{p}-e\mathbf{A}_\mathrm{m})^2}{2m_0^2c^2}\right)(\mathbf{p}-e\mathbf{A}_\mathrm{m})_z \\
&\qquad + (\mathbf{p}-e\mathbf{A}_\mathrm{m})_z\left(1-\frac{(\mathbf{p}-e\mathbf{A}_\mathrm{m})^2}{2m_0^2c^2}\right)\Bigg]\frac{\partial}{\partial z}\Bigg\} \\
&\quad \times\Bigg[\left(1-\frac{(\mathbf{p}-e\mathbf{A}_\mathrm{m})^2}{2m_0^2c^2}\right)[A_{ez}(\mathbf{p}-e\mathbf{A}_\mathrm{m})_y - A_{ey}(\mathbf{p}-e\mathbf{A}_\mathrm{m})_z] \\
&\quad + [A_{ez}(\mathbf{p}-e\mathbf{A}_\mathrm{m})_y - A_{ey}(\mathbf{p}-e\mathbf{A}_\mathrm{m})_z]\left(1-\frac{(\mathbf{p}-e\mathbf{A}_\mathrm{m})^2}{2m_0^2c^2}\right)\Bigg]dx
\end{aligned}
\tag{24}
$$

Having explained the terms Q and $\mathfrak{L}_{cx}$ in Eq.(11) we may turn to writing the simpler Eq.(10) in the matrix form of Eq.(15) as well as substituting the operators of Eq.(3):

$$\left[\begin{pmatrix} \left(\frac{\hbar}{i}\frac{\partial}{\partial x} - eA_{\mathrm{m}x}\right)^2 & 0 & 0 \\ 0 & \left(\frac{\hbar}{i}\frac{\partial}{\partial y} - eA_{\mathrm{m}y}\right)^2 & 0 \\ 0 & 0 & \left(\frac{\hbar}{i}\frac{\partial}{\partial z} - eA_{\mathrm{m}z}\right)^2 \end{pmatrix}\right.$$

$$-\frac{1}{c^2}\begin{pmatrix} \left(\frac{\hbar}{i}\frac{\partial}{\partial t} + e\phi_{\mathrm{e}}\right)^2 & 0 & 0 \\ 0 & \left(\frac{\hbar}{i}\frac{\partial}{\partial t} + e\phi_{\mathrm{e}}\right)^2 & 0 \\ 0 & 0 & \left(\frac{\hbar}{i}\frac{\partial}{\partial t} + e\phi_{\mathrm{e}}\right)^2 \end{pmatrix}$$

$$\left. + m_0^2c^2\begin{pmatrix} 1 & 0 & 0 \\ 0 & 1 & 0 \\ 0 & 0 & 1 \end{pmatrix}\right]\begin{pmatrix} \Psi_{\mathrm{x}0x} & 0 & 0 \\ 0 & \Psi_{\mathrm{x}0y} & 0 \\ 0 & 0 & \Psi_{\mathrm{x}0z} \end{pmatrix} = 0 \tag{25}$$

Since only the terms in the main diagonals are not zero this matrix equation is essentially three times Eq.(4). The summation sign of Eq.(4) has disappeared but the index j retains the values $j = 1, 2, 3$ that it has in Eq.(4):

$$\left[\left(\frac{\hbar}{i}\frac{\partial}{\partial x_j} - eA_{\mathrm{m}x_j}\right)^2 - \frac{1}{c^2}\left(\frac{\hbar}{i}\frac{\partial}{\partial t} + e\phi_{\mathrm{e}}\right)^2 + m_0^2c^2\right]\Psi_{\mathrm{x}0x_j} = 0 \tag{26}$$

Equation (11) can be written in matrix form like Eq.(25) with $\Psi_{\mathrm{x}0x}$, $\Psi_{\mathrm{x}0y}$, $\Psi_{\mathrm{x}0z}$ replaced by $\Psi_{\mathrm{x}1x}$, $\Psi_{\mathrm{x}1y}$, $\Psi_{\mathrm{x}1z}$ and another matrix instead of 0 on the right side. The shorter notation of Eq.(26) is a better choice. We begin by rewriting the first term on the right side of Eq.(11) with the help of Eq.(18):

$$\begin{aligned} &\frac{4\lambda_{\mathrm{C}}A_{\mathrm{e}}}{e}[(\mathbf{p} - e\mathbf{A}_{\mathrm{m}})^2 + m_0^2c^2]Q \\ &\qquad \doteq \frac{4\lambda_{\mathrm{C}}A_{\mathrm{e}}}{e}[1 + (\mathbf{p} - e\mathbf{A}_{\mathrm{m}})^2/m_0^2c^2]\frac{(\mathbf{p} - e\mathbf{A}_{\mathrm{m}})^2}{[1 + (\mathbf{p} - e\mathbf{A}_{\mathrm{m}}^2)/m_0^2c^2]^{3/2}} \\ &\qquad \doteq \frac{4\lambda_{\mathrm{C}}A_{\mathrm{e}}}{e}(\mathbf{p} - e\mathbf{A}_{\mathrm{m}})^2\left(1 - \frac{1}{2}\frac{(\mathbf{p} - e\mathbf{A}_{\mathrm{m}})^2}{m_0^2c^2}\right) \end{aligned} \tag{27}$$

Equation (11) may now be rewritten in analogy to Eq.(26). Note the change of the variable $\mathcal{L}_{\mathrm{cx}}/\alpha$ to $\mathfrak{L}_{\mathrm{cx}}$ explained in detail in the following Eq.(29):

$$\left[\left(\frac{\hbar}{i}\frac{\partial}{\partial x_j} - eA_{\mathrm{m}x_j}\right)^2 - \frac{1}{c^2}\left(\frac{\hbar}{i}\frac{\partial}{\partial t} + e\phi_{\mathrm{e}}\right)^2 + m_0^2c^2\right]\Psi_{\mathrm{x}1x_j}$$
$$= -\left\{\frac{4\lambda_{\mathrm{C}}A_{\mathrm{e}}}{e}\left(\frac{\hbar}{i}\frac{\partial}{\partial x_j} - eA_{\mathrm{m}x_j}\right)^2\left[1 - \frac{1}{2m_0^2c^2}\left(\frac{\hbar}{i}\frac{\partial}{\partial x_j} - eA_{\mathrm{m}x_j}\right)^2\right]\right.$$
$$+ \frac{1}{c^2}\left[\left(\frac{\hbar}{i}\frac{\partial}{\partial t} + e\phi_{\mathrm{e}}\right)(\mathfrak{L}_{\mathrm{cx}1j} + \mathfrak{L}_{\mathrm{cx}2j} + \mathfrak{L}_{\mathrm{cx}3j} + \mathfrak{L}_{\mathrm{cx}4j} + \mathfrak{L}_{\mathrm{cx}5j})\right.$$
$$\left.\left. + (\mathfrak{L}_{\mathrm{cx}1j} + \mathfrak{L}_{\mathrm{cx}2j} + \mathfrak{L}_{\mathrm{cx}3j} + \mathfrak{L}_{\mathrm{cx}4j} + \mathfrak{L}_{\mathrm{cx}5j})\left(\frac{\hbar}{i}\frac{\partial}{\partial t} + e\phi_{\mathrm{e}}\right) + \alpha\mathfrak{L}_{\mathrm{cx}}^2\right]\right\}\Psi_{\mathrm{x}0x_j} \tag{28}$$

The last term $\alpha\mathfrak{L}_{\mathrm{cx}}^2 = \alpha(\mathcal{L}_{\mathrm{cx}}/\alpha)^2$ may be dropped because of the factor α. The operators $\mathfrak{L}_{\mathrm{cx}1}$ to $\mathfrak{L}_{\mathrm{cx}5}$ introduced here follow from Eqs.(20) to (24) with the operators of Eq.(3) and the substitution

$$\frac{1}{\alpha}\mathcal{L}_{\mathrm{cx}kj} \rightarrow \mathfrak{L}_{\mathrm{cx}kj},\ k = 1,\ 2,\ 3,\ 4,\ 5 \tag{29}$$

The font Euler Script Medium is used for the symbol $\mathcal{L}$ but Euler Fractur Medium is used for the symbol $\mathfrak{L}$. The matrix $\boldsymbol{\mathfrak{L}}_{\mathrm{cx}1}$, using Euler Fractur Bold, has the terms $\mathfrak{L}_{\mathrm{cx}1j}$ along its main diagonal and zeroes everywhere else:

$$\mathfrak{L}_{\mathrm{cx}1j} = \frac{Ze}{2\alpha m_0^2c}\left\{\left[A_{\mathrm{e}z}\left(\frac{\hbar}{i}\frac{\partial}{\partial y} - eA_{\mathrm{m}y}\right) - A_{\mathrm{e}y}\left(\frac{\hbar}{i}\frac{\partial}{\partial z} - eA_{\mathrm{m}z}\right)\right]\right.$$
$$\times\left(\frac{\hbar}{i}\frac{\partial}{\partial x} - eA_{\mathrm{m}x}\right)\left[1 - \frac{1}{m_0^2c^2}\left(\frac{\hbar}{i}\frac{\partial}{\partial x_j} - eA_{\mathrm{m}x_j}\right)^2\right]$$
$$+ \left[1 - \frac{1}{m_0^2c^2}\left(\frac{\hbar}{i}\frac{\partial}{\partial x_j} - eA_{\mathrm{m}x_j}\right)^2\right]\left[A_{\mathrm{e}z}\left(\frac{\hbar}{i}\frac{\partial}{\partial y} - eA_{\mathrm{m}y}\right)\right.$$
$$\left.\left. - A_{\mathrm{e}y}\left(\frac{\hbar}{i}\frac{\partial}{\partial z} - eA_{\mathrm{m}z}\right)\right]\left(\frac{\hbar}{i}\frac{\partial}{\partial x} - eA_{\mathrm{m}x}\right)\right\}$$
$$j = 1,\ 2,\ 3;\quad x_1 = x,\ x_2 = y,\ x_3 = z \tag{30}$$

The mixed notation x, y, z, and x_j in Eq.(30) requires clarification. The terms with x, y, z

$$A_{\mathrm{e}z}\left(\frac{\hbar}{i}\frac{\partial}{\partial y} - eA_{\mathrm{m}y}\right),\ A_{\mathrm{e}y}\left(\frac{\hbar}{i}\frac{\partial}{\partial z} - eA_{\mathrm{m}z}\right),\ \left(\frac{\hbar}{i}\frac{\partial}{\partial x} - eA_{\mathrm{m}x}\right) \tag{31}$$

form matrices of rank 3 with equal values for all elements in the main diagonal like the second and third matrix in Eq.(25). On the other hand, the terms

$$\left(\frac{\hbar}{i}\frac{\partial}{\partial x_j} - eA_{\mathrm{m}x_j}\right)^2 \tag{32}$$

are the terms of a matrix with rank 3 like the first matrix in Eq.(25) with the terms along the main diagonal varying according to $j = 1, 2, 3$. The remaining expressions for $\mathfrak{L}_{\mathrm{cx}2j}$ to $\mathfrak{L}_{\mathrm{cx}5j}$ assume the following form:

$$\begin{aligned}\mathfrak{L}_{\mathrm{cx}2j} = \frac{Ze}{2\alpha m_0 c}\int \Bigg\{ & \left[\frac{\partial\phi_{\mathrm{m}}}{\partial y}\left(\frac{\hbar}{i}\frac{\partial}{\partial z} - eA_{\mathrm{m}z}\right) - \frac{\partial\phi_{\mathrm{m}}}{\partial z}\left(\frac{\hbar}{i}\frac{\partial}{\partial y} - eA_{\mathrm{m}y}\right)\right] \\ \times & \left[1 - \frac{1}{2m_0^2c^2}\left(\frac{\hbar}{i}\frac{\partial}{\partial x_j} - eA_{\mathrm{m}x_j}\right)^2\right] + \left[1 - \frac{1}{2m_0^2c^2}\left(\frac{\hbar}{i}\frac{\partial}{\partial x_j} - eA_{\mathrm{m}x_j}\right)^2\right] \\ & \times \left[\frac{\partial\phi_{\mathrm{m}}}{\partial y}\left(\frac{\hbar}{i}\frac{\partial}{\partial z} - eA_{\mathrm{m}z}\right) - \frac{\partial\phi_{\mathrm{m}}}{\partial z}\left(\frac{\hbar}{i}\frac{\partial}{\partial y} - eA_{\mathrm{m}y}\right)\right]\Bigg\} dx \end{aligned} \tag{33}$$

$$\begin{aligned}\mathfrak{L}_{\mathrm{cx}3j} = \frac{Ze}{2\alpha m_0 c}\int \Bigg(& A_{\mathrm{e}z}\frac{\partial}{\partial t}\Bigg\{\left[1 - \frac{1}{2m_0^2c^2}\left(\frac{\hbar}{i}\frac{\partial}{\partial x_j} - eA_{\mathrm{m}x_j}\right)^2\right]\left(\frac{\hbar}{i}\frac{\partial}{\partial y} - eA_{\mathrm{m}y}\right) \\ & + \left(\frac{\hbar}{i}\frac{\partial}{\partial y} - eA_{\mathrm{m}y}\right)\left[1 - \frac{1}{2m_0^2c^2}\left(\frac{\hbar}{i}\frac{\partial}{\partial x_j} - eA_{\mathrm{m}x_j}\right)^2\right]\Bigg\} \\ & - A_{\mathrm{e}z}\frac{\partial}{\partial t}\Bigg\{\left[1 - \frac{1}{2m_0^2c^2}\left(\frac{\hbar}{i}\frac{\partial}{\partial x_j} - eA_{\mathrm{m}x_j}\right)^2\right]\left(\frac{\hbar}{i}\frac{\partial}{\partial z} - eA_{\mathrm{m}z}\right) \\ & + \left(\frac{\hbar}{i}\frac{\partial}{\partial z} - eA_{\mathrm{m}z}\right)\left[1 - \frac{1}{2m_0^2c^2}\left(\frac{\hbar}{i}\frac{\partial}{\partial x_j} - eA_{\mathrm{m}x_j}\right)^2\right]\Bigg\}\Bigg) dx \end{aligned} \tag{34}$$

$$\mathfrak{L}_{\mathrm{cx}4j} = \frac{Zec}{\alpha}\int\left(\frac{\partial A_{\mathrm{e}y}}{\partial z} - \frac{\partial A_{\mathrm{e}z}}{\partial y}\right) dx \tag{35}$$

$$\begin{aligned}\mathfrak{L}_{\mathrm{cx}5j} = \frac{Ze}{4\alpha m_0^2 c}\int \Bigg(& \Bigg\{\left[1 - \frac{1}{2m_0^2c^2}\left(\frac{\hbar}{i}\frac{\partial}{\partial x_j} - eA_{\mathrm{m}x_j}\right)^2\right]\left(\frac{\hbar}{i}\frac{\partial}{\partial y} - eA_{\mathrm{m}y}\right) \\ & + \left(\frac{\hbar}{i}\frac{\partial}{\partial y} - eA_{\mathrm{m}y}\right)\left[1 - \frac{1}{2m_0^2c^2}\left(\frac{\hbar}{i}\frac{\partial}{\partial x_j} - eA_{\mathrm{m}x_j}\right)^2\right]\Bigg\}\frac{\partial}{\partial y} \\ & + \Bigg\{\left[1 - \frac{1}{2m_0^2c^2}\left(\frac{\hbar}{i}\frac{\partial}{\partial x_j} - eA_{\mathrm{m}x_j}\right)^2\left(\frac{\hbar}{i}\frac{\partial}{\partial z} - eA_{\mathrm{m}z}\right)\right. \\ & + \left(\frac{\hbar}{i}\frac{\partial}{\partial z} - eA_{\mathrm{m}z}\right)\left[1 - \frac{1}{2m_0^2c^2}\left(\frac{\hbar}{i}\frac{\partial}{\partial x_j} - eA_{\mathrm{m}x_j}\right)^2\right]\Bigg\}\frac{\partial}{\partial z}\Bigg) \end{aligned}$$

$$\times\left\{\left[1-\frac{1}{2m_0^2c^2}\left(\frac{\hbar}{i}\frac{\partial}{\partial x_j}-eA_{\mathrm{m}x_j}\right)^2\right]\right.$$
$$\times\left[A_{\mathrm{e}x}\left(\frac{\hbar}{i}\frac{\partial}{\partial y}-eA_{\mathrm{m}y}\right)-A_{\mathrm{e}y}\left(\frac{\hbar}{i}\frac{\partial}{\partial z}-eA_{\mathrm{m}z}\right)\right]$$
$$+\left[A_{\mathrm{e}x}\left(\frac{\hbar}{i}\frac{\partial}{\partial y}-eA_{\mathrm{m}y}\right)-A_{\mathrm{e}y}\left(\frac{\hbar}{i}\frac{\partial}{\partial z}-eA_{\mathrm{m}z}\right)\right]$$
$$\left.\times\left[1-\frac{1}{2m_0^2c^2}\left(\frac{\hbar}{i}\frac{\partial}{\partial x_j}-eA_{\mathrm{m}x_j}\right)^2\right]\right\}dx \quad (36)$$

Equations (30) to (36) make all terms in Eq.(28) defined for known values of the potentials $\mathbf{A}_{\mathrm{m}}$, $\mathbf{A}_{\mathrm{e}}$, ϕ_{e} and the rest mass m_0 of a charged particle. The equations are unique provided we accept Eq.(1.1-47) as necessary. Only the success of the derived results can determine that. The complexity of Eqs.(30) and (33) to (36) makes one hesitant to replace Eq.(1.1-47) by a more complex relation.

Equation (26) can be solved for certain initial and boundary conditions. The solution of Eq.(28) requires a particular solution of the inhomogeneous equation that contains the solution $\Psi_{\mathrm{x}0x_j}$ of Eq.(26). The homogeneous part of Eq.(28) equals Eq.(26).

When the notation of Eqs.(25) and (26) is used one must rewrite Eq.(8) with a longer subscript

$$\Psi_{\mathrm{x}x_j}=\Psi_{\mathrm{x}0x_j}+\alpha\Psi_{\mathrm{x}1x_j} \quad (37)$$

as was already mentioned in the text following Eq.(26). These long subscripts will often be shortened as follows:

$$\Psi_{\mathrm{x}0x_j}=\Psi_0,\quad \Psi_{\mathrm{x}1x_j}=\Psi_1 \quad (38)$$

2.2 Modified Klein-Gordon Difference Equation

For the derivation of the Klein-Gordon difference equation including the modified Maxwell equations we start from Eq.(2.1-2):

$$(\mathbf{p}-e\mathbf{A}_{\mathrm{m}})^2-\frac{1}{c^2}(\mathcal{H}-e\phi_{\mathrm{e}})^2=-m_0^2c^2$$
$$(p_x-eA_{\mathrm{m}x})^2+(p_y-eA_{\mathrm{m}y})^2+(p_z-eA_{\mathrm{m}z})^2-\frac{1}{c^2}(\mathcal{H}-e\phi_{\mathrm{e}})^2=-m_0^2c^2 \quad (1)$$

Instead of the differential operators of Eq.(2.1-3) we use difference operators. Since the transition from first- to second-order difference operators is not as simple as in the case of differential operators we must define them both according to Eqs.(1.2-1), (1.2-3), and (1.2-7):

$$p_{x_j}\Psi \to \frac{\hbar}{i}\frac{\tilde{\Delta}\Psi}{\tilde{\Delta}x_j} = \frac{\hbar}{i}\frac{\Psi(x_j+\Delta x_j)-\Psi(x_j-\Delta x_j)}{2\Delta x_j} \tag{2}$$

$$\mathcal{H}\Psi \to -\frac{\hbar}{i}\frac{\tilde{\Delta}\Psi}{\tilde{\Delta}t} = -\frac{\hbar}{i}\frac{\Psi(t+\Delta t)-\Psi(t-\Delta t)}{2\Delta t} \tag{3}$$

$$p_{x_j}^2\Psi \to -\hbar^2\frac{\tilde{\Delta}^2\Psi}{\tilde{\Delta}x_j^2} = -\hbar^2\frac{\Psi(x_j+\Delta x_j)-2\Psi(x_j)+\Psi(x_j-\Delta x_j)}{(\Delta x_j)^2} \tag{4}$$

$$\mathcal{H}^2\Psi \to -\hbar^2\frac{\tilde{\Delta}^2\Psi}{\tilde{\Delta}t^2} = -\hbar^2\frac{\Psi(t+\Delta t)-2\Psi(t)+\Psi(t-\Delta t)}{(\Delta t)^2} \tag{5}$$

For the normalized variables

$$\begin{gathered}\theta = t/\Delta t,\ \zeta_j = x_j/c\Delta t,\ N = T/\Delta t\\ 0 \le t \le T,\ 0 \le x_j \le cT,\ 0 \le \theta \le N,\ 0 \le \zeta_j \le N\end{gathered} \tag{6}$$

we follow the simplified notation of Eqs.(1.2-3) and (1.2-7):

$$\frac{\hbar}{i}\frac{\tilde{\Delta}\Psi}{\tilde{\Delta}\zeta_j} = \frac{\hbar}{2i}[\Psi(\zeta_j+1)-\Psi(\zeta_j-1)] \tag{7}$$

$$-\frac{\hbar}{i}\frac{\tilde{\Delta}\Psi}{\tilde{\Delta}\theta} = -\frac{\hbar}{2i}[\Psi(\theta+1)-\Psi(\theta-1)] \tag{8}$$

$$-\hbar^2\frac{\tilde{\Delta}^2\Psi}{\tilde{\Delta}\zeta_j^2} = -\hbar^2[\Psi(\zeta_j+1)-2\Psi(\zeta_j)+\Psi(\zeta_j-1)] \tag{9}$$

$$-\hbar^2\frac{\tilde{\Delta}^2\Psi}{\tilde{\Delta}\theta^2} = -\hbar^2[\Psi(\theta+1)-2\Psi(\theta)+\Psi(\theta-1)] \tag{10}$$

These definitions suffice to rewrite all equations in Section 2.1 from Eq.(2.1-3) to Eq.(2.1-20). From Eq.(2.1-21) on we encounter integrations in addition to differentiations. To remain consistent we have no choice but to substitute summations for integrations as shown in Table 1.2-1:

$$\int \varphi(x)dx = \int^x \varphi(\nu)d\nu \to \mathop{\mathrm{S}}^{x} \varphi(\nu+1)\Delta\nu \tag{11}$$

We may write the matrix equation (2.1-25) by substituting $\tilde{\Delta}/\tilde{\Delta}x$ for $\partial/\partial x$ to $\tilde{\Delta}/\tilde{\Delta}t$ for $\partial/\partial t$, but must observe that the squares $\tilde{\Delta}^2/\tilde{\Delta}x^2$ to $\tilde{\Delta}^2/\tilde{\Delta}t^2$ are replaced according to Eqs.(4) and (5). Equation (2.1-8), $\Psi_\mathrm{x} = \Psi_{\mathrm{x}0} + \alpha\Psi_{\mathrm{x}1}$, applies from here on.

Using the notation x_j rather than x, y, z we rewrite the differential matrix equation of Eq.(2.1-25) with the help of Eqs.(2) to (5):

$$\left[\begin{pmatrix} \left(\frac{\hbar}{i}\frac{\tilde{\Delta}}{\tilde{\Delta}x_1} - eA_{\mathrm{m}x_1}\right)^2 & 0 & 0 \\ 0 & \left(\frac{\hbar}{i}\frac{\tilde{\Delta}}{\tilde{\Delta}x_2} - eA_{\mathrm{m}x_2}\right)^2 & 0 \\ 0 & 0 & \left(\frac{\hbar}{i}\frac{\tilde{\Delta}}{\tilde{\Delta}x_3} - eA_{\mathrm{m}x_3}\right)^2 \end{pmatrix}\right.$$

$$-\frac{1}{c^2}\begin{pmatrix} \left(\frac{\hbar}{i}\frac{\tilde{\Delta}}{\tilde{\Delta}t} + e\phi_{\mathrm{e}}\right)^2 & 0 & 0 \\ 0 & \left(\frac{\hbar}{i}\frac{\tilde{\Delta}}{\tilde{\Delta}t} + e\phi_{\mathrm{e}}\right)^2 & 0 \\ 0 & 0 & \left(\frac{\hbar}{i}\frac{\tilde{\Delta}}{\tilde{\Delta}t} + e\phi_{\mathrm{e}}\right)^2 \end{pmatrix}$$

$$\left.+\, m_0^2c^2\begin{pmatrix} 1 & 0 & 0 \\ 0 & 1 & 0 \\ 0 & 0 & 1 \end{pmatrix}\right]\begin{pmatrix} \Psi_{\mathrm{x}0x_1} & 0 & 0 \\ 0 & \Psi_{\mathrm{x}0x_2} & 0 \\ 0 & 0 & \Psi_{\mathrm{x}0x_3} \end{pmatrix} = 0 \quad (12)$$

The three equations represented by Eq.(2.1-26) assume the following form as difference equations:

$$\left[\left(\frac{\hbar}{i}\frac{\tilde{\Delta}}{\tilde{\Delta}x_j} - eA_{\mathrm{m}x_j}\right)^2 - \frac{1}{c^2}\left(\frac{\hbar}{i}\frac{\tilde{\Delta}}{\tilde{\Delta}t} + e\phi_{\mathrm{e}}\right)^2 + m_0^2c^2\right]\Psi_{\mathrm{x}0x_j} = 0 \quad (13)$$

We shorten $\Psi_{\mathrm{x}0x_j}$ to Ψ_0 as well as x_j to x and rewrite this equation explicitly rather than in operator form. To simplify the notation further we do not write a variable that is not changed, e.g., we write $\Psi_0(x + \Delta x) - \Psi_0(x - \Delta x)$ rather than $\Psi_0(x + \Delta x, t) - \Psi_0(x - \Delta x, t)$. The first term in Eq.(13) becomes:

$$\Psi_0 = \Psi_{\mathrm{x}0x_j}, \quad x = x_j$$

$$\left(\frac{\hbar}{i}\frac{\tilde{\Delta}}{\tilde{\Delta}x} - eA_{\mathrm{m}x}\right)^2\Psi_0$$

$$= -\hbar^2\frac{\tilde{\Delta}^2\Psi_0}{\tilde{\Delta}x^2} + 2i\hbar eA_{\mathrm{m}x}\frac{\tilde{\Delta}\Psi_0}{\tilde{\Delta}x} + \left(e^2A_{\mathrm{m}x}^2 + i\hbar e\frac{\tilde{\Delta}A_{\mathrm{m}x}}{\tilde{\Delta}x}\right)\Psi_0$$

$$= -\hbar^2 \frac{\Psi_0(x+\Delta x) - 2\Psi_0(x) + \Psi_0(x-\Delta x)}{(\Delta x)^2} + 2i\hbar e A_{\mathrm{m}x} \frac{\Psi_0(x+\Delta x) - \Psi_0(x-\Delta x)}{2\Delta x}$$
$$+ \left(e^2 A_{\mathrm{m}x}^2 + i\hbar e \frac{A_{\mathrm{m}x}(x+\Delta x) - A_{\mathrm{m}x}(x-\Delta x)}{2\Delta x} \right) \Psi_0 \quad (14)$$

The second term of Eq.(13) assumes the form:

$$\Psi_0 = \Psi_{\mathrm{x}0x_j}$$

$$\left(\frac{\hbar}{i} \frac{\tilde{\Delta}}{\tilde{\Delta} t} + e\phi_{\mathrm{e}} \right)^2 \Psi_0$$
$$= -\hbar^2 \frac{\tilde{\Delta}^2 \Psi_0}{\tilde{\Delta} t^2} - 2i\hbar e \phi_{\mathrm{e}} \frac{\tilde{\Delta}\Psi_0}{\tilde{\Delta} t} + \left(e^2 \phi_{\mathrm{e}}^2 - i\hbar e \frac{\tilde{\Delta}\phi_{\mathrm{e}}}{\tilde{\Delta} t} \right) \Psi_0$$
$$= -\hbar^2 \frac{\Psi_0(t+\Delta t) - 2\Psi_0(t) + \Psi_0(t-\Delta t)}{(\Delta t)^2} - 2i\hbar e \phi_{\mathrm{e}} \frac{\Psi_0(t+\Delta t) - \Psi_0(t-\Delta t)}{2\Delta t}$$
$$+ \left(e^2 \phi_{\mathrm{e}}^2 - i\hbar e \frac{\phi_{\mathrm{e}}(t+\Delta t) - \phi_{\mathrm{e}}(t-\Delta t)}{2\Delta t} \right) \Psi_0 \quad (15)$$

Substitution of Eqs.(14) and (15) into Eq.(13) yields a difference equation instead of the differential equation (2.1-26). We write here both variables x_j and t of Ψ_0, $A_{\mathrm{m}x_j}$, ϕ_{e}:

$$\Psi_0 = \Psi_{\mathrm{x}0x_j}$$

$$\frac{\Psi_0(x_j+\Delta x_j,t) - 2\Psi_0(x_j,t) + \Psi_0(x_j-\Delta x_j,t)}{(\Delta x_j)^2}$$
$$- \frac{1}{c^2} \frac{\Psi_0(x_j,t+\Delta t) - 2\Psi_0(x_j,t) + \Psi_0(x_j,t-\Delta t)}{(\Delta t)^2}$$
$$- 2i\frac{e}{\hbar} \left(A_{\mathrm{m}x_j}(x_j,t) \frac{\Psi_0(x_j+\Delta x_j,t) - \Psi_0(x_j-\Delta x_j,t)}{2\Delta x_j} \right.$$
$$\left. + \frac{1}{c^2} \phi_{\mathrm{e}}(x_j,t) \frac{\Psi_0(x_j,t+\Delta t) - \Psi_0(x_j,t-\Delta t)}{2\Delta t} \right)$$
$$- \frac{e^2}{\hbar^2} \left[A_{\mathrm{m}x_j}^2(x_j,t) - \frac{1}{c^2}\phi_{\mathrm{e}}^2(x_j,t) + \frac{\hbar}{e} \left(\frac{A_{\mathrm{m}x_j}(x_j+\Delta x_j,t) - A_{\mathrm{m}x_j}(x_j-\Delta x_j,t)}{2\Delta x_j} \right. \right.$$
$$\left. \left. + \frac{1}{c^2} \frac{\phi_{\mathrm{e}}(x_j,t+\Delta t) - \phi_{\mathrm{e}}(x_j,t-\Delta t)}{2\Delta t} \right) + \frac{m_0^2 c^2}{e^2} \right] \Psi_0(x_j,t) = 0 \quad (16)$$

Turning to Eq.(2.1-11) we avoid again writing a matrix equation and use the space-saving notation of Eq.(13). Equation (2.1-27) remains unchanged since the operators **p** and $\mathcal{H}$ can stand for differential or difference operators, but Eq.(2.1-28) must be rewritten for difference operators:

$$\left[\left(\frac{\hbar}{i}\frac{\tilde{\Delta}}{\tilde{\Delta}x_j} - eA_{\mathrm{m}x_j}\right)^2 - \frac{1}{c^2}\left(\frac{\hbar}{i}\frac{\tilde{\Delta}}{\tilde{\Delta}t} + e\phi_{\mathrm{e}}\right)^2 + m_0^2c^2\right]\Psi_{\mathrm{x}1x_j}$$
$$= -\left\{\frac{4\lambda_{\mathrm{C}}A_{\mathrm{e}}}{e}\left(\frac{\hbar}{i}\frac{\tilde{\Delta}}{\tilde{\Delta}x_j} - eA_{\mathrm{m}x_j}\right)^2\left[1 - \frac{1}{2m_0^2c^2}\left(\frac{\hbar}{i}\frac{\tilde{\Delta}}{\tilde{\Delta}x_j} - eA_{\mathrm{m}x_j}\right)^2\right]\right.$$
$$+ \frac{1}{c^2}\left[\left(\frac{\hbar}{i}\frac{\tilde{\Delta}}{\tilde{\Delta}t} + e\phi_{\mathrm{e}}\right)(\mathsf{L}_{\mathrm{cx}1j} + \mathsf{L}_{\mathrm{cx}2j} + \mathsf{L}_{\mathrm{cx}3j} + \mathsf{L}_{\mathrm{cx}4j} + \mathsf{L}_{\mathrm{cx}5j})\right.$$
$$\left.\left. + (\mathsf{L}_{\mathrm{cx}1j} + \mathsf{L}_{\mathrm{cx}2j} + \mathsf{L}_{\mathrm{cx}3j} + \mathsf{L}_{\mathrm{cx}4j} + \mathsf{L}_{\mathrm{cx}5j})\left(\frac{\hbar}{i}\frac{\tilde{\Delta}}{\tilde{\Delta}t} + e\phi_{\mathrm{e}}\right) + \alpha\mathsf{L}_{\mathrm{cx}}^2\right]\right\}\Psi_{\mathrm{x}0x_j} \quad (17)$$

As in the case of Eq.(2.1-28) we drop the last term $\alpha\mathsf{L}_{\mathrm{cx}}^2$ because of the factor α. The operators $\mathsf{L}_{\mathrm{cx1}}$ to $\mathsf{L}_{\mathrm{cx5}}$ (Euler Roman Medium font) introduced here follow from Eqs.(2.1-20) to (2.1-24) with the operators of Eqs.(2), (3), and (1.2-23) as well as the following substitution based on Eq.(2.1-29):

$$\frac{1}{\alpha_{\mathrm{e}}}\mathfrak{L}_{\mathrm{cxk}j} \to \mathsf{L}_{\mathrm{cxk}j}, \quad k = 1,\ 2,\ 3,\ 4,\ 5 \quad (18)$$

The matrix $\mathbf{L}_{\mathrm{cxk}}$, using the Euler Roman Bold font, has the terms $\mathsf{L}_{\mathrm{cxk}j}$ along the main diagonal and zeros everywhere else:

$$\mathbf{L}_{\mathrm{cxk}} = \begin{pmatrix} \mathsf{L}_{\mathrm{cxk1}} & 0 & 0 \\ 0 & \mathsf{L}_{\mathrm{cxk2}} & 0 \\ 0 & 0 & \mathsf{L}_{\mathrm{cxk3}} \end{pmatrix}, \qquad k = 1,\ 2,\ 3,\ 4,\ 5 \quad (19)$$

In analogy to Eqs.(2.1-30) and (2.1-33) to (2.1-36) we obtain:

$$\mathsf{L}_{\mathrm{cx}1j} = \frac{Ze}{2\alpha m_0^2 c}\left\{\left[A_{\mathrm{e}z}\left(\frac{\hbar}{i}\frac{\tilde{\Delta}}{\tilde{\Delta}y} - eA_{\mathrm{m}y}\right) - A_{\mathrm{e}y}\left(\frac{\hbar}{i}\frac{\tilde{\Delta}}{\tilde{\Delta}z} - eA_{\mathrm{m}z}\right)\right]\right.$$
$$\times\left(\frac{\hbar}{i}\frac{\tilde{\Delta}}{\tilde{\Delta}x} - eA_{\mathrm{m}x}\right)\left[1 - \frac{1}{m_0^2c^2}\left(\frac{\hbar}{i}\frac{\tilde{\Delta}}{\tilde{\Delta}x_j} - eA_{\mathrm{m}x_j}\right)^2\right]$$
$$+\left[1 - \frac{1}{m_0^2c^2}\left(\frac{\hbar}{i}\frac{\tilde{\Delta}}{\tilde{\Delta}x_j} - eA_{\mathrm{m}x_j}\right)^2\right]\left[A_{\mathrm{e}z}\left(\frac{\hbar}{i}\frac{\tilde{\Delta}}{\tilde{\Delta}y} - eA_{\mathrm{m}y}\right)\right.$$
$$\left.\left. - A_{\mathrm{e}y}\left(\frac{\hbar}{i}\frac{\tilde{\Delta}}{\tilde{\Delta}z} - eA_{\mathrm{m}z}\right)\right]\left(\frac{\hbar}{i}\frac{\tilde{\Delta}}{\tilde{\Delta}x} - eA_{\mathrm{m}x}\right)\right\}$$
$$j = 1,\ 2,\ 3;\quad x_1 = x,\ x_2 = y,\ x_3 = z \quad (20)$$

As previously in the case of Eqs.(2.1-31) and (2.1-32) the terms with the variables x, y, z

$$A_{\mathrm{e}z}\left(\frac{\hbar}{i}\frac{\tilde{\Delta}}{\tilde{\Delta}y}-eA_{\mathrm{m}y}\right),\ A_{\mathrm{e}y}\left(\frac{\hbar}{i}\frac{\tilde{\Delta}}{\tilde{\Delta}z}-eA_{\mathrm{m}z}\right),\ \left(\frac{\hbar}{i}\frac{\tilde{\Delta}}{\tilde{\Delta}x}-eA_{\mathrm{m}x}\right) \tag{21}$$

form matrices of rank 3 with equal values for all elements in the main diagonal like the second and third matrix in Eq.(12). But the terms

$$\left(\frac{\hbar}{i}\frac{\tilde{\Delta}}{\tilde{\Delta}x_j}-eA_{\mathrm{m}x_j}\right)^2 \tag{22}$$

are the terms of a matrix with rank 3 like the first matrix in Eq.(12) with $x_j = x_1, x_2, x_3 = x, y, z$. The four expressions for $\mathsf{L}_{\mathrm{cx}2j}$ to $\mathsf{L}_{\mathrm{cx}5j}$ are written according to Eqs.(2.1-33) to (2.1-36) with the help of Eq.(1.2-23):

$$\begin{aligned}\mathsf{L}_{\mathrm{cx}2j} = \frac{Ze}{2\alpha m_0 c}\mathop{\mathfrak{S}}_{c}^{x}\Bigg\{&\left[\frac{\tilde{\Delta}\phi_{\mathrm{m}}}{\tilde{\Delta}y}\left(\frac{\hbar}{i}\frac{\tilde{\Delta}}{\tilde{\Delta}z}-eA_{\mathrm{m}z}\right)-\frac{\tilde{\Delta}\phi_{\mathrm{m}}}{\tilde{\Delta}z}\left(\frac{\hbar}{i}\frac{\tilde{\Delta}}{\tilde{\Delta}y}-eA_{\mathrm{m}y}\right)\right]\\ \times&\left[1-\frac{1}{2m_0^2c^2}\left(\frac{\hbar}{i}\frac{\tilde{\Delta}}{\tilde{\Delta}x_j}-eA_{\mathrm{m}x_j}\right)^2\right]+\left[1-\frac{1}{2m_0^2c^2}\left(\frac{\hbar}{i}\frac{\tilde{\Delta}}{\tilde{\Delta}x_j}-eA_{\mathrm{m}x_j}\right)^2\right]\\ &\times\left[\frac{\tilde{\Delta}\phi_{\mathrm{m}}}{\tilde{\Delta}y}\left(\frac{\hbar}{i}\frac{\tilde{\Delta}}{\tilde{\Delta}z}-eA_{\mathrm{m}z}\right)-\frac{\tilde{\Delta}\phi_{\mathrm{m}}}{\tilde{\Delta}z}\left(\frac{\hbar}{i}\frac{\tilde{\Delta}}{\tilde{\Delta}y}-eA_{\mathrm{m}y}\right)\right]\Bigg\}\Delta x\end{aligned} \tag{23}$$

$$\begin{aligned}\mathsf{L}_{\mathrm{cx}3j} = \frac{Ze}{2\alpha m_0 c}\mathop{\mathfrak{S}}_{c}^{x}\Bigg(&A_{\mathrm{e}z}\frac{\tilde{\Delta}}{\tilde{\Delta}t}\left\{\left[1-\frac{1}{2m_0^2c^2}\left(\frac{\hbar}{i}\frac{\tilde{\Delta}}{\tilde{\Delta}x_j}-eA_{\mathrm{m}x_j}\right)^2\right]\left(\frac{\hbar}{i}\frac{\tilde{\Delta}}{\tilde{\Delta}y}-eA_{\mathrm{m}y}\right)\right.\\ &\left.+\left(\frac{\hbar}{i}\frac{\tilde{\Delta}}{\tilde{\Delta}y}-eA_{\mathrm{m}y}\right)\left[1-\frac{1}{2m_0^2c^2}\left(\frac{\hbar}{i}\frac{\tilde{\Delta}}{\tilde{\Delta}x_j}-eA_{\mathrm{m}x_j}\right)^2\right]\right\}\\ &-A_{\mathrm{e}z}\frac{\tilde{\Delta}}{\tilde{\Delta}t}\left\{\left[1-\frac{1}{2m_0^2c^2}\left(\frac{\hbar}{i}\frac{\tilde{\Delta}}{\tilde{\Delta}x_j}-eA_{\mathrm{m}x_j}\right)^2\right]\left(\frac{\hbar}{i}\frac{\tilde{\Delta}}{\tilde{\Delta}z}-eA_{\mathrm{m}z}\right)\right.\\ &\left.+\left(\frac{\hbar}{i}\frac{\tilde{\Delta}}{\tilde{\Delta}z}-eA_{\mathrm{m}z}\right)\left[1-\frac{1}{2m_0^2c^2}\left(\frac{\hbar}{i}\frac{\tilde{\Delta}}{\tilde{\Delta}x_j}-eA_{\mathrm{m}x_j}\right)^2\right]\right\}\Bigg)\Delta x\end{aligned} \tag{24}$$

$$\mathsf{L}_{\mathrm{cx}4j} = \frac{Zec}{\alpha}\mathop{\mathfrak{S}}_{c}^{x}\left(\frac{\tilde{\Delta}A_{\mathrm{e}y}}{\tilde{\Delta}z}-\frac{\tilde{\Delta}A_{\mathrm{e}z}}{\tilde{\Delta}y}\right)\Delta x \tag{25}$$

$$\begin{aligned}
\mathsf{L}_{\mathrm{cx}5j} = \frac{Ze}{4\alpha m_0^2 c} \oint_c^x \Bigg(\Bigg\{ \bigg[1 - \frac{1}{2m_0^2c^2} \Big(\frac{\hbar}{i}\frac{\tilde{\Delta}}{\tilde{\Delta} x_j} - eA_{\mathrm{m}x_j} \Big)^2 \bigg] \Big(\frac{\hbar}{i}\frac{\tilde{\Delta}}{\tilde{\Delta} y} - eA_{\mathrm{m}y} \Big) \\
+ \Big(\frac{\hbar}{i}\frac{\tilde{\Delta}}{\tilde{\Delta} y} - eA_{\mathrm{m}y} \Big) \bigg[1 - \frac{1}{2m_0^2c^2} \Big(\frac{\hbar}{i}\frac{\tilde{\Delta}}{\tilde{\Delta} x_j} - eA_{\mathrm{m}x_j} \Big)^2 \bigg] \Bigg\} \frac{\tilde{\Delta}}{\tilde{\Delta} y} \\
+ \Bigg\{ \bigg[1 - \frac{1}{2m_0^2c^2} \Big(\frac{\hbar}{i}\frac{\tilde{\Delta}}{\tilde{\Delta} x_j} - eA_{\mathrm{m}x_j} \Big)^2 \Big(\frac{\hbar}{i}\frac{\tilde{\Delta}}{\tilde{\Delta} z} - eA_{\mathrm{m}z} \Big) \\
+ \Big(\frac{\hbar}{i}\frac{\tilde{\Delta}}{\tilde{\Delta} z} - eA_{\mathrm{m}z} \Big) \bigg[1 - \frac{1}{2m_0^2c^2} \Big(\frac{\hbar}{i}\frac{\tilde{\Delta}}{\tilde{\Delta} x_j} - eA_{\mathrm{m}x_j} \Big)^2 \bigg] \Bigg\} \frac{\tilde{\Delta}}{\tilde{\Delta} z} \Bigg) \\
\times \Bigg\{ \bigg[1 - \frac{1}{2m_0^2c^2} \Big(\frac{\hbar}{i}\frac{\tilde{\Delta}}{\tilde{\Delta} x_j} - eA_{\mathrm{m}x_j} \Big)^2 \bigg] \\
\times \bigg[A_{\mathrm{e}x} \Big(\frac{\hbar}{i}\frac{\tilde{\Delta}}{\tilde{\Delta} y} - eA_{\mathrm{m}y} \Big) - A_{\mathrm{e}y} \Big(\frac{\hbar}{i}\frac{\tilde{\Delta}}{\tilde{\Delta} z} - eA_{\mathrm{m}z} \Big) \bigg] \\
+ \bigg[A_{\mathrm{e}x} \Big(\frac{\hbar}{i}\frac{\tilde{\Delta}}{\tilde{\Delta} y} - eA_{\mathrm{m}y} \Big) - A_{\mathrm{e}y} \Big(\frac{\hbar}{i}\frac{\tilde{\Delta}}{\tilde{\Delta} z} - eA_{\mathrm{m}z} \Big) \bigg] \\
\times \bigg[1 - \frac{1}{2m_0^2c^2} \Big(\frac{\hbar}{i}\frac{\tilde{\Delta}}{\tilde{\Delta} x_j} - eA_{\mathrm{m}x_j} \Big)^2 \bigg] \Bigg\} \Delta x \qquad (26)
\end{aligned}$$

Equations (20) to (26) make all terms in Eq.(17) defined for known values of the potentials $\mathbf{A}_{\mathrm{m}}$, $\mathbf{A}_{\mathrm{e}}$, ϕ_{e} and the rest mass m_0 of a charged particle. The equations are unique if one accepts Eq.(1.1-47). Equation (16) can be solved for certain initial and boundary conditions. The solution of Eq.(17) calls for a particular solution of the inhomogeneous equation that contains the solution $\Psi_{\mathrm{x}0x_j}$ of Eq.(16). The homogeneous part of Eq.(17) is equal to Eqs.(13) or (16).

Our calculation contains so far the simplification of the Hamilton function and the term $\alpha(2\lambda_{\mathrm{C}}A_{\mathrm{e}}/e)Q$ in Eq.(2.1-5) that go back to Eqs.(1.1-42) to (1.1-46) rather than the exact Hamilton function of Eq.(1.1-37) without the restriction $v \ll c$. A second simplification was introduced by Eq.(2.1-9) for small values of α. But our equations are still too complicated and we introduce further simplifications for the potentials $\mathbf{A}_{\mathrm{m}}$ and ϕ_{e}:

$$\begin{aligned}
\mathbf{A}_{\mathrm{m}} = \mathbf{A}_{\mathrm{m}0} + \alpha \mathbf{A}_{\mathrm{m}1}(\mathbf{r},t), \quad A_{\mathrm{m}x_j} = A_{\mathrm{m}0x_j} + \alpha A_{\mathrm{m}1x_j}(x_j,t) \\
\phi_{\mathrm{e}} = \phi_{\mathrm{e}0} + \alpha\phi_{\mathrm{e}1}(x_j,t), \quad x_j = x,\ y,\ z; \quad \Psi_{\mathrm{x}} = \Psi_{\mathrm{x}0} + \alpha\Psi_{\mathrm{x}1} \qquad (27)
\end{aligned}$$

We first rewrite Eqs.(2.1-10) and (2.1-11) with the approximations for $\mathbf{A}_{\mathrm{m}}$ and ϕ_{e}:

$$\left((\mathbf{p}-e\mathbf{A}_{\mathrm{m0}})^2-\frac{1}{c^2}(\mathcal{H}_x-e\phi_{\mathrm{e0}})^2+m_0^2c^2\right)\Psi_{\mathrm{x0}}=0 \tag{28}$$

$$\begin{aligned}&\left((\mathbf{p}-e\mathbf{A}_{\mathrm{m0}})^2-\frac{1}{c^2}(\mathcal{H}_x-e\phi_{\mathrm{e0}})^2+m_0^2c^2\right)\Psi_{\mathrm{x1}}\\&=-\Bigg[\frac{4\lambda_{\mathrm{C}}A_{\mathrm{e}}}{e}[(\mathbf{p}-e\mathbf{A}_{\mathrm{m0}})^2+m_0^2c^2]Q\\&-\frac{1}{c^2}\left((\mathcal{H}_x-e\phi_{\mathrm{e0}})\frac{\mathcal{L}_{\mathrm{cx}}}{\alpha}+\frac{\mathcal{L}_{\mathrm{cx}}}{\alpha}(\mathcal{H}_x-e\phi_{\mathrm{e0}})+\frac{\mathcal{L}_{\mathrm{cx}}^2}{\alpha}\right)-e[\mathbf{A}_{\mathrm{m1}}\cdot(\mathbf{p}-e\mathbf{A}_{\mathrm{m0}})\\&+(\mathbf{p}-e\mathbf{A}_{\mathrm{m0}})\ \ \mathbf{A}_{\mathrm{m1}}]+\frac{e}{c^2}[\phi_{\mathrm{e1}}(\mathcal{H}_x-e\phi_{\mathrm{e0}})+(\mathcal{H}_x-e\phi_{\mathrm{e0}})\phi_{\mathrm{e1}}]\Bigg]\Psi_{\mathrm{x0}}\end{aligned} \tag{29}$$

Equation (28) corresponds to Eq.(2.1-2). If we replace $\mathbf{A}_{\mathrm{m}}$ by $\mathbf{A}_{\mathrm{m0}}$ we must replace $A_{\mathrm{m}x_j}$ by $A_{\mathrm{m0}x_j}$ in Eqs.(2.1-4) and (2.1-28). This implies the same replacement of $A_{\mathrm{m}x_j}$ in Eqs.(13) to (16). Hence, Eq.(16) is reduced to the following form by the approximation of Eq.(27):

$$\begin{aligned}&\frac{\Psi_{\mathrm{x0}x_j}(x_j+\Delta x_j,t)-2\Psi_{\mathrm{x0}x_j}(x_j,t)+\Psi_{\mathrm{x0}x_j}(x_j-\Delta x_j,t)}{(\Delta x_j)^2}\\&\qquad-\frac{1}{c^2}\frac{\Psi_{\mathrm{x0}x_j}(x_j,t+\Delta t)-2\Psi_{\mathrm{x0}x_j}(x_j,t)+\Psi_{\mathrm{x0}x_j}(x_j,t-\Delta t)}{(\Delta t)^2}\\&-2i\frac{e}{\hbar}\Bigg(A_{\mathrm{m0}x_j}\frac{\Psi_{\mathrm{x0}x_j}(x_j+\Delta x_j,t)-\Psi_{\mathrm{x0}x_j}(x_j-\Delta x_j,t)}{2\Delta x_j}\\&\qquad+\frac{1}{c^2}\phi_{\mathrm{e0}}\frac{\Psi_{\mathrm{x0}x_j}(x_j,t+\Delta t)-\Psi_{\mathrm{x0}x_j}(x_j,t-\Delta t)}{2\Delta t}\Bigg)\\&\qquad-\frac{e^2}{\hbar^2}\left(A_{\mathrm{m0}x_j}^2-\frac{1}{c^2}\phi_{\mathrm{e0}}^2+\frac{m_0^2c^2}{e^2}\right)\Psi_{\mathrm{x0}x_j}(x_j,t)=0\end{aligned} \tag{30}$$

The difference equation representing Eq.(29) must be written in a more compact form by means of the definitions of Eqs.(2) and (3):

$$\begin{aligned}&\left[\left(\frac{\hbar}{i}\frac{\tilde{\Delta}}{\tilde{\Delta}x_j}-eA_{\mathrm{m0}x_j}\right)^2-\frac{1}{c^2}\left(\frac{\hbar}{i}\frac{\tilde{\Delta}}{\tilde{\Delta}t}+e\phi_{\mathrm{e0}}\right)^2+m_0^2c^2\right]\Psi_{\mathrm{x1}x_j}(x_j,t)\\&=-\Bigg\{\frac{4\lambda_{\mathrm{C}}A_{\mathrm{e}}}{e}\left(\frac{\hbar}{i}\frac{\tilde{\Delta}}{\tilde{\Delta}x_j}-eA_{\mathrm{m0}x_j}\right)^2\left[1-\frac{1}{2m_0^2c^2}\left(\frac{\hbar}{i}\frac{\tilde{\Delta}}{\tilde{\Delta}x_j}-eA_{\mathrm{m0}x_j}\right)^2\right]\\&\quad+\frac{1}{c^2}\Bigg[\left(\frac{\hbar}{i}\frac{\tilde{\Delta}}{\tilde{\Delta}t}+e\phi_{\mathrm{e0}}\right)(\mathsf{L}_{\mathrm{cx}1j}+\mathsf{L}_{\mathrm{cx}2j}+\mathsf{L}_{\mathrm{cx}3j}+\mathsf{L}_{\mathrm{cx}4j}+\mathsf{L}_{\mathrm{cx}5j})\\&\quad+(\mathsf{L}_{\mathrm{cx}1j}+\mathsf{L}_{\mathrm{cx}2j}+\mathsf{L}_{\mathrm{cx}3j}+\mathsf{L}_{\mathrm{cx}4j}+\mathsf{L}_{\mathrm{cx}5j})\left(\frac{\hbar}{i}\frac{\tilde{\Delta}}{\tilde{\Delta}t}+e\phi_{\mathrm{e0}}\right)+\alpha\mathsf{L}_{\mathrm{cx}}^2\Bigg]\end{aligned}$$

$$- e\left[A_{\mathrm{m}1x_j}\left(\frac{\hbar}{i}\frac{\tilde{\Delta}}{\tilde{\Delta}x_j} - eA_{\mathrm{m}0x_j}\right) + \left(\frac{\hbar}{i}\frac{\tilde{\Delta}}{\tilde{\Delta}x_j} - eA_{\mathrm{m}0x_j}\right)A_{\mathrm{m}1x_j}\right]$$
$$- \frac{e}{c^2}\left[\phi_{\mathrm{e}1}\left(\frac{\hbar}{i}\frac{\tilde{\Delta}}{\tilde{\Delta}t} + e\phi_{\mathrm{e}0}\right) + \left(\frac{\hbar}{i}\frac{\tilde{\Delta}}{\tilde{\Delta}t} + e\phi_{\mathrm{e}0}\right)\phi_{\mathrm{e}1}\right]\Bigg\}\Psi_{\mathrm{x}0x_j}(x_j,t) \quad (31)$$

As in Eq.(17) we ignore the term $\alpha \mathsf{L}_{\mathrm{cx}}^2$ because of the factor α. The terms $\mathsf{L}_{\mathrm{cx1j}}$ to $\mathsf{L}_{\mathrm{cx5j}}$ still have to be rewritten to correspond to the simplifications of Eq.(27). We have to add two more approximations:

$$\mathbf{A}_{\mathrm{e}} = \mathbf{A}_{\mathrm{e}0} + \alpha A_{\mathrm{e}1}(\mathbf{r},t), \quad A_{\mathrm{e}x_j} = A_{\mathrm{e}0x_j} + \alpha A_{\mathrm{e}1x_j}(x_j,t)$$
$$\phi_{\mathrm{m}} = \phi_{\mathrm{m}0} + \alpha\phi_{\mathrm{m}1}(x_j,t) \quad (32)$$

Only the terms $\mathbf{A}_{\mathrm{e}0}$ and $\phi_{\mathrm{m}0}$ are needed in Eq.(31); the terms multiplied by α are neglected. The term $\mathsf{L}_{\mathrm{cx}1j}$ in Eq.(20) is not changed much optically, but all terms $A_{\mathrm{e}0}$, $A_{\mathrm{e}0.}$, $A_{\mathrm{m}0.}$ are now constants:

$$\mathsf{L}_{\mathrm{cx}1j} = \frac{Ze}{2\alpha m_0^2 c}\Bigg\{\left[A_{\mathrm{e}0z}\left(\frac{\hbar}{i}\frac{\tilde{\Delta}}{\tilde{\Delta}y} - eA_{\mathrm{m}0y}\right) - A_{\mathrm{e}0y}\left(\frac{\hbar}{i}\frac{\tilde{\Delta}}{\tilde{\Delta}z} - eA_{\mathrm{m}0z}\right)\right]$$
$$\times\left(\frac{\hbar}{i}\frac{\tilde{\Delta}}{\tilde{\Delta}x} - eA_{\mathrm{m}0x}\right)\left[1 - \frac{1}{m_0^2c^2}\left(\frac{\hbar}{i}\frac{\tilde{\Delta}}{\tilde{\Delta}x_j} - eA_{\mathrm{m}0x_j}\right)^2\right]$$
$$+\left[1 - \frac{1}{m_0^2c^2}\left(\frac{\hbar}{i}\frac{\tilde{\Delta}}{\tilde{\Delta}x_j} - eA_{\mathrm{m}0x_j}\right)^2\right]\left[A_{\mathrm{e}0z}\left(\frac{\hbar}{i}\frac{\tilde{\Delta}}{\tilde{\Delta}y} - eA_{\mathrm{m}0y}\right)\right.$$
$$\left. - A_{\mathrm{e}0y}\left(\frac{\hbar}{i}\frac{\tilde{\Delta}}{\tilde{\Delta}z} - eA_{\mathrm{m}0z}\right)\right]\left(\frac{\hbar}{i}\frac{\tilde{\Delta}}{\tilde{\Delta}x} - eA_{\mathrm{m}0x}\right)\Bigg\}$$
$$j = 1,\ 2,\ 3;\quad x_1 = x,\ x_2 = y,\ x_3 = z \quad (33)$$

The term $\mathsf{L}_{\mathrm{cx}2j}$ of Eq.(23) vanishes since the differences $\tilde{\Delta}\phi_{\mathrm{m}0}/\tilde{\Delta}y$, etc. of a constant is zero. This implies that our results will not depend on the existence or nonexistence of a magnetic charge according to Eq.(1.1-22):

$$\mathsf{L}_{\mathrm{cx}2j} = 0 \quad (34)$$

For $\mathsf{L}_{\mathrm{cx}3j}$ of Eq.(24) apply the same comments as for Eq.(20):

$$\mathsf{L}_{\mathrm{cx}3j} = \frac{Ze}{2\alpha m_0 c}\mathop{\mathcal{S}}_{c}^{x}\Bigg(A_{\mathrm{e}0z}\frac{\tilde{\Delta}}{\tilde{\Delta}t}\Bigg\{\left[1 - \frac{1}{2m_0^2c^2}\left(\frac{\hbar}{i}\frac{\tilde{\Delta}}{\tilde{\Delta}x_j} - eA_{\mathrm{m}0x_j}\right)^2\right]\left(\frac{\hbar}{i}\frac{\tilde{\Delta}}{\tilde{\Delta}y} - eA_{\mathrm{m}0y}\right)$$
$$+\left(\frac{\hbar}{i}\frac{\tilde{\Delta}}{\tilde{\Delta}y} - eA_{\mathrm{m}0y}\right)\left[1 - \frac{1}{2m_0^2c^2}\left(\frac{\hbar}{i}\frac{\tilde{\Delta}}{\tilde{\Delta}x_j} - eA_{\mathrm{m}0x_j}\right)^2\right]\Bigg\}$$

$$
- A_{e0z}\frac{\tilde{\Delta}}{\tilde{\Delta}t}\left\{\left[1-\frac{1}{2m_0^2c^2}\left(\frac{\hbar}{i}\frac{\tilde{\Delta}}{\tilde{\Delta}x_j}-eA_{m0x_j}\right)^2\right]\left(\frac{\hbar}{i}\frac{\tilde{\Delta}}{\tilde{\Delta}z}-eA_{m0z}\right)\right.
+\left.\left(\frac{\hbar}{i}\frac{\tilde{\Delta}}{\tilde{\Delta}z}-eA_{m0z}\right)\left[1-\frac{1}{2m_0^2c^2}\left(\frac{\hbar}{i}\frac{\tilde{\Delta}}{\tilde{\Delta}x_j}-eA_{m0x_j}\right)^2\right]\right\}\Bigg)\Delta x \quad (35)
$$

$$
\mathsf{L}_{cx4j}=0,\quad \tilde{\Delta}A_{e0x}/\tilde{\Delta}x=0,\ \tilde{\Delta}A_{e0z}/\tilde{\Delta}y=0 \quad (36)
$$

$$
\begin{aligned}
\mathsf{L}_{cx5j}={}&\frac{Ze}{4\alpha m_0^2c}\oint_c^x\Bigg(\left\{\left[1-\frac{1}{2m_0^2c^2}\left(\frac{\hbar}{i}\frac{\tilde{\Delta}}{\tilde{\Delta}x_j}-eA_{m0x_j}\right)^2\right]\left(\frac{\hbar}{i}\frac{\tilde{\Delta}}{\tilde{\Delta}y}-eA_{m0y}\right)\right.\\
&+\left.\left(\frac{\hbar}{i}\frac{\tilde{\Delta}}{\tilde{\Delta}y}-eA_{m0y}\right)\left[1-\frac{1}{2m_0^2c^2}\left(\frac{\hbar}{i}\frac{\tilde{\Delta}}{\tilde{\Delta}x_j}-eA_{m0x_j}\right)^2\right]\right\}\frac{\tilde{\Delta}}{\tilde{\Delta}y}\\
&+\left\{\left[1-\frac{1}{2m_0^2c^2}\left(\frac{\hbar}{i}\frac{\tilde{\Delta}}{\tilde{\Delta}x_j}-eA_{m0x_j}\right)^2\left(\frac{\hbar}{i}\frac{\tilde{\Delta}}{\tilde{\Delta}z}-eA_{m0z}\right)\right.\right.\\
&+\left.\left(\frac{\hbar}{i}\frac{\tilde{\Delta}}{\tilde{\Delta}z}-eA_{m0z}\right)\left[1-\frac{1}{2m_0^2c^2}\left(\frac{\hbar}{i}\frac{\tilde{\Delta}}{\tilde{\Delta}x_j}-eA_{m0x_j}\right)^2\right]\right\}\frac{\tilde{\Delta}}{\tilde{\Delta}z}\Bigg)\\
&\times\left\{\left[1-\frac{1}{2m_0^2c^2}\left(\frac{\hbar}{i}\frac{\tilde{\Delta}}{\tilde{\Delta}x_j}-eA_{m0x_j}\right)^2\right]\right.\\
&\qquad\times\left[A_{e0x}\left(\frac{\hbar}{i}\frac{\tilde{\Delta}}{\tilde{\Delta}y}-eA_{m0y}\right)-A_{e0y}\left(\frac{\hbar}{i}\frac{\tilde{\Delta}}{\tilde{\Delta}z}-eA_{m0z}\right)\right]\\
&+\left[A_{e0x}\left(\frac{\hbar}{i}\frac{\tilde{\Delta}}{\tilde{\Delta}y}-eA_{m0y}\right)-A_{e0y}\left(\frac{\hbar}{i}\frac{\tilde{\Delta}}{\tilde{\Delta}z}-eA_{m0z}\right)\right]\\
&\qquad\times\left.\left[1-\frac{1}{2m_0^2c^2}\left(\frac{\hbar}{i}\frac{\tilde{\Delta}}{\tilde{\Delta}x_j}-eA_{m0x_j}\right)^2\right]\right\}\Delta x \quad (37)
\end{aligned}
$$

2.3 Solution of the Difference Equation of Ψ_{x0}

We start from Eq.(2.2-30). The normalized variables θ and ζ are used instead of the nonnormalized variables t and x_j. The number of spatial variables is reduced to one:

$$
\theta=t/\Delta t,\quad \zeta=x_j/c\Delta t,\quad 0\le\theta\le N,\quad 0\le\zeta\le N \quad (1)
$$

The function Ψ_{x0x_j} is shortened to Ψ_0. The potentials $\mathbf{A}_{e0x_j}$ and $\mathbf{A}_{m0x_j}$ have the components A_{e0x}, A_{e0y}, A_{e0z} and A_{m0x}, A_{m0y}, A_{m0z} which are constants. They are not affected by the normalizations of Eq.(1). In the special case of Eq.(2.2-30) we could write $A_{m0\zeta}$ and not decide whether $A_{m0\zeta}$ represents A_{m0x}, A_{m0y}, or A_{m0z} since the choice of ζ or $x_j=x$, y, z in Eq.(1) will decide what

$A_{\mathrm{m}0\zeta}$ stands for. This simplification would not be possible when we advance from Eq.(2.2-30) to Eq.(2.2-31) in Chapter 3. Hence, we will not use it.

The following normalized partial difference equation with one spatial variable is obtained from Eq.(2.2-30):

$$\Psi_{\mathrm{x}0x_j} = \Psi_{\mathrm{x}0} = \Psi_0, \; x_j = x$$

$$[\Psi_0(\zeta+1,\theta)-2\Psi_0(\zeta,\theta)+\Psi_0(\zeta-1,\theta)]-[\Psi_0(\zeta,\theta+1)-2\Psi_0(\zeta,\theta)+\Psi_0(\zeta,\theta-1)]$$
$$-i\lambda_1\{[\Psi_0(\zeta+1,\theta)-\Psi_0(\zeta-1,\theta)]+\lambda_3[\Psi_0(\zeta,\theta+1)-\Psi_0(\zeta,\theta-1)]\}-\lambda_2^2\Psi_0(\zeta,\theta)=0$$

$$\lambda_1 = \frac{ecTA_{\mathrm{m}0x}}{N\hbar}, \quad \lambda_3 = \frac{\phi_{\mathrm{e}0}}{cA_{\mathrm{m}0x}}, \quad \lambda_1\lambda_3 = \frac{eT\phi_{\mathrm{e}0}}{N\hbar}, \quad N = \frac{T}{\Delta t}$$
$$\lambda_2^2=\lambda_1^2\left(1-\frac{\phi_{\mathrm{e}0}^2}{c^2A_{\mathrm{m}0x}^2}+\frac{m_0^2c^2}{e^2A_{\mathrm{m}0x}^2}\right) = \left(\frac{T}{N\hbar}\right)^2(m_0^2c^4-e^2\phi_{\mathrm{e}0}^2+e^2c^2A_{\mathrm{m}0x}^2)$$
$$\lambda_2^2-\lambda_1^2 = (T/N\hbar)^2(m_0^2c^4-e^2\phi_{\mathrm{e}0}^2) > 0 \text{ for } m_0c^2 > e\phi_{\mathrm{e}0}, \; \lambda_2^2 \gg \lambda_1^2$$
$$\lambda_1^2\lambda_3^2+\lambda_2^2=(T/N\hbar)^2(m_0^2c^4+e^2c^2A_{\mathrm{m}0x}^2), \; \lambda_1^2\lambda_3^2+\lambda_2^2-\lambda_1^2=(m_0c^2T/N\hbar)^2$$
$$\lambda_1^2\lambda_3^2-\lambda_2^2 = -(T/N\hbar)^2(m_0^2c^4-2e^2\phi_{\mathrm{e}0}^2+e^2c^2A_{\mathrm{m}0x}^2) \tag{2}$$

We look for a solution excited by an exponential ramp function as boundary condition. The exponential ramp function avoids the instant jump of a step function and is thus better suited for equations that describe particles with a finite mass m_0. The constant ι cannot be chosen. It will turn out to be imaginary:

$$\begin{aligned}\Psi_0(0,\theta) = \Psi_{00}S(\theta)(1-e^{-\iota\theta}) &= 0 && \text{for } \theta < 0\\ &= \Psi_{00}(1-e^{-\iota\theta}) && \text{for } \theta \geq 0\end{aligned} \tag{3}$$

An initial condition is needed for the time $\theta = 0$ and all values of ζ. We choose it to be

$$\Psi_0(\zeta,\theta) = 0 \quad \text{for } \theta = 0, \; \zeta \geq 0 \tag{4}$$

Since Eq.(2) is a second-order difference equation with respect to θ we need a second initial condition in addition to Eq.(4). We proceed as follows. Equation (2) requires $\Psi_0(\zeta,0)$, and $\Psi_0(\zeta,1)$ to compute $\Psi_0(\zeta,\theta)$ for $\theta \geq 2$. Since $\Psi_0(\zeta,0)$ is defined by Eq.(4) we may choose a second condition

$$\Psi_0(\zeta,1) = d' \tag{5}$$

but the connection with $\partial\Psi_0(\zeta,\theta)/\partial\theta$ for $\theta = 0$ of the differential theory becomes clear if we use the "right" difference quotient of Eq.(1.3-1) instead of Eq.(5):

$$\Psi_0(\zeta,1) - \Psi_0(\zeta,0) = d \tag{6}$$

We use the following ansatz (Habermann 1987) to find a general solution from a steady-state solution $F(\zeta)$ multiplied by $(1 - e^{-\iota\theta})$ plus a deviation $u(\zeta,\theta)$ from $(1 - e^{-\iota\theta})F(\zeta)$:

$$\Psi_0(\zeta,\theta) = \Psi_{00}[(1 - e^{-\iota\theta})F(\zeta) + u(\zeta,\theta)], \quad \zeta,\ \theta \geq 0 \tag{7}$$

Substitution of $\Psi_{00}(1 - e^{-\iota\theta})F(\zeta)$ into Eq.(2) produces the following relations

$$\Psi_0(\zeta,\theta) = \Psi_{00}(1 - e^{-\iota\theta})F(\zeta),\ \Psi_0(\zeta,\theta \pm 1) = \Psi_{00}(1 - e^{-\iota(\theta\pm1)})F(\zeta)$$
$$\Psi_0(\zeta \pm 1,\theta) = \Psi_{00}(1 - e^{-\iota\theta})F(\zeta \pm 1)$$

and we obtain:

$$\begin{aligned}(1 - e^{-\iota\theta})[F(\zeta+1) - 2F(\zeta) + F(\zeta-1)] + e^{-\iota\theta}(e^{-\iota} - 2 + e^{\iota})F(\zeta)\\ - i\lambda_1\{(1 - e^{-\iota\theta})[F(\zeta+1) - F(\zeta-1)] - \lambda_3 e^{-\iota\theta}(e^{-\iota} - e^{\iota})F(\zeta)\}\\ - \lambda_2^2(1 - e^{-\iota\theta})F(\zeta) = 0\end{aligned} \tag{8}$$

We collect all terms multiplied with $1 - e^{-\iota\theta}$ on the left side and the others on the right side:

$$\begin{aligned}(1-e^{-\iota\theta})\{F(\zeta+1) - 2F(\zeta) + F(\zeta-1) - i\lambda_1[F(\zeta+1) - F(\zeta-1)] - \lambda_2^2F(\zeta)\}\\ = -e^{-\iota\theta}[e^{-\iota} - 2 + e^{\iota} + i\lambda_1\lambda_3(e^{-\iota} - e^{\iota})]F(\zeta)\end{aligned}$$

This equation must hold for all values of θ, which is possible only if both sides are zero:

$$F(\zeta+1) - 2F(\zeta) + F(\zeta-1) - i\lambda_1[F(\zeta+1) - F(\zeta-1)] - \lambda_2^2F(\zeta) = 0 \tag{9}$$

$$e^{-\iota} - 2 + e^{\iota} + i\lambda_1\lambda_3(e^{-\iota} - e^{\iota}) = 0 \tag{10}$$

For the solution of Eq.(9) we may use the ansatz

$$F(\zeta) = A_1 v^\zeta, \; F(\zeta \pm 1) = A_1 v^{\zeta \pm 1} \tag{11}$$

and obtain an equation for v:

$$\begin{aligned}
&(1 - i\lambda_1)v^2 - (2 + \lambda_2^2)v + 1 + i\lambda_1 = 0 \\
v_{1,2} &= \frac{1 + i\lambda_1}{2(1 + \lambda_1^2)}\{2 + \lambda_2^2 \pm [4(\lambda_2^2 - \lambda_1^2) + \lambda_2^4]^{1/2}\} \\
&\doteq (1 + i\lambda_1)[1 \pm (\lambda_2^2 - \lambda_1^2)^{1/2}], \quad \lambda_2^2, \; \lambda_1^2 \ll 1
\end{aligned} \tag{12}$$

$$\begin{aligned}
v_1 &\doteq (1 + i\lambda_1)[1 - (\lambda_2^2 - \lambda_1^2)^{1/2}] \\
&\doteq \exp[-(\lambda_2^2 - \lambda_1^2)^{1/2}]e^{i\lambda_1}
\end{aligned} \tag{13}$$

$$\begin{aligned}
v_2 &\doteq (1 + i\lambda_1)[1 + (\lambda_2^2 - \lambda_1^2)^{1/2}] \\
&\doteq \exp[(\lambda_2^2 - \lambda_1^2)^{1/2}]e^{i\lambda_1}
\end{aligned} \tag{14}$$

We obtain for $F(\zeta)$ in Eq.(11):

$$F(\zeta) = A_{10}\exp\{[(\lambda_2^2 - \lambda_1^2)^{1/2} + i\lambda_1]\zeta\} + A_{11}\exp\{[-(\lambda_2^2 - \lambda_1^2)^{1/2} + i\lambda_1]\zeta\} \tag{15}$$

According to Eq.(2) the difference $\lambda_2^2 - \lambda_1^2$ will be positive except for extremely large values of ϕ_{e0}^2. We restrict the calculation to this case $\lambda_2^2 - \lambda_1^2 > 0$. In addition, we want $F(0)$ to be 1, which is achieved by the choice $A_{11} = 1$ and $A_{10} = 0$. The function $F(\zeta)$ becomes:

$$\begin{aligned}
&F(\zeta) = \exp[-(\lambda_2^2 - \lambda_1^2)^{1/2}\zeta]e^{i\lambda_1\zeta} \\
&\lambda_2^2 - \lambda_1^2 > 0, \; F(0) = 1, \; A_{10} = 0, \; A_{11} = 1
\end{aligned} \tag{16}$$

Equation (10) yields a quadratic equation for e^ι:

$$\begin{aligned}
&(1 - i\lambda_1\lambda_3)e^{2\iota} - 2e^\iota + 1 + i\lambda_1\lambda_3 = 0 \\
e^\iota &= \frac{(1 + i\lambda_1\lambda_3)(1 \pm i\lambda_1\lambda_3)}{1 + \lambda_1^2\lambda_3^2} \doteq 1 + i(\lambda_1\lambda_3 \pm \lambda_1\lambda_3)
\end{aligned} \tag{17}$$

A trivial solution $e^\iota = 1$, $\iota = 0$ and a nontrivial solution are obtained:

$$e^\iota = 1 + 2i\lambda_1\lambda_3, \; \iota \doteq 2i\lambda_1\lambda_3 \tag{18}$$

Equation (7) must satisfy the boundary condition of Eq.(3). Since Eq.(16) yields $F(0) = 1$ we get:

$$\Psi_{00}[1 - e^{-\iota\theta} + u(0,\theta)] = \Psi_{00}(1 - e^{-\iota\theta}), \quad \theta \geq 0$$
$$u(0,\theta) = 0 \tag{19}$$

Hence, $u(\zeta,\theta)$ has a homogeneous boundary condition, which is the goal of this method of solution. The initial conditions of Eqs.(4) and (6) yield with $F(0) = 1$:

$$\Psi_0(\zeta,0) = \Psi_{00}u(\zeta,0) = 0, \quad \zeta > 0 \tag{20}$$

$$\Psi_{00}[(1 - e^{-\iota})F(\zeta) + u(\zeta,1)] - \Psi_{00}u(\zeta,0) = 0$$
$$u(\zeta,1) - u(\zeta,0) = -(1 - e^{-\iota})F(\zeta), \quad \zeta > 0 \tag{21}$$

Substitution of $u(\zeta,\theta)$ into Eq.(2) yields the same equation with Ψ_0 replaced by u:

$$[u(\zeta+1,\theta) - 2u(\zeta,\theta) + u(\zeta-1,\theta)] - [u(\zeta,\theta+1) - 2u(\zeta,\theta) + u(\zeta,\theta-1)]$$
$$- i\lambda_1\{[u(\zeta+1,\theta) - u(\zeta-1,\theta)] + \lambda_3[u(\zeta,\theta+1) - u(\zeta,\theta-1)]\} - \lambda_2^2 u(\zeta,\theta) = 0 \tag{22}$$

Particular solutions $u_\kappa(\zeta,\theta)$ of this equation can be obtained by the extension of Bernoulli's product method for the separation of variables from partial differential equations to partial difference equations:

$$u_\kappa(\zeta,\theta) = \phi_\kappa(\zeta)\psi_\kappa(\theta) = \phi(\zeta)\psi(\theta) \tag{23}$$

We observe that the notation $\phi(\zeta)\psi(\theta)$ instead of $\phi_\kappa(\zeta)\psi_\kappa(\theta)$ is a generally made simplification that will be acceptable in Chapter 2. The substitution of $u_\kappa(\zeta,\theta)$ for $u(\zeta,\theta)$ in Eq.(22) yields:

$$[\phi(\zeta+1)\psi(\theta) - 2\phi(\zeta)\psi(\theta) + \phi(\zeta-1)\psi(\theta)]$$
$$- [\phi(\zeta)\psi(\theta+1) - 2\phi(\zeta)\psi(\theta) + \phi(\zeta)\psi(\theta-1)]$$
$$- i\lambda_1\Big\{[\phi(\zeta+1)\psi(\theta) - \phi(\zeta-1)\psi(\theta] + \lambda_3[\phi(\zeta)\psi(\theta+1) - \phi(\zeta)\psi(\theta-1)]\Big\}$$
$$- \lambda_2^2\phi(\theta)\psi(\theta) = 0 \tag{24}$$

In analogy to the procedure used for partial differential equations we multiply with $1/\phi(\zeta)\psi(\theta)$ and separate the variables:

$$\frac{1}{\phi(\zeta)}\Big\{[\phi(\zeta+1)-2\phi(\zeta)+\phi(\zeta-1)]-i\lambda_1[\phi(\zeta+1)-\phi(\zeta-1)]\Big\}$$
$$=\frac{1}{\psi(\theta)}\Big\{[\psi(\theta+1)-2\psi(\theta)+\psi(\theta-1)]+i\lambda_1\lambda_3[\psi(\theta+1)-\psi(\theta-1)]\Big\}+\lambda_2^2$$
$$=-(2\pi\rho_\kappa/N)^2 \quad (25)$$

A constant $-(2\pi\rho_\kappa/N)^2$ is written at the end of Eq.(25) since a function of ζ can be equal to a function of θ for any ζ and θ only if they are equal to a constant. The divisor N will permit the use of an orthogonality interval of length N rather than 1 later on. Equation (25) yields two ordinary difference equations for $\phi(\zeta)$ and $\psi(\theta)$:

$$[\phi(\zeta+1)-2\phi(\zeta)+\phi(\zeta-1)]-i\lambda_1[\phi(\zeta+1)-\phi(\zeta-1)]$$
$$+(2\pi\rho_\kappa/N)^2\phi(\zeta)=0 \quad (26)$$

$$[\psi(\theta+1)-2\psi(\theta)+\psi(\theta-1)]+i\lambda_1\lambda_3[\psi(\theta+1)-\psi(\theta-1)]$$
$$+[(2\pi\rho_\kappa/N)^2+\lambda_2^2]\psi(\theta)=0 \quad (27)$$

Equation (26) can be solved like Eq.(9) for $F(\zeta)$ by using an ansatz equivalent to Eq.(11):

$$\phi(\zeta)=A_2v^\zeta,\ \phi(\zeta\pm 1)=A_2v^{\zeta\pm 1} \quad (28)$$

The following equation is obtained for v:

$$(1-i\lambda_1)v^2-[2-(2\pi\rho_\kappa/N)^2]v+1+i\lambda_1=0$$
$$v_{3,4}=\frac{1+i\lambda_1}{2(1+\lambda_1^2)}\Bigg\{2-\left(\frac{2\pi\kappa\rho_\kappa}{N}\right)^2$$
$$\pm 2i\frac{2\pi\rho_\kappa}{N}\left[1-\frac{1}{4}\left(\frac{2\pi\rho_\kappa}{N}\right)^2+\frac{\lambda_1^2}{(2\pi\rho_\kappa/N)^2}\right]^{1/2}\Bigg\} \quad (29)$$

We shall need $v_{3,4}$ only for real values of the square root. Furthermore, we need $v_{3,4}$ only up to the first order of Δt. Let us note that λ_1^2 in the square root is of order $O(\Delta t)^2$ according to Eq.(2):

$$\text{for } (2\pi\rho_\kappa/N)^2 < 4 + \frac{4\lambda_1^2}{(2\pi\rho_\kappa/N)^2} = 4 + O(\Delta t)^2$$

$$v_3 = 1 - \frac{1}{2}\left(\frac{2\pi\rho_\kappa}{N}\right)^2 - \lambda_1\frac{2\pi\rho_\kappa}{N}\left(1 - \frac{(2\pi\rho_\kappa/N)^2}{4}\right)^{1/2}$$
$$+i\left[\frac{2\pi\rho_\kappa}{N}\left(1 - \frac{(2\pi\rho_\kappa/N)^2}{4}\right)^{1/2} + \lambda_1\left(1 - \frac{(2\pi\rho_\kappa/N)^2}{4}\right)\right] + O(\Delta t)^2 \quad (30)$$

$$v_4 = 1 - \frac{1}{2}\left(\frac{2\pi\rho_\kappa}{N}\right)^2 + \lambda_1\frac{2\pi\rho_\kappa}{N}\left(1 - \frac{(2\pi\rho_\kappa/N)^2}{4}\right)^{1/2}$$
$$-i\left[\frac{2\pi\rho_\kappa}{N}\left(1 - \frac{(2\pi\rho_\kappa/N)^2}{4}\right)^{1/2} - \lambda_1\left(1 - \frac{(2\pi\rho_\kappa/N)^2}{4}\right)\right] + O(\Delta t)^2 \quad (31)$$

We use Eqs.(29)–(31) to write v_3^ζ and v_4^ζ as follows, ignoring terms of order $O(\Delta t)^2$ or higher:

$$\text{for } (2\pi\rho_\kappa/N)^2 < 4 + \frac{4\lambda_1^2}{(2\pi\rho_\kappa/N)^2} = 4 + O(\Delta)^2$$

$$v_{3,4} = (1 + i\lambda_1)\left\{1 - \frac{1}{2}\left(\frac{2\pi\rho_\kappa}{N}\right)^2 \pm i\frac{2\pi\rho_\kappa}{N}\left[1 - \frac{1}{4}\left(\frac{2\pi\rho_\kappa}{N}\right)^2\right]^{1/2}\right\} \quad (32)$$

$$v_3^\zeta = (1+i\lambda_1)^\zeta\left\{\left[1 - \frac{1}{2}\left(\frac{2\pi\rho_\kappa}{N}\right)^2\right]^2 + \left(\frac{2\pi\rho_\kappa}{N}\right)^2\left[1 - \frac{1}{4}\left(\frac{2\pi\rho_\kappa}{N}\right)^2\right]\right\}^{\zeta/2} e^{i\varphi_\kappa\zeta}$$
$$= (1 + i\lambda_1)^\zeta \{1\}^{\zeta/2} e^{i\varphi_\kappa\zeta} \doteq (1 + i\lambda_1\zeta)e^{i\varphi_\kappa\zeta} \doteq e^{i(\varphi_\kappa+\lambda_1)\zeta}$$
$$= \cos(\varphi_\kappa + \lambda_1)\zeta + i\sin(\varphi_\kappa + \lambda_1)\zeta \quad (33)$$

$$v_4^\zeta = (1 + i\lambda_1)^\zeta e^{-i\varphi_\kappa\zeta} = e^{-i(\varphi_\kappa-\lambda_1)\zeta}$$
$$= \cos(\varphi_\kappa - \lambda_1)\zeta - i\sin(\varphi_\kappa - \lambda_1)\zeta \quad (34)$$

$$\varphi_\kappa = \operatorname{arctg}\frac{(2\pi\rho_\kappa/N)[1 - (2\pi\rho_\kappa/N)^2/4]^{1/2}}{1 - (2\pi\rho_\kappa/N)^2/2}, \quad \lambda_1 = \frac{ecA_{m0x}\Delta t}{\hbar} \quad (35)$$

The function $\phi(\zeta)$ is written in complex form using Eqs.(33) and (34) with λ_1 and φ_κ defined by Eq.(35):

$$\phi(\zeta) = A_{30}v_3^\zeta + A_{31}v_4^\zeta = A_{30}e^{i(\varphi_\kappa+\lambda_1)\zeta} + A_{31}e^{-i(\varphi_\kappa-\lambda_1)\zeta} \quad (36)$$

The boundary condition $u(0,\theta) = 0$ in Eq.(19) requires in Eq.(36) the following relation between the constants:

$$A_{31} = -A_{30}$$
$$\phi(\zeta) = A_{30}\left(e^{i(\varphi_\kappa+\lambda_1)\zeta} - e^{-i(\varphi_\kappa-\lambda_1)\zeta}\right) = 2iA_{30}e^{i\lambda_1\zeta}\sin\varphi_\kappa\zeta \tag{37}$$

Equations (36) and (37) will be used with a Fourier series. This requires that we choose $\varphi_\kappa = \varphi_\kappa(\rho_\kappa)$ so that we get an orthogonal system of sine and cosine functions in an interval $0 \le \zeta \le N = T/\Delta t$ with a maximum of N periods. We must choose φ_κ as follows:

$$\begin{gathered}\varphi_\kappa = 0,\ \pm 1\cdot 2\pi,\ \pm 2\cdot 2\pi,\ \ldots,\ \pm\frac{N}{2}\cdot 2\pi \\ \varphi_\kappa = 2\pi\kappa/N, \quad \kappa = 0,\ \pm 1,\ \pm 2,\ \ldots,\ \pm N/2 \\ 0 \le \zeta \le N = T/\Delta t\end{gathered} \tag{38}$$

The variable φ_κ can assume $N+1$ values. There are N orthogonal sine functions, N orthogonal cosine functions, and a constant defined by $\kappa = 0$. The function $\phi(\zeta) = \text{constant}$ is a solution of Eq.(26) for $\varphi_\kappa = 2\pi\rho_\kappa/N = 0$.

In order to obtain the eigenvalues $(2\pi\rho_\kappa/N)^2$ associated with the angles φ_κ we must solve Eq.(35) for $(2\pi\rho_\kappa/N)^2$. With the relation

$$1 + \operatorname{tg}^2\varphi_\kappa = \frac{1}{\cos^2\varphi_\kappa} \tag{39}$$

we obtain from Eq.(35) two solutions:

$$\left(\frac{2\pi\rho_\kappa}{N}\right)^2 = 2(1+\cos\varphi_\kappa) = 4\cos^2\frac{\varphi_\kappa}{2} \tag{40}$$
$$= 2(1-\cos\varphi_\kappa) = 4\sin^2\frac{\varphi_\kappa}{2} \tag{41}$$

In order to see which solution to use we take the inverse of Eq.(41):

$$\varphi_\kappa = 2\arcsin\frac{2\pi\rho_\kappa/N}{2} \tag{42}$$

At least for small values of φ_κ we are back to Eq.(35). Hence, we use Eq.(41) to define ρ_κ with the help of φ_κ:

$$\begin{aligned}\rho_\kappa &= \frac{N}{\pi}\sin\frac{\varphi_\kappa}{2} = \frac{N}{\pi}\sin\frac{\pi\kappa}{N}, \quad \kappa = 0,\ \pm 1,\ \pm 2,\ \ldots,\ \pm N/2 \\ &\doteq \frac{N}{2\pi}\varphi_\kappa = \kappa \quad \text{for } \varphi_\kappa \ll 1\end{aligned} \tag{43}$$

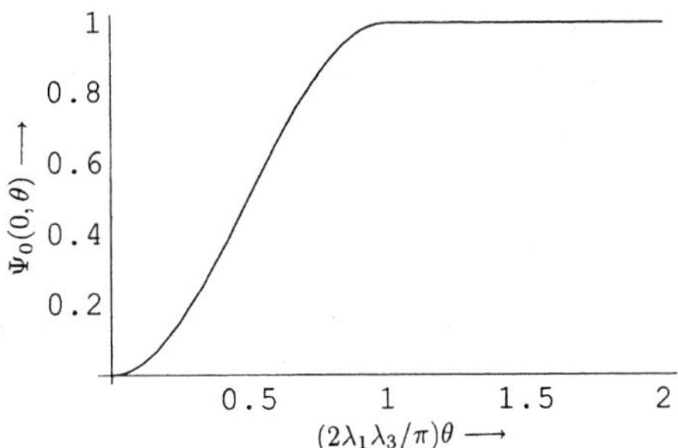

FIG.2.3-1. Plot of the excitation function $\Psi_0(0,\theta)$ according to Eq.(46).

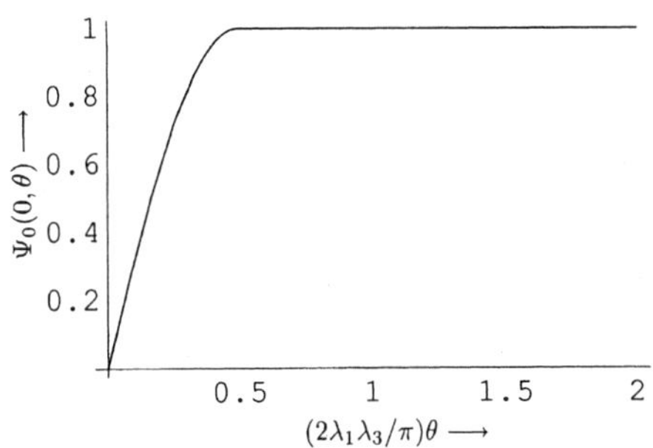

FIG.2.3-2. Plot of the excitation function $\Psi_0(0,\theta)$ according to Eq.(47).

We want to derive a plot of the boundary condition or excitation function of Eq.(3). Substitution of ι from Eq.(18) produces a complex function. Either the real or the imaginary component represents an acceptable boundary condition:

$$\Psi_0(0,\theta) = \Psi_{00} S(\theta)(1 - \cos 2\lambda_1\lambda_3\theta) \tag{44}$$

$$\Psi_0(0,\theta) = \Psi_{00} S(\theta) \sin 2\lambda_1\lambda_3\theta \tag{45}$$

These are (causal) cosinusoidal and sinusoidal functions that are zero for $\theta \leq 0$. One may readily derive step functions with a smoothed step from them:

$$\Psi_0(0,\theta) = \Psi_{00}\{S(\theta)(1 - \cos 2\lambda_1\lambda_3\theta) + S(\theta - \pi/2\lambda_1\lambda_3)[1 - \cos(2\lambda_1\lambda_3\theta - \pi)]\} \tag{46}$$

$$\Psi_0(0,\theta) = \Psi_{00}\{S(\theta) \sin 2\lambda_1\lambda_3\theta + S(\theta - \pi/4\lambda_1\lambda_3)[1 - \sin(2\lambda_1\lambda_3\theta - \pi/2)]\} \tag{47}$$

Plots of Eqs.(46) and (47) are shown in Figs.2.3-1 and 2.3-2. The tangent at $\theta = 0$ is zero in Fig.2.3-1 and larger than zero in Fig.2.3-2. There is no jump or kink at $(2\lambda_1\lambda_3/\pi)\theta = 1$ in Fig.2.3-1 or at $(2\lambda_1\lambda_3/\pi)\theta = 0.5$ in Fig.2.3-2.

Boundary conditions with any time variation other than those of Eqs.(46) and (47) must be represented by a superposition of these inherent boundary conditions. The choice $d = 0$ in Eq.(6) yields Fig.2.3-1, while $d = (2\lambda_1\lambda_3/\pi)\Psi_{00}$ yields Fig.2.3-2.

2.4 Time-Dependent Solutions of $\Psi_{x0}(\zeta, \theta)$

Equation (2.3-27) is the time-dependent difference equation of $\psi(\theta)$. The separation variable ρ_κ is defined in Eq.(2.3-43):

$$\begin{aligned}
\psi(\theta+1) - 2\psi(\theta) + \psi(\theta-1) + i\lambda_1\lambda_3[\psi(\theta+1) - \psi(\theta-1)]& \\
+[(2\pi\rho_\kappa/N)^2 + \lambda_2^2]\psi(\theta) &= 0 \\
\varphi_\kappa = \frac{2\pi\kappa}{N}, \quad \kappa = 0,\ \pm1,\ \pm2, \ldots,\ \pm\frac{N}{2}& \\
\left(\frac{2\pi\rho_\kappa}{N}\right)^2 = 4\sin^2\frac{\varphi_\kappa}{2}, \quad \rho_\kappa = \frac{N}{\pi}\sin\frac{\varphi_\kappa}{2}& \qquad (1)
\end{aligned}$$

Substitution of $\psi(\theta) = A_2 v^\theta$ into Eq.(1) yields an equation for v

$$(1 + i\lambda_1\lambda_3)v^2 - [2 - (2\pi\rho_\kappa/N)^2 - \lambda_2^2]v + 1 - i\lambda_1\lambda_3 = 0 \qquad (2)$$

with the solutions

$$\begin{aligned}
v_{5,6} = \frac{1 - i\lambda_1\lambda_3}{1 + \lambda_1^2\lambda_3^2}\Biggl(1 - \frac{1}{2}\left[\left(\frac{2\pi\rho_\kappa}{N}\right)^2 + \lambda_2^2\right]& \\
\pm\frac{1}{2}\left\{\left[\left(\frac{2\pi\rho_\kappa}{N}\right)^2 + \lambda_2^2\right]^2 - 4\left[\left(\frac{2\pi\rho_\kappa}{N}\right)^2 + \lambda_2^2\right] - 4\lambda_1^2\lambda_3^2\right\}^{1/2}\Biggr)& \qquad (3)
\end{aligned}$$

The square root is imaginary for

$$\left[\left(\frac{2\pi\rho_\kappa}{N}\right)^2 + \lambda_2^2\right]^2 - 4\left[\left(\frac{2\pi\rho_\kappa}{N}\right)^2 + \lambda_2^2\right] - 4\lambda_1^2\lambda_3^2 < 0 \qquad (4)$$

If we replace the sign $<$ by the equality sign $=$ we obtain a quadratic equation with the solutions

$$\begin{aligned}
(2\pi\rho_\kappa/N)^2 &= 2 - \lambda_2^2 \pm 2(1 + \lambda_1^2\lambda_3^3)^{1/2} \doteq 2 - \lambda_2^2 \pm (2 + \lambda_1^2\lambda_3^2) \\
&\doteq 4 - \lambda_2^2 + \lambda_1^2\lambda_3^2 \qquad \text{for } +(2 + \lambda_1^2\lambda_3^2) \\
&\doteq -\lambda_2^2 - \lambda_1^2\lambda_3^2 \qquad \text{for } -(2 + \lambda_1^2\lambda_3^3) \qquad (5)
\end{aligned}$$

Using Eq.(2.3-2) we see that the square root in Eq.(3) is imaginary if the following two conditions are satisfied:

$$
\begin{gathered}
4-\lambda_2^2+\lambda_1^2\lambda_3^2>(2\pi\rho_\kappa/N)^2=4\sin^2(\varphi_\kappa/2)\\
4-\left(\frac{\Delta t}{\hbar}\right)^2(m_0^2c^4-2e^2\phi_{e0}^2+e^2c^2A_{m0x}^2)>(2\pi\rho_\kappa/N)^2
\end{gathered}
\tag{6}
$$

We may rewrite these equations as follows:

$$
\begin{aligned}
\varphi_\kappa&=\frac{2\pi\kappa}{N}<2\arcsin\left(1-\frac{\lambda_2^2-\lambda_1^2\lambda_3^2}{4}\right)^{1/2}\\
\kappa<\kappa_0&=\frac{N}{\pi}\arcsin\left(1-\frac{\lambda_2^2-\lambda_1^2\lambda_3^2}{4}\right)^{1/2}
\end{aligned}
\tag{7}
$$

In addition to Eq.(6) we get from Eq.(5) a second condition:

$$
\begin{gathered}
-\lambda_2^2-\lambda_1^2\lambda_3^2<(2\pi\rho_\kappa/N)^2=4\sin^2(\varphi_\kappa/2)\\
-\left(\frac{\Delta t}{\hbar}\right)^2(m_0^2c^4+e^2c^2A_{m0x}^2)<(2\pi\rho_\kappa/N)^2
\end{gathered}
\tag{8}
$$

The condition of Eq.(8) is always satisfied since the range of interest of $(2\pi\rho_\kappa/N)^2=4\sin^2(\varphi_\kappa/2)$ is restricted as follows:

$$
-N/2\le\kappa\le N/2,\quad -1\le\varphi_\kappa/\pi\le+1,\quad 0\le4\sin^2(\varphi_\kappa/2)\le4 \tag{9}
$$

On the other hand the condition of Eq.(7) may not be satisfied. Both the real and the imaginary root of $v_{5,6}$ must be analyzed. We consider the imaginary root.

$$
\begin{aligned}
v_{5,6}&=\frac{1-i\lambda_1\lambda_3}{1+\lambda_1^2\lambda_3^2}\Bigg(1-\frac{1}{2}\left[\left(\frac{2\pi\rho_\kappa}{N}\right)^2+\lambda_2^2\right]\pm i\left[\left(\frac{2\pi\rho_\kappa}{N}\right)^2+\lambda_2^2\right]^{1/2}\\
&\qquad\times\left\{1-\frac{1}{4}\left[\left(\frac{2\pi\rho_\kappa}{N}\right)^2+\lambda_2^2\right]+\frac{\lambda_1^2\lambda_3^2}{(2\pi\rho_\kappa/N)^2+\lambda_2^2}\right\}^{1/2}\Bigg)\\
&=\frac{1-i\lambda_1\lambda_3}{1+\lambda_1^2\lambda_3^2}(1+\lambda_1^2\lambda_3^2)^{1/2}e^{\pm i\beta_\kappa}\doteq(1-i\lambda_1\lambda_3)e^{\pm i\beta_\kappa}\doteq e^{-i\lambda_1\lambda_3}e^{\pm i\beta_\kappa}
\end{aligned}
\tag{10}
$$

$$
\operatorname{tg}\beta_\kappa=\frac{[(2\pi\rho_\kappa/N)^2+\lambda_2^2]^{1/2}\{1-[(2\pi\rho_\kappa/N)^2+\lambda_2^2]/4+\lambda_1^2\lambda_3^2/[(2\pi\rho_\kappa/N)^2+\lambda_2^2]\}^{1/2}}{1-[(2\pi\rho_\kappa/N)^2+\lambda_2^2]/2}
\tag{11}
$$

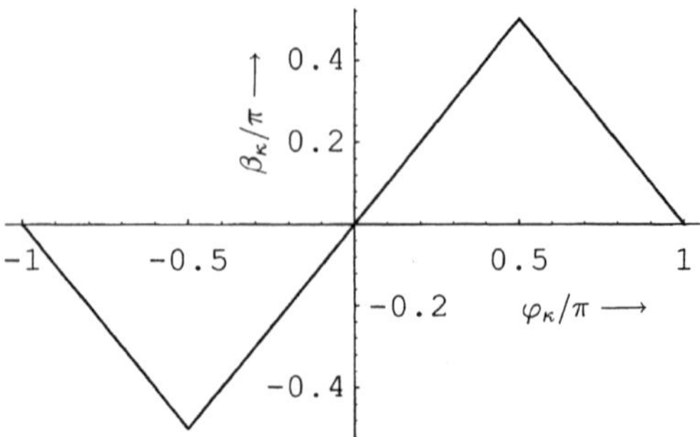

FIG.2.4-1. Plot of β_κ/π according to Eq.(14).

Equation (11) for β_κ can be simplified with the help of the following identities

$$1 + \mathrm{tg}^2\,\beta_\kappa = \frac{1}{\cos^2\beta_\kappa}, \quad \sin\beta_\kappa = (1-\cos^2\beta_\kappa)^{1/2} \tag{12}$$

which are solved for $\cos\beta_\kappa$ and $\sin\beta_\kappa$. Substitution of $2\pi\rho_\kappa/N$ from Eq.(1) brings:

$$\begin{gathered}\frac{\beta_\kappa}{\pi} = \frac{1}{\pi}\arcsin\left(1 - \frac{[1-\lambda_2^2/2 - 2\sin^2(\varphi_\kappa/2)]^2}{1+\lambda_1^2\lambda_3^2}\right)^{1/2} \\ \varphi_\kappa = 2\pi\kappa/N, \quad \kappa = 0,\ \pm 1,\ \ldots,\ \pm N/2\end{gathered} \tag{13}$$

For small values of λ_1^2 and λ_2^2 one obtains

$$\frac{\beta_\kappa}{\pi} = \frac{1}{\pi}\arcsin\left[1 - \left(1 - 2\sin^2\frac{\pi}{2}\frac{\varphi_\kappa}{\pi}\right)^2\right]^{1/2}, \quad \lambda_1^2\lambda_3^2,\ \lambda_2^2 \ll 1 \tag{14}$$

A plot of this equation is shown in Fig.2.4-1.

The plot of Fig.2.4-1 needs some details close to $\varphi_\kappa/\pi = 0,\ \pm 1$ where the simplification $\lambda_2^2 \ll 1$ is not justified since $\sin^2(\varphi_\kappa/2)$ approaches zero. For φ_κ/π equal to zero we get from Eq.(13):

$$\begin{aligned}\frac{\beta_\kappa}{\pi} &= \frac{1}{\pi}\arcsin\left(1 - \frac{(1-\lambda_2^2/2)^2}{1+\lambda_1^2\lambda_3^2}\right)^{1/2}, \quad \varphi_\kappa = 0 \\ &= 0.0225 \quad \text{for } \lambda_2^2 = 0.005,\ \lambda_1^2\lambda_3^2 = 0\end{aligned} \tag{15}$$

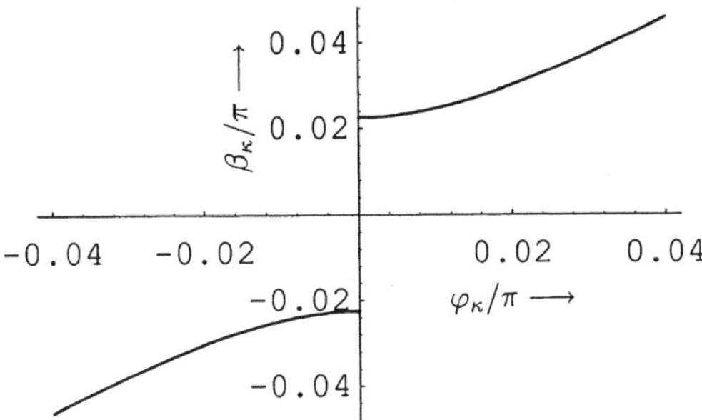

FIG.2.4-2. Plot of β_κ/π according to Eq.(13) in the neighborhood of $\varphi_\kappa/\pi = 0$ for $\lambda_2^2 = 0.005$, $\lambda_1^2\lambda_3^2 = 0$.

Figure 2.4-2 shows a plot of β_κ/π for $\lambda_2^2 = 0.005$, $\lambda_1^2\lambda_3^2 = 0$ in the interval $-0.04 < \varphi_\kappa/\pi < 0.04$ according to Eq.(13). The important feature of Fig.2.4-2 is that β_κ does not become zero for $\varphi_\kappa/\pi = 0$ but assumes a value defined by Eq.(15):

$$\left|\frac{\beta_\kappa}{\pi}\right| = \frac{1}{\pi}\arcsin\left(1 - \frac{(1-\lambda_2^2/2)^2}{1+\lambda_1^2\lambda_3^2}\right)^{1/2} \doteq \frac{\lambda_2}{\pi} \tag{16}$$

For the neighborhood of $\varphi_\kappa/\pi = 1$ we obtain from Eq.(7)

$$\begin{aligned}
2\sin\frac{\varphi_\kappa}{2} &< (4 - \lambda_2^2 + \lambda_1^2\lambda_3^2)^{1/2} \doteq 2\left[1 - \frac{1}{8}(\lambda_2^2 - \lambda_1^2\lambda_3^2)\right] \\
\frac{\varphi_\kappa}{2} &< \arcsin\left[1 - \frac{1}{8}(\lambda_2^2 - \lambda_1^2\lambda_3^2)\right] \\
\frac{\varphi_\kappa}{\pi} &< \frac{2}{\pi}\arcsin\left[1 - \frac{1}{8}(\lambda_2^2 - \lambda_1^2\lambda_3^2)\right] = 0.9775 \\
&\qquad \text{for } \lambda_2^2 = 0.005,\ \lambda_1^2\lambda_3^2 = 0
\end{aligned} \tag{17}$$

Plots of β_κ/π close to $\varphi_\kappa/\pi = \pm 1$ are shown according to Eq.(13) in Figs.2.4-3a and b.

The function $v_{5,6}^\theta$ according to Eq.(10) can be written in the following form if the conditions of Eqs.(7) and (8) are satisfied:

$$v_5^\theta = e^{-i(\lambda_1\lambda_3+\beta_\kappa)\theta} \tag{18}$$

$$v_6^\theta = e^{-i(\lambda_1\lambda_3-\beta_\kappa)\theta} \tag{19}$$

The function $\psi(\theta)$ of Eq.(1) becomes:

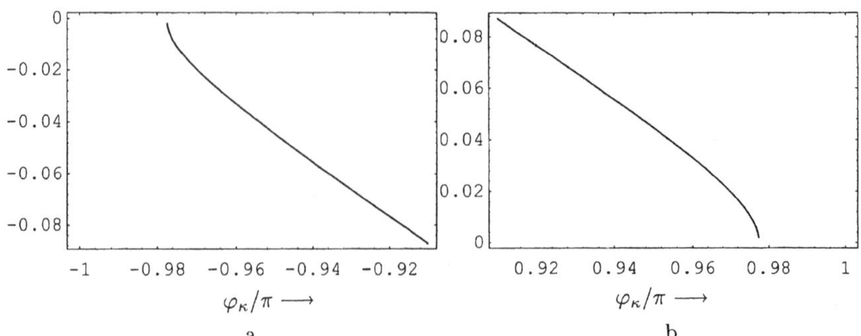

FIG.2.4-3. Plots of β_κ/π according to Eq.(13) in the neighborhood of $\varphi_\kappa/\pi = -1$ (a) and $\varphi_\kappa/\pi = +1$ (b) for $\lambda_2^2 = 0.005$, $\lambda_1^2\lambda_3^2 = 0$.

$$\psi(\theta) = A_{40}v_5^\theta + A_{41}v_6^\theta = A_{40}e^{-i(\lambda_1\lambda_3+\beta_\kappa)\theta} + A_{41}e^{-i(\lambda_1\lambda_3-\beta_\kappa)\theta} \tag{20}$$

The particular solution $u_\kappa(\zeta,\theta)$ of Eq.(2.3-23) assumes the following form with the help of Eqs.(2.3-37), (18), and (19):

$$u_\kappa(\zeta,\theta) = \Big\{A_1\exp[-i(\lambda_1\lambda_3+\beta_\kappa)\theta] + A_2\exp[-i(\lambda_1\lambda_3-\beta_\kappa)\theta]\Big\}e^{i\lambda_1\zeta}\sin\varphi_\kappa\zeta \tag{21}$$

The solution $u_\kappa(\zeta,\theta)$ is usually generalized by making A_1 and A_2 functions of a real variable κ and integrating over all values of κ from zero to infinity. This would imply nondenumerably many oscillators or photons. It is usual in quantum field theory to reduce the nondenumerably many oscillators to denumerably many, using *box normalization* to accomplish this reduction. We follow the spirit of this approach but (a) the box normalization is avoided since it is not based on a physical principle but strictly on the success of the calculation and (b) the denumerably many oscillators or photons are replaced by a finite number since we have no means to produce or observe more.

We introduce arbitrarily large but finite time and space intervals $0 \le t \le T$ and $0 \le x_j \le cT$. The normalized variables θ and ζ cover the intervals

$$0 \le \theta = t/\Delta t \le T/\Delta t,\ \ 0 \le \zeta = y/c\Delta t \le T/\Delta t,\ \ T/\Delta t = N \gg 1 \tag{22}$$

Equation (21) is not generalized by a Fourier integral but by a Fourier sum (Harmuth et al. 2001). The notation $\kappa > -\kappa_0$ in the following equation means the smallest integer larger than $-\kappa_0$ while $\kappa < \kappa_0$ means the largest integer smaller than κ_0:

$$u(\zeta,\theta) = \sum_{\kappa>-\kappa_0}^{<\kappa_0} \left\{A_1(\kappa)\exp[-i(\lambda_1\lambda_3+\beta_\kappa)\theta] + A_2(\kappa)\exp[-i(\lambda_1\lambda_3-\beta_\kappa)\theta]\right\} \times e^{i\lambda_1\zeta}\sin\varphi_\kappa\zeta \quad (23)$$

The summation is symmetric over negative and positive values of κ, while the differential theory always yields nonsymmetric sums over positive values of κ. Substitution of Eq.(23) for $u(\zeta,0)$ and $u(\zeta,1)$ in Eqs.(2.3-20) and (2.3-21) yields:

$$u(\zeta,0) = \sum_{\kappa>-\kappa_0}^{<\kappa_0} [A_1(\kappa)+A_2(\kappa)]e^{i\lambda_1\zeta}\sin\varphi_\kappa\zeta = 0 \quad (24)$$

$$u(\zeta,1) - u(\zeta,0) = \sum_{\kappa>-\kappa_0}^{<\kappa_0} \left[A_1(\kappa)\left(e^{-i(\lambda_1\lambda_3+\beta_\kappa)}-1\right)+A_2(\kappa)\left(e^{-i(\lambda_1\lambda_3-\beta_\kappa)}-1\right)\right] \times e^{i\lambda_1\zeta}\sin\varphi_\kappa\zeta = -(1-e^{-2i\lambda_1\lambda_3})F(\zeta) \doteq -2i\lambda_1\lambda_3 F(\zeta) \quad (25)$$

To solve these equations for $A_1(\kappa)$ and $A_2(\kappa)$ we use the Fourier series in the form of the sine series:

$$g_s(\kappa) = \frac{2}{N}\int_0^N f_s(\zeta)\sin\frac{2\pi\kappa\zeta}{N}d\zeta, \quad f_s(\zeta) = \sum_{\kappa=-N/2+1}^{N/2-1} g_s(\kappa)\sin\frac{2\pi\kappa\zeta}{N} \quad (26)$$

The factor $e^{i\lambda_1\zeta}$ in Eqs.(24) and (25) can be taken in front of the summation sign. With $\varphi_\kappa = 2\pi\kappa/N$ the substitution of Eq.(26) into Eqs.(24) and (25) yields:

$$A_1(\kappa) + A_2(\kappa) = 0 \quad (27)$$

$$A_1(\kappa)\left(e^{-i(\lambda_1\lambda_3+\beta_\kappa)}-1\right) + A_2(\kappa)\left(e^{-i(\lambda_1\lambda_3-\beta_\kappa)}-1\right) = -\frac{4i\lambda_1\lambda_3}{N}\int_0^N F(\zeta)e^{-i\lambda_1\zeta}\sin\frac{2\pi\kappa\zeta}{N}d\zeta = -2i\lambda_1\lambda_3 I_T(\kappa/N) \quad (28)$$

$$I_T(\kappa/N) = \frac{2}{N}\int_0^N F(\zeta)e^{-i\lambda_1\zeta}\sin\frac{2\pi\kappa\zeta}{N}d\zeta = \frac{2}{N}\int_0^N \exp[-(\lambda_2^2-\lambda_1^2)^{1/2}\zeta]\sin\frac{2\pi\kappa\zeta}{N}d\zeta = \frac{2}{N}\frac{(2\pi\kappa/N)\{1-\exp[-(\lambda_2^2-\lambda_1^2)^{1/2}N]\}}{\lambda_2^2-\lambda_1^2+(2\pi\kappa/N)^2} \quad (29)$$

2.5 Hamilton Function and Quantization

The Klein-Gordon differential equation defines a wave whose energy density is given by the term T_{00} of the energy-impulse tensor[1]:

$$T_{00} = \frac{1}{c^2}\frac{\partial\Psi^*}{\partial t}\frac{\partial\Psi}{\partial t} + \nabla\Psi^* \cdot \nabla\Psi + \frac{m_0^2c^2}{\hbar^2}\Psi^*\Psi \tag{1}$$

To produce the dimension J/m^3 for T_{00} the dimension of $\Psi^*\Psi$ must be J/m or VAs/m in electromagnetic units. For a planar wave propagating in the direction x we have $\nabla = \partial/\partial x$.

The Fourier series expansion in Eqs.(2.4-31) and (2.4-32) permits an arbitrarily large but finite time T and a corresponding spatial distance cT in the direction of x, using the intervals $0 \le t \le T$ and $0 \le x \le cT$. In the direction y and z we use the intervals $-L/2 \le y \le L/2$ and $-L/2 \le z \le L/2$ with $L = cT$. The energy U of the Klein-Gordon wave in this interval equals

$$U = \int_{-L/2}^{L/2}\int_{-L/2}^{L/2}\left[\int_0^{cT}\left(\frac{1}{c^2}\frac{\partial\Psi^*}{\partial t}\frac{\partial\Psi}{\partial t} + \frac{\partial\Psi^*}{\partial x}\frac{\partial\Psi}{\partial x} + \frac{m_0^2c^2}{\hbar^2}\Psi^*\Psi\right)dx\right]dy\,dz \tag{2}$$

The dimension of U is VAs. Equation (2) is rewritten with the normalized variables $\theta = t/\Delta t$, $\zeta = x/c\Delta t$ of Eq.(2.3-1):

$$\begin{aligned}U &= c\Delta t\int_{-L/2c\Delta t}^{L/2c\Delta t}\int_{-L/2c\Delta t}^{L/2c\Delta t}\left[\int_0^N\left(\frac{\partial\Psi^*}{\partial\theta}\frac{\partial\Psi}{\partial\theta} + \frac{\partial\Psi^*}{\partial\zeta}\frac{\partial\Psi}{\partial\zeta} + \frac{m_0^2c^4(\Delta t)^2}{\hbar^2}\Psi^*\Psi\right)d\zeta\right] \\ &\qquad\times d\left(\frac{y}{c\Delta t}\right)d\left(\frac{z}{c\Delta t}\right) \\ &= \frac{L^2}{c\Delta t}\int_0^N\left(\frac{\partial\Psi^*}{\partial\theta}\frac{\partial\Psi}{\partial\theta} + \frac{\partial\Psi^*}{\partial\zeta}\frac{\partial\Psi}{\partial\zeta} + \frac{m_0^2c^4(\Delta t)^2}{\hbar^2}\Psi^*\Psi\right)d\zeta\end{aligned} \tag{3}$$

We obtain from Eq.(2.4-32) for $\Psi = \Psi_0$:

$$\Psi_0^*\Psi_0 = \Psi_{00}^2[(1 - e^{2i\lambda_1\lambda_3\theta})F^*(\zeta) + u^*(\zeta,\theta)][(1 - e^{-2i\lambda_1\lambda_3\theta})F(\zeta) + u(\zeta,\theta)]$$

[1] Berestezki et al. 1970, 1982; § 10, Eq. 13.

$$
\begin{aligned}
&= 4\Psi_{00}^2\Bigg[\frac{1}{2}(1-\cos 2\lambda_1\lambda_3\theta)\exp[-2(\lambda_2^2-\lambda_1^2)^{1/2}\zeta] \\
&\qquad - 2\lambda_1\lambda_3\cos\lambda_1\lambda_3\sin\lambda_1\lambda_3\theta\exp[-(\lambda_2^2-\lambda_1^2)^{1/2}\zeta] \\
&\qquad\qquad \times \sum_{\kappa=-N/2+1}^{N/2-1}\frac{I_{\mathrm{T}}(\kappa/N)}{\sin\beta_\kappa}\sin\beta_\kappa\theta\sin\frac{2\pi\kappa\zeta}{N} \\
&\qquad + \lambda_1^2\lambda_3^2\Bigg(\sum_{\kappa=-N/2+1}^{N/2-1}\frac{I_{\mathrm{T}}(\kappa/N)}{\sin\beta_\kappa}\sin\beta_\kappa\theta\sin\frac{2\pi\kappa\zeta}{N}\Bigg)^2\Bigg]
\end{aligned}
\tag{4}
$$

The differentiation of $\Psi(\zeta,\theta)=\Psi_0(\zeta,\theta)$ of Eq.(2.4-32) with respect to θ or ζ produces the following results with the help of Eqs.(2.3-16) and (2.4-31):

$$
\frac{\partial\Psi}{\partial\theta}=\frac{\partial\Psi_0}{\partial\theta}=\Psi_{00}\left(2i\lambda_1\lambda_3 e^{-2i\lambda_1\lambda_3\theta}e^{i\lambda_1\zeta}\exp[-(\lambda_2^2-\lambda_1^2)^{1/2}\zeta]+\frac{\partial u}{\partial\theta}\right) \tag{5}
$$

$$
\begin{aligned}
\frac{\partial u}{\partial\theta} &= -2\lambda_1\lambda_3 e^{-i\lambda_1\lambda_3(\theta-1)}e^{i\lambda_1\zeta} \\
&\quad\times\sum_{\kappa=-N/2+1}^{N/2-1}\frac{I_{\mathrm{T}}(\kappa/N)}{\sin\beta_\kappa}(\lambda_1\lambda_3\sin\beta_\kappa\theta+i\beta_\kappa\cos\beta_\kappa\theta)\sin\frac{2\pi\kappa\zeta}{N}
\end{aligned}
\tag{6}
$$

$$
\begin{aligned}
\frac{\partial\Psi}{\partial\zeta}=\frac{\partial\Psi_0}{\partial\zeta} &= \Psi_{00}\Bigg((1-e^{-2i\lambda_1\lambda_3\theta})[-(\lambda_2^2-\lambda_1^2)^{1/2}+i\lambda_1] \\
&\qquad \times\exp[-(\lambda_2^2-\lambda_1^2)^{1/2}\zeta]e^{i\lambda_1\zeta}+\frac{\partial u}{\partial\zeta}\Bigg)
\end{aligned}
\tag{7}
$$

$$
\begin{aligned}
\frac{\partial u}{\partial\zeta} &= 2\lambda_1\lambda_3 e^{-i\lambda_1\lambda_3(\theta-1)}e^{i\lambda_1\zeta}\sum_{\kappa=-N/2+1}^{N/2-1}\frac{I_{\mathrm{T}}(\kappa/N)}{\sin\beta_\kappa}\sin\beta_\kappa\theta \\
&\qquad\times\left(\lambda_1\sin\frac{2\pi\kappa\zeta}{N}-\frac{2\pi i\kappa}{N}\cos\frac{2\pi\kappa\zeta}{N}\right)
\end{aligned}
\tag{8}
$$

The first term in Eq.(3) becomes:

$$
\begin{aligned}
\frac{\partial\Psi^*}{\partial\theta}\frac{\partial\Psi}{\partial\theta} &= \frac{\partial\Psi_0^*}{\partial\theta}\frac{\partial\Psi_0}{\partial\theta} \\
&= \Psi_{00}^2\left(-2i\lambda_1\lambda_3 e^{2i\lambda_1\lambda_3\theta}e^{-i\lambda_1\zeta}\exp[-(\lambda_2^2-\lambda_1^2)^{1/2}\zeta]+\frac{\partial u^*}{\partial\theta}\right) \\
&\quad\times\left(2i\lambda_1\lambda_3 e^{-2i\lambda_1\lambda_3\theta}e^{i\lambda_1\zeta}\exp[-(\lambda_2^2-\lambda_1^2)^{1/2}\zeta]+\frac{\partial u}{\partial\theta}\right)
\end{aligned}
$$

$$
\begin{aligned}
&= 4\lambda_1^2\lambda_3^2\Psi_{00}^2\Bigg[\exp[-2(\lambda_2^2-\lambda_1^2)^{1/2}\zeta] \\
&-2\exp[-(\lambda_2^2-\lambda_1^2)^{1/2}\zeta]\sum_{\kappa=-N/2+1}^{N/2-1}\frac{I_{\mathrm{T}}(\kappa/N)}{\sin\beta_\kappa}[\lambda_1\lambda_3\sin\lambda_1\lambda_3(\theta+1)\sin\beta_\kappa\theta \\
&\qquad +\beta_\kappa\cos\lambda_1\lambda_3(\theta+1)\cos\beta_\kappa\theta]\sin\frac{2\pi\kappa\zeta}{N} \\
&+\Bigg(\sum_{\kappa=-N/2+1}^{N/2-1}\frac{I_{\mathrm{T}}(\kappa/N)\lambda_1\lambda_3}{\sin\beta_\kappa}\sin\beta_\kappa\theta\sin\frac{2\pi\kappa\zeta}{N}\Bigg)^2 \\
&+\Bigg(\sum_{\kappa=-N/2+1}^{N/2-1}\frac{I_{\mathrm{T}}(\kappa/N)\beta_\kappa}{\sin\beta_\kappa}\cos\beta_\kappa\theta\sin\frac{2\pi\kappa\zeta}{N}\Bigg)^2\Bigg] \qquad (9)
\end{aligned}
$$

For the second term in Eq.(3) we get:

$$
\begin{aligned}
\frac{\partial\Psi^*}{\partial\zeta}\frac{\partial\Psi}{\partial\zeta} &= \frac{\partial\Psi_0^*}{\partial\zeta}\frac{\partial\Psi_0}{\partial\zeta} = \Psi_{00}^2\Big((1-e^{2i\lambda_1\lambda_3\theta})e^{-i\lambda_1\zeta}[-(\lambda_2^2-\lambda_1^2)^{1/2}-i\lambda_1] \\
&\qquad \times\exp[-(\lambda_2^2-\lambda_1^2)^{1/2}\zeta]+\frac{\partial u^*}{\partial\zeta}\Big) \\
&\times\Big((1-e^{-2i\lambda_1\lambda_3\theta})e^{i\lambda_1\zeta}[-(\lambda_2^2-\lambda_1^2)^{1/2}+i\lambda_1] \\
&\qquad \times\exp[-(\lambda_2^2-\lambda_1^2)^{1/2}\zeta]+\frac{\partial u}{\partial\zeta}\Big) \qquad (10)
\end{aligned}
$$

After considerable time one obtains the following result:

$$
\begin{aligned}
\frac{\partial\Psi_0^*}{\partial\zeta}\frac{\partial\Psi_0}{\partial\zeta} &= \Psi_{00}^2\Big\{2\lambda_2^2(1-\cos2\lambda_1\lambda_3\theta)\exp[-2(\lambda_2^2-\lambda_1^2)^{1/2}\zeta] \\
&-8\lambda_1\lambda_3\sin\lambda_1\lambda_3\theta\exp[-(\lambda_2^2-\lambda_1^2)^{1/2}\zeta]\sum_{\kappa=-N/2+1}^{N/2-1}\frac{I_{\mathrm{T}}(\kappa/N)}{\sin\beta_\kappa}\sin\beta_\kappa\theta \\
&\times\Big[\cos\lambda_1\lambda_3\Big(\lambda_1^2\sin\frac{2\pi\kappa\zeta}{N}-\frac{2\pi\kappa(\lambda_2^2-\lambda_1^2)^{1/2}}{N}\cos\frac{2\pi\kappa\zeta}{N}\Big) \\
&+\sin\lambda_1\lambda_3\Big(\lambda_1(\lambda_2^2-\lambda_1^2)^{1/2}\sin\frac{2\pi\kappa\zeta}{N}+\frac{2\pi\kappa\lambda_1}{N}\cos\frac{2\pi\kappa\zeta}{N}\Big)\Big] \\
&+4\lambda_1^2\lambda_3^2\Big[\Big(\sum_{\kappa=-N/2+1}^{N/2-1}\frac{\lambda_1 I_{\mathrm{T}}(\kappa/N)}{\sin\beta_\kappa}\sin\beta_\kappa\theta\sin\frac{2\pi\kappa\zeta}{N}\Big)^2
\end{aligned}
$$

$$+\left(\sum_{\kappa=-N/2+1}^{N/2-1}\frac{2\pi\kappa I_{\mathrm{T}}(\kappa/N)}{N\sin\beta_\kappa}\sin\beta_\kappa\theta\cos\frac{2\pi\kappa\zeta}{N}\right)^2\Bigg]\Bigg\} \qquad (11)$$

Equations (9), (11), and (4) have to be substituted into Eq.(3). The integration with respect to ζ is straightforward but lengthy (Harmuth and Meffert 2003, Sec. 6.8). The energy U of Eq.(3) is separated into a constant part U_{c} and a variable part $U_{\mathrm{v}}(\theta)$ that is composed of sinusoidal functions of θ and has the time-average zero:

$$U = U_{\mathrm{c}} + U_{\mathrm{v}}(\theta) \qquad (12)$$

We obtain the following three constant parts from Eqs.(4), (9), and (11):

$$U_{\mathrm{c}1} = \frac{L^2}{c\Delta t}\left(\int_0^N \frac{m_0^2c^4(\Delta t)^2}{\hbar^2}\Psi_0^*\Psi_0\,d\zeta\right)_c = \frac{L^2}{c\Delta t}\frac{m_0^2c^4(\Delta t)^2}{\hbar^2}\Psi_{00}^2 N$$
$$\times\left(2\frac{1-\exp[-2(\lambda_2^2-\lambda_1^2)^{1/2}N]}{2\,(\lambda_2^2-\lambda_1^2)^{1/2}\,N} + \lambda_1^2\lambda_3^2\sum_{\kappa=-N/2+1}^{N/2-1}\frac{I_{\mathrm{T}}^2(\kappa/N)}{\sin^2\beta_\kappa}\right)$$
$$\doteq \frac{L^2}{c\Delta t}\frac{m_0^2c^4(\Delta t)^2}{\hbar^2}\Psi_{00}^2 N\lambda_1^2\lambda_3^2\sum_{\kappa=-N/2+1}^{N/2-1}\left(\frac{I_{\mathrm{T}}(\kappa/N)}{\sin\beta_\kappa}\right)^2 \text{ for } N\gg 1,\ \lambda_1^2\neq\lambda_2^2 \qquad (13)$$

$$U_{\mathrm{c}2} = \frac{L^2}{c\Delta t}\left(\int_0^N \frac{\partial\Psi_0^*}{\partial\theta}\frac{\partial\Psi_0}{\partial\theta}d\zeta\right)_c$$
$$\doteq \frac{L^2}{c\Delta t}\Psi_{00}^2 N\lambda_1^2\lambda_3^2\sum_{\kappa=-N/2+1}^{N/2-1}\left(\frac{I_{\mathrm{T}}(\kappa/N)}{\sin\beta_\kappa}\right)^2(\lambda_1^2\lambda_3^2+\beta_\kappa^2), \quad N\gg 1 \qquad (14)$$

$$U_{\mathrm{c}3} = \frac{L^2}{c\Delta t}\left(\int_0^N \frac{\partial\Psi_0^*}{\partial\zeta}\frac{\partial\Psi_0}{\partial\zeta}d\zeta\right)_{\mathrm{c}}$$
$$\doteq \frac{L^2}{c\Delta t}\Psi_{00}^2 N\lambda_1^2\lambda_3^2\sum_{\kappa=-N/2+1}^{N/2-1}\left(\frac{I_{\mathrm{T}}(\kappa/N)}{\sin\beta_\kappa}\right)^2\left[\lambda_1^2+\left(\frac{2\pi\kappa}{N}\right)^2\right], \quad N\gg 1 \qquad (15)$$

The sum of these three components yields U_{c}:

$$
\begin{aligned}
U_{\mathrm{c}} &= U_{\mathrm{c}2} + U_{\mathrm{c}3} + U_{\mathrm{c}1} \\
&= \frac{L^2}{c\Delta t}\Psi_{00}^2 N\lambda_1^2\lambda_3^2 \sum_{\kappa=-N/2+1}^{N/2-1} \left(\frac{I_{\mathrm{T}}(\kappa/N)}{\sin\beta_\kappa}\right)^2 \\
&\qquad \times \left[\beta_\kappa^2 + \lambda_1^2\lambda_3^2 + \lambda_1^2 + \left(\frac{2\pi\kappa}{N}\right)^2 + \frac{m_0^2c^4(\Delta t)^2}{\hbar^2}\right] = \sum_{\kappa=-N/2+1}^{N/2-1} U_{\mathrm{c}\kappa}(\kappa)
\end{aligned}
$$

$$\lambda_1,\ \lambda_2,\ \lambda_3\ \text{Eq.(2.3-2)},\quad I_{\mathrm{T}}(\kappa/N)\ \text{Eq.(2.4-29)},\quad \beta_\kappa\ \text{Eq.(2.4-13)} \tag{16}$$

For the derivation of the Hamilton function $\mathcal{H}$ we need only the constant energy U_{c} since the average of the variable energy $U_{\mathrm{v}}(\theta)$ is zero. We normalize U_{c} in Eq.(16):

$$\frac{c\Delta t U_{\mathrm{c}}}{L^2\Psi_{00}^2 N} = \frac{cT\Delta t U_{\mathrm{c}}}{L^2\Psi_{00}^2 TN} = \frac{cTU_{\mathrm{c}}}{L^2\Psi_{00}^2 N^2} = \frac{\mathcal{H}}{N^2} \tag{17}$$

$$\frac{cTU_{\mathrm{c}}}{L^2\Psi_{00}^2} = \mathcal{H} = \sum_{\kappa=-N/2+1}^{N/2-1} \mathcal{H}_\kappa = \sum_{\kappa=-N/2+1}^{N/2-1} d^2(\kappa)$$

$$d(\kappa) = \lambda_1\lambda_2 N\frac{I_{\mathrm{T}}(\kappa/N)}{\sin\beta_\kappa}\left[\beta_\kappa^2 + \lambda_1^2\lambda_3^2 + \lambda_1^2 + \left(\frac{2\pi\kappa}{N}\right)^2 + \frac{m_0^2c^4(\Delta t)^2}{\hbar^2}\right]^{1/2} \tag{18}$$

The component $\mathcal{H}_\kappa$ of the sum of Eq.(18) may be rewritten as follows:

$$
\begin{aligned}
\mathcal{H}_\kappa &= (2\pi\kappa)^2\frac{d(\kappa)}{2\pi\kappa}(\sin 2\pi\kappa\theta - i\cos 2\pi\kappa\theta)\frac{d(\kappa)}{2\pi\kappa}(\sin 2\pi\kappa\theta + i\cos 2\pi\kappa\theta) \\
&= -2\pi i\kappa\frac{2\pi\kappa d(\kappa)}{\sqrt{2\pi i\kappa}}e^{2\pi i\kappa\theta}\frac{2\pi\kappa d(\kappa)}{\sqrt{2\pi i\kappa}}e^{-2\pi i\kappa\theta} = -2\pi i\kappa p_\kappa(\theta)q_\kappa(\theta)
\end{aligned} \tag{19}
$$

$$p_\kappa(\theta) = \frac{2\pi\kappa d(\kappa)}{\sqrt{2\pi i\kappa}}e^{2\pi i\kappa\theta} \tag{20}$$

$$
\begin{aligned}
\dot{p}_\kappa &= \frac{\tilde{\Delta} p_\kappa}{\tilde{\Delta}\theta} = \frac{2\pi\kappa d(\kappa)}{\sqrt{2\pi i\kappa}}e^{2\pi i\kappa\theta}\frac{1}{2\Delta\theta}(e^{2\pi i\kappa\Delta\theta} - e^{-2\pi i\kappa\Delta\theta}) \\
&= \frac{2\pi\kappa d(\kappa)}{\sqrt{2\pi i\kappa}}e^{2\pi i\kappa\theta}\frac{i\sin 2\pi\kappa\Delta\theta}{\Delta\theta} = p_\kappa\frac{i\sin 2\pi\kappa\Delta\theta}{\Delta\theta} \doteq 2\pi i\kappa p_\kappa
\end{aligned} \tag{21}
$$

$$q_\kappa(\theta) = \frac{2\pi\kappa d(\kappa)}{\sqrt{2\pi i\kappa}}e^{-2\pi i\kappa\theta} \tag{22}$$

$$
\begin{aligned}
\dot{q}_\kappa &= \frac{\tilde{\Delta} q_\kappa}{\tilde{\Delta}\theta} = -\frac{2\pi\kappa d(\kappa)}{\sqrt{2\pi i\kappa}}e^{-2\pi i\kappa\theta}\frac{i\sin 2\pi\kappa\Delta\theta}{\Delta\theta} \\
&= -q_\kappa\frac{i\sin 2\pi\kappa\Delta\theta}{\Delta\theta} \doteq -2\pi i\kappa q_\kappa(\theta)
\end{aligned} \tag{23}
$$

The relations for the finite derivatives $\tilde{\Delta}\mathcal{H}_\kappa/\tilde{\Delta}q_\kappa$ and $\tilde{\Delta}\mathcal{H}_\kappa/\tilde{\Delta}p_\kappa$ have the form known from the differential theory but use finite differences instead of differentials:

$$\frac{\tilde{\Delta}\mathcal{H}_\kappa}{\tilde{\Delta}q_\kappa} = -2\pi i\kappa p_\kappa \frac{(q_\kappa + \Delta q_\kappa) - (q_\kappa - \Delta q_\kappa)}{2\Delta q_\kappa} = -2\pi i\kappa p_\kappa \doteq -\dot{p}_\kappa \tag{24}$$

$$\frac{\tilde{\Delta}\mathcal{H}_\kappa}{\tilde{\Delta}p_\kappa} = -2\pi i q_\kappa \doteq +\dot{q}_\kappa \tag{25}$$

We introduce the definitions

$$a_\kappa = \frac{d(\kappa)}{2\pi\kappa}e^{2\pi i\kappa\theta}, \quad a_\kappa^* = \frac{d(\kappa)}{2\pi\kappa}e^{-2\pi i\kappa\theta} \tag{26}$$

to rewrite the Hamilton function of Eq.(18):

$$\mathcal{H} = -i\sum_{\kappa=0}^{N} 2\pi\kappa p_\kappa q_\kappa = \sum_{\kappa=0}^{N}(2\pi\kappa)^2 a_\kappa a_\kappa^* = \sum_{\kappa=0}^{N}\frac{2\pi\kappa}{T}\hbar b_\kappa b_\kappa^* = \sum_{\kappa=0}^{N}\mathcal{H}_\kappa \tag{27}$$

$$b_\kappa = \left(\frac{2\pi\kappa T}{\hbar}\right)^{1/2} a_\kappa, \quad b_\kappa^* = \left(\frac{2\pi\kappa T}{\hbar}\right)^{1/2} a_\kappa^*, \quad T = N\Delta t \tag{28}$$

Equations (27) and (28) are written in a form that permits one to follow the standard ways of quantization. It can be used both for the Heisenberg and the Schrödinger approach. Let us consider the Heisenberg approach first. The terms b_κ and b_κ^* are replaced by operators b_κ^- and b_κ^+:

$$b_\kappa^* \to b_\kappa^+ = \frac{1}{\sqrt{2}}\left(\alpha\zeta - \frac{1}{\alpha}\frac{\partial}{\partial\zeta}\right), \quad b_\kappa \to b_\kappa^- = \frac{1}{\sqrt{2}}\left(\alpha\zeta + \frac{1}{\alpha}\frac{\partial}{\partial\zeta}\right) \tag{29}$$

Alternately one may interchange b_κ^* and b_κ:

$$b_\kappa \to b_\kappa^+ = \frac{1}{\sqrt{2}}\left(\alpha\zeta - \frac{1}{\alpha}\frac{\partial}{\partial\zeta}\right), \quad b_\kappa^* \to b_\kappa^- = \frac{1}{\sqrt{2}}\left(\alpha\zeta + \frac{1}{\alpha}\frac{\partial}{\partial\zeta}\right) \tag{30}$$

The two ways of quantization are a well-known ambiguity of the theory (Becker 1963, 1964, vol. 2, § 52; Heitler 1954, p. 57). If we use first Eq.(29) we obtain from Eq.(27):

$$b_\kappa^- b_\kappa^+ = \frac{\mathcal{H}_\kappa T}{2\pi\kappa\hbar} \equiv \frac{\mathsf{E}_\kappa T}{2\pi\kappa\hbar} \tag{31}$$

which leads to the eigenvalues

$$\mathsf{E}_\kappa = \mathsf{E}_{\kappa n} = \frac{2\pi\kappa\hbar}{T}\left(n + \frac{1}{2}\right), \quad n = 0,\ 1,\ \ldots,\ N \tag{32}$$

as shown in some detail by Becker (1964, vol. II, § 15). On the other hand, Eq.(30) leads to

$$b_\kappa^+ b_\kappa^- = \frac{\mathsf{E}_\kappa T}{2\pi\kappa\hbar} \tag{33}$$

and the eigenvalues

$$\mathsf{E}_\kappa = \mathsf{E}_{\kappa n} = \frac{2\pi\kappa\hbar}{T}\left(n - \frac{1}{2}\right), \quad n = 0,\ 1,\ \ldots,\ N \tag{34}$$

We prefer here the use of the Schrödinger approach mainly because it leads to differential equations that we may readily replace by difference equations. We replace the differential operators of Eq.(29) by difference operators:

$$b_\kappa^* \to b_\kappa^+ = \frac{1}{\sqrt{2}}\left(\alpha\zeta - \frac{1}{\alpha}\frac{\tilde{\Delta}}{\tilde{\Delta}\zeta}\right), \quad b_\kappa \to b_\kappa^- = \frac{1}{\sqrt{2}}\left(\alpha\zeta + \frac{\tilde{\Delta}}{\tilde{\Delta}\zeta}\right) \tag{35}$$

The products $b_\kappa^+ b_\kappa^-$ or $b_\kappa^- b_\kappa^+$ produce a second-order difference operator:

$$b_\kappa^- b_\kappa^+ = b_\kappa^+ b_\kappa^- = \frac{1}{2}\left(\alpha^2\zeta^2 - \frac{1}{\alpha^2}\frac{\tilde{\Delta}^2}{\tilde{\Delta}\zeta^2}\right) \tag{36}$$

Substituting into Eq.(27) and applying the operator to a function Φ yields in analogy to Eq.(31):

$$\frac{1}{2}\left(\alpha^2\zeta^2 - \frac{1}{\alpha^2}\frac{\tilde{\Delta}^2}{\tilde{\Delta}\zeta^2}\right)\Phi = \frac{\mathsf{E}_\kappa T}{2\pi\kappa\hbar}\Phi = \lambda_\kappa\Phi, \quad \lambda_\kappa = \frac{\mathsf{E}_\kappa T}{2\pi\kappa\hbar} \tag{37}$$

For the solution of Eq.(37) one makes first the substitution

$$\xi = \alpha\zeta,\ \zeta = \frac{1}{\alpha}\xi,\ \frac{\tilde{\Delta}^2}{\tilde{\Delta}\zeta^2} = \alpha^2\frac{\tilde{\Delta}^2}{\tilde{\Delta}\xi^2} \tag{38}$$

that produces a difference equation with a variable coefficient ξ:

$$\frac{\tilde{\Delta}^2\Phi}{\tilde{\Delta}\xi^2} + (2\lambda_\kappa - \xi^2)\Phi = 0 \tag{39}$$

The substitution[2]

[2]For details of the calculation see Harmuth and Meffert 2003, Sec. 3.6.

$$\Phi = e^{-\xi^2/2}\chi(\xi) \tag{40}$$

leads to the equation

$$\frac{\tilde{\Delta}^2\chi(\xi)}{\tilde{\Delta}\xi^2} - 2\xi\frac{\tilde{\Delta}\chi(\xi)}{\tilde{\Delta}\xi} + (2\lambda_\kappa - 1)\chi(\xi) = 0 \tag{41}$$

which can be simplified by the further substitution

$$x = \xi/\Delta\xi, \ \Delta\xi = 1 \tag{42}$$

to the equation

$$\chi(x+1) - 2\chi(x) + \chi(x-1) - x\,[\chi(x+1) - \chi(x-1)] + (2\lambda_\kappa - 1)\chi(\xi) = 0 \tag{43}$$

This difference equation with variable coefficient x can be solved by a factorial series (of the second kind)

$$\chi(x) = b_0 + b_1\frac{x-1}{1!} + b_2\frac{(x-1)(x-2)}{2!} + \ldots + b_j\frac{(x-1)(x-2)\ldots(x-j)}{j!} + \ldots \tag{44}$$

whose coefficients b_j are determined by a recursion formula with three terms:

$$b_{j+2} = \frac{j+1}{(j+2)^2+1}\{(2\lambda_\kappa - 1 - 2j)b_j + [2\lambda_\kappa - 1 - 3(j+2)]b_{j+1}\} \tag{45}$$

A polynomial solution with highest term b_n exists if the following two conditions are satisfied:

$$b_{n+1} = \frac{n}{(n+1)^2+1}\{[2\lambda_\kappa - 1 - 2(n-1)]b_{n-1} + [2\lambda_\kappa - 1 - 3(n+1)]b_n\} = 0 \tag{46}$$

$$2n = 2\lambda_\kappa - 1, \quad n = 0,\ 1,\ 2,\ \ldots,\ N \tag{47}$$

Substitution of Eq.(47) into Eq.(37) yields the eigenvalues known from the differential theory or from Eq.(32) of the Heisenberg approach, except for the limit N of n:

$$\mathsf{E}_\kappa = \mathsf{E}_{\kappa n} = \frac{2\pi\kappa\hbar}{T}\left(n + \frac{1}{2}\right), \quad n = 0,\ 1,\ 2,\ \ldots,\ N \tag{48}$$

The recursion formula of Eq.(45) is a difference equation of second order. Hence, two of the coefficients b_j in Eq.(44) can be chosen. Our first choice is b_0, but we leave the value of b_0 undefined so that it remains available for normalization. The second choice is *not* b_1 but $b_{n+1} = 0$, which is Eq.(46).

The eigenvalues of Eq.(48) are the same as in the differential theory except for the finite value of N. The associated eigenfunctions in the differential theory are the Hermite polynomials that are defined by a recursion formula with two terms. Equation (45) is a recursion formula with three terms. The eigenfunctions are thus different from the Hermite polynomials.

2.6 Plots for First-Order Approximation

In the differential theory we had derived the probability $p(\kappa)$ of a photon with a period number κ (Harmuth and Meffert 2003, Eq.4.5-14):

$$p(\kappa) = 1/1.20205\kappa^3 \tag{1}$$

This function is plotted for the interval $1 \leq \kappa \leq 49$ in Fig.2.6-1 and for the interval $1 \leq \kappa \leq 500$ in Fig.2.6-2 using a logarithmic scale for $p(\kappa)$. The plots hold for integer values of κ.

In the difference theory the energy $U_{\mathrm{c}\kappa}(\kappa)$ for the period number κ is defined by Eq.(2.5-16):

$$U_{\mathrm{c}\kappa}(\kappa) = \frac{L^2\Psi_{00}^2 N^2 \lambda_1^2\lambda_3^2}{cT}\frac{I_{\mathrm{T}}^2(\kappa/N)}{\sin^2\beta_\kappa}\left[\beta_\kappa^2 + \lambda_1^2\lambda_3^2 + \lambda_1^2 + \left(\frac{2\pi\kappa}{N}\right)^2 + \left(\frac{m_0c^2\Delta t}{\hbar}\right)^2\right]$$

$$\beta_\kappa = \arcsin\left(1 - \frac{[1-\lambda_2^2/2 - 2\sin^2(\pi\kappa/N)]^2}{1+\lambda_1^2\lambda_3^2}\right)^{1/2}$$

$$\sin^2\beta_\kappa = 1 - \frac{[1-\lambda_2^2/2 - 2\sin^2(\pi\kappa/N)]^2}{1+\lambda_1^2\lambda_3^2}$$

$$I_{\mathrm{T}}(\kappa/N) = \frac{2}{N}\frac{(2\pi\kappa/N)\{1-\exp[-(\lambda_2^2-\lambda_1^2)^{1/2}N]\}}{\lambda_2^2-\lambda_1^2+(2\pi\kappa/N)^2}$$

$$\lambda_2^2 = \lambda_1^2\left(1 - \frac{\phi_{\mathrm{e}0}^2}{c^2A_{\mathrm{m}0x}^2} + \frac{m_0^2c^2}{e^2A_{\mathrm{m}0x}^2}\right) = \frac{(\Delta t)^2}{\hbar^2}\left(m_0^2c^4 - e^2\phi_{\mathrm{e}0}^2 + e^2c^2A_{\mathrm{m}0x}^2\right)$$

$$\lambda_2^2 - \lambda_1^2 = (\Delta t)^2(m_0^2c^4 - e^2\phi_{\mathrm{e}0}^2)/\hbar,\quad \lambda_1^2 = e^2c^2(\Delta t)^2A_{\mathrm{m}0x}^2/\hbar^2$$

$$\lambda_1^2\lambda_3^2 = e^2(\Delta t)^2\phi_{\mathrm{e}0}^2/\hbar^2,\quad 0 \leq \kappa < \kappa_0 = \frac{N}{\pi}\arcsin\left(1 - \frac{\lambda_2^2-\lambda_1^2\lambda_3^2}{4}\right)^{1/2}$$

$$\lambda_2^2 - \lambda_1^2\lambda_3^2 = (\Delta t)^2(m_0^2c^4 - 2e^2\phi_{\mathrm{e}0}^2 + e^2c^2A_{\mathrm{m}0x}^2)/\hbar^2,\quad N = T/\Delta t \tag{2}$$

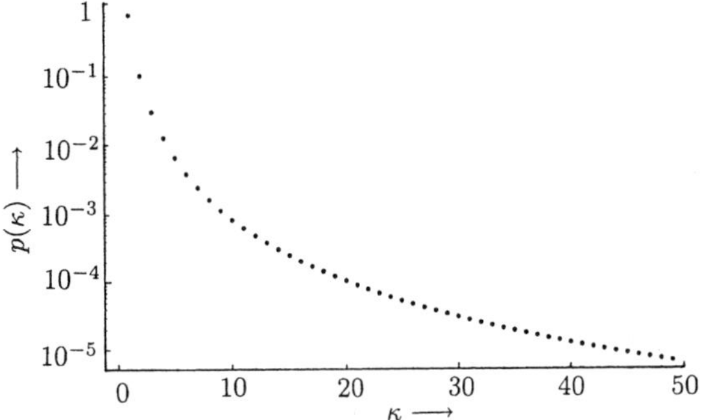

FIG.2.6-1. Probability $p(\kappa)$ of a photon with period number κ according to Eq.(1) of the differential theory in the interval $1 \le \kappa \le 49$.

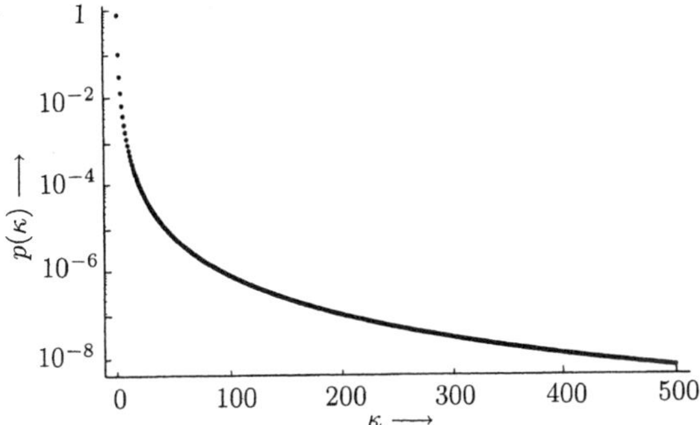

FIG.2.6-2. Probability $p(\kappa)$ of a photon with period number κ according to Eq.(1) of the differential theory in the interval $1 \le \kappa \le 500$.

The energy of a photon with period number κ and a certain value of n is defined by Eq.(2.5-48):

$$\mathsf{E}_{\kappa n} = \frac{2\pi\kappa\hbar}{T}\left(n + \frac{1}{2}\right), \quad n = 0,\ 1,\ \ldots,\ N \tag{3}$$

The average value of $\mathsf{E}_{\kappa n}$ for all $N+1$ values of n becomes:

$$\mathsf{E}_{\kappa} = \frac{2\pi\kappa\hbar}{T}\frac{1}{N+1}\sum_{\kappa=0}^{N}\left(n + \frac{1}{2}\right) = \frac{1}{2}(N+1)\frac{2\pi\kappa\hbar}{T} \doteq \frac{1}{2}N\frac{2\pi\kappa\hbar}{T} \tag{4}$$

For a specific value of n the energy $U_{c\kappa}(\kappa)$ requires the number $U_{c\kappa}(\kappa)/\mathsf{E}_{\kappa n}$of photons:

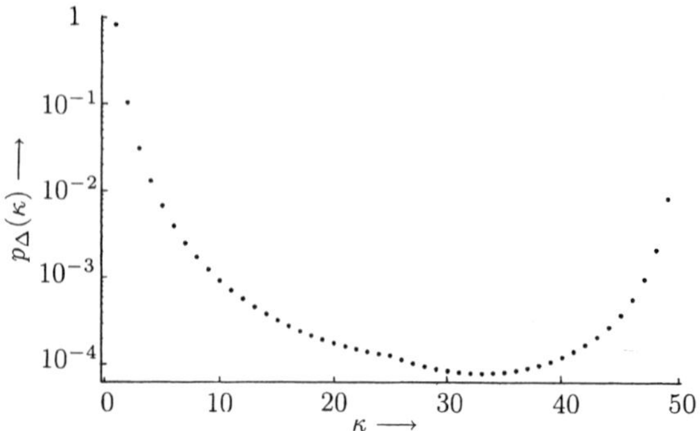

FIG.2.6-3. Probability $p_\Delta(\kappa)$ according to Eqs.(9) and (14) for $N = 100$, $\lambda_2 = 10^{-4}$, $\lambda_1\lambda_3 = 0.1\lambda_2$ in the interval $1 \le \kappa \le 49$.

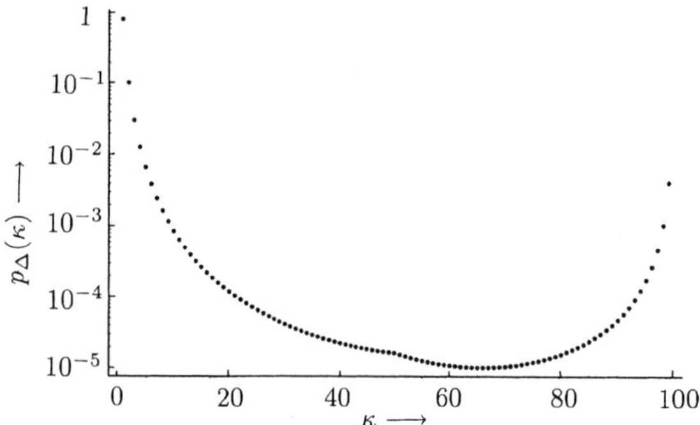

FIG.2.6-4. Probability $p_\Delta(\kappa)$ according to Eqs.(9) and (14) for $N = 200$, $\lambda_2 = 5 \times 10^{-5}$, $\lambda_1\lambda_3 = 0.1\lambda_2$ in the interval $1 \le \kappa \le 99$.

$$\frac{U_{\mathrm{c}\kappa}(\kappa)}{\mathsf{E}_{\kappa n}} = D_{\mathrm{r}n} r_\Delta(\kappa)$$

$$r_\Delta(\kappa) = \frac{I_{\mathrm{T}}^2(\kappa/N)}{\sin^2\beta_\kappa}\frac{1}{2\pi\kappa}\left[\beta_\kappa^2 + \lambda_1^2\lambda_3^2 + \lambda_1^2 + \left(\frac{2\pi\kappa}{N}\right)^2 + \frac{m_0^2c^4(\Delta t)^2}{\hbar^2}\right]$$

$$D_{\mathrm{r}n} = \frac{L^2\Psi_1^2N^2\lambda_1^2\lambda_3^2}{c\hbar}\left(n + \frac{1}{2}\right)^{-1} \tag{5}$$

If photons with various values of n are equally frequent we obtain the following relation in place of Eq.(5):

$$\frac{U_{\mathrm{c}\kappa}(\kappa}{\mathsf{E}_\kappa} = D_{\mathrm{r}} r_\Delta(\kappa)$$

$$D_{\mathrm{r}} = \frac{2L^2\Psi_1^2N\lambda_1^2\lambda_3^2}{c\hbar} \tag{6}$$

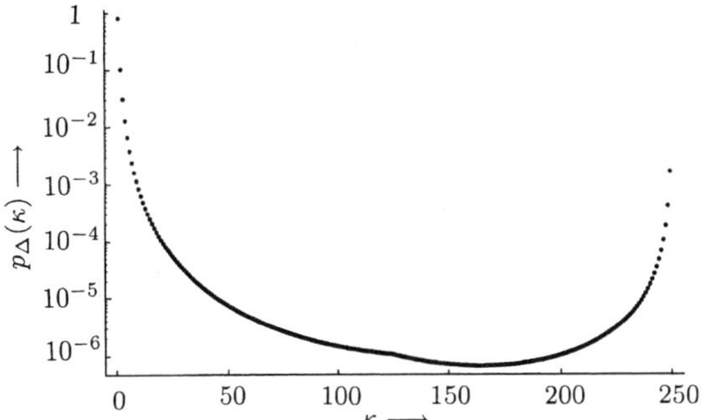

FIG.2.6-5. Probability $p_\Delta(\kappa)$ according to Eqs.(9) and (14) for $N = 500$, $\lambda_2 = 2 \times 10^{-5}$, $\lambda_1\lambda_3 = 0.1\lambda_2$ in the interval $1 \leq \kappa \leq 249$.

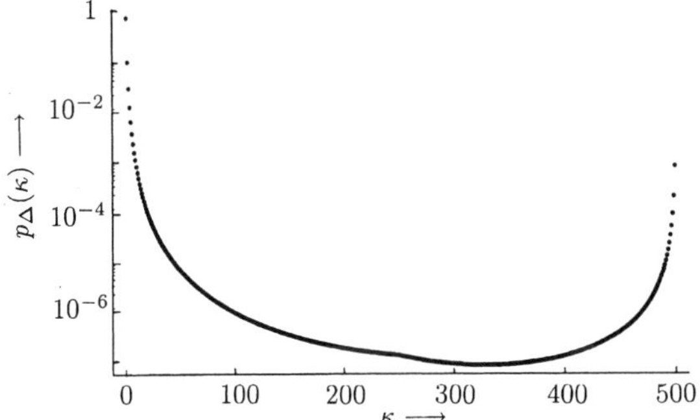

FIG.2.6-6. Probability $p_\Delta(\kappa)$ according to Eqs.(9) and (14) for $N = 1000$, $\lambda_2 = 10^{-5}$, $\lambda_1\lambda_3 = 0.1\lambda_2$ in the interval $1 \leq \kappa \leq 499$.

The total number of photons equals the sum over κ in Eqs.(5) or (6):

$$s_{rn}^{\Delta} = \sum_{\kappa=0}^{N/2-1} \frac{U_{c\kappa}(\kappa)}{\mathsf{E}_{c\kappa}} = D_{rn} \sum_{\kappa=0}^{N/2-1} r_\Delta(\kappa) \tag{7}$$

$$s_{\mathrm{r}}^{\Delta} = \sum_{\kappa=0}^{N/2-1} \frac{U_{c\kappa}(\kappa)}{\mathsf{E}_{\kappa}} = D_{\mathrm{r}} \sum_{\kappa=0}^{N/2-1} r_\Delta(\kappa) \tag{8}$$

The probability $p_\Delta(\kappa)$ of a photon with period number κ is the same for Eqs.(5) and (6) since the constants D_{rn} and D_{r} drop out:

$$p_\Delta(\kappa) = \frac{r_\Delta(\kappa)}{\sum_{\kappa=0}^{N/2-1} r_\Delta(\kappa)} \tag{9}$$

For the computation of $r_\Delta(\kappa)$ and $p_\Delta(\kappa)$ one can make several simplifications in Eqs.(5) and (9). First we deal with the special case $\kappa = 0$. We see from Fig.2.4-2 that β_κ and thus $\sin\beta_\kappa$ is not zero at $\varphi_\kappa = 2\pi\kappa/N = 0$ or $\kappa = 0$. The function $I_T^2(\kappa/N)$ decreases like κ^2, the factor $1/2\pi\kappa$ increases like $1/\kappa$, and the terms in brackets in Eq.(5) become constant. Hence, $r_\Delta(0)$ and $p_\Delta(0)$ approach zero. For discrete values of κ we get $I_T(\kappa/N)^2/2\pi\kappa = 0^2/0$ for $\kappa = 0$, which makes $r_\Delta(0)$ and $p_\Delta(0)$ undefined.

This was an argument in terms of differential mathematics. In terms of mathematics of finite differences one has to state that β_κ in Fig.2.4-2 is defined for $\kappa = \ldots, 3, 2, 1$ but not for intermediate values and particularly not for $\kappa \to 0$ or $\kappa = 0$. In terms of physics a photon with period number $\kappa = 0$ or $\kappa \to 0$ would be a strange thing indeed. It would have trouble producing interference phenomena.

We leave out the terms $\kappa = 0$ in Eqs.(7) to (9). The following approximations hold for $1 \leq \kappa < N/2$:

$$\kappa_0 = \frac{N}{\pi}\arcsin\left(1 - \frac{\lambda_2^2 - \lambda_1^2\lambda_3^2}{4}\right)^{1/2} < \frac{N}{\pi}\frac{\pi}{2} = \frac{N}{2}, \quad \text{Eq.(2.4-7)}$$

$$1 \leq \kappa \leq N/2 - 1 < \kappa_0 \quad \text{for } \lambda_2^2 - \lambda_1^2\lambda_3^2 \ll 1 \tag{10}$$

$$\lambda_2^2 - \lambda_1^2\lambda_3^2 \ll 1 \quad \text{for } \Delta t \ll \hbar/m_0c^2 \tag{11}$$

$$(\lambda_1^2 - \lambda_2^2)^{1/2}N = N\Delta t(m_0^2c^4 - e^2\phi_{e0}^2)^{1/2}/\hbar \gg 1, \quad \text{for } T \gg \hbar/m_0c^2 \tag{12}$$

$$r_\Delta(\kappa) = \left(\frac{2\pi\kappa/N}{\lambda_2^2 + (2\pi\kappa/N)^2}\right)^2 \frac{1}{\sin^2\beta_\kappa}\frac{1}{2\pi\kappa}\left[\beta_\kappa^2 + \left(\frac{2\pi\kappa}{N}\right)^2\right] \tag{13}$$

We obtain the following values for $\lambda_1\lambda_3 = 0.1\lambda_2$:

$$\begin{aligned}\sum_{\kappa=1}^{N/2-1} r_\Delta(\kappa) &= 98.743 \quad &&\text{for} \quad N = 100,\ \lambda_2 = 10^{-4}\\ &= 390.914 && N = 200,\ \lambda_2 = 5\times10^{-5}\\ &= 2430.32 && N = 500,\ \lambda_2 = 2\times10^{-5}\\ &= 9706.03 && N = 1000,\ \lambda_2 = 10^{-5}\end{aligned} \tag{14}$$

Equation (11) states that Δt can be arbitrarily small but finite while Eq.(12) states that T can be arbitrarily large but finite. This is in line with our assumptions. We have the numerical value $h/m_0c^2 = 2.95241 \times 10^{-23}$ s for pions π^+ and π^-.

Figures 2.6-3 to 2.6-6 show $p_\Delta(\kappa)$ plotted with a logarithmic scale for the four cases $N = 100, 200, 500, 1000$ of Eq.(14). The plots drop rapidly with increasing values of κ, similarly to the plots of Figs.2.6-1 and 2.6-2. The conspicuous difference is the increase of $p_\Delta(\kappa)$ when κ approaches $N/2$. To show the practical identity for $\kappa < 0.2N$ of the plots of $p_\Delta(\kappa)$ for $N = 1000$ in Fig.2.6-6 with the plots of $p(\kappa)$ in Figs.2.6-1 and 2.6-2 we plot the difference $p_\Delta(\kappa) - p(\kappa)$ in Figs.2.6-7 and 2.6-8. According to Fig.2.6-7 the function $p_\Delta(\kappa)$

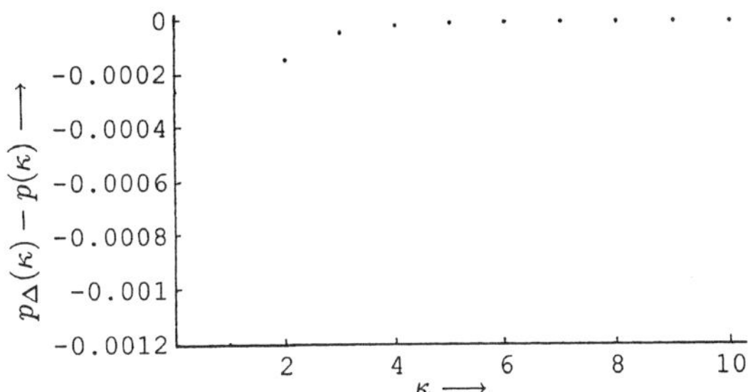

FIG.2.6-7. Difference $p_\Delta(\kappa) - p(\kappa)$ between the plots of Fig.2.6-6 for $N = 100$ and Fig.2.6-1 for the interval $1 \leq \kappa \leq 10$. The point at $\kappa = 1$ is barely visible since it has the value -0.0012.

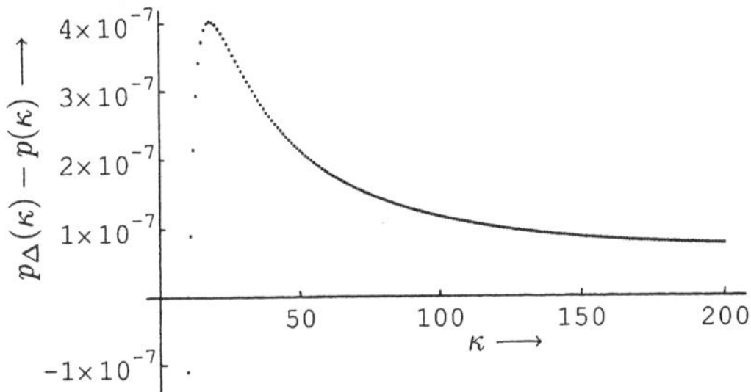

FIG.2.6-8. Difference $p_\Delta(\kappa) - p(\kappa)$ between the plots of Fig.2.6-6 for $N = 100$ and Fig.2.6-1 for the interval $10 \leq \kappa \leq 200$.

is slightly smaller than $p(\kappa)$ in the interval $1 \leq \kappa \leq 10$. For larger values of κ we get $p(\kappa) > p_\Delta(\kappa)$ in the interval $11 \leq \kappa < 200$ according to Fig.2.6-8. But the difference is of the order 10^{-3} or less in Fig.2.6-7 and of the order 10^{-7} in Fig.2.6-8.

The increase of the plots of Figs.2.6-3 to 2.6-6 for $\kappa \to N/2$ is a typical result of a finite difference theory. By limiting the interval of interest to a maximum value of $\kappa = 49$, 99, 249, 499 we introduce a boundary condition at the upper limit of κ that causes $p_\Delta(\kappa)$ to increase when approaching this upper limit. We see in Fig.2.6-3 at the upper limit $p_\Delta(49) \doteq 10^{-2}$ while Fig.2.6-6 shows $p_\Delta(499) \doteq 10^{-3}$. One will expect that this value decreases to zero for $N \to \infty$ in analogy to Figs.2.6-1 and 2.6-2. In a finite difference theory one must specify that one wants the k most significant digits of $p_\Delta(\kappa)$. Then one must choose $N > 2\kappa$ so large that a further increase of N does not change these k digits[1].

[1]In a differential theory we always have $\kappa \ll \infty$ if we make a plot as function of κ. The corresponding condition $\kappa \ll N$ is introduced by Eqs.(2.5-21) and (2.5-22) for the difference theorie. We hope to discuss the matter in more detail in a book on the Dirac difference equations.

3 Inhomogeneous Difference Equation

3.1 Inhomogeneous Term of Eq.(2.2-31)

We have developed solutions of the homogeneous difference equation (2.2-30) for $\Psi_{\mathrm{x}0x_j}(x_j,t)$ in Sections 2.3 to 2.6. The next step is to find a particular solution of the inhomogeneous Eq.(2.2-31). The homogeneous part of that equation equals Eq.(2.2-30) as is readily seen from Eqs.(2.2-28) and (2.2-29). Hence, the homogeneous solution is obtained by replacing $\Psi_{\mathrm{x}0x_j} = \Psi_0$ in Section 2.3 by $\Psi_{\mathrm{x}1x_j} = \Psi_1$. This implies that Eq.(2.4-32) is both the solution of Eqs.(2.2-28), (2.2-30) and of the homogeneous part of Eq.(2.2-29), (2.2-31) if we replace $\Psi_{\mathrm{x}0x_j}(x_j,t) = \Psi_0$ by $\Psi_{\mathrm{x}1x_j}(x_j,t) = \Psi_1$. We copy Ψ_0 for $x_j \to \zeta$ and $t \to \theta$ from Eq.(2.4-32). The notation $\Psi_{1\mathrm{H}}(\zeta,\theta)$ means that this is only the homogeneous solution of $\Psi_1(\zeta,\theta)$:

$$\begin{aligned}\Psi_0(\zeta,\theta) &= \Psi_{1\mathrm{H}}(\zeta,\theta) = \Psi_{00}[(1-e^{-2i\lambda_1\lambda_3\theta})F(\zeta)+u(\zeta,\theta)]\\ &= \Psi_{00}\Bigg[(1-e^{-2i\lambda_1\lambda_3\theta})\exp[-(\lambda_2^2-\lambda_1^2)^{1/2}\zeta]e^{i\lambda_1\zeta}\\ &\quad -2i\lambda_1\lambda_3 e^{i\lambda_1\lambda_3}e^{i\lambda_1(\zeta-\lambda_3\theta)}\sum_{\kappa=-N/2+1}^{N/2-1}\frac{I_{\mathrm{T}}(\kappa/N)}{\sin\beta_\kappa}\sin\beta_\kappa\theta\sin\frac{2\pi\kappa\zeta}{N}\Bigg]\\ &\qquad I_{\mathrm{T}}(\kappa/N)\ \text{see Eq.(2.4-29)}\end{aligned} \tag{1}$$

The next task is to write Eq.(2.2-31) with the normalization of Eq.(2.3-1) and reduce the number of spatial variables from three to one:

$$\begin{gathered}\theta = t/\Delta t,\quad \zeta = x_j/\Delta x,\ j=1,\ x_j = x\\ \theta\Delta t = t,\ \zeta\Delta x = x_j,\quad \frac{\tilde{\Delta}}{\tilde{\Delta}t}\to\frac{1}{\Delta t}\frac{\tilde{\Delta}}{\tilde{\Delta}\theta},\ \frac{\tilde{\Delta}}{\tilde{\Delta}x_j}\to\frac{1}{\Delta x}\frac{\tilde{\Delta}}{\tilde{\Delta}\zeta}\end{gathered} \tag{2}$$

Equation (2.2-31) assumes the following form:

$$\left[\left(\frac{\hbar}{i\Delta x}\frac{\tilde{\Delta}}{\tilde{\Delta}\zeta}-eA_{\mathrm{m}0x}\right)^2-\frac{1}{c^2}\left(\frac{\hbar}{i\Delta t}\frac{\tilde{\Delta}}{\tilde{\Delta}\theta}-e\phi_{\mathrm{e}0}\right)^2+m_0^2c^2\right]\Psi_{\mathrm{x}1x}(\zeta,\theta)$$

ISSN 1076-5670/05
DOI: 10.1016/S1076-5670(05)37003-0

$$
\begin{aligned}
= -\Bigg\{ & \frac{4\lambda_{\mathrm{C}} A_{\mathrm{e}0}}{e}\left(\frac{\hbar}{i\Delta x}\frac{\tilde{\Delta}}{\tilde{\Delta}\zeta} - eA_{\mathrm{m}0x}\right)^2\left[1 - \frac{1}{2m_0^2c^2}\left(\frac{\hbar}{i\Delta x}\frac{\tilde{\Delta}}{\tilde{\Delta}\zeta} - eA_{\mathrm{m}0x}\right)^2\right] \\
& + \frac{1}{c^2}\left[\left(\frac{\hbar}{i\Delta t}\frac{\tilde{\Delta}}{\tilde{\Delta}\theta} + e\phi_{\mathrm{e}0}\right)(\mathsf{L}_{\mathrm{cx}11} + \mathsf{L}_{\mathrm{cx}21} + \mathsf{L}_{\mathrm{cx}31} + \mathsf{L}_{\mathrm{cx}41} + \mathsf{L}_{\mathrm{cx}51})\right. \\
& \left. + (\mathsf{L}_{\mathrm{cx}11} + \mathsf{L}_{\mathrm{cx}21} + \mathsf{L}_{\mathrm{cx}31} + \mathsf{L}_{\mathrm{cx}41} + \mathsf{L}_{\mathrm{cx}51})\left(\frac{\hbar}{i\Delta t}\frac{\tilde{\Delta}}{\tilde{\Delta}\theta} + e\phi_{\mathrm{e}0}\right)\right] \\
& - e\left[A_{\mathrm{m}1x}\left(\frac{\hbar}{i\Delta x}\frac{\tilde{\Delta}}{\tilde{\Delta}\zeta} - eA_{\mathrm{m}0x}\right) + \left(\frac{\hbar}{i\Delta x}\frac{\tilde{\Delta}}{\tilde{\Delta}\zeta} - eA_{\mathrm{m}0x}\right)A_{\mathrm{m}1x}\right] \\
& - \frac{e}{c^2}\left[\phi_{\mathrm{e}1}\left(\frac{\hbar}{i\Delta t}\frac{\tilde{\Delta}}{\tilde{\Delta}\theta} + e\phi_{\mathrm{e}0}\right) + \left(\frac{\hbar}{i\Delta t}\frac{\tilde{\Delta}}{\tilde{\Delta}\theta} + e\phi_{\mathrm{e}0}\right)\phi_{\mathrm{e}1}\right]\Bigg\}\Psi_{\mathrm{x}0x}(\zeta,\theta) \qquad (3)
\end{aligned}
$$

$$
\begin{aligned}
\mathbf{A}_{\mathrm{e}} &= \mathbf{A}_{\mathrm{e}0} + \alpha\mathbf{A}_{\mathrm{e}1}(\zeta,\theta), \quad A_{\mathrm{e}x} = A_{\mathrm{e}0x} + \alpha A_{\mathrm{e}1x}(\zeta,\theta) \\
\mathbf{A}_{\mathrm{m}} &= \mathbf{A}_{\mathrm{m}0} + \alpha\mathbf{A}_{\mathrm{m}1}(\zeta,\theta), \quad A_{\mathrm{m}x} = A_{\mathrm{m}0x} + \alpha A_{\mathrm{m}1x}(\zeta,\theta) \\
\phi_{\mathrm{e}} &= \phi_{\mathrm{e}0} + \alpha\phi_{\mathrm{e}1}(\zeta,\theta), \quad \phi_{\mathrm{m}} = \phi_{\mathrm{m}0} + \alpha\phi_{\mathrm{m}1}(\zeta,\theta) \qquad (4)
\end{aligned}
$$

Using Eqs.(2.2-14), (2.2-15), and (2.3-2) we may write the homogeneous equation defined by the first line of Eq.(3) as follows:

$$
\begin{aligned}
&\left[\left(\frac{\hbar}{i\Delta x}\frac{\tilde{\Delta}}{\tilde{\Delta}\zeta} - eA_{\mathrm{m}0x}\right)^2 - \frac{1}{c^2}\left(\frac{\hbar}{i\Delta t}\frac{\tilde{\Delta}}{\tilde{\Delta}\theta} - e\phi_{\mathrm{e}0}\right)^2 + m_0^2c^2\right]\Psi_{\mathrm{x}1x}(\zeta,\theta) \\
&\quad = -\frac{\hbar^2}{(\Delta x)^2}\frac{\Psi_{\mathrm{x}1x}(\zeta+\Delta\zeta,\theta) - 2\Psi_{\mathrm{x}1x}(\zeta,\theta) + \Psi_{\mathrm{x}1x}(\zeta-\Delta\zeta,\theta)}{(\Delta\zeta)^2} \\
&\quad + \frac{\hbar^2}{(c\Delta t)^2}\frac{\Psi_{\mathrm{x}1x}(\zeta,\theta+\Delta\theta) - 2\Psi_{\mathrm{x}1x}(\zeta,\theta) + \Psi_{\mathrm{x}1x}(\zeta,\theta-\Delta\theta)}{(\Delta\theta)^2} \\
&\quad + \frac{2ie\hbar}{\Delta x}A_{\mathrm{m}0x}\frac{\Psi_{\mathrm{x}1x}(\zeta+\Delta\zeta,\theta) - \Psi_{\mathrm{x}1x}(\zeta-\Delta\zeta,\theta)}{2\Delta\zeta} \\
&\quad - \frac{2ie\hbar}{c^2\Delta t}\phi_{\mathrm{e}0}\frac{\Psi_{\mathrm{x}1x}(\zeta,\theta+\Delta\theta) - \Psi_{\mathrm{x}1x}(\zeta,\theta-\Delta\theta)}{2\Delta\theta} \\
&\quad - e^2\left(A_{\mathrm{m}0x}^2 - \frac{1}{c^2}\phi_{\mathrm{e}0}^2 + \frac{m_0^2c^2}{e^2}\right)\Psi_{\mathrm{x}1x}(\zeta,\theta) = 0 \qquad (5)
\end{aligned}
$$

With $\Delta x = c\Delta t$ and $\Delta\zeta = \Delta\theta = 1$ we get:

$$
\begin{aligned}
\Bigg(&\big[\Psi_1(\zeta+1,\theta) - 2\Psi_1(\zeta,\theta) + \Psi_1(\zeta-1,\theta)\big] - \big[\Psi_1(\zeta,\theta+1) - 2\Psi_1(\zeta,\theta) + \Psi_1(\zeta,\theta-1)\big] \\
& - i\lambda_1\Big\{\big[\Psi_1(\zeta+1,\theta) - \Psi_1(\zeta-1,\theta)\big] + \lambda_3\big[\Psi_1(\zeta,\theta+1) - \Psi_1(\zeta,\theta-1)\big]\Big\} \\
& - \lambda_2^2\Psi_1(\zeta,\theta)\Bigg)\left(-\frac{\hbar^2}{(\Delta x)^2}\right) = 0
\end{aligned}
$$

$$
\Psi_1 = \Psi_{\mathrm{x}1x}, \quad \lambda_1,\ \lambda_2,\ \lambda_3 \quad \text{see Eq.(2.3-2)} \qquad (6)
$$

The second line in Eq.(3) is rewritten as follows:

$$\frac{4\lambda_C A_{e0}}{e}\left[\left(\frac{\hbar}{i\Delta x}\frac{\tilde{\Delta}}{\tilde{\Delta}\zeta} - eA_{m0x}\right)^2\left[1 - \frac{1}{2m_0^2c^2}\left(\frac{\hbar}{i\Delta x}\frac{\tilde{\Delta}}{\tilde{\Delta}\zeta} - eA_{m0x}\right)^2\right]\right.$$
$$= -\frac{4\lambda_C A_{e0}\hbar^2}{e(\Delta x)^2}\left(\frac{\tilde{\Delta}}{\tilde{\Delta}\zeta} - \frac{ie\Delta x}{\hbar}A_{m0x}\right)^2$$
$$\times\left[1 + \frac{1}{2}\left(\frac{\hbar}{m_0c\Delta x}\right)^2\left(\frac{\tilde{\Delta}}{\tilde{\Delta}\zeta} - \frac{ie\Delta x}{\hbar}A_{m0x}\right)^2\right] \quad (7)$$

If Δx is significantly larger than the Compton wavelength

$$\lambda_C = \frac{2\pi\hbar}{m_0c} = \frac{h}{m_0c} \quad (8)$$

we may ignore the second term in brackets on the right side of Eq.(7). For pions π^+ and π^- we have

$$\lambda_C = \frac{h}{m_0c} = \frac{6.626 \times 10^{-34}}{273.2 \times 9.109 \times 10^{-31} \times 2.998 \times 10^8} = 8.88 \times 10^{-15}\,[\mathrm{m}] \quad (9)$$

If the condition $\Delta x \gg h/m_0c$ is satisfied we get a second-order difference equation according to the first term on the right side of Eq.(7), otherwise we get a fourth-order difference equation according to the whole right side of Eq.(7). With the help of Eq.(2.2-14) we may rewrite the first term on the right side of Eq.(7) as follows with $\Psi_{x0x}(\zeta,\theta) = \Psi_0(\zeta,\theta)$:

$$\Delta x \gg \lambda_C = h/m_0c,\ \Psi_0 = \Psi_{x0x}$$

$$\frac{4\lambda_C A_{e0}\hbar^2}{e(\Delta x)^2}\left(\frac{\tilde{\Delta}}{\tilde{\Delta}\zeta} - \frac{ie\Delta x}{\hbar}A_{m0x}\right)^2\Psi_0(\zeta,\theta)$$
$$= \frac{4\lambda_C A_{e0}\hbar^2}{e(\Delta x)^2}\left[\frac{\Psi_0(\zeta+\Delta\zeta,\theta) - 2\Psi_0(\zeta,\theta) + \Psi_0(\zeta-\Delta\zeta,\theta)}{(\Delta\zeta)^2}\right.$$
$$\left. - \frac{2ie\Delta x}{\hbar}A_{m0x}\frac{\Psi_0(\zeta+\Delta\zeta,\theta) - \Psi_0(\zeta-\Delta\zeta,\theta)}{2\Delta\zeta} + \left(\frac{e\Delta x}{\hbar}A_{m0x}\right)^2\Psi_0(\zeta,\theta)\right]$$
$$= \frac{4\lambda_C A_{e0}\hbar^2}{e(\Delta x)^2}\Big[\Psi_0(\zeta+1,\theta) - 2\Psi_0(\zeta,\theta) + \Psi_0(\zeta-1,\theta)$$
$$- \frac{ie\Delta x}{\hbar}A_{m0x}[\Psi_0(\zeta+1,\theta) - \Psi_0(\zeta-1,\theta)] + \left(\frac{e\Delta x}{\hbar}A_{m0x}\right)^2\Psi_0(\zeta,\theta)\Big] \quad (10)$$

For the other extreme, $\Delta x \ll h/m_0 c$, we ignore the term 1 on the right side of Eq.(7) and use only the last term, which is multiplied by $(\hbar/m_0 c\Delta x)^2$:

$$\Delta x \ll \lambda_C = h/m_0 c, \quad \Psi_0 = \Psi_{x0x}$$

$$\begin{aligned}
&\frac{4\lambda_C A_{e0}\hbar^2}{e(\Delta x)^2}\left(\frac{\tilde{\Delta}}{\tilde{\Delta}\zeta} - \frac{ie\Delta x}{\hbar}A_{m0x}\right)^2 \\
&\qquad \times \frac{1}{2}\left(\frac{\hbar}{m_0 c\Delta x}\right)^2\left(\frac{\tilde{\Delta}}{\tilde{\Delta}\zeta} - \frac{ie\Delta x}{\hbar}A_{m0x}\right)^2 \Psi_0(\zeta,\theta) \\
&\doteq \frac{2\lambda_C A_{e0}}{e}\left(\frac{\hbar}{\Delta x}\right)^2\left(\frac{\hbar}{m_0 c\Delta x}\right)^2\left(\frac{\tilde{\Delta}}{\tilde{\Delta}\zeta} - \frac{ie\Delta x}{\hbar}A_{m0x}\right)^4 \Psi_0(\zeta,\theta)
\end{aligned} \tag{11}$$

Since we have not yet shown explicitly difference operators $\tilde{\Delta}^n/\tilde{\Delta}\zeta^n$ for n larger than 2 we must postpone the further analysis of Eq.(11) to Section 6.1. However, comparison of the first line of Eq.(10) with the last line of Eq.(11) shows that completely different equations are obtained for $\Delta x \gg h/m_0 c$ and $\Delta x \ll h/m_0 c$. Nothing comparable can exist in a differential theory since dx is never large or small compared with λ_C but is always infinitesimal.

We return to Eq.(3). The last two lines may be rewritten in the following form:

$$\begin{aligned}
&\Bigg\{e\left[A_{m1x}\left(\frac{\hbar}{i\Delta x}\frac{\tilde{\Delta}}{\tilde{\Delta}\zeta} - eA_{m0x}\right) + \left(\frac{\hbar}{i\Delta x}\frac{\tilde{\Delta}}{\tilde{\Delta}\zeta} - eA_{m0x}\right)A_{m1x}(\zeta,\theta)\right] \\
&\quad + \frac{e}{c^2}\left[\phi_{e1}\left(\frac{\hbar}{i\Delta t}\frac{\tilde{\Delta}}{\tilde{\Delta}\theta} + e\phi_{e0}\right) + \left(\frac{\hbar}{i\Delta t}\frac{\tilde{\Delta}}{\tilde{\Delta}\theta} + e\phi_{e0}\right)\phi_{e1}(\zeta,\theta)\right]\Bigg\}\Psi_0(\zeta,\theta) \\
&= e\Bigg[A_{m1x}(\zeta,\theta)\left(\frac{\hbar}{i\Delta x}\frac{\Psi_0(\zeta+\Delta\zeta,\theta) - \Psi_0(\zeta-\Delta\zeta,\theta)}{2\Delta\zeta} - eA_{m0x}\Psi_0(\zeta,\theta)\right) \\
&\quad + \frac{\hbar}{i\Delta x}\frac{\Psi_0(\zeta+\Delta\zeta,\theta)A_{m1x}(\zeta+\Delta\zeta,\theta) - \Psi_0(\zeta-\Delta\zeta,\theta)A_{m1x}(\zeta-\Delta\zeta,\theta)}{2\Delta\zeta} \\
&\qquad\qquad - eA_{m0x}A_{m1x}(\zeta,\theta)\Psi_0(\zeta,\theta)\Bigg] \\
&+ \frac{e}{c^2}\Bigg[\phi_{e1}(\zeta,\theta)\left(\frac{\hbar}{i\Delta t}\frac{\Psi_0(\zeta,\theta+\Delta\theta) + \Psi_0(\zeta,\theta-\Delta\theta)}{2\Delta\theta} + e\phi_{e0}\Psi_0(\zeta,\theta)\right) \\
&\quad + \frac{\hbar}{i\Delta t}\frac{\Psi_0(\zeta,\theta+\Delta\theta)\phi_{e1}(\zeta,\theta+\Delta\theta) - \Psi_0(\zeta,\theta-\Delta\theta)\phi_{e1}(\zeta,\theta-\Delta\theta)}{2\Delta\theta} \\
&\qquad\qquad + e\phi_{e0}\phi_{e1}(\zeta,\theta)\Psi_0(\zeta,\theta)\Bigg]
\end{aligned}$$

$$
\begin{aligned}
= e\Bigg[& A_{\mathrm{m}1x}(\zeta,\theta)\left(\frac{\hbar}{2i\Delta x}[\Psi_0(\zeta+1,\theta)-\Psi_0(\zeta-1,\theta)] - eA_{\mathrm{m}0x}\Psi_0(\zeta,\theta)]\right) \\
& + \frac{\hbar}{2i\Delta x}[\Psi_0(\zeta+1,\theta)A_{\mathrm{m}1x}(\zeta+1,\theta)-\Psi_0(\zeta-1,\theta)A_{\mathrm{m}1x}(\zeta-1,\theta)] \\
& - 2eA_{\mathrm{m}0x}A_{\mathrm{m}1x}(\zeta,\theta)\Psi_0(\zeta,\theta)\Bigg] \\
+ \frac{e}{c^2}\Bigg[& \phi_{\mathrm{e}1}(\zeta,\theta)\left(\frac{\hbar}{2i\Delta t}[\Psi_0(\zeta,\theta+1)-\Psi_0(\zeta,\theta-1)] + e\phi_{\mathrm{e}0}\Psi_0(\zeta,\theta)\right) \\
& + \frac{\hbar}{2i\Delta t}[\Psi_0(\zeta,\theta+1)\phi_{\mathrm{e}1}(\zeta,\theta+1)-\Psi_0(\zeta,\theta-1)\phi_{\mathrm{e}1}(\zeta,\theta-1)] \\
& + 2e\phi_{\mathrm{e}0}\phi_{\mathrm{e}1}(\zeta,\theta)\Psi_0(\zeta,\theta)\Bigg] \qquad (12)
\end{aligned}
$$

Consider next the terms $\mathsf{L}_{\mathrm{cx}11}$ to $\mathsf{L}_{\mathrm{cx}51}$ in Eq.(3). They are shown for $j = 1$ by Eqs.(2.2-33) to (2.2-37). We rewrite Eq.(2.2-33) for one space variable $x_j = x$, $j = 1$:

$$
\begin{aligned}
\mathsf{L}_{\mathrm{cx}11} = \frac{Ze}{2\alpha m_0^2 c}\Bigg\{ & \left[A_{\mathrm{e}0z}\left(\frac{\hbar}{i}\frac{\tilde{\Delta}}{\tilde{\Delta}y} - eA_{\mathrm{m}0y}\right) - A_{\mathrm{e}0y}\left(\frac{\hbar}{i}\frac{\tilde{\Delta}}{\tilde{\Delta}z} - eA_{\mathrm{m}0z}\right)\right] \\
& \times \left(\frac{\hbar}{i}\frac{\tilde{\Delta}}{\tilde{\Delta}x} - eA_{\mathrm{m}0x}\right)\left[1 - \frac{1}{m_0^2c^2}\left(\frac{\hbar}{i}\frac{\tilde{\Delta}}{\tilde{\Delta}x} - eA_{\mathrm{m}0x}\right)^2\right] \\
& + \left[1 - \frac{1}{m_0^2c^2}\left(\frac{\hbar}{i}\frac{\tilde{\Delta}}{\tilde{\Delta}x} - eA_{\mathrm{m}0x}\right)^2\right]\left[eA_{\mathrm{e}0z}\left(\frac{\hbar}{i}\frac{\tilde{\Delta}}{\tilde{\Delta}y} - eA_{\mathrm{m}0y}\right)\right. \\
& \left. - A_{\mathrm{e}0y}\left(\frac{\hbar}{i}\frac{\tilde{\Delta}}{\tilde{\Delta}z} - eA_{\mathrm{m}0z}\right)\right]\left(\frac{\hbar}{i}\frac{\tilde{\Delta}}{\tilde{\Delta}x} - eA_{\mathrm{m}0x}\right)\Bigg\} \qquad (13)
\end{aligned}
$$

The differences $\tilde{\Delta}/\tilde{\Delta}y$ and $\tilde{\Delta}/\tilde{\Delta}z$ are zero but the constants $A_{\mathrm{m}0x}$, $A_{\mathrm{m}0y}$, $A_{\mathrm{m}0z}$ are generally not zero:

$$
\begin{aligned}
\mathsf{L}_{\mathrm{cx}11} = \frac{Ze}{2\alpha m_0^2 c}\Bigg\{ & e(-A_{\mathrm{e}0z}A_{\mathrm{m}0y} + A_{\mathrm{e}0y}A_{\mathrm{m}0z}) \\
& \times \left(\frac{\hbar}{i}\frac{\tilde{\Delta}}{\tilde{\Delta}x} - eA_{\mathrm{m}0x}\right)\left[1 - \frac{1}{m_0^2c^2}\left(\frac{\hbar}{i}\frac{\tilde{\Delta}}{\tilde{\Delta}x} - eA_{\mathrm{m}0x}\right)^2\right] \\
& + \left[1 - \frac{1}{m_0^2c^2}\left(\frac{\hbar}{i}\frac{\tilde{\Delta}}{\tilde{\Delta}x} - eA_{\mathrm{m}0x}\right)^2\right]e(-A_{\mathrm{e}0z}A_{\mathrm{m}0y} + A_{\mathrm{e}0y}A_{\mathrm{m}0z}) \\
& \times \left(\frac{\hbar}{i}\frac{\tilde{\Delta}}{\tilde{\Delta}x} - eA_{\mathrm{m}0x}\right)\Bigg\} \qquad (14)
\end{aligned}
$$

We note that the terms in lines 3 and 4 can be written in reverse order since this requires only that $(-i\hbar\tilde{\Delta}/\tilde{\Delta}x - eA_{\mathrm{m}0x})$ commutes with the constant 1 and itself. We obtain for $x = \Delta x\,\zeta$ and $\Delta x = \Delta x \Delta\zeta$:

$$\begin{aligned}\mathsf{L}_{\mathrm{cx}11} &= i\frac{Ze^2}{\alpha m_0}\frac{\hbar}{m_0 c\Delta x}(A_{\mathrm{e}0z}A_{\mathrm{m}0y} - A_{\mathrm{e}0y}A_{\mathrm{m}0z})\\ &\times\left(\frac{\tilde{\Delta}}{\tilde{\Delta}\zeta} - \frac{ie\Delta x}{\hbar}A_{\mathrm{m}0x}\right)\left[1 + \left(\frac{\hbar}{m_0 c\Delta x}\right)^2\left(\frac{\tilde{\Delta}}{\tilde{\Delta}\zeta} - \frac{ie\Delta x}{\hbar}A_{\mathrm{m}0x}\right)^2\right] \qquad (15)\end{aligned}$$

Substitution into Eq.(2.2-31) yields with $\theta = t/\Delta t$ and thus $\Delta\theta = \Delta t/\Delta t = 1$:

$$\begin{aligned}&\frac{1}{c^2}\left(\frac{\hbar}{i\Delta t}\frac{\tilde{\Delta}}{\tilde{\Delta}\theta} + e\phi_{\mathrm{e}0}\right)\mathsf{L}_{\mathrm{cx}11} = \frac{1}{c^2}\mathsf{L}_{\mathrm{cx}11}\left(\frac{\hbar}{i\Delta t}\frac{\tilde{\Delta}}{\tilde{\Delta}\theta} + e\phi_{\mathrm{e}0}\right)\\ &= \frac{Ze^2}{\alpha}\left(\frac{\hbar}{m_0 c\Delta x}\right)^2(A_{\mathrm{e}0z}A_{\mathrm{m}0y} - A_{\mathrm{e}0y}A_{\mathrm{m}0z})\left(\frac{\tilde{\Delta}}{\tilde{\Delta}\theta} + \frac{ie\Delta t}{\hbar}\phi_{\mathrm{e}0}\right)\\ &\times\left(\frac{\tilde{\Delta}}{\tilde{\Delta}\zeta} - \frac{ie\Delta x}{\hbar}A_{\mathrm{m}0x}\right)\left[1 + \left(\frac{\hbar}{m_0 c\Delta x}\right)^2\left(\frac{\tilde{\Delta}}{\tilde{\Delta}\zeta} - \frac{ie\Delta x}{\hbar}A_{\mathrm{m}0x}\right)^2\right] \qquad (16)\end{aligned}$$

As in Eq.(7) we obtain again difference operators of higher than second order. As before we may distinguish between $\Delta x \gg h/m_0 c$ and $\Delta x \ll h/m_0 c$:

$$\Delta x \gg h/m_0 c$$

$$\begin{aligned}&\left[\frac{1}{c^2}\left(\frac{\hbar}{i\Delta t}\frac{\tilde{\Delta}}{\tilde{\Delta}\theta} + e\phi_{\mathrm{e}0}\right)\mathsf{L}_{\mathrm{cx}11} + \frac{1}{c^2}\mathsf{L}_{\mathrm{cx}11}\left(\frac{\hbar}{i\Delta t}\frac{\tilde{\Delta}}{\tilde{\Delta}\theta} + e\phi_{\mathrm{e}0}\right)\right]\Psi_0(\zeta,\theta)\\ &\doteq 2\frac{Ze^2}{\alpha}\left(\frac{\hbar}{m_0 c\Delta x}\right)^2(A_{\mathrm{e}0z}A_{\mathrm{m}0y} - A_{\mathrm{e}0y}A_{\mathrm{m}0z})\\ &\times\left(\frac{\tilde{\Delta}}{\tilde{\Delta}\theta}\frac{\tilde{\Delta}}{\tilde{\Delta}\zeta} + \frac{ie\Delta t}{\hbar}\phi_{\mathrm{e}0}\frac{\tilde{\Delta}}{\tilde{\Delta}\zeta} - \frac{ice\Delta t}{\hbar}A_{\mathrm{m}0x}\frac{\tilde{\Delta}}{\tilde{\Delta}\theta} + \frac{ce^2(\Delta t)^2}{\hbar^2}\phi_{\mathrm{e}0}A_{\mathrm{m}0x}\right)\Psi_0(\zeta,\theta)\end{aligned}$$

$$\begin{aligned}&= 2\frac{Ze^2}{\alpha}\left(\frac{\hbar}{m_0 c\Delta x}\right)^2(A_{\mathrm{e}0z}A_{\mathrm{m}0y} - A_{\mathrm{e}0y}A_{\mathrm{m}0z})\\ &\times\left(\frac{1}{4}\left\{[\Psi_0(\zeta+1,\theta+1) - \Psi_0(\zeta+1,\theta-1)] + [\Psi_0(\zeta-1,\theta+1) - \Psi_0(\zeta-1,\theta-1)]\right\}\right.\\ &\qquad + \frac{1}{2}\frac{ie\Delta t}{\hbar}\phi_{\mathrm{e}0}[\Psi_0(\zeta+1,\theta) - \Psi_0(\zeta-1,\theta)]\\ &\left. - \frac{1}{2}\frac{ie\Delta x}{\hbar}A_{\mathrm{m}0x}[\Psi_0(\zeta,\theta+1) - \Psi_0(\zeta,\theta-1)] + \frac{ce^2(\Delta t)^2}{\hbar^2}\phi_{\mathrm{e}0}A_{\mathrm{m}0x}\Psi_0(\zeta,\theta)\right) \qquad (17)\end{aligned}$$

$$\Delta x \ll h/m_0c$$

$$\left[\frac{1}{c^2}\left(\frac{\hbar}{i\Delta t}\frac{\tilde{\Delta}}{\tilde{\Delta}\theta}+e\phi_{e0}\right)\mathsf{L}_{cx11}+\frac{1}{c^2}\mathsf{L}_{cx11}\left(\frac{\hbar}{i\Delta t}\frac{\tilde{\Delta}}{\tilde{\Delta}\theta}+e\phi_{e0}\right)\right]\Psi_0(\zeta,\theta)$$
$$\doteq 2\frac{Ze^2}{\alpha}\left(\frac{\hbar}{m_0c\Delta x}\right)^4(A_{e0z}A_{m0y}-A_{e0y}A_{m0z})$$
$$\times\left(\frac{\tilde{\Delta}}{\tilde{\Delta}\theta}+\frac{ie\Delta t}{\hbar}\phi_{e0}\right)\left(\frac{\tilde{\Delta}}{\tilde{\Delta}\zeta}-\frac{ie\Delta x}{\hbar}A_{m0x}\right)^3\Psi_0(\zeta,\theta) \quad (18)$$

As in Eq.(11) we obtain difference operators of higher than second order and refer to Section 6.1 for further analysis.

The terms L_{cx21} in Eq.(3) are zero according to Eq.(2.2-34) and we get in Eq.(3):

$$\left[\frac{1}{c^2}\left(\frac{\hbar}{i\Delta t}\frac{\tilde{\Delta}}{\tilde{\Delta}\theta}+e\phi_{e0}\right)\mathsf{L}_{cx21}+\frac{1}{c^2}\mathsf{L}_{cx21}\left(\frac{\hbar}{i\Delta t}\frac{\tilde{\Delta}}{\tilde{\Delta}\theta}+e\phi_{e0}\right)\right]\Psi_0(\zeta,\theta)=0 \quad (19)$$

The term L_{cx31} is shown for $j=1$ by Eq.(2.2-35). We rewrite Eq.(2.2-35) for one space variable $x_j=x$, $j=1$:

$$\mathsf{L}_{cx31}=\frac{Ze}{2\alpha m_0c}\mathop{\mathbb{S}}_{c}^{x}\left(A_{e0z}\frac{\tilde{\Delta}}{\tilde{\Delta}t}\left\{\left[1-\frac{1}{2m_0^2c^2}\left(\frac{\hbar}{i}\frac{\tilde{\Delta}}{\tilde{\Delta}x}-eA_{m0x}\right)^2\right]\left(\frac{\hbar}{i}\frac{\tilde{\Delta}}{\tilde{\Delta}y}-eA_{m0y}\right)\right.\right.$$
$$\left.+\left(\frac{\hbar}{i}\frac{\tilde{\Delta}}{\tilde{\Delta}y}-eA_{m0y}\right)\left[1-\frac{1}{2m_0^2c^2}\left(\frac{\hbar}{i}\frac{\tilde{\Delta}}{\tilde{\Delta}x}-eA_{m0x}\right)^2\right]\right\}$$
$$-A_{e0z}\frac{\tilde{\Delta}}{\tilde{\Delta}t}\left\{\left[1-\frac{1}{2m_0^2c^2}\left(\frac{\hbar}{i}\frac{\tilde{\Delta}}{\tilde{\Delta}x}-eA_{m0x}\right)^2\right]\left(\frac{\hbar}{i}\frac{\tilde{\Delta}}{\tilde{\Delta}z}-eA_{m0z}\right)\right.$$
$$\left.\left.+\left(\frac{\hbar}{i}\frac{\tilde{\Delta}}{\tilde{\Delta}z}-eA_{m0z}\right)\left[1-\frac{1}{2m_0^2c^2}\left(\frac{\hbar}{i}\frac{\tilde{\Delta}}{\tilde{\Delta}x}-eA_{m0x}\right)^2\right]\right\}\right)\Delta x \quad (20)$$

Again, the differences $\tilde{\Delta}/\tilde{\Delta}y$ and $\tilde{\Delta}/\tilde{\Delta}z$ are zero but the constants A_{m0x}, A_{m0y}, A_{m0z} are generally not zero. We obtain with $x=\zeta\Delta x$:

$$
\begin{aligned}
\mathsf{L}_{\mathrm{cx}31}\Psi_0(\zeta,\theta) = \frac{Ze^2}{2\alpha m_0 c} \oint_c^{\zeta\Delta x} \Bigg\{ & - A_{\mathrm{e}0z}A_{\mathrm{m}0y}\frac{\tilde{\Delta}}{\tilde{\Delta} t}\left[1 - \frac{1}{2m_0^2c^2}\left(\frac{\hbar}{i}\frac{\tilde{\Delta}}{\tilde{\Delta} x} - eA_{\mathrm{m}0x}\right)^2\right] \\
& - A_{\mathrm{e}0z}A_{\mathrm{m}0y}\frac{\tilde{\Delta}}{\tilde{\Delta} t}\left[1 - \frac{1}{2m_0^2c^2}\left(\frac{\hbar}{i}\frac{\tilde{\Delta}}{\tilde{\Delta} x} - eA_{\mathrm{m}0x}\right)^2\right] \\
& + A_{\mathrm{e}0z}A_{\mathrm{m}0z}\frac{\tilde{\Delta}}{\tilde{\Delta} t}\left[1 - \frac{1}{2m_0^2c^2}\left(\frac{\hbar}{i}\frac{\tilde{\Delta}}{\tilde{\Delta} x} - eA_{\mathrm{m}0x}\right)^2\right] \\
& + A_{\mathrm{e}0z}A_{\mathrm{m}0z}\frac{\tilde{\Delta}}{\tilde{\Delta} t}\left[1 - \frac{1}{2m_0^2c^2}\left(\frac{\hbar}{i}\frac{\tilde{\Delta}}{\tilde{\Delta} x} - eA_{\mathrm{m}0x}\right)^2\right]\Bigg\}\Psi_0(x,\theta)\Delta x \\
= \frac{Ze^2}{2\alpha m_0 c} \oint_c^{\zeta\Delta x} & 2A_{\mathrm{e}0z}(A_{\mathrm{m}0z} - A_{\mathrm{m}0y}) \\
& \times \frac{\tilde{\Delta}}{\tilde{\Delta} t}\left[1 - \frac{1}{2m_0^2c^2}\left(\frac{\hbar}{i}\frac{\tilde{\Delta}}{\tilde{\Delta} x} - eA_{\mathrm{m}0x}\right)^2\right]\Psi_0(x,\theta)\Delta x \qquad (21)
\end{aligned}
$$

We could write Eq.(21) in the form

$$
L_{\mathrm{cx}31}\Psi_0(\zeta,\theta) = \frac{Ze^2}{2\alpha m_0 c}V'_{\mathrm{cx}31}(\zeta,\theta) \qquad (22)
$$

but this way of writing would give $V'_{\mathrm{cx}31}(\zeta,\theta)$ a physical dimension. We prefer to use a dimension-free function $V_{\mathrm{cx}31}(\zeta,\theta)$. To achieve this we note that the square of the mechanical momentum m_0c in the first line of Eq.(3) has the electromagnetic dimension $(\mathrm{VAs}^2/\mathrm{m})^2$. Since the factor $\hbar/c^2\Delta t$ of the terms $L_{\mathrm{cx}i1}$ in that equation has the dimension $\mathrm{VAs}^3/\mathrm{m}^2$ we infer that the terms $L_{\mathrm{cx}i1}$ have the electromagnetic dimension VAs. The factor m_0c^2 has the dimension VAs. Hence, we rewrite Eq.(22) as follows

$$
L_{\mathrm{cx}31}\Psi_0(\zeta,\theta) = \frac{m_0c^2}{2\alpha}V_{\mathrm{cx}31}(\zeta,\theta) = \frac{Ze^2}{2\alpha m_0 c}V'_{\mathrm{cx}31}(\zeta,\theta) \qquad (23)
$$

and obtain

$$
\begin{aligned}
V_{\mathrm{cx}31}(\zeta,\theta) &= \frac{Ze^2}{m_0^2c^3}V'_{\mathrm{cx}31}(\zeta,\theta) \\
&= \frac{Ze^2}{m_0^2c^3}\oint_c^{\zeta\Delta x} 2A_{\mathrm{e}0z}(A_{\mathrm{m}0z} - A_{\mathrm{m}0y}) \\
&\qquad \times \frac{\tilde{\Delta}}{\tilde{\Delta} t}\left[1 - \frac{1}{m_0^2c^2}\left(\frac{\hbar}{i}\frac{\tilde{\Delta}}{\tilde{\Delta} x} - eA_{\mathrm{m}0x}\right)^2\right]\Psi_0(x,\theta)\Delta x \qquad (24)
\end{aligned}
$$

Equation (24) written explicitly assumes the form

$$V_{\mathrm{cx31}}(\zeta,\theta)=\frac{Ze^2}{m_0^2c^3}\oint_c^{\zeta\Delta x}2A_{\mathrm{e0}z}(A_{\mathrm{m0}z}-A_{\mathrm{m0}y})\frac{\tilde{\Delta}}{\tilde{\Delta}t}\left[1+\frac{1}{2m_0^2c^2}\right.$$
$$\times\left(-\hbar^2\frac{\Psi_0(x+\Delta x,\theta)-2\Psi_0(x,\theta)+\Psi_0(x-\Delta x,\theta)}{(\Delta x)^2}\right.$$
$$\left.\left.+2i\hbar eA_{\mathrm{m0}x}\frac{\Psi_0(x+\Delta x,\theta)-\Psi_0(x-\Delta x,\theta)}{2\Delta x}+e^2A_{\mathrm{m0}x}^2\Psi_0(x,\theta)\right)\right]\Delta x \quad (25)$$

From Eqs.(3) and (23) we obtain the following relation for $\mathsf{L}_{\mathrm{cx31}}\Psi_0(\zeta,\theta)$ and $V_{\mathrm{cx31}}(\zeta,\theta)$

$$\frac{1}{c^2}\left(\frac{\hbar}{i\Delta t}\frac{\tilde{\Delta}}{\tilde{\Delta}\theta}+e\phi_{\mathrm{e0}}\right)\mathsf{L}_{\mathrm{cx31}}\Psi_0(\zeta,\theta)=\frac{1}{c^2}\mathsf{L}_{\mathrm{cx31}}\left(\frac{\hbar}{i\Delta t}\frac{\tilde{\Delta}}{\tilde{\Delta}\theta}+e\phi_{\mathrm{e0}}\right)\Psi_0(\zeta,\theta)$$
$$=-i\frac{m_0\hbar}{2\alpha\Delta t}V_{\mathrm{cx31}}(\zeta,\theta)\left(\frac{\tilde{\Delta}}{\tilde{\Delta}\theta}+\frac{ie\Delta t}{\hbar}\phi_{\mathrm{e0}}\right)\Psi_0(\zeta,\theta) \quad (26)$$

which yields:

$$\left[\frac{1}{c^2}\left(\frac{\hbar}{i\Delta t}\frac{\tilde{\Delta}}{\tilde{\Delta}\theta}+e\phi_{\mathrm{e0}}\right)\mathsf{L}_{\mathrm{cx31}}+\frac{1}{c^2}\mathsf{L}_{\mathrm{cx31}}\left(\frac{\hbar}{i\Delta t}\frac{\tilde{\Delta}}{\tilde{\Delta}\theta}+e\phi_{\mathrm{e0}}\right)\right]\Psi_0(\zeta,\theta)$$
$$=-i\frac{m_0\hbar}{\alpha\Delta t}\left(\frac{\tilde{\Delta}}{\tilde{\Delta}\theta}+\frac{ie\Delta t}{\hbar}\phi_{\mathrm{e0}}\right)\Psi_0(\zeta,\theta)$$
$$=-i\frac{m_0\hbar}{\alpha\Delta t}V_{\mathrm{cx31}}(\zeta,\theta)$$
$$\times\left(\frac{1}{2}[\Psi_0(\zeta,\theta+1)-\Psi_0(\zeta,\theta-1)]+\frac{ie\Delta t}{\hbar}\phi_{\mathrm{e0}}\Psi_0(\zeta,\theta)\right) \quad (27)$$

According to Eq.(2.2-36) the term $\mathsf{L}_{\mathrm{cx41}}$ is zero and we get in Eq.(3):

$$\left[\frac{1}{c^2}\left(\frac{\hbar}{i\Delta t}\frac{\tilde{\Delta}}{\tilde{\Delta}\theta}+e\phi_{\mathrm{e0}}\right)\mathsf{L}_{\mathrm{cx41}}+\frac{1}{c^2}\mathsf{L}_{\mathrm{cx41}}\left(\frac{\hbar}{i\Delta t}\frac{\tilde{\Delta}}{\tilde{\Delta}\theta}+e\phi_{\mathrm{e0}}\right)\right]\Psi_0(\zeta,\theta)=0 \quad (28)$$

The final term $\mathsf{L}_{\mathrm{cx51}}$ in Eq.(3) is defined for $j=1$ by Eq.(2.2-37):

$$\mathsf{L}_{\mathrm{cx51}}=\frac{Ze}{4\alpha m_0^2c}\oint_c^x\left(\left\{\left[1-\frac{1}{2m_0^2c^2}\left(\frac{\hbar}{i}\frac{\tilde{\Delta}}{\tilde{\Delta}x}-eA_{\mathrm{m0}x}\right)^2\right]\left(\frac{\hbar}{i}\frac{\tilde{\Delta}}{\tilde{\Delta}y}-eA_{\mathrm{m0}y}\right)\right.\right.$$
$$\left.+\left(\frac{\hbar}{i}\frac{\tilde{\Delta}}{\tilde{\Delta}y}-eA_{\mathrm{m0}y}\right)\left[1-\frac{1}{2m_0^2c^2}\left(\frac{\hbar}{i}\frac{\tilde{\Delta}}{\tilde{\Delta}x}-eA_{\mathrm{m0}x}\right)^2\right]\right\}\frac{\tilde{\Delta}}{\tilde{\Delta}y}$$

$$
\begin{aligned}
&+\left\{\left[1-\frac{1}{2m_0^2c^2}\left(\frac{\hbar}{i}\frac{\tilde{\Delta}}{\tilde{\Delta}x}-eA_{\mathrm{m0}x}\right)^2\left(\frac{\hbar}{i}\frac{\tilde{\Delta}}{\tilde{\Delta}z}-eA_{\mathrm{m0}z}\right)\right.\right.\\
&\left.\left.+\left(\frac{\hbar}{i}\frac{\tilde{\Delta}}{\tilde{\Delta}z}-eA_{\mathrm{m0}z}\right)\left[1-\frac{1}{2m_0^2c^2}\left(\frac{\hbar}{i}\frac{\tilde{\Delta}}{\tilde{\Delta}x}-eA_{\mathrm{m0}x}\right)^2\right]\right\}\frac{\tilde{\Delta}}{\tilde{\Delta}z}\right)\\
&\times\left\{\left[1-\frac{1}{2m_0^2c^2}\left(\frac{\hbar}{i}\frac{\tilde{\Delta}}{\tilde{\Delta}x}-eA_{\mathrm{m0}x}\right)^2\right]\right.\\
&\qquad\times\left[A_{\mathrm{e0}x}\left(\frac{\hbar}{i}\frac{\tilde{\Delta}}{\tilde{\Delta}y}-eA_{\mathrm{m0}y}\right)-A_{\mathrm{e0}y}\left(\frac{\hbar}{i}\frac{\tilde{\Delta}}{\tilde{\Delta}z}-eA_{\mathrm{m0}z}\right)\right]\\
&+\left[A_{\mathrm{e0}x}\left(\frac{\hbar}{i}\frac{\tilde{\Delta}}{\tilde{\Delta}y}-eA_{\mathrm{m0}y}\right)-A_{\mathrm{e0}y}\left(\frac{\hbar}{i}\frac{\tilde{\Delta}}{\tilde{\Delta}z}-eA_{\mathrm{m0}x}\right)\right]\\
&\qquad\left.\times\left[1-\frac{1}{2m_0^2c^2}\left(\frac{\hbar}{i}\frac{\tilde{\Delta}}{\tilde{\Delta}x}-eA_{\mathrm{m0}x}\right)^2\right]\right\}\Delta x
\end{aligned}
\tag{29}
$$

The differences $\tilde{\Delta}/\tilde{\Delta}y$ and $\tilde{\Delta}/\tilde{\Delta}z$ are again zero and all terms multiplied by them are removed. This means primarily lines 1–4 but as a consequence lines 5–8 too, which leaves a summation constant C'_{cx51}. In analogy to Eq.(23) we get:

$$
L_{\mathrm{cx51}}=\frac{m_0c^2}{4\alpha}C_{\mathrm{cx51}}=\frac{Ze}{4\alpha m_0^2c}C'_{\mathrm{cx51}}
\tag{30}
$$

We obtain from Eqs.(3) and (30):

$$
\begin{aligned}
&\left[\frac{1}{c^2}\left(\frac{\hbar}{i\Delta t}\frac{\tilde{\Delta}}{\tilde{\Delta}\theta}+e\phi_{\mathrm{e0}}\right)\mathrm{L}_{\mathrm{cx51}}+\frac{1}{c^2}\mathrm{L}_{\mathrm{cx51}}\left(\frac{\hbar}{i\Delta t}\frac{\tilde{\Delta}}{\tilde{\Delta}\theta}+e\phi_{\mathrm{e0}}\right)\right]\Psi_0(\zeta,\theta)\\
&\quad=-i\frac{m_0\hbar}{2\alpha\Delta t}C_{\mathrm{cx51}}\left(\frac{\tilde{\Delta}}{\tilde{\Delta}\theta}+\frac{ie\Delta t}{\hbar}\phi_{\mathrm{e0}}\right)\Psi_0(\zeta,\theta)\\
&\quad=-i\frac{m_0\hbar}{2\alpha\Delta t}C_{\mathrm{cx51}}\left(\frac{1}{2}[\Psi_0(\zeta,\theta+1)-\Psi_0(\zeta,\theta-1)]+\frac{ie\Delta t}{\hbar}\phi_{\mathrm{e0}}\Psi_0(\zeta,\theta)\right)
\end{aligned}
\tag{31}
$$

With the help of Eqs.(6), (10), (17), (28), (31), and (12) we may rewrite Eq.(3). The lines are numbered for easier reference later on. We remove the factor $-\hbar^2/(\Delta x)^2$ of Eq.(6) by multiplying Eq.(3) with $-(\Delta x)^2/\hbar^2$:

$$
\Delta x\gg\lambda_{\mathrm{C}}=h/m_0c
$$

$$[\Psi_1(\zeta+1,\theta)-2\Psi_1(\zeta,\theta)+\Psi_1(\zeta-1,\theta)]-[\Psi_1(\zeta,\theta+1)-2\Psi_1(\zeta,\theta)+\Psi_1(\zeta,\theta-1)] \qquad 1$$

$$-i\lambda_1\Big\{[\Psi_1(\zeta+1,\theta)-\Psi_1(\zeta-1,\theta)]+\lambda_3[\Psi_1(\zeta,\theta+1)-\Psi_1(\zeta,\theta-1)]\Big\}-\lambda_2^2\Psi_1(\zeta,\theta) \qquad 2$$

$$
\begin{aligned}
&= \frac{4\lambda_{\mathrm{C}} A_{\mathrm{e}}(\zeta,\theta)}{e}\Big[\Psi_0(\zeta+1,\theta) - 2\Psi_0(\zeta,\theta) + \Psi_0(\zeta-1,\theta) \\
&\qquad - i\frac{e\Delta x}{\hbar}A_{\mathrm{m}0x}[\Psi_0(\zeta+1,\theta) - \Psi_0(\zeta-1,\theta)] + \left(\frac{e\Delta x}{\hbar}A_{\mathrm{m}0x}\right)^2\Psi_0(\zeta,\theta)\Big] \\
&+ \frac{2Ze^2}{\alpha m_0^2c^2}(A_{\mathrm{e}0z}A_{\mathrm{m}0y} - A_{\mathrm{e}0y}A_{\mathrm{m}0z}) \\
&\times\Big(\frac{1}{4}\big\{[\Psi_0(\zeta+1,\theta+1) - \Psi_0(\zeta+1,\theta-1)] + [\Psi_0(\zeta-1,\theta+1) - \Psi_0(\zeta-1,\theta-1)]\big\} \\
&\qquad + \frac{1}{2}\frac{ie\Delta t}{\hbar}\phi_{\mathrm{e}0}[\Psi_0(\zeta+1,\theta) - \Psi_0(\zeta-1,\theta)] \\
&\qquad - \frac{ie\Delta x}{2\hbar}A_{\mathrm{m}0x}[\Psi_0(\zeta,\theta+1) - \Psi_0(\zeta,\theta-1)] + \frac{ce^2(\Delta t)^2}{\hbar^2}\phi_{\mathrm{e}0}A_{\mathrm{m}0x}\Psi_0(\zeta,\theta)\Big) \\
&- i\frac{m_0(\Delta x)^2}{\alpha\hbar\Delta t}V_{\mathrm{cx}31}(\zeta,\theta)\left(\frac{1}{2}[\Psi_0(\zeta,\theta+1) - \Psi_0(\zeta,\theta-1)] + \frac{ie\Delta t}{\hbar}\phi_{\mathrm{e}0}\Psi_0(\zeta,\theta)\right) \\
&- i\frac{m_0(\Delta x)^2}{2\alpha\hbar\Delta t}C_{\mathrm{cx}51}\left(\frac{1}{2}[\Psi_0(\zeta,\theta+1) - \Psi_0(\zeta,\theta-1)] + \frac{ie\Delta t}{\hbar}\phi_{\mathrm{e}0}\Psi_0(\zeta,\theta)\right) \\
&- e\left(\frac{\Delta x}{\hbar}\right)^2\Big[A_{\mathrm{m}1x}(\zeta,\theta)\left(\frac{\hbar}{2i\Delta x}[\Psi_0(\zeta+1,\theta) - \Psi_0(\zeta-1,\theta)] - eA_{\mathrm{m}0x}\Psi_0(\zeta,\theta)]\right) \\
&\qquad + \frac{\hbar}{2i\Delta x}[\Psi_0(\zeta+1,\theta)A_{\mathrm{m}1x}(\zeta+1,\theta) - \Psi_0(\zeta-1,\theta)A_{\mathrm{m}1x}(\zeta-1,\theta)] \\
&\qquad\qquad - 2eA_{\mathrm{m}0x}A_{\mathrm{m}1x}(\zeta,\theta)\Psi_0(\zeta,\theta)\Big] \\
&- \frac{e}{c^2}\left(\frac{\Delta x}{\hbar}\right)^2\Big[\phi_{\mathrm{e}1}(\zeta,\theta)\left(\frac{\hbar}{2i\Delta t}[\Psi_0(\zeta,\theta+1) - \Psi_0(\zeta,\theta-1)] + e\phi_{\mathrm{e}0}\Psi_0(\zeta,\theta)\right) \\
&\qquad + \frac{\hbar}{2i\Delta t}[\Psi_0(\zeta,\theta+1)\phi_{\mathrm{e}1}(\zeta,\theta+1) - \Psi_0(\zeta,\theta-1)\phi_{\mathrm{e}1}(\zeta,\theta-1)] \\
&\qquad\qquad + 2e\phi_{\mathrm{e}0}\phi_{\mathrm{e}1}(\zeta,\theta)\Psi_0(\zeta,\theta)\Big] \qquad (32)
\end{aligned}
$$

3.2 Evaluation of Eq.(3.1-32)

Equation (3.1-32) needs to be simplified before attempting to find a solution. A first step is to make as many as possible of the constant components of $\mathbf{A}_{\mathrm{e}}$, $\mathbf{A}_{\mathrm{m}}$, ϕ_{e}, and ϕ_{m} in Eq.(3.1-4) zero. The two constants $A_{\mathrm{m}0x}$ and $\phi_{\mathrm{e}0}$ must be retained to keep Eq.(2.3-2) valid but the following components may be chosen equal to zero:

$$A_{\mathrm{e}0x} = A_{\mathrm{e}0y} = A_{\mathrm{e}0z} = 0, \quad A_{\mathrm{m}0y} = A_{\mathrm{m}0z} = 0, \quad \phi_{\mathrm{m}0} = 0 \qquad (1)$$

The first two lines of Eq.(3.1-32) remain unchanged. We note that they are equal to the left side of Eq.(2.3-2) except that Ψ_0 is replaced by Ψ_1.

Lines 3 and 4 are multiplied by $A_e(\zeta,\theta)$, which according to Eqs.(3.1-4) becomes $\alpha A_{e1}(\zeta,\theta)$. These two lines are of order $O(\alpha)$, they are left out. Lines 5-8 are zero because of Eq.(1).

Line 9 is zero according to Eqs.(3.1-25) and (1). Line 10 remains unchanged. We shorten the coefficient C_{cx51} to C_{cx}:

$$C_{cx} = C_{cx51} \tag{2}$$

The remaining lines 11 to 16 do not permit any significant change. Hence, our first simplification of Eq.(3.1-32) leads to the elimination only of lines 3 to 9:

$$\begin{aligned}
&[\Psi_1(\zeta+1,\theta)-2\Psi_1(\zeta,\theta)+\Psi_1(\zeta-1,\theta)]-[\Psi_1(\zeta,\theta+1)-2\Psi_1(\zeta,\theta)+\Psi_1(\zeta,\theta-1)]\\
&-i\lambda_1\{[\Psi_1(\zeta+1,\theta)-\Psi_1(\zeta-1,\theta)]+\lambda_3[\Psi_1(\zeta,\theta+1)-\Psi_1(\zeta,\theta-1)]\}-\lambda_2^2\Psi_1(\zeta,\theta)\\
&\qquad = G_1(\zeta,\theta)
\end{aligned} \tag{3}$$

$$\begin{aligned}
G_1(\zeta,\theta) = &-i\frac{m_0(\Delta x)^2C_{cx}}{2\alpha\hbar\Delta t}\left(\frac{1}{2}[\Psi_0(\zeta,\theta+1)-\Psi_0(\zeta,\theta-1)]+\frac{ie\Delta t\phi_{e0}}{\hbar}\Psi_0(\zeta,\theta)\right)\\
&-i\frac{e\Delta x}{2\hbar}\{[A_{m1x}(\zeta+1,\theta)+A_{m1x}(\zeta,\theta)]\Psi_0(\zeta+1,\theta)\\
&\qquad -[A_{m1x}(\zeta,\theta)+A_{m1x}(\zeta-1,\theta)]\Psi_0(\zeta-1,\theta)\}\\
&\qquad\qquad -2\left(\frac{e\Delta x}{\hbar}\right)^2 A_{m0x}A_{m1x}(\zeta,\theta)\Psi_0(\zeta,\theta)\\
&-i\frac{e\Delta x}{2c\hbar}\{[\phi_{e1}(\zeta,\theta+1)+\phi_{e1}(\zeta,\theta)]\Psi_0(\zeta,\theta+1)\\
&\qquad -[\phi_{e1}(\zeta,\theta)+\phi_{e1}(\zeta,\theta-1)]\Psi_0(\zeta,\theta-1)\}\\
&\qquad\qquad +2\left(\frac{e\Delta x}{c\hbar}\right)^2\phi_{e0}\phi_{e1}(\zeta,\theta)\Psi_0(\zeta,\theta)
\end{aligned} \tag{4}$$

We need the general solution of the homogeneous Eq.(3) plus a particular solution of the inhomogeneous equation. The boundary condition of Eq.(2.3-3) is used again:

$$\begin{aligned}
\Psi_1(0,\theta) = \Psi_{00}S(\theta)(1-e^{-\iota\theta}) &= 0 && \text{for } \theta < 0\\
&= \Psi_{00}(1-e^{-\iota\theta}) && \text{for } \theta \geq 0
\end{aligned} \tag{5}$$

The initial conditions of Eqs.(2.3-4) and (2.3-6) are also used for $\Psi_1(\zeta,0)$:

$$\Psi_1(\zeta,\theta) = 0 \quad \text{for } \theta = 0,\ \zeta > 0 \tag{6}$$
$$\Psi_1(\zeta,\theta+1) - \Psi_1(\zeta,\theta) = 0 \quad \text{or } \Psi_1(\zeta,1) = 0 \quad \text{for } \theta = 0,\ \zeta > 0 \tag{7}$$

The ansatz of Eq.(2.3-7) is also used once more but we write $\hat{u}(\zeta,\theta)$ to emphasize the difference with $u(\zeta,\theta)$ in Eq.(2.3-7) or (2.4-23):

$$\Psi_1(\zeta,\theta) = \Psi_{00}[(1 - e^{-\iota\theta})F(\zeta) + \hat{u}(\zeta,\theta)] \tag{8}$$

The determination of $F(\zeta)$ and ι follows the calculation from Eq.(2.3-8) to (2.3-18):

$$F(\zeta) = \exp[-(\lambda_2^2 - \lambda_1^2)^{1/2}\zeta]e^{i\lambda_1\zeta} \tag{9}$$
$$\iota = 2i\lambda_1\lambda_3 \tag{10}$$

The boundary and initial conditions for $\hat{u}(\zeta,\theta)$ are the same as for $u(\zeta,\theta)$ in Eqs.(2.3-19) to (2.3-21):

$$\hat{u}(0,\theta) = 0 \qquad \theta \geq 0 \tag{11}$$
$$\hat{u}(\zeta,0) = 0 \qquad \zeta > 0 \tag{12}$$
$$\hat{u}(\zeta,1) - \hat{u}(\zeta,0) = -(1 - e^{-\iota})F(\zeta) \qquad \zeta > 0 \tag{13}$$

Substitution of $\hat{u}(\zeta,\theta)$ of Eq.(8) into Eq.(3) yields the same equation with $\Psi_1(\zeta,\theta)$ replaced by $\hat{u}(\zeta,\theta)$:

$$\begin{aligned}&[\hat{u}(\zeta+1,\theta) - 2\hat{u}(\zeta,\theta) + \hat{u}(\zeta-1,\theta)] - [\hat{u}(\zeta,\theta+1) - 2\hat{u}(\zeta,\theta) + \hat{u}(\zeta,\theta-1)]\\ &- i\lambda_1\{[\hat{u}(\zeta+1,\theta) - \hat{u}(\zeta-1,\theta)] + \lambda_3[\hat{u}(\zeta,\theta+1) - \hat{u}(\zeta,\theta-1)]\} - \lambda_2^2\hat{u}(\zeta,\theta)\\ &= G_1(\zeta,\theta)\end{aligned} \tag{14}$$

We write the solution of Eq.(14) in the form

$$\hat{u}(\zeta,\theta) = u(\zeta,\theta) + \hat{v}(\zeta,\theta) \tag{15}$$

where $u(\zeta,\theta)$ is the solution of the homogeneous equation and $\hat{v}(\zeta,\theta)$ is a particular solution of the inhomogeneous equation. The homogeneous equation equals Eq.(2.3-22). We solved it with Bernoulli's product method:

$$u_\kappa(\zeta,\theta) = \phi_\kappa(\zeta)\psi_\kappa(\theta) \tag{16}$$

which yielded Eq.(2.3-36) for $\phi(\zeta) = \phi_\kappa(\zeta)$

$$\phi_\kappa(\zeta) = e^{i\lambda_1\zeta}(A_{30}e^{i\varphi_\kappa\zeta} + A_{31}e^{-i\varphi_\kappa\zeta}) = e^{i\lambda_1\zeta}(A_{32}\cos\varphi_\kappa\zeta + iA_{33}\sin\varphi_\kappa\zeta)$$
$$\varphi_\kappa = 2\pi\kappa/N,\ \kappa = 0,\ \pm1,\ \ldots,\ \pm N/2 \tag{17}$$

and Eq.(2.4-20) for $\psi(\theta) = \psi_\kappa(\theta)$:

$$\psi_\kappa(\theta) = e^{-i\lambda_1\lambda_3\theta}(A_{41}e^{i\beta_\kappa\theta} + A_{40}e^{-i\beta_\kappa\theta}) \tag{18}$$

Using the approximation $\beta_\kappa = \varphi_\kappa$ of Fig.2.4-1 in the interval $-\pi/2 < \varphi_\kappa < \pi/2$ we may rewrite Eq.(18) in the form

$$\begin{aligned}\psi_\kappa(\theta) &= e^{-i\lambda_1\lambda_3\theta}(A_{41}e^{i\varphi_\kappa\theta} + A_{40}e^{-i\varphi_\kappa\theta}) \\ &= e^{-i\lambda_1\lambda_3\theta}(A_{42}\cos\varphi_\kappa\theta + iA_{43}\sin\varphi_\kappa\theta)\end{aligned} \tag{19}$$

We turn to the inhomogeneous solution $\hat{v}(\zeta,\theta)$. Let $\hat{v}(\zeta,\theta)$ be substituted for $\hat{u}(\zeta,\theta)$ in Eq.(14). Writing in some detail we obtain in analogy to Eqs.(16)–(19):

$$\begin{aligned}\hat{v}_\kappa(\zeta,\theta) = \hat{\phi}_\kappa(\zeta)\hat{\psi}_\kappa(\theta) &= e^{i\lambda_1\zeta}(\hat{A}_{32}\cos\varphi_\kappa\zeta + i\hat{A}_{33}\sin\varphi_\kappa\zeta) \\ &\quad\times e^{-i\lambda_1\lambda_3\theta}(\hat{A}_{42}\cos\varphi_\kappa\theta + i\hat{A}_{43}\sin\varphi_\kappa\theta) \\ &= e^{-i\lambda_1\lambda_3\theta}(\hat{A}_{32}\hat{A}_{42}\cos\varphi_\kappa\theta + i\hat{A}_{32}\hat{A}_{43}\sin\varphi_\kappa\theta)e^{i\lambda_1\zeta}\cos\varphi_\kappa\zeta \\ &\quad - e^{-i\lambda_1\lambda_3\theta}(\hat{A}_{33}\hat{A}_{43}\sin\varphi_\kappa\theta - i\hat{A}_{33}\hat{A}_{42}\cos\varphi_\kappa\theta)e^{i\lambda_1\zeta}\sin\varphi_\kappa\zeta\end{aligned}$$

In the spirit of the method of the variation of the constant we make the constants $\hat{A}_{42}$ and $\hat{A}_{43}$ functions of κ and write $\hat{v}_\kappa(\zeta,\theta)$ in a shorter form:

$$\hat{v}_\kappa(\zeta,\theta) = S_\kappa(\theta)e^{i\lambda_1\zeta}\cos\frac{2\pi\kappa\zeta}{N} + T_\kappa(\theta)e^{i\lambda_1\zeta}\sin\frac{2\pi\kappa\zeta}{N} \tag{20}$$

For a more general solution[1] $\hat{v}(\zeta,\theta)$ we may sum over all values of κ. To shorten a number of formulas we use the notation $\phi_c(\zeta)$ and $\phi_s(\zeta)$ for $e^{i\lambda_1\zeta}\cos(2\pi\kappa\zeta/N)$ and $e^{i\lambda_1\zeta}\sin(2\pi\kappa\zeta/N)$:

$$\begin{aligned}\hat{v}(\zeta,\theta) &= \sum_{\kappa>-\kappa_0}^{<\kappa_0}\left(S_\kappa(\theta)e^{i\lambda_1\zeta}\cos\frac{2\pi\kappa\zeta}{N} + T_\kappa(\theta)e^{i\lambda_1\zeta}\sin\frac{2\pi\kappa\zeta}{N}\right) \\ &= \sum_{\kappa=-N/2+1}^{N/2-1}[S_\kappa(\theta)\phi_c(\zeta) + T_\kappa(\theta)\phi_s(\zeta)]\end{aligned} \tag{21}$$

---

[1]See Smirnov (1961), vol. II, § 169 for the use of this method for partial differential equations.

Substitution of $\hat{v}(\zeta,\theta)$ for $\hat{u}(\zeta,\theta)$ into Eq.(14) yields:

$$\sum_{\kappa=-N/2+1}^{N/2-1} \Big[S_\kappa(\theta)[\phi_c(\zeta+1)-2\phi_c(\zeta)+\phi_c(\zeta-1)]+T_\kappa(\theta)[\phi_s(\zeta+1)-2\phi_s(\zeta)+\phi_s(\zeta-1)]$$
$$-\phi_c(\zeta)[S_\kappa(\theta+1)-2S_\kappa(\theta)+S_\kappa(\theta-1)]-\phi_s(\zeta)[T_\kappa(\theta+1)-2T_\kappa(\theta)+T_\kappa(\theta-1)]$$
$$-i\lambda_1\Big(S_\kappa(\theta)[\phi_c(\zeta+1)-\phi_c(\zeta-1)]+T_\kappa(\theta)[\phi_s(\zeta+1)-\phi_s(\zeta-1)]$$
$$+\lambda_3\{\phi_c(\zeta)[S_\kappa(\theta+1)-S_\kappa(\theta-1)]+\phi_s(\zeta)[T_\kappa(\theta+1)-T_\kappa(\theta-1)]\}\Big)$$
$$-\lambda_2^2[S_\kappa(\theta)\phi_c(\zeta)+T_\kappa(\theta)\phi_s(\zeta)]\Big] = G_1(\zeta,\theta) \quad (22)$$

With

$$\phi_c(\zeta\pm 1) = e^{\pm i\lambda_1} e^{i\lambda_1\zeta}\left(\cos\frac{2\pi\kappa}{N}\cos\frac{2\pi\kappa\zeta}{N} \mp \sin\frac{2\pi\kappa}{N}\sin\frac{2\pi\kappa\zeta}{N}\right) \quad (23)$$

$$\phi_s(\zeta\pm 1) = e^{\pm i\lambda_1} e^{i\lambda_1\zeta}\left(\cos\frac{2\pi\kappa}{N}\sin\frac{2\pi\kappa\zeta}{N} \pm \sin\frac{2\pi\kappa}{N}\cos\frac{2\pi\kappa\zeta}{N}\right) \quad (24)$$

we get

$$\phi_c(\zeta+1)-\phi_c(\zeta-1) = 2e^{i\lambda_1\zeta}\Big(-\cos\lambda_1\sin\frac{2\pi\kappa}{N}\sin\frac{2\pi\kappa\zeta}{N}$$
$$+i\sin\lambda_1\cos\frac{2\pi\kappa}{N}\cos\frac{2\pi\kappa\zeta}{N}\Big) \quad (25)$$

$$\phi_s(\zeta+1)-\phi_s(\zeta-1) = 2e^{i\lambda_1\zeta}\Big(\cos\lambda_1\sin\frac{2\pi\kappa}{N}\cos\frac{2\pi\kappa\zeta}{N}$$
$$+i\sin\lambda_1\cos\frac{2\pi\kappa}{N}\sin\frac{2\pi\kappa\zeta}{N}\Big) \quad (26)$$

$$\phi_c(\zeta+1)-2\phi_c(\zeta)+\phi_c(\zeta-1) = 2e^{i\lambda_1\zeta}\Big[\Big(\cos\lambda_1\cos\frac{2\pi\kappa}{N}-1\Big)\cos\frac{2\pi\kappa\zeta}{N}$$
$$-i\sin\lambda_1\sin\frac{2\pi\kappa}{N}\sin\frac{2\pi\kappa\zeta}{N}\Big] \quad (27)$$

$$\phi_s(\zeta+1)-2\phi_s(\zeta)+\phi_s(\zeta-1)=2e^{i\lambda_1\zeta}\Bigg[\left(\cos\lambda_1\cos\frac{2\pi\kappa}{N}-1\right)\sin\frac{2\pi\kappa\zeta}{N} + i\sin\lambda_1\sin\frac{2\pi\kappa}{N}\cos\frac{2\pi\kappa\zeta}{N}\Bigg] \quad (28)$$

Equation (22) assumes the following form:

$$\begin{aligned}
&\sum_{\kappa=-N/2+1}^{N/2-1}\\
&\Bigg\{2S_\kappa(\theta)e^{i\lambda_1\zeta}\Bigg[\left(\cos\lambda_1\cos\frac{2\pi\kappa}{N}-1\right)\cos\frac{2\pi\kappa\zeta}{N}-i\sin\lambda_1\sin\frac{2\pi\kappa}{N}\sin\frac{2\pi\kappa\zeta}{N}\Bigg]\\
&+2T_\kappa(\theta)e^{i\lambda_1\zeta}\Bigg[\left(\cos\lambda_1\cos\frac{2\pi\kappa}{N}-1\right)\sin\frac{2\pi\kappa\zeta}{N}+i\sin\lambda_1\sin\frac{2\pi\kappa}{N}\cos\frac{2\pi\kappa\zeta}{N}\Bigg]\\
&\qquad -e^{i\lambda_1\zeta}\cos\frac{2\pi\kappa\zeta}{N}[S_\kappa(\theta+1)-2S_\kappa(\theta)+S_\kappa(\theta-1)]\\
&\qquad -e^{i\lambda_1\zeta}\sin\frac{2\pi\kappa\zeta}{N}[T_\kappa(\theta+1)-2T_\kappa(\theta)+T_\kappa(\theta-1)]\\
&-i\lambda_1\Bigg[2S_\kappa(\theta)e^{i\lambda_1\zeta}\left(-\cos\lambda_1\sin\frac{2\pi\kappa}{N}\sin\frac{2\pi\kappa\zeta}{N}+i\sin\lambda_1\cos\frac{2\pi\kappa}{N}\cos\frac{2\pi\kappa\zeta}{N}\right)\\
&\quad +2T_\kappa(\theta)e^{i\lambda_1\zeta}\left(\cos\lambda_1\sin\frac{2\pi\kappa}{N}\cos\frac{2\pi\kappa\zeta}{N}+i\sin\lambda_1\cos\frac{2\pi\kappa}{N}\sin\frac{2\pi\kappa\zeta}{N}\right)\\
&+\lambda_3\left(e^{i\lambda_1\zeta}\cos\frac{2\pi\kappa\zeta}{N}[S_\kappa(\theta+1)-S_\kappa(\theta-1)]+e^{i\lambda_1\zeta}\sin\frac{2\pi\kappa\zeta}{N}[T_\kappa(\theta+1)-T_\kappa(\theta-1)]\right)\Bigg]\\
&\quad -\lambda_2^2\left(S_\kappa(\theta)e^{i\lambda_1\zeta}\cos\frac{2\pi\kappa\zeta}{N}+T_\kappa(\theta)e^{i\lambda_1\zeta}\sin\frac{2\pi\kappa\zeta}{N}\right)\Bigg\}=G_1(\zeta,\theta) \quad (29)
\end{aligned}$$

This equation may be reduced to the following form:

$$\begin{aligned}
&\sum_{\kappa=-N/2+1}^{N/2-1}\Bigg[-(1+i\lambda_1\lambda_3)S_\kappa(\theta+1)+\left(2(\cos\lambda_1+\lambda_1\sin\lambda_1)\cos\frac{2\pi\kappa}{N}-\lambda_2^2\right)S_\kappa(\theta)\\
&-(1-i\lambda_1\lambda_3)S_\kappa(\theta-1)+2i(\sin\lambda_1-\lambda_1\cos\lambda_1)\sin\frac{2\pi\kappa}{N}T_\kappa(\theta)\Bigg]e^{i\lambda_1\zeta}\cos\frac{2\pi\kappa\zeta}{N}\\
&+\sum_{\kappa=-N/2+1}^{N/2-1}\Bigg[-(1+i\lambda_1\lambda_3)T_\kappa(\theta+1)+\left(2(\cos\lambda_1+\lambda_1\sin\lambda_1)\cos\frac{2\pi\kappa}{N}-\lambda_2^2\right)T_\kappa(\theta)\\
&-(1-i\lambda_1\lambda_3)T_\kappa(\theta-1)-2i(\sin\lambda_1-\lambda_1\cos\lambda_1)\sin\frac{2\pi\kappa}{N}S_\kappa(\theta)\Bigg]e^{i\lambda_1\zeta}\sin\frac{2\pi\kappa\zeta}{N}\\
&=G_1(\zeta,\theta) \quad (30)
\end{aligned}$$

The left side of Eq.(30) is essentially a Fourier series in terms of ζ. The right side may be represented in the same form:

$$G_1(\zeta,\theta) = \sum_{\kappa=-N/2+1}^{N/2-1} \left(G_{s\kappa}(\theta,\kappa) e^{i\lambda_1\zeta} \sin\frac{2\pi\kappa\zeta}{N} + G_{c\kappa}(\theta,\kappa) e^{i\lambda_1\zeta} \cos\frac{2\pi\kappa\zeta}{N} \right) \quad (31)$$

Multiplication with $2N^{-1}e^{-i\lambda_1\zeta}\sin(2\pi\nu\zeta/N)$ or $2N^{-1}e^{-i\lambda_1\zeta}\cos(2\pi\nu\zeta/N)$ and integration over the orthogonality interval $0 \le \zeta \le N$ yields $G_{s\kappa}(\theta,\kappa)$ and $G_{c\kappa}(\theta,\kappa)$:

$$G_{c\kappa}(\theta,\kappa) = \frac{2}{N}\int_0^N G_1(\zeta,\theta) e^{-i\lambda_1\zeta} \cos\frac{2\pi\kappa\zeta}{N} d\zeta \quad \text{for } \nu = \kappa \quad (32)$$

$$G_{s\kappa}(\theta,\kappa) = \frac{2}{N}\int_0^N G_1(\zeta,\theta) e^{-i\lambda_1\zeta} \sin\frac{2\pi\kappa\zeta}{N} d\zeta \quad \text{for } \nu = \kappa \quad (33)$$

Equations (30) and (31) yield for every component κ of the two sums the following two equations:

$$\begin{aligned} &-(1+i\lambda_1\lambda_3)S_\kappa(\theta+1) + \left(2(\cos\lambda_1 + \lambda_1\sin\lambda_1)\cos\frac{2\pi\kappa}{N} - \lambda_2^2\right)S_\kappa(\theta) \\ &\quad -(1-i\lambda_1\lambda_3)S_\kappa(\theta-1) - 2i(\sin\lambda_1 - \lambda_1\cos\lambda_1)\sin\frac{2\pi\kappa}{N}T_\kappa(\theta) = G_{c\kappa}(\theta,\kappa) \end{aligned} \quad (34)$$

$$\begin{aligned} &-(1+i\lambda_1\lambda_3)T_\kappa(\theta+1) + \left(2(\cos\lambda_1 + \lambda_1\sin\lambda_1)\cos\frac{2\pi\kappa}{N} - \lambda_2^2\right)T_\kappa(\theta) \\ &\quad -(1-i\lambda_1\lambda_3)T_\kappa(\theta-1) - 2i(\sin\lambda_1 - \lambda_1\cos\lambda_1)\sin\frac{2\pi\kappa}{N}S_\kappa(\theta) = G_{s\kappa}(\theta,\kappa) \end{aligned} \quad (35)$$

The following substitutions and simplifications are made:

$$\begin{aligned} \lambda_1 &= ecTA_{m0x}/N\hbar,\ \cos\lambda_1 = 1 + O(\Delta t)^2,\ \lambda_1\sin\lambda_1 = O(\Delta t)^2 \\ \lambda_2^2 &= (T/N\hbar)^2(m_0^2c^4 - e\phi_{e0}^2 + e^2c^2A_{m0x}^2) = O(\Delta t)^2 \quad (36) \\ \lambda_1^2\lambda_3^2 &= (eT\phi_{e0}/N\hbar)^2 = O(\Delta t)^2, \quad T = N\Delta t \quad (37) \\ 1 + i\lambda_1\lambda_3 &\doteq e^{i\lambda_1\lambda_3},\ 1 - i\lambda_1\lambda_3 \doteq e^{-i\lambda_1\lambda_3}, \quad (38) \end{aligned}$$

$$2(\cos\lambda_1 + \lambda_1 \sin\lambda_1)\cos\frac{2\pi\kappa}{N} - \lambda_2^2 \doteq 2\cos\frac{2\pi\kappa}{N} \tag{39}$$

$$2(\sin\lambda_1 - \lambda_1\cos\lambda_1)\sin\frac{2\pi\kappa}{N} = O(\Delta t) \tag{40}$$

The small value of Eq.(40) tempts one to leave out the fourth term with $T_\kappa(\theta)$ in Eq.(34) and with $S_\kappa(\theta)$ in Eq.(35). This is not justified mathematically since we do not know how large the sum of the remaining four terms—including the inhomogeneous term—is. From the standpoint of physics one would eliminate the coupling of the two processes represented by Eqs.(34) and (35).

We may solve Eq.(34) for $T_\kappa(\theta)$ and substitute $T_\kappa(\theta)$, $T_\kappa(\theta\pm1)$ into Eq.(35) to obtain an ordinary, inhomogeneous difference equation for $S_\kappa(\theta)$:

$$\begin{aligned}
&e^{2i\lambda_1\lambda_3}S_\kappa(\theta+2) - 4\cos\frac{2\pi\kappa}{N}e^{i\lambda_1\lambda_3}S_\kappa(\theta+1) + 2\left(2+\cos\frac{4\pi\kappa}{N}\right)S_\kappa(\theta)\\
&\qquad - 4\cos\frac{2\pi\kappa}{N}e^{-i\lambda_1\lambda_3}S_\kappa(\theta-1) + e^{-2i\lambda_1\lambda_3}S_\kappa(\theta-2)\\
&= -e^{i\lambda_1\lambda_3}G_{c\kappa}(\theta+1,\kappa) + 2\cos\frac{2\pi\kappa}{N}G_{c\kappa}(\theta,\kappa) - e^{-i\lambda_1\lambda_3}G_{c\kappa}(\theta-1,\kappa)
\end{aligned} \tag{41}$$

Similarly, we may solve Eq.(35) for $S_\kappa(\theta)$ and substitute $S_\kappa(\theta)$, $S_\kappa(\theta\pm1)$ into Eq.(34) to obtain an ordinary, inhomogeneous difference equation for $T_\kappa(\theta)$:

$$\begin{aligned}
&e^{2i\lambda_1\lambda_3}T_\kappa(\theta+2) - 4\cos\frac{2\pi\kappa}{N}e^{i\lambda_1\lambda_3}T_\kappa(\theta+1) + 2\left(2+\cos\frac{4\pi\kappa}{N}\right)T_\kappa(\theta)\\
&\qquad - 4\cos\frac{2\pi\kappa}{N}e^{-i\lambda_1\lambda_3}T_\kappa(\theta-1) + e^{-2i\lambda_1\lambda_3}T_\kappa(\theta-2)\\
&= -e^{i\lambda_1\lambda_3}G_{s\kappa}(\theta+1,\kappa) + 2\cos\frac{2\pi\kappa}{N}G_{s\kappa}(\theta,\kappa) - e^{-i\lambda_1\lambda_3}G_{s\kappa}(\theta-1,\kappa)
\end{aligned} \tag{42}$$

The homogeneous parts of Eqs.(41) and (42) are equal; their only differences are the subscripts c and s of the inhomogeneous terms.

For the solution of the homogeneous Eqs.(41) and (42) we make the usual ansatz

$$S_\kappa(\theta) = c_\kappa v_\kappa^\theta \quad \text{or} \quad T_\kappa(\theta) = d_\kappa v_\kappa^\theta \tag{43}$$

and obtain an equation for v_κ:

$$\begin{aligned}
&e^{2i\lambda_1\lambda_3}v_\kappa^2 - 4\cos\frac{2\pi\kappa}{N}e^{i\lambda_1\lambda_3}v_\kappa + 2\left(2+\cos\frac{4\pi\kappa}{N}\right)\\
&\qquad\qquad - 4\cos\frac{2\pi\kappa}{N}e^{-i\lambda_1\lambda_3}v_\kappa^{-1} + e^{-2i\lambda_1\lambda_3}v_\kappa^{-2} = 0
\end{aligned} \tag{44}$$

This is a fourth-order equation. Because of its symmetry it can be reduced to two equations of second order:

$$(e^{i\lambda_1\lambda_3}v_\kappa + e^{-i\lambda_1\lambda_3}v_\kappa^{-1})^2 - 4\cos\frac{2\pi\kappa}{N}(e^{i\lambda_1\lambda_3}v_\kappa + e^{-i\lambda_1\lambda_3}v_\kappa^{-1}) + 2\left(1+\cos\frac{4\pi\kappa}{N}\right) = 0 \quad (45)$$

$$\begin{aligned}(e^{i\lambda_1\lambda_3}v_\kappa + e^{-i\lambda_1\lambda_3}v_\kappa^{-1}) &= 2\cos\frac{2\pi\kappa}{N} \pm \left[4\cos^2\frac{2\pi\kappa}{N} - 2\left(1+\cos\frac{4\pi\kappa}{N}\right)\right]^{1/2} \\ &= 2\cos\frac{2\pi\kappa}{N} \qquad \text{double root} \end{aligned} \quad (46)$$

The second quadratic equation

$$\begin{aligned} e^{i\lambda_1\lambda_3}v_\kappa + e^{-i\lambda_1\lambda_3}v_\kappa^{-1} - 2\cos\frac{2\pi\kappa}{N} &= 0 \\ e^{2i\lambda_1\lambda_3}v_\kappa^2 - 2\cos\frac{2\pi\kappa}{N}e^{i\lambda_1\lambda_3}v_\kappa + 1 &= 0 \end{aligned} \quad (47)$$

yields two single roots

$$e^{i\lambda_1\lambda_3}v_\kappa = \cos\frac{2\pi\kappa}{N} \pm \left(\cos^2\frac{2\pi\kappa}{N} - 1\right)^{1/2}$$

$$v_{\kappa 1} = e^{-i\lambda_1\lambda_3}\left(\cos\frac{2\pi\kappa}{N} + i\sin\frac{2\pi\kappa}{N}\right) = e^{-i\lambda_1\lambda_3}e^{2\pi i\kappa/N} \quad (48)$$

$$v_{\kappa 2} = e^{-i\lambda_1\lambda_3}\left(\cos\frac{2\pi\kappa}{N} - i\sin\frac{2\pi\kappa}{N}\right) = e^{-i\lambda_1\lambda_3}e^{-2\pi i\kappa/N} \quad (49)$$

and we obtain for v_κ^θ in Eq.(43) the first two solutions:

$$v_{\kappa 1}^\theta = e^{i(2\pi\kappa/N - \lambda_1\lambda_3)\theta} = \cos\left(\frac{2\pi\kappa}{N} - \lambda_1\lambda_3\right)\theta + i\sin\left(\frac{2\pi\kappa}{N} - \lambda_1\lambda_3\right)\theta \quad (50)$$

$$v_{\kappa 2}^\theta = e^{-i(2\pi\kappa/N + \lambda_1\lambda_3)\theta} = \cos\left(\frac{2\pi\kappa}{N} + \lambda_1\lambda_3\right)\theta - i\sin\left(\frac{2\pi\kappa}{N} + \lambda_1\lambda_3\right)\theta \quad (51)$$

The double root of Eq.(46) calls for two more solutions:

$$\begin{aligned}
\theta v_{\kappa 1}^{\theta} &= \theta e^{i(2\pi\kappa/N-\lambda_1\lambda_3)\theta} \\
&= \theta\left[\cos\left(\frac{2\pi\kappa}{N}-\lambda_1\lambda_3\right)\theta + i\sin\left(\frac{2\pi\kappa}{N}-\lambda_1\lambda_3\right)\theta\right] \qquad (52)\\
\theta v_{\kappa 2}^{\theta} &= \theta e^{-i(2\pi\kappa/N+\lambda_1\lambda_3)\theta} \\
&= \theta\left[\cos\left(\frac{2\pi\kappa}{N}+\lambda_1\lambda_3\right)\theta - i\sin\left(\frac{2\pi\kappa}{N}+\lambda_1\lambda_3\right)\theta\right] \qquad (53)
\end{aligned}$$

The general solution $S_\kappa(\theta)$ in Eq.(43) becomes:

$$\begin{aligned}
S_\kappa(\theta) &= c_{\kappa 1}v_{\kappa 1}^{\theta} + c_{\kappa 2}v_{\kappa 2}^{\theta} + c_{\kappa 3}\theta v_{\kappa 1}^{\theta} + c_{\kappa 4}\theta v_{\kappa 2}^{\theta} \\
&= c_{\kappa 1}e^{i(2\pi\kappa/N-\lambda_1\lambda_3)\theta} + c_{\kappa 2}e^{-i(2\pi\kappa/N+\lambda_1\lambda_3)\theta} \\
&\quad + c_{\kappa 3}\theta e^{i(2\pi\kappa/N-\lambda_1\lambda_3)\theta} + c_{\kappa 4}\theta e^{-i(2\pi\kappa/N+\lambda_1\lambda_3)\theta} \qquad (54)
\end{aligned}$$

For $T_\kappa(\theta)$ we obtain from Eq.(43):

$$\begin{aligned}
T_\kappa(\theta) &= d_{\kappa 1}v_{\kappa 1}^{\theta} + d_{\kappa 2}v_{\kappa 2}^{\theta} + d_{\kappa 3}\theta v_{\kappa 1}^{\theta} + d_{\kappa 4}\theta v_{\kappa 2}^{\theta} \\
&= d_{\kappa 1}e^{i(2\pi\kappa/N-\lambda_1\lambda_3)\theta} + d_{\kappa 2}e^{-i(2\pi\kappa/N+\lambda_1\lambda_3)\theta} \\
&\quad + d_{\kappa 3}\theta e^{i(2\pi\kappa/N-\lambda_1\lambda_3)\theta} + d_{\kappa 4}\theta e^{-i(2\pi\kappa/N+\lambda_1\lambda_3)\theta} \qquad (55)
\end{aligned}$$

We turn to the inhomogeneous solution of Eqs.(41) and (42). A general method for this task goes back to Lagrange (1736–1813). It is discussed in the available books[1] but we present it in Section 6.3 in more detail. The constants $c_{\kappa i}$ and $d_{\kappa i}$, with $i = 1, 2, 3, 4$, in Eqs.(54) and (55) are replaced by variables $c_{\kappa i}(\theta)$ and $d_{\kappa i}(\theta)$:

$$S_\kappa(\theta) = c_{\kappa 1}(\theta)v_{\kappa 1}^{\theta} + c_{\kappa 2}(\theta)v_{\kappa 2}^{\theta} + c_{\kappa 3}(\theta)\theta v_{\kappa 1}^{\theta} + c_{\kappa 4}(\theta)\theta v_{\kappa 2}^{\theta} \qquad (56)$$

$$T_\kappa(\theta) = d_{\kappa 1}(\theta)v_{\kappa 1}^{\theta} + d_{\kappa 2}(\theta)v_{\kappa 2}^{\theta} + d_{\kappa 3}(\theta)\theta v_{\kappa 1}^{\theta} + d_{\kappa 4}(\theta)\theta v_{\kappa 2}^{\theta} \qquad (57)$$

Equations (6.3-23) and (6.3-30) define $d_{\kappa i}(\theta)$ and $c_{\kappa i}(\theta)$ in a form suited for computer use.

The solution derived here for Eq.(3) is extremely general. The only significant requirement is that the function $G_1(\zeta,\theta)$ of Eq.(4) permits a cosine and sine transformation in terms of ζ according to Eqs.(32) and (33). Convergence of the Fourier series of Eq.(31) is not required since κ does not approach infinity but only $\pm(N/2-1)$.

[1] Nörlund 1924, p. 396; 1929, p. 22, 125; Milne-Thomson 1951, p. 374.

Substitution of Eqs.(56) and (57) into Eq.(21) yields the particular solution of the inhomogeneous equation (14). We must still satisfy the boundary and initial conditions of Eqs.(11) to (13). Using Eqs.(2.3-19) to (2.3-21) and Eq.(15) we obtain the following boundary and initial conditions for $\hat{v}(\zeta,\theta)$ from Eqs.(11) to (13):

$$\hat{v}(0,\theta) = 0 \quad \text{for } \theta \geq 0 \tag{58}$$

$$\hat{v}(\zeta,0) = 0 \quad \text{for } \zeta > 0 \tag{59}$$

$$\hat{v}(\zeta,1) - \hat{v}(\zeta,0) = 0 \quad \text{for } \zeta > 0 \tag{60}$$

The boundary condition of Eq.(58) is satisfied if we discard the first term in Eq.(21):

$$\hat{v}(\zeta,\theta) = \sum_{\kappa=-N/2+1}^{N/2-1} T_\kappa(\theta) e^{i\lambda_1\zeta} \sin\frac{2\pi\kappa\zeta}{N} \tag{61}$$

The initial condition of Eq.(59) is satisfied for the third and fourth term of $T_\kappa(\theta)$ in Eq.(57) due to the factor θ. The first and second term have the coefficients $d_{\kappa i}(0)$ for $\theta = 0$:

$$\hat{v}(\zeta,0) = \sum_{\kappa=-N/2+1}^{N/2-1} [d_{\kappa 1}(0) + d_{\kappa 2}(0)] e^{i\lambda_1\zeta} \sin\frac{2\pi\kappa\zeta}{N} = 0 \tag{62}$$

From Eqs.(57) and (61) we get with Eqs.(48), (49), (60), and (59) the relation

$$\begin{aligned}\hat{v}(\zeta,1) = &\sum_{\kappa=-N/2+1}^{N/2-1} \{[d_{\kappa 1}(1) + d_{\kappa 3}(1)]v_{\kappa 1} + [d_{\kappa 2}(1) + d_{\kappa 4}(1)]v_{\kappa 2}\} \\ &\times e^{i\lambda_1\zeta} \sin\frac{2\pi\kappa\zeta}{N} = 0 \\ = e^{-i\lambda_1\lambda_3} &\sum_{\kappa=-N/2+1}^{N/2-1} \Big([d_{\kappa 1}(1) + d_{\kappa 2}(1) + d_{\kappa 3}(1) + d_{\kappa 4}(1)] \cos\frac{2\pi\kappa}{N} \\ &+ i[d_{\kappa 1}(1) - d_{\kappa 2}(1) + d_{\kappa 3}(1) - d_{\kappa 4}(1)] \sin\frac{2\pi\kappa}{N}\Big) e^{i\lambda_1\zeta} \sin\frac{2\pi\kappa\zeta}{N} = 0\end{aligned} \tag{63}$$

We multiply Eq.(62) with $N^{-1}e^{-i\lambda_1\zeta}\sin(2\pi\nu\zeta/N)$ and integrate over the orthogonality interval $0 \leq \zeta \leq N$:

$$d_{\kappa 1}(0) + d_{\kappa 2}(0) = 0 \qquad \text{for } \nu = \kappa \tag{64}$$

Equation (63) is multiplied with $N^{-1}e^{-i\lambda_1\zeta}\sin(2\pi\nu\zeta/N)$ and integrated over the interval $0 \leq \zeta \leq N$. This time we obtain two equations:

$$d_{\kappa 1}(1) + d_{\kappa 2}(1) + d_{\kappa 3}(1) + d_{\kappa 4}(1) = 0 \quad \text{for } \nu = \kappa \tag{65}$$
$$d_{\kappa 1}(1) - d_{\kappa 2}(1) + d_{\kappa 3}(1) - d_{\kappa 4}(1) = 0 \tag{66}$$

We get from Eqs.(64) to (66):

$$d_{\kappa 1}(0) = -d_{\kappa 2}(0) \tag{67}$$
$$d_{\kappa 1}(1) = -d_{\kappa 3}(1) \tag{68}$$
$$d_{\kappa 2}(1) = -d_{\kappa 4}(1) \tag{69}$$

Equation (6.3-24) yields:

$$d_{\kappa 1}(1) = d_{\kappa 1}(1) - d_{\kappa 1}(0) + d_{\kappa 1}, \quad d_{\kappa 1}(0) = d_{\kappa 1} \tag{70}$$
$$d_{\kappa 2}(1) = d_{\kappa 2}(1) - d_{\kappa 2}(0) + d_{\kappa 2}, \quad d_{\kappa 2}(0) = d_{\kappa 2} \tag{71}$$

From Eqs.(67), (70), and (71) we get

$$d_{\kappa 2} = -d_{\kappa 1} \tag{72}$$

There is no corresponding link between $d_{\kappa 3}$ and $d_{\kappa 4}$ due to the factor θ in the third and fourth term of Eq.(57).

From Eqs.(6.3-23) and (6.3-22) as well as Eqs.(6.3-17) to (6.3-21) we obtain the following relations for $d_{\kappa 1}(1)$ and $d_{\kappa 3}(1)$:

$$d_{\kappa 1}(1) = \Delta d_{\kappa 1}(0) + d_{\kappa 1} = \frac{D_{\kappa 1}(1)}{D_{\kappa 0}(1)} + d_{\kappa 1} \tag{73}$$
$$d_{\kappa 3}(1) = \Delta d_{\kappa 3}(0) + d_{\kappa 3} = \frac{D_{\kappa 3}(1)}{D_{\kappa 0}(1)} + d_{\kappa 3} \tag{74}$$

Substitution into Eq.(68) yields:

$$d_{\kappa 3} = -d_{\kappa 1} - \frac{D_{\kappa 1}(1) + D_{\kappa 3}(1)}{D_{\kappa 0}(1)} \tag{75}$$

Similarly we obtain for $d_{\kappa 2}(1)$ and $d_{\kappa 4}(1)$ of Eq.(69):

$$d_{\kappa 2}(1) = \Delta d_{\kappa 2}(0) + d_{\kappa 2} = \frac{D_{\kappa 2}(1)}{D_{\kappa 0}(1)} + d_{\kappa 2} \tag{76}$$

$$d_{\kappa 4}(1) = \Delta d_{\kappa 4}(0) + d_{\kappa 4} = \frac{D_{\kappa 4}(1)}{D_{\kappa 0}(1)} + d_{\kappa 4} \tag{77}$$

$$d_{\kappa 4} = -d_{\kappa 2} - \frac{D_{\kappa 2}(1) + D_{\kappa 4}(1)}{D_{\kappa 0}(1)} \tag{78}$$

Substitution of Eq.(72) brings:

$$d_{\kappa 4} = d_{\kappa 1} - \frac{D_{\kappa 2}(1) + D_{\kappa 4}(1)}{D_{\kappa 0}(1)} \tag{79}$$

The constants $d_{\kappa 4}$, $d_{\kappa 3}$, and $d_{\kappa 2}$ are now expressed in terms of $d_{\kappa 1}$. The sums $d_{\kappa 1} + d_{\kappa 3}$ and $d_{\kappa 2} + d_{\kappa 4}$ are further developed from Eq.(6.3-53) on in Section 6.3, after the evaluation of the determinants $D_{\kappa 1}(\theta)$–$D_{\kappa 4}(\theta)$ and $D_{\kappa 0}(\theta)$.

We may now write the solution for $\Psi_1(\zeta,\theta)$ of Eq.(3). Starting with Eqs.(8), (9), (10), and (15) we obtain the following expressions:

$$\Psi_1(\zeta,\theta) = \Psi_{00}\Big\{(1 - e^{-2i\lambda_1\lambda_3\theta})\exp[-(\lambda_2^2 - \lambda_1^2)^{1/2}\zeta]e^{i\lambda_1\zeta} + u(\zeta,\theta) + \hat{v}(\zeta,\theta)\Big\} \tag{80}$$

$$u(\zeta,\theta) = -2i\lambda_1\lambda_3 e^{i\lambda_1(\zeta+\lambda_3\theta)} \sum_{\kappa=-N/2+1}^{N/2-1} \frac{I_{\mathrm{T}}(\kappa/N)}{\sin\beta_\kappa}\sin\beta_\kappa\theta\sin\frac{2\pi\kappa\zeta}{N} \tag{81}$$

$I_{\mathrm{T}}(\kappa/N)$ see Eq.(2.4-29); λ_1, λ_2, λ_3 see Eqs.(36), (37)

$$\hat{v}(\zeta,\theta) = \sum_{\kappa=-N/2+1}^{N/2-1} T_\kappa(\theta)e^{i\lambda_1\zeta}\sin\frac{2\pi\kappa\zeta}{N} \tag{82}$$

$$T_\kappa(\theta) = [d_{\kappa 1}(\theta) + \theta d_{\kappa 3}(\theta)]v_{\kappa 1}^\theta + [d_{\kappa 2}(\theta) + \theta d_{\kappa 4}(\theta)]v_{\kappa 3}^\theta \tag{83}$$

$v_{\kappa 1}$, $v_{\kappa 2}$ see Eqs.(48), (49); $d_{\kappa i}(\theta)$ see Eqs.(6.3-23), (6.3-47)–(6.3-50)

$d_{\kappa 2}$, $d_{\kappa 3}$, $d_{\kappa 4}$ see Eqs.(72), (75), (79)

Comparison of Eq.(80) with Eq.(2.4-32) shows that $\Psi_1(\zeta,\theta)$ can also be written in the following form:

$$\Psi_1(\zeta,\theta) = \Psi_0(\zeta,\theta) + \Psi_{00}\hat{v}(\zeta,\theta)$$

$$\hat{v}(\zeta,\theta) = \sum_{\kappa=-N/2+1}^{N/2-1}\Big([d_{\kappa 1}(\theta) + \theta d_{\kappa 3}(\theta)]e^{i(2\pi\kappa/N - \lambda_1\lambda_3)\theta} + [d_{\kappa 2}(\theta) + \theta d_{\kappa 4}(\theta)]e^{-i(2\pi\kappa/N + \lambda_1\lambda_3)\theta}\Big)e^{i\lambda_1\zeta}\sin\frac{2\pi\kappa\zeta}{N} \tag{84}$$

3.3 Quantization of the Solution for $\Delta x \gg h/m_0c$

In Section 2.5 we derived the Hamilton function of $\Psi_0 = \Psi_{\mathrm{x}0x_j}$ according to Eqs.(2.1-37), (2.1-38) and carried out the quantization with Eq.(2.5-29) or (2.5-30). We follow the process used there, but with $\Psi_1 = \Psi_{\mathrm{x}1x_j}$ of Eq.(3.2-80) we carry the approximation to the next order in α. Using Eq.(2.1-38) we write

$$\Psi = \Psi_0 + \alpha\Psi_1 = \Psi_{\mathrm{x}0x_j} + \alpha\Psi_{\mathrm{x}1x_j} \tag{1}$$

Equations (2.5-1) to (2.5-3) remain unchanged but from Eq.(2.5-4) on we must replace $\Psi = \Psi_0$ with $\Psi = \Psi_0 + \alpha\Psi_1$:

$$\Psi^*\Psi = (\Psi_0^* + \alpha\Psi_1^*)(\Psi_0 + \alpha\Psi_1) = \Psi_0^*\Psi_0 + \alpha(\Psi_0^*\Psi_1 + \Psi_1^*\Psi_0) + O(\alpha^2) \tag{2}$$

$$\frac{\partial\Psi^*}{\partial\theta}\frac{\partial\Psi}{\partial\theta} = \frac{\partial\Psi_0^*}{\partial\theta}\frac{\partial\Psi_0}{\partial\theta} + \alpha\left(\frac{\partial\Psi_0^*}{\partial\theta}\frac{\partial\Psi_1}{\partial\theta} + \frac{\partial\Psi_1^*}{\partial\theta}\frac{\partial\Psi_0}{\partial\theta}\right) + O(\alpha^2) \tag{3}$$

$$\frac{\partial\Psi^*}{\partial\zeta}\frac{\partial\Psi}{\partial\zeta} = \frac{\partial\Psi_0^*}{\partial\zeta}\frac{\partial\Psi_0}{\partial\zeta} + \alpha\left(\frac{\partial\Psi_0^*}{\partial\zeta}\frac{\partial\Psi_1}{\partial\zeta} + \frac{\partial\Psi_1^*}{\partial\zeta}\frac{\partial\Psi_0}{\partial\zeta}\right) + O(\alpha^2) \tag{4}$$

The function $\Psi_0(\zeta,\theta)$ is defined in Eq.(2.4-32) and used in Eq.(2.5-4):

$$\begin{aligned}\Psi_0(\zeta,\theta) &= \Psi_{00}[(1 - e^{-2i\lambda_1\lambda_3\theta}F(\zeta) + u(\zeta,\theta)]\\ &= \Psi_{00}\Bigg[(1 - e^{-2i\lambda_1\lambda_3\theta})\exp[-(\lambda_2^2 - \lambda_1^2)^{1/2}\zeta]e^{i\lambda_1\zeta}\\ &\quad -2i\lambda_1\lambda_3 e^{i\lambda_1\lambda_3}e^{i\lambda_1\zeta}e^{-i\lambda_1\lambda_3\theta}\sum_{\kappa=-N/2+1}^{N/2-1}\frac{I_{\mathrm{T}}(\kappa/N)}{\sin\beta_\kappa}\sin\beta_\kappa\theta\sin\frac{2\pi\kappa\zeta}{N}\Bigg]\end{aligned} \tag{5}$$

The function $\Psi_1(\zeta,\theta)$ is written in the form of Eq.(3.2-84):

$$\Psi_1(\zeta,\theta) = \Psi_0(\zeta,\theta) + \Psi_{00}\hat{v}(\zeta,\theta) \tag{6}$$

$$\begin{aligned}\hat{v}(\zeta,\theta) = e^{i\lambda_1(\zeta-\lambda_3\theta)}\sum_{\kappa=-N/2+1}^{N/2-1}&\{[d_{\kappa1}(\theta) + \theta d_{\kappa3}(\theta)]e^{2\pi i\kappa\theta/N}\\ &+ [d_{\kappa2}(\theta) + \theta d_{\kappa4}(\theta)]e^{-2\pi i\kappa\theta/N}\}\frac{\sin 2\pi\kappa\zeta}{N}\end{aligned} \tag{7}$$

With the help of Eq.(6) we may rewrite Eqs.(2)–(4) as follows, using the notation $\mathcal{R}e(\dots)$ for the real part of the expression in parentheses:

$$\begin{aligned}\Psi^*\Psi &= \Psi_0^*\Psi_0 + \alpha[2\Psi_0^*\Psi_0 + \Psi_{00}(\Psi_0^*\hat{v} + \hat{v}^*\Psi_0)]\\ &= \Psi_0^*\Psi_0 + 2\alpha[\Psi_0^*\Psi_0 + \Psi_{00}\mathcal{R}e(\Psi_0^*\hat{v})]\end{aligned} \tag{8}$$

$$\begin{aligned}\frac{\partial\Psi^*}{\partial\theta}\frac{\partial\Psi}{\partial\theta} &= \frac{\partial\Psi_0^*}{\partial\theta}\frac{\partial\Psi_0}{\partial\theta} + \alpha\left[2\frac{\partial\Psi_0^*}{\partial\theta}\frac{\partial\Psi_0}{\partial\theta} + \Psi_{00}\left(\frac{\partial\Psi_0^*}{\partial\theta}\frac{\partial\hat{v}}{\partial\theta} + \frac{\partial\hat{v}^*}{\partial\theta}\frac{\partial\Psi_0}{\partial\theta}\right)\right]\\ &= \frac{\partial\Psi_0^*}{\partial\theta}\frac{\partial\Psi_0}{\partial\theta} + 2\alpha\left[\frac{\partial\Psi_0^*}{\partial\theta}\frac{\partial\Psi_0}{\partial\theta} + \Psi_{00}\Re e\left(\frac{\partial\Psi_0^*}{\partial\theta}\frac{\partial\hat{v}}{\partial\theta}\right)\right] \end{aligned} \quad (9)$$

$$\begin{aligned}\frac{\partial\Psi^*}{\partial\zeta}\frac{\partial\Psi}{\partial\zeta} &= \frac{\partial\Psi_0^*}{\partial\zeta}\frac{\partial\Psi_0}{\partial\zeta} + \alpha\left[2\frac{\partial\Psi_0^*}{\partial\zeta}\frac{\partial\Psi_0}{\partial\zeta} + \Psi_{00}\left(\frac{\partial\Psi_0^*}{\partial\zeta}\frac{\partial\hat{v}}{\partial\zeta} + \frac{\partial\hat{v}^*}{\partial\zeta}\frac{\partial\Psi_0}{\partial\zeta}\right)\right]\\ &= \frac{\partial\Psi_0^*}{\partial\zeta}\frac{\partial\Psi_0}{\partial\zeta} + 2\alpha\left[\frac{\partial\Psi_0^*}{\partial\zeta}\frac{\partial\Psi_0}{\partial\zeta} + \Psi_{00}\Re e\left(\frac{\partial\Psi_0^*}{\partial\zeta}\frac{\partial\hat{v}}{\partial\zeta}\right)\right] \end{aligned} \quad (10)$$

The terms $\Psi_0^*\Psi_0$, $(\partial\Psi_0^*/\partial\theta)(\partial\Psi_0/\partial\theta)$, and $(\partial\Psi_0^*/\partial\zeta)(\partial\Psi_0/\partial\zeta)$ are shown in Eqs.(2.5-4), (2.5-9), and (2.5-11). We still must evaluate the additional terms from $\Psi_0^*\hat{v}$ to $(\partial\Psi_0^*/\partial\zeta)(\partial\hat{v}/\partial\zeta)$ in Eqs.(8)–(10) with the help of Eqs.(5) and (7). Since we are interested in the effect of the inhomogeneous term $G_1(\zeta,\theta)$ in Eq.(3.2-4) we must show more clearly how $\hat{v}(\zeta,\theta)$ depends on $G_1(\zeta,\theta)$. From Eqs.(6.3-1) and (3.2-33) we obtain:

$$H_{s\kappa}(\theta,\kappa) = \frac{2}{N}\int_0^N\left[-e^{i\lambda_1\lambda_3}G_1(\zeta,\theta+1) + 2\cos\frac{2\pi\kappa}{N}G_1(\zeta,\theta) - e^{-i\lambda_1\lambda_3}G_1(\zeta,\theta-1)\right]$$
$$\times\, e^{-i\lambda_1\zeta}\sin\frac{2\pi\kappa\zeta}{N}d\zeta, \quad \theta = 1,\ 2,\ \ldots,\ N-2 \quad (11)$$

The four functions $d_{\kappa1}(\theta)$ to $d_{\kappa4}(\theta)$ in Eq.(7) are defined by Eqs.(6.3-22) and (6.6-23). We use them in the form of Eqs.(6.3-59)–(6.3-62):

$$d_{\kappa1}(\theta) = \sum_{n=0}^{\theta-1}H_{s\kappa}(n,\kappa)\{nF_1(n,\kappa) + F_3(n,\kappa) + i[nF_5(n,\kappa) + F_7(n,\kappa)] + d_{\kappa1}\} \quad (12)$$

$$d_{\kappa2}(\theta) = \sum_{n=0}^{\theta-1}H_{s\kappa}(n,\kappa)\{nF_2(n,\kappa) + F_4(n,\kappa) + i[nF_6(n,\kappa) + F_8(n,\kappa)] + d_{\kappa2}\} \quad (13)$$

$$d_{\kappa3}(\theta) = -\sum_{n=0}^{\theta-1}H_{s\kappa}(n,\kappa)\{F_1(n,\kappa) + iF_5(n,\kappa) + d_{\kappa3}\} \quad (14)$$

$$d_{\kappa4}(\theta) = -\sum_{n=0}^{\theta-1}H_{s\kappa}(n,\kappa)\{F_2(n,\kappa) + iF_6(n,\kappa) + d_{\kappa4}\} \quad (15)$$

The functions $F_1(n,\kappa)$ to $F_8(n,\kappa)$ are defined by Eqs.(6.3-51)–(6.3-58). With the help of these equations we may rewrite the terms

$$[d_{\kappa1}(\theta) + \theta d_{\kappa3}(\theta)]e^{2\pi i\kappa\theta/N} + [d_{\kappa2}(\theta) + \theta d_{\kappa4}(\theta)]e^{-2\pi i\kappa\theta/N} \quad (16)$$

of $\hat{v}(\zeta,\theta)$ in Eq.(7) into the form of Eq.(6.3-75) and obtain the expression

$$\hat{v}(\zeta,\theta) = e^{i\lambda_1(\zeta-\lambda_3\theta)} \sum_{\kappa=-N/2+1}^{N/2-1} [J_{\rm r}(\theta,\kappa) + iJ_{\rm i}(\theta,\kappa)] \sin\frac{2\pi\kappa\zeta}{N}$$

$$J_{\rm r}(\theta,\kappa) = [J_1(\theta,\kappa) - \theta J_2(\theta,\kappa)]\cos\frac{2\pi\kappa\theta}{N} - [J_3(\theta,\kappa) - \theta J_4(\theta,\kappa)]\sin\frac{2\pi\kappa\theta}{N}$$

$$J_{\rm i}(\theta,\kappa) = [J_5(\theta,\kappa) - \theta J_6(\theta,\kappa)]\sin\frac{2\pi\kappa\theta}{N} + [J_7(\theta,\kappa) - \theta J_8(\theta,\kappa)]\cos\frac{2\pi\kappa\theta}{N} \quad (17)$$

The sums over n shown in Eqs.(12)–(15) have been separated in Eq.(6.3-75).

We obtain $\Psi_0^*(\zeta,\theta)$ from Eq.(5) by changing the sign of i. The product $\Psi_{00}\Psi_0^*(\zeta,\theta)\hat{v}(\zeta,\theta)$ becomes:

$$\Psi_{00}\Psi_0^*(\zeta,\theta)\hat{v}(\zeta,\theta) = \Psi_{00}^2 e^{-i\lambda_1\lambda_3\theta}\Bigg((1 - e^{2i\lambda_1\lambda_3\theta})\exp[-(\lambda_2^2 - \lambda_1^2)^{1/2}\zeta]$$
$$+ 2i\lambda_1\lambda_3 e^{i\lambda_1\lambda_3(\theta-1)} \sum_{\kappa=-N/2+1}^{N/2-1} \frac{I_{\rm T}(\kappa/N)}{\sin\beta_\kappa}\sin\beta_\kappa\theta\sin\frac{2\pi\kappa\zeta}{N}\Bigg)$$
$$\times \sum_{\kappa=-N/2+1}^{N/2-1} [J_{\rm r}(\theta,\kappa) + iJ_{\rm i}(\theta,\kappa)]\sin\frac{2\pi\kappa\zeta}{N} \quad (18)$$

We have a product of two sums. The terms $\sin(2\pi\kappa\zeta/N)$ are the important parts as will soon become clear. We write ν for κ in the second sum and separate the products $\sin(2\pi\kappa\zeta/N)\sin(2\pi\nu\zeta)/N$ for $\kappa=\nu$ and for $\kappa\neq\nu$:

$$\Psi_{00}\Psi_0^*(\zeta,\theta)\hat{v}(\zeta,\theta) = \Psi_{00}^2 e^{-i\lambda_1\lambda_3\theta}\Bigg[(1 - e^{2i\lambda_1\lambda_3\theta})$$
$$\times \sum_{\kappa=-N/2+1}^{N/2-1} [J_{\rm r}(\theta,\kappa) + iJ_{\rm i}(\theta,\kappa)]\exp[-(\lambda_2^2 - \lambda_1^2)^{1/2}\zeta]\sin\frac{2\pi\kappa\zeta}{N}$$
$$+ 2i\lambda_1\lambda_3 e^{i\lambda_1\lambda_3(\theta-1)}\Bigg(\sum_{\kappa=-N/2+1}^{N/2-1} \frac{I_{\rm T}(\kappa/N)}{\sin\beta_\kappa}\sin(\beta_\kappa\theta)[J_{\rm r}(\theta,\kappa)+iJ_{\rm i}(\theta,\kappa)]\sin^2\frac{2\pi\kappa\zeta}{N}$$
$$+ \sum_{\kappa=-N/2+1}^{N/2-1,\neq\nu} \sum_{\nu=-N/2+1}^{N/2-1} \frac{I_{\rm T}(\kappa/N)}{\sin\beta_\kappa}\sin\beta_\kappa\theta$$
$$\times [J_{\rm r}(\theta,\nu) + iJ_{\rm i}(\theta,\nu)]\sin\frac{2\pi\nu\zeta}{N}\sin\frac{2\pi\kappa\zeta}{N}\Bigg)\Bigg] \quad (19)$$

The integral over ζ of this expression taken from 0 to N will be needed according to Eqs.(2.5-3) and (8). Only the terms

$$\int_0^N \exp[-(\lambda_2^2-\lambda_1^2)^{1/2}\zeta]\sin\frac{2\pi\kappa\zeta}{N}\,d\zeta \quad \text{and} \quad \int_0^N \sin^2\frac{2\pi\kappa\zeta}{N}\,d\zeta$$

contribute to the integral:

$$\begin{aligned}
&\Psi_{00}\int_0^N \Psi_0^*(\zeta,\theta)\hat{v}(\zeta,\theta)d\zeta \\
&\quad = \Psi_{00}^2 \sum_{\kappa=-N/2+1}^{N/2-1}\Bigg(2\sin(\lambda_1\lambda_3\theta)\frac{(2\pi\kappa/N)\{1-\exp[-(\lambda_2^2-\lambda_1^2)^{1/2}N]\}}{\lambda_2^2-\lambda_1^2+(2\pi\kappa/N)^2} \\
&\qquad\qquad \times [J_{\rm i}(\theta,\kappa)-iJ_{\rm r}(\theta,\kappa)] \\
&\qquad + N\lambda_1\lambda_3\frac{I_{\rm T}(\kappa/N)}{\sin\beta_\kappa}\sin(\beta_\kappa\theta)\{J_{\rm r}(\theta,\kappa)\sin\lambda_1\lambda_3 - J_{\rm i}(\theta,\kappa)\cos\lambda_1\lambda_3 \\
&\qquad\qquad + i[J_{\rm r}(\theta,\kappa)\cos\lambda_1\lambda_3 + J_{\rm i}(\theta,\kappa)\sin\lambda_1\lambda_3]\}\Bigg)
\end{aligned} \tag{20}$$

We turn to Eq.(9). It requires $(\partial\Psi_0^*/\partial\theta)(\partial\Psi_0/\partial\theta)$, which is found in Eq.(2.5-9). For the term $\partial\hat{v}/\partial\theta$ we obtain from Eq.(17):

$$\begin{aligned}
\frac{\partial\hat{v}(\zeta,\theta)}{\partial\theta} &= e^{i\lambda_1\zeta}\sum_{\kappa=-N/2+1}^{N/2-1}[J_{\rm R}(\theta,\kappa)+iJ_{\rm I}(\theta,\kappa)]\sin\frac{2\pi\kappa\zeta}{N} \\
J_{\rm R}(\theta,\kappa) &= \left(\frac{\partial J_{\rm r}(\theta,\kappa)}{\partial\theta}+\lambda_1\lambda_3 J_{\rm i}(\theta,\kappa)\right)\cos\lambda_1\lambda_3\theta \\
&\quad + \left(\frac{\partial J_{\rm i}(\theta,\kappa)}{\partial\theta}-\lambda_1\lambda_3 J_{\rm r}(\theta,\kappa)\right)\sin\lambda_1\lambda_3\theta \\
J_{\rm I}(\theta,\kappa) &= -\left(\frac{\partial J_{\rm r}(\theta,\kappa)}{\partial\theta}+\lambda_1\lambda_3 J_{\rm i}(\theta,\kappa)\right)\sin\lambda_1\lambda_3\theta \\
&\quad + \left(\frac{\partial J_{\rm i}(\theta,\kappa)}{\partial\theta}-\lambda_1\lambda_3 J_{\rm r}(\theta,\kappa)\right)\cos\lambda_1\lambda_3\theta
\end{aligned} \tag{21}$$

The factor $e^{i\lambda_1\zeta}$ is not written as $\cos\lambda_1\zeta + i\sin\lambda_1\zeta$ since it will be cancelled in the following equation for $(\partial\Psi_0^*/\partial\theta)(\partial\hat{v}/\partial\theta)$ by $e^{-i\lambda_1\zeta}$. The term $\partial\Psi_0^*/\partial\theta$ is obtained with the help of Eqs.(2.5-5) and (2.5-6):

$$\Psi_{00}\frac{\partial\Psi_0^*(\zeta,\theta)}{\partial\theta}\frac{\partial\hat{v}(\zeta,\theta)}{\partial\theta} = \Psi_{00}^2\Bigg[-2i\lambda_1\lambda_3 e^{2i\lambda_1\lambda_3\theta}$$

$$\times\sum_{\kappa=-N/2+1}^{N/2-1}[J_{\mathrm{R}}(\theta,\kappa)+iJ_{\mathrm{I}}(\theta,\kappa)]\exp[-(\lambda_2^2-\lambda_1^2)^{1/2}\zeta]\sin\frac{2\pi\kappa\zeta}{N}$$

$$-2\lambda_1\lambda_3 e^{i\lambda_1\lambda_3(\theta-1)}\Bigg(\sum_{\kappa=-N/2+1}^{N/2-1}\frac{I_{\mathrm{T}}(\kappa/N)}{\sin\beta_\kappa}(\lambda_1\lambda_3\sin\beta_\kappa\theta$$

$$-i\beta_\kappa\cos\beta_\kappa\theta)[J_{\mathrm{R}}(\theta,\kappa)+iJ_{\mathrm{I}}(\theta,\kappa)]\sin^2\frac{2\pi\kappa\zeta}{N}$$

$$+\sum_{\kappa=-N/2+1}^{N/2-1,\neq\nu}\sum_{\nu=-N/2+1}^{N/2-1}\frac{I_{\mathrm{T}}(\kappa/N)}{\sin\beta_\kappa}[J_{\mathrm{R}}(\theta,\nu)+iJ_{\mathrm{I}}(\theta,\nu)]$$

$$\times(\lambda_1\lambda_3\sin\beta_\kappa\theta-i\beta_\kappa\cos\beta_\kappa\theta)\sin\frac{2\pi\kappa\zeta}{N}\sin\frac{2\pi\nu\zeta}{N}\Bigg)\Bigg]$$

$$J_{\mathrm{R}}(\theta,\nu)=J_{\mathrm{R}}(\theta,\kappa),\ J_{\mathrm{I}}(\theta,\nu)=J_{\mathrm{I}}(\theta,\kappa)\quad\text{for } \kappa\to\nu \tag{22}$$

As in Eq.(20) we integrate this expression over ζ:

$$\Psi_{00}\int_0^N\frac{\partial\Psi_0^*(\zeta,\theta)}{\partial\theta}\frac{\partial\hat{v}(\zeta,\theta)}{\partial\theta}d\zeta = \Psi_{00}^2\Bigg(2\lambda_1\lambda_3\sum_{\kappa=-N/2+1}^{N/2-1}\{[J_{\mathrm{R}}(\theta,\kappa)+iJ_{\mathrm{I}}(\theta,\kappa)]\sin 2\lambda_1\lambda_3\theta$$

$$+[J_{\mathrm{I}}(\theta,\kappa)-iJ_{\mathrm{R}}(\theta,\kappa)]\cos 2\lambda_1\lambda_3\theta\}\frac{(2\pi\kappa/N)\{1-\exp[-(\lambda_2^2-\lambda_1^2)^{1/2}N]\}}{\lambda_2^2-\lambda_1^2+(2\pi\kappa/N)^2}$$

$$-\lambda_1\lambda_3 N\sum_{\kappa=-N/2+1}^{N/2-1}\frac{I_{\mathrm{T}}(\kappa/N)}{\sin\beta_\kappa}\{[J_{\mathrm{R}}(\theta,\kappa)+iJ_{\mathrm{I}}(\theta,\kappa)]$$

$$\times[\lambda_1\lambda_3\sin\beta_\kappa\theta\cos\lambda_1\lambda_3(\theta-1)+\beta_\kappa\cos\beta_\kappa\theta\sin\lambda_1\lambda_3(\theta-1)]$$

$$-[J_{\mathrm{I}}(\theta,\kappa)-iJ_{\mathrm{R}}(\theta,\kappa)]$$

$$\times[\lambda_1\lambda_3\sin\beta_\kappa\theta\sin\lambda_1\lambda_3(\theta-1)-\beta_\kappa\cos\beta_\kappa\theta\cos\lambda_1\lambda_3(\theta-1)]\}\Bigg) \tag{23}$$

Turning to Eq.(10) we recognize the need for the term $\partial\hat{v}/\partial\zeta$. Equation (17) yields:

$$\frac{\partial \hat{v}(\zeta,\theta)}{\partial \zeta} = e^{i\lambda_1(\zeta-\lambda_3\theta)} \sum_{\kappa=-N/2+1}^{N/2-1} [J_{\mathrm{r}}(\theta,\kappa) + iJ_{\mathrm{i}}(\theta,\kappa)] \times \left(\frac{2\pi\kappa}{N} \cos \frac{2\pi\kappa\zeta}{N} + i\lambda_1 \sin \frac{2\pi\kappa\zeta}{N} \right) \quad (24)$$

The term $(\partial\Psi_0^*/\partial\zeta)(\partial\hat{v}/\partial\zeta)$ in Eq.(10) is obtained with the help of Eqs.(2.5-7) and (2.5-8):

$$\Psi_{00} \frac{\partial \Psi_0^*(\zeta,\theta)}{\partial \zeta} \frac{\partial \hat{v}(\zeta,\theta)}{\partial \zeta} = \Psi_{00}^2 \Bigg\{ -2[\lambda_1 - i(\lambda_2^2 - \lambda_1^2)^{1/2}]$$
$$\times \sum_{\kappa=-N/2+1}^{N/2-1} [J_r(\theta,\kappa) + iJ_{\mathrm{i}}(\theta,\kappa)] \bigg(\frac{2\pi\kappa}{N} \exp[-(\lambda_2^2-\lambda_1^2)^{1/2}\zeta] \cos \frac{2\pi\kappa\zeta}{N}$$
$$+ i\lambda_1 \exp[-(\lambda_2^2 - \lambda_1^2)^{1/2}\zeta] \sin \frac{2\pi\kappa\zeta}{N} \bigg) \sin \lambda_1\lambda_3\theta$$
$$+ 2i\lambda_1\lambda_3 e^{-i\lambda_1\lambda_3} \sum_{\kappa=-N/2+1}^{N/2-1} \frac{I_{\mathrm{T}}(\kappa/N)}{\sin\beta_\kappa} [J_{\mathrm{r}}(\theta,\kappa) + iJ_{\mathrm{i}}(\theta,\kappa)] \sin\beta_\kappa\theta$$
$$\times \left[\lambda_1^2 \sin^2 \frac{2\pi\kappa\zeta}{N} + \left(\frac{2\pi\kappa}{N}\right)^2 \cos^2 \frac{2\pi\kappa\zeta}{N} \right]$$
$$+ 2\lambda_1\lambda_3 e^{-i\lambda_1\lambda_3} \sum_{\kappa=-N/2+1}^{N/2-1,\neq\nu} \sum_{\nu=-N/2+1}^{N/2-1} \frac{I_{\mathrm{T}}(\kappa/N)}{\sin\beta_\kappa} [J_{\mathrm{r}}(\theta,\nu) + iJ_{\mathrm{i}}(\theta,\nu)] \sin\beta_\kappa\theta$$
$$\times \left(\lambda_1 \sin\frac{2\pi\kappa\zeta}{N} + i\frac{2\pi\kappa}{N} \cos\frac{2\pi\kappa\zeta}{N} \right) \left(\frac{2\pi\nu}{N} \cos\frac{2\pi\nu\zeta}{N} + i\lambda_1 \sin\frac{2\pi\nu\zeta}{N} \right) \Bigg\} \quad (25)$$

This expression is integrated over ζ following Eqs.(20) and (23):

$$\Psi_{00} \int_0^N \frac{\partial \Psi_0^*(\zeta,\theta)}{\partial \zeta} \frac{\partial \hat{v}(\zeta,\theta)}{\partial \zeta} d\zeta = \Psi_{00}^2 \Bigg\{ 2\lambda_2^2 \sin\lambda_1\lambda_3\theta$$
$$\times \sum_{\kappa=-N/2+1}^{N/2-1} [J_{\mathrm{i}}(\theta,\kappa) - iJ_{\mathrm{r}}(\theta,\kappa)] \frac{(2\pi\kappa/N)\{1-\exp[-(\lambda_2^2-\lambda_1^2)^{1/2}N]\}}{\lambda_2^2 - \lambda_1^2 + (2\pi\kappa/N)^2}$$
$$-N\lambda_1\lambda_3 \sum_{\kappa=-N/2+1}^{N/2-1} \left[\lambda_1^2 + \left(\frac{2\pi\kappa}{N}\right)^2 \right] \{ J_{\mathrm{i}}(\theta,\kappa) \cos\lambda_1\lambda_3 - J_{\mathrm{r}}(\theta,\kappa) \sin\lambda_1\lambda_3$$
$$- i[J_{\mathrm{r}}(\theta,\kappa) \cos\lambda_1\lambda_3 + J_{\mathrm{i}}(\theta,\kappa) \sin\lambda_1\lambda_3]\} \frac{I_{\mathrm{T}}(\kappa/N)}{\sin\beta_\kappa} \sin\beta_\kappa\theta \Bigg\} \quad (26)$$

Let us turn to the text following Eq.(2.5-11). In order to allow for the generalization of the terms of Eq.(2.5-3) by Eqs.(8)–(10) we define the energy $\hat{U}$ by the sum of the following three components

$$\hat{U} = \hat{U}_1 + \hat{U}_2 + \hat{U}_3 = \hat{U}_c + \hat{U}_v(\theta) \tag{27}$$

where $\hat{U}_1$ to $\hat{U}_3$ are obtained by the substitution of Eqs.(8)–(10) into Eq.(2.5-3). We use the notation $\hat{U}$ to $\hat{U}_v(\theta)$ to distinguish the terms from the approximations U to $U_v(\theta)$ in Eq.(2.5-12):

$$\hat{U}_1 = \frac{L^2}{c\Delta t}\frac{m_0^2c^4(\Delta t)^2}{\hbar^2}\int_0^N \Psi^*\Psi d\zeta = \frac{L^2}{c\Delta t}\frac{m_0^2c^4(\Delta t)^2}{\hbar^2}\bigg((1+2\alpha)\int_0^N \Psi_0^*\Psi_0 d\zeta$$
$$+ 2\alpha\Psi_{00}\int_0^N \Re e(\Psi_0^*\hat{v})d\zeta\bigg) \tag{28}$$

$$\hat{U}_2 = \frac{L^2}{c\Delta t}\int_0^N \frac{\partial\Psi^*}{\partial\theta}\frac{\partial\Psi}{\partial\theta}d\zeta = \frac{L^2}{c\Delta t}\bigg[(1+2\alpha)\int_0^N \frac{\partial\Psi_0^*}{\partial\theta}\frac{\partial\Psi_0}{\partial\theta}d\zeta$$
$$+ 2\alpha\Psi_{00}\int_0^N \Re e\left(\frac{\partial\Psi_0^*}{\partial\theta}\frac{\partial\hat{v}}{\partial\theta}\right)d\zeta\bigg] \tag{29}$$

$$\hat{U}_3 = \frac{L^2}{c\Delta t}\int_0^N \frac{\partial\Psi^*}{\partial\zeta}\frac{\partial\Psi}{\partial\zeta}d\zeta = \frac{L^2}{c\Delta t}\bigg[(1+2\alpha)\int_0^N \frac{\partial\Psi_0^*}{\partial\zeta}\frac{\partial\Psi_0}{\partial\zeta}d\zeta$$
$$+ 2\alpha\Psi_{00}\int_0^N \Re e\left(\frac{\partial\Psi_0^*}{\partial\zeta}\frac{\partial\hat{v}}{\partial\zeta}\right)d\zeta\bigg] \tag{30}$$

As in Eqs.(2.5-13)–(2.5-15) we want the time-invariant part

$$\hat{U}_c = \hat{U}_{c1} + \hat{U}_{c2} + \hat{U}_{c3} \tag{31}$$

of $\hat{U}_1$, $\hat{U}_2$, $\hat{U}_3$ and we ignore the time-variable part $\hat{U}_v(\theta)$. Again we write $\hat{U}_c$ to $\hat{U}_{c3}$ to distinguish the terms from the approximations U_c to U_{c3} in Eq.(2.5-16). When deriving Eqs.(2.5-13)–(2.5-15) we simply left out terms containing $\sin\beta_\kappa\theta$ or $\cos\beta_\kappa\theta$. The time variations of Eqs.(28), (29), (30) are not so obvious and we must write integrals over θ. Using Eqs.(2.5-13)–(2.5-15) we may write the time invariant part of Eqs.(28)–(30) as follows:

$$\hat{U}_{c1} = \int_0^N \hat{U}_1 d\theta = \frac{L^2}{c\Delta t}\frac{m_0^2 c^4(\Delta t)^2}{\hbar^2}\Bigg[(1+2\alpha)\Psi_{00}^2 N\lambda_1^2\lambda_3^2 \sum_{\kappa=-N/2+1}^{N/2-1}\left(\frac{I_{\rm T}(\kappa/N)}{\sin\beta_\kappa}\right)^2 + 2\alpha\Psi_{00}\int_0^N\left(\int_0^N \mathcal{R}e(\Psi_0^*\hat{v})d\zeta\right)d\theta\Bigg] \tag{32}$$

$$\hat{U}_{c2} = \int_0^N \hat{U}_2 d\theta = \frac{L^2}{c\Delta t}\Bigg\{(1+2\alpha)\Psi_{00}^2 N\lambda_1^2\lambda_3^2 \sum_{\kappa=-N/2+1}^{N/2-1}\left(\frac{I_{\rm T}(\kappa/N)}{\sin\beta_\kappa}\right)^2(\lambda_1^2\lambda_3^2+\beta_\kappa^2) + 2\alpha\Psi_{00}\int_0^N\left[\int_0^N \mathcal{R}e\left(\frac{\partial\Psi_0^*}{\partial\theta}\frac{\partial\hat{v}}{\partial\theta}\right)d\zeta\right]d\theta\Bigg\} \tag{33}$$

$$\hat{U}_{c3} = \int_0^N \hat{U}_3 d\theta = \frac{L^2}{c\Delta t}\Bigg\{(1+2\alpha)\Psi_{00}^2 N\lambda_1^2\lambda_3^2 \sum_{\kappa=-N/2+1}^{N/2-1}\left(\frac{I_{\rm T}(\kappa/N)}{\sin\beta_\kappa}\right)^2\left[\lambda_1^2+\left(\frac{2\pi\kappa}{N}\right)^2\right] + 2\alpha\Psi_{00}\int_0^N\left[\int_0^N \mathcal{R}e\left(\frac{\partial\Psi_0^*}{\partial\zeta}\frac{\partial\hat{v}}{\partial\zeta}\right)d\zeta\right]d\theta\Bigg\} \tag{34}$$

We recognize on the right side of Eqs.(32)–(34) the energies U_{c1}, U_{c2}, U_{c3} of Eqs.(2.5-13)–(2.5-15) multiplied by $1+2\alpha \doteq 1.0146$. This difference of 1.5% is barely visible in a plot. The interesting terms are the integrals multiplied by $2\alpha\Psi_{00}$. We shall analyze them in the following Section 3.4.

A comparison of Eqs.(31) and (2.5-16) shows that the only difference is the symbol ˆ. Hence, we may use the results of Section 2.5 from Eq.(2.5-16) on. The energies E_κ of Eqs.(2.5-32), (2.5-34), and (2.5-48) are obtained. One could write $\hat{\mathsf{E}}_\kappa$ instead of E_κ and obtain the result $\hat{\mathsf{E}}_\kappa = \mathsf{E}_\kappa$, but there is little incentive to do so.

3.4 Evaluation of the Energy $\hat{U}_c$

We start from the energies $\hat{U}_{c1}$, $\hat{U}_{c2}$, $\hat{U}_{c3}$ in Eqs.(3.3-32)–(3.3-34). Their first terms equal the energies U_{c1}, U_{c2}, U_{c3} in Eqs.(2.5-13)–(2.5-15), except for the factor $1+2\alpha$. The results of Chapter 2 apply to these terms. Here we are interested in the second terms that are multiplied by 2α:

$$\hat{U}_{\alpha 1} = 2\alpha\Psi_{00}\frac{L^2}{c\Delta t}\left(\frac{m_0c^2\Delta t}{\hbar}\right)^2\int_0^N\left(\int_0^N \mathcal{R}e(\Psi_0^*\hat{v})d\zeta\right)d\theta \tag{1}$$

$$\hat{U}_{\alpha 2} = 2\alpha\Psi_{00}\frac{L^2}{c\Delta t}\int_0^N\left[\int_0^N \mathcal{R}e\left(\frac{\partial\Psi_0^*}{\partial\theta}\frac{\partial\hat{v}}{\partial\theta}\right)d\zeta\right]d\theta \tag{2}$$

$$\hat{U}_{\alpha 3} = 2\alpha\Psi_{00}\frac{L^2}{c\Delta t}\int_0^N\left[\int_0^N \mathcal{R}e\left(\frac{\partial\Psi_0^*}{\partial\zeta}\frac{\partial\hat{v}}{\partial\zeta}\right)d\zeta\right]d\theta \tag{3}$$

The sum $\hat{U}_{\alpha c}$ of these terms is ultimately wanted as function $\hat{U}_{\alpha\kappa}(\kappa)$ of κ as shown by Eq.(2.5-16)

$$\hat{U}_{\alpha c} = \hat{U}_{\alpha 1} + \hat{U}_{\alpha 2} + \hat{U}_{\alpha 3} = \sum_{\kappa=-N/2+1}^{N/2-1}\hat{U}_{\alpha\kappa}(\kappa) \tag{4}$$

$$\hat{U}_{\alpha 1} = \sum_{\kappa=-N/2+1}^{N/2-1}\hat{U}_{1\kappa}(\kappa),\quad \hat{U}_{\alpha 2} = \sum_{\kappa=-N/2+1}^{N/2-1}\hat{U}_{2\kappa}(\kappa),\quad \hat{U}_{\alpha 3} = \sum_{\kappa=-N/2+1}^{N/2-1}\hat{U}_{3\kappa}(\kappa) \tag{5}$$

and Eqs.(1)–(3) must be written correspondingly. We write $\hat{U}_{1\kappa}(\kappa)$ with the index κ after the integration over ζ that indicates that the summation over κ in Eq.(3.3-20) was not carried out:

$$\hat{U}_{1\kappa}(\kappa) = 2\alpha\Psi_{00}\frac{L^2}{c\Delta t}\left(\frac{m_0c^2\Delta t}{\hbar}\right)^2\int_0^N \mathcal{R}e\left(\int_0^N \Psi_0^*(\zeta,\theta)\hat{v}(\zeta,\theta)\,d\zeta\right)_\kappa d\theta$$

$$\begin{aligned}\mathcal{H}_{1\kappa}(\kappa) &= \frac{\hat{U}_{1\kappa}(\kappa)}{2\alpha\Psi_{00}^2(L^2/c\Delta t)(m_0c^2\Delta t/\hbar)^2}\\ &= \frac{1}{\Psi_{00}}\int_0^N \mathcal{R}e\left(\int_0^N \Psi_0^*(\zeta,\theta)\hat{v}(\zeta,\theta)d\zeta\right)_\kappa d\theta\end{aligned} \tag{6}$$

Dividing Eq.(3.3-20) by Ψ_{00}^2 and leaving out the summation sign yields the following expression for $\mathcal{H}_{1\kappa}(\kappa)$:

$$\begin{aligned}\mathcal{H}_{1\kappa}(\kappa) = &\,2\frac{(2\pi\kappa/N)\{1-\exp[-(\lambda_2^2-\lambda_1^2)^{1/2}N]\}}{\lambda_2^2-\lambda_1^2+(2\pi\kappa/N)^2}\\ &\times\int_0^N \mathcal{R}e[J_{\mathrm{i}}(\theta,\kappa)-iJ_{\mathrm{r}}(\theta,\kappa)]\sin\lambda_1\lambda_3\theta\,d\theta\\ &+N\lambda_1\lambda_3\frac{I_{\mathrm{T}}(\kappa/N)}{\sin\beta_\kappa}\int_0^N \mathcal{R}e\big\{J_{\mathrm{r}}(\theta,\kappa)\sin\lambda_1\lambda_3 - J_{\mathrm{i}}(\theta,\kappa)\cos\lambda_1\lambda_3\\ &+i[J_{\mathrm{r}}(\theta,\kappa)\cos\lambda_1\lambda_3+iJ_{\mathrm{i}}(\theta,\kappa)\sin\lambda_1\lambda_3]\big\}\sin\beta_\kappa\theta\,d\theta\end{aligned} \tag{7}$$

TABLE 3.4-1

FUNCTIONS β_κ, $I_T(\kappa/N)$, ... AND THE EQUATIONS EQ.(2.4-13), EQ.(2.4-29), ... DEFINING THEM, WHICH ARE REQUIRED FOR THE CALCULATION OF $\mathcal{H}_{1\kappa}(\kappa)$, $\mathcal{H}_{2\kappa}(\kappa)$, AND $\mathcal{H}_{3\kappa}(\kappa)$ OF EQS.(28), (39), AND (45). THE NUMERICAL VALUES OBTAINED FOR EQUATIONS SHOWN WITH BRACKETS, E.G. EQ.[2.4-13], SHOULD BE REMEMBERED BY THE COMPUTER TO REDUCE THE COMPUTING TIME.

β_κ	Eq.[2.4-13]	$I_T(\kappa/N)$	Eq.[2.4-29]	$\Psi_0(\zeta,\theta_n)$	Eq.[3.4-8]
$A_{m1x}(\zeta,\theta_n)$	Eq.(3.4-14)	$\phi_{e1}(\zeta,\theta_n)$	Eq.(3.4-15)	ϕ_{e0}	Eq.(3.4-10)
A_{m0x}	Eq.(3.4-9)	$G_1(\zeta,\theta_n)$	Eq.[3.4-18]	$G_{1-}(\zeta,\theta_n)$	Eq.[3.4-19]
$G_{1+}(\zeta,\theta_n)$	Eq.[3.4-20]	$G_{s\kappa 1}(\theta_n,\kappa)$	Eq.[3.4-21]	$G_{s\kappa 1-}(\theta_n,\kappa)$	Eq.[3.4-22]
$G_{s\kappa 1+}(\theta_n,\kappa)$	Eq.[3.4-23]	$H_{s\kappa}(\theta_n,\kappa)$	Eq.[3.4-24]	$F_1(n,\kappa)$	Eq.(6.3-51)
$F_2(n,\kappa)$	Eq.(6.3-52)	$F_3(n,\kappa)$	Eq.(6.3-53)	$F_4(n,\kappa)$	Eq.(6.3-54)
$F_5(n,\kappa)$	Eq.(6.3-55)	$F_6(n,\kappa)$	Eq.(6.3-56)	$F_7(n,\kappa)$	Eq.(6.3-57)
$F_8(n,\kappa)$	Eq.(6.3-58)	$d_{3r}(\kappa)$	Eq.(6.3-73)	$d_{3i}(\kappa)$	Eq.(6.3-74)
$J_1(\theta,\kappa)$	Eq.[6.3-75]	$J_2(\theta,\kappa)$	Eq.[6.3-75]	$J_3(\theta,\kappa)$	Eq.[6.3-75]
$J_4(\theta,\kappa)$	Eq.[6.3-75]	$J_5(\theta,\kappa)$	Eq.[6.3-75]	$J_6(\theta,\kappa)$	Eq.[6.3-75]
$J_7(\theta,\kappa)$	Eq.[6.3-75]	$J_8(\theta,\kappa)$	Eq.[6.3-75]	$J_r(\theta,\kappa)$	Eq.[3.3-17]
$J_i(\theta,\kappa)$	Eq.[3.3-17]	$A_{1\kappa}(\kappa)$	Eq.[3.4-25]	$B_{1\kappa}(\kappa)$	Eq.[3.4-26]
$I_{11}(\theta,\kappa)$	Eq.(3.4-27)	$\mathcal{H}_{1\kappa}(\kappa)$	Eq.[3.4-28]		
$K_1(\theta,\kappa)$	Eq.[6.3-75]	$K_2(\theta,\kappa)$	Eq.[6.3-75]	$K_3(\theta,\kappa)$	Eq.[6.3-75]
$K_4(\theta,\kappa)$	Eq.[6.3-75]	$K_5(\theta,\kappa)$	Eq.[6.3-75]	$K_6(\theta,\kappa)$	Eq.[6.3-75]
$K_7(\theta,\kappa)$	Eq.[6.3-75]	$K_8(\theta,\kappa)$	Eq.[6.3-75]	$\partial J_r/\partial\theta$	Eq.[3.4-36]
$\partial J_i/\partial\theta$	Eq.[3.4-37]	$J_R(\theta,\kappa)$	Eq.[3.3-21]	$J_I(\theta,\kappa)$	Eq.[3.3-21]
$I_{21}(\theta,\kappa)$	Eq.(3.4-32)	$I_{22}(\theta,\kappa)$	Eq.(3.4-33)	$I_{23}(\theta,\kappa)$	Eq.(3.4-34)
$I_{24}(\theta,\kappa)$	Eq.(3.4-35)	$\mathcal{H}_{2\kappa}(\kappa)$	Eq.[3.4-39]		
$I_{31}(\theta,\kappa)$	Eq.(3.4-43)	$I_{32}(\theta,\kappa)$	Eq.(3.4-44)	$\mathcal{H}_{3\kappa}(\kappa)$	Eq.[3.4-45]

We observe that $J_r(\theta,\kappa)$ and $J_i(\theta,\kappa)$ are not real but complex functions. An attempt to separate real and imaginary components analytically creates unnecessary complications since the computer instruction $\mathcal{R}e$ will do this separation numerically.

In order to integrate Eq.(7) over θ we need to evaluate integrals containing $J_r(\theta,\kappa)$ and $J_i(\theta,\kappa)$. For the determination of $J_r(\theta,\kappa)$ and $J_i(\theta,\kappa)$ we start with Eq.(3.1-1) for $\Psi_0(\zeta,\theta)$ and choose $\Psi_{00}=1$. The variable θ of Eq.(3.1-1) will become the variable n in $H_{s\kappa}(n,\kappa)$ in Eqs.(6.3-47)–(6.3-50). For this reason we rewrite Eq.(3.1-1) with θ replaced by θ_n and shorten θ_n to either θ or n as required:

$$\Psi_0(\zeta,\theta_n) = (1-e^{-2i\lambda_1\lambda_3\theta_n})\exp[-(\lambda_2^2-\lambda_1^2)^{1/2}\zeta]e^{i\lambda_1\zeta}$$
$$-2i\lambda_1\lambda_3 e^{i\lambda_1\lambda_3}e^{i\lambda_1(\zeta-\lambda_3\theta_n)}\sum_{\kappa=-N/2+1}^{N/2-1}\frac{I_T(\kappa/N)}{\sin\beta_\kappa}\sin\beta_\kappa\theta_n\sin\frac{2\pi\kappa\zeta}{N} \quad (8)$$

Table 3.4-1 lists the sequence of functions required to obtain $\mathcal{H}_{1\kappa}(\kappa)$ and the equations that define them. The function $\Psi_0(\zeta,\theta_n)$ is needed for $G_1(\zeta,\theta)$ of Eq.(3.2-4). A number of definitions is introduced to make Eq.(3.2-4) suitable for a computer:

$$A_{m0x} = \frac{e\Delta x}{\hbar} A_{m0x} = \frac{ec\Delta t}{\hbar} A_{m0x} \tag{9}$$

$$\phi_{e0} = \frac{e\Delta t}{\hbar}\phi_{e0} = \frac{e\Delta x}{c\hbar}\phi_{e0} \tag{10}$$

$$A_{m1x}(\zeta,\theta_n) = \frac{e\Delta x}{\hbar} A_{m1x}(\zeta,\theta_n) = \frac{ec\Delta t}{\hbar} A_{m1x}(\zeta,\theta_n) \tag{11}$$

$$\phi_{e1}(\zeta,\theta_n) = \frac{e\Delta x}{c\hbar}\phi_{e1}(\zeta,\theta_n) = \frac{e\Delta t}{\hbar}\phi_{e1}(\zeta,\theta_n) \tag{12}$$

$$\begin{aligned} G_1(\zeta,\theta_n) = &-i\frac{A_{m0x}C_{cx}}{\alpha}\bigg(\frac{1}{2}[\Psi_0(\zeta,\theta_n+1) - \Psi_0(\zeta,\theta_n-1)] \\ &\qquad + i\phi_{e0}\Psi_0(\zeta,\theta_n)\bigg) \\ &- \frac{i}{2}\{[A_{m1x}(\zeta+1,\theta_n) + A_{m1x}(\zeta,\theta_n)]\Psi_0(\zeta+1,\theta_n) \\ &\quad - [A_{m1x}(\zeta,\theta_n) + A_{m1x}(\zeta-1,\theta_n)]\Psi_0(\zeta-1,\theta_n)\} \\ &\qquad - 2A_{m0x}A_{m1x}(\zeta,\theta_n)\Psi_0(\zeta,\theta_n) \\ &- \frac{i}{2}\{[\phi_{e1}(\zeta,\theta_n+1) + \phi_{e1}(\zeta,\theta_n)]\Psi_0(\zeta,\theta_n+1) \\ &\quad - [\phi_{e1}(\zeta,\theta_n) + \phi_{e1}(\zeta,\theta_n-1)]\Psi_0(\zeta,\theta_n-1)\} \\ &\qquad + 2\phi_{e0}\phi_{e1}(\zeta,\theta_n)\Psi_0(\zeta,\theta_n) \end{aligned} \tag{13}$$

The space and time variation of the potentials $A_{m1x}(\zeta,\theta_n)$ and $\phi_{e1}(\zeta,\theta_n)$ must still be defined. A particularly simple choice is to let the potentials be independent of time and vary linearly with the space variable ζ:

$$A_{m1x}(\zeta,\theta_n) = A_{m1x}(\zeta) = A_{1m}\zeta/N \tag{14}$$

$$\phi_{e1}(\zeta,\theta_n) = \phi_{e1}(\zeta) = \phi_{1e}\zeta/N \tag{15}$$

Here A_{1m} and ϕ_{1e} are dimension-free constants.

The two constants ϕ_{e0} and A_{m0x} in Eqs.(10) and (9) may be expressed in terms of λ_1 and λ_3 according to Eq.(2.3-2):

$$\phi_{e0} = \frac{e\Delta t}{\hbar}\phi_{e0} = \lambda_1\lambda_3 \tag{16}$$

$$A_{m0x} = \frac{ec\Delta t}{\hbar}A_{m0x} = \lambda_1 \tag{17}$$

The constant C_{cx} poses a problem. Its value is arbitrary and we may thus choose it to be zero without causing a mathematical mistake. We do so here but point out that this is not a particularly satisfying solution of the problem. The simplified Eq.(13) becomes:

$$\begin{aligned}
G_1(\zeta,\theta_n) = -\frac{i}{2}\{[A_{\mathrm{m}1x}(\zeta+1) + A_{\mathrm{m}1x}(\zeta)]\Psi_0(\zeta+1,\theta_n) \\
- [A_{\mathrm{m}1x}(\zeta) + A_{\mathrm{m}1x}(\zeta-1)]\Psi_0(\zeta-1,\theta_n)\} - 2A_{\mathrm{m}0x}A_{\mathrm{m}1x}(\zeta)\Psi_0(\zeta,\theta_n) \\
- i\phi_{\mathrm{e}1}(\zeta)[\Psi_0(\zeta,\theta_n+1) - \Psi_0(\zeta,\theta_n-1)] + 2\phi_{\mathrm{e}0}\phi_{\mathrm{e}1}(\zeta)\Psi_0(\zeta,\theta_n) \quad (18)
\end{aligned}$$

We shall also need $G_1(\zeta,\theta_n-1)$ and $G_1(\zeta,\theta_n+1)$. A not very elegant but lucid method to produce these time shifted functions is to write Eq.(18) for $\theta_n \to \theta_n - 1$ and $\theta_n \to \theta_n + 1$:

$$\begin{aligned}
G_{1-}(\zeta,\theta_n) = -\frac{i}{2}\{[A_{\mathrm{m}1x}(\zeta+1)+A_{\mathrm{m}1x}(\zeta)]\Psi_0(\zeta+1,\theta_n-1) \\
- [A_{\mathrm{m}1x}(\zeta)+A_{\mathrm{m}1x}(\zeta-1)]\Psi_0(\zeta-1,\theta_n-1)\}-2A_{\mathrm{m}0x}A_{\mathrm{m}1x}(\zeta)\Psi_0(\zeta,\theta_n-1) \\
- i\phi_{\mathrm{e}1}(\zeta)[\Psi_0(\zeta,\theta_n) - \Psi_0(\zeta,\theta_n-2)] + 2\phi_{\mathrm{e}0}\phi_{\mathrm{e}1}(\zeta)\Psi_0(\zeta,\theta_n-1) \quad (19)
\end{aligned}$$

$$\begin{aligned}
G_{1+}(\zeta,\theta_n) = -\frac{i}{2}\{[A_{\mathrm{m}1x}(\zeta+1) + A_{\mathrm{m}1x}(\zeta)]\Psi_0(\zeta+1,\theta_n+1) \\
-[A_{\mathrm{m}1x}(\zeta)+A_{\mathrm{m}1x}(\zeta-1)]\Psi_0(\zeta-1,\theta_n+1)\}-2A_{\mathrm{m}0x}A_{\mathrm{m}1x}(\zeta)\Psi_0(\zeta,\theta_n+1) \\
- i\phi_{\mathrm{e}1}(\zeta)[\Psi_0(\zeta,\theta_n+2) - \Psi_0(\zeta,\theta_n)] + 2\phi_{\mathrm{e}0}\phi_{\mathrm{e}1}(\zeta)\Psi_0(\zeta,\theta_n+1) \quad (20)
\end{aligned}$$

We can compute $G_{\mathrm{s}\kappa}(\theta_n,\kappa)$ of Eq.(3.2-33) once $G_1(\zeta,\theta_n)$ of Eq.(18) is obtained. The notation $G_{\mathrm{s}\kappa}(\theta_n,\kappa)$ is changed to $G_{\mathrm{s}\kappa1}(\theta_n,\kappa)$ to make the use of Eqs.(19) and (20) more lucid. Let us observe that Eq.(3.2-33) treats the function $G_1(\zeta,\theta)$ as a step function with step width 1. The integral may readily be replaced by a sum of $G_1(\zeta,\theta)$ taken at $\zeta = 0,\ 1,\ \dots,\ N-1$:

$$\begin{aligned}
G_{\mathrm{s}\kappa1}(\theta_n,\kappa) = G_{\mathrm{s}\kappa}(\theta_n,\kappa) &= \frac{2}{N}\int_0^N G_1(\zeta,\theta_n)e^{-i\lambda_1\zeta}\sin\frac{2\pi\kappa\zeta}{N}d\zeta \\
&= \frac{2}{N}\sum_{\zeta=0}^{N-1} G_1(\zeta,\theta_n)e^{-i\lambda_1\zeta}\sin\frac{2\pi\kappa\zeta}{N} \quad (21)
\end{aligned}$$

The functions $G_{\mathrm{s}\kappa1-}(\theta_n,\kappa)$ and $G_{\mathrm{s}\kappa1+}(\theta_n,\kappa)$ follow from Eqs.(19) and (20):

$$G_{s\kappa 1-}(\theta_n,\kappa) = G_{s\kappa}(\theta_n - 1,\kappa) = \frac{2}{N}\sum_{\zeta=0}^{N-1} G_{1-}(\zeta,\theta_n)e^{-i\lambda_1\zeta}\sin\frac{2\pi\kappa\zeta}{N} \qquad (22)$$

$$G_{s\kappa 1+}(\theta_n,\kappa) = G_{s\kappa}(\theta_n + 1,\kappa) = \frac{2}{N}\sum_{\zeta=0}^{N-1} G_{1+}(\zeta,\theta_n)e^{-i\lambda_1\zeta}\sin\frac{2\pi\kappa\zeta}{N} \qquad (23)$$

With Eqs.(21)–(23) we obtain the function $H_{s\kappa}(\theta_n,\kappa)$ from Eq.(6.3-1):

$$\begin{aligned}H_{s\kappa}(\theta_n,\kappa) &= -e^{i\lambda_1\lambda_3}G_{s\kappa}(\theta_n+1,\kappa) + 2\cos\frac{2\pi\kappa}{N}G_{s\kappa}(\theta_n,\kappa) - e^{-i\lambda_1\lambda_3}G_{s\kappa}(\theta_n-1,\kappa)\\ &= -e^{i\lambda_1\lambda_3}G_{s\kappa 1+}(\theta_n,\kappa) + 2\cos\frac{2\pi\kappa}{N}G_{s\kappa 1}(\theta_n,\kappa) - e^{-i\lambda_1\lambda_3}G_{s\kappa 1-}(\theta_n,\kappa) \qquad (24)\end{aligned}$$

Eight products of $H_{s\kappa}(n,\kappa)$ with certain combinations of the functions $F_1(n,\kappa)$ to $F_8(n,\kappa)$, defined by Eqs.(6.3-51) to (6.3-58), are listed in Eq.(6.3-75) and denoted $K_1(n,\kappa)$ to $K_8(n,\kappa)$. The functions $J_i(\theta,\kappa)$ and $J_r(\theta,\kappa)$ required in Eq.(7) are derived from $J_1(\theta,\kappa)$ to $J_8(\theta,\kappa)$ in Eq.(6.3-75); they are shown in Eq.(3.3-17). We are now ready to evaluate Eq.(7). The following definitions are made to write it in a more compact form:

$$A_{1\kappa}(\kappa) = \frac{2(2\pi\kappa/N)\{1-\exp[-(\lambda_2^2-\lambda_1^2)^{1/2}N]\}}{\lambda_2^2-\lambda_1^2+(2\pi\kappa/N)^2} \qquad (25)$$

$$B_{1\kappa}(\kappa) = N\lambda_1\lambda_3\frac{I_T(\kappa/N)}{\sin\beta_\kappa} \qquad (26)$$

Equation (7) without the integration signs may be written as follows:

$$\begin{aligned}I_{11}(\theta,\kappa) &= A_{1\kappa}(\kappa)\mathcal{R}e[J_i(\theta,\kappa) - iJ_r(\theta,\kappa)]\sin\lambda_1\lambda_3\theta\\ &\quad + B_{1\kappa}(\kappa)\sin(\beta_\kappa\theta)\,\mathcal{R}e\{J_r(\theta,\kappa)\sin\lambda_1\lambda_3 - J_i(\theta,\kappa)\cos\lambda_1\lambda_3\\ &\qquad + i[J_r(\theta,\kappa)\cos\lambda_1\lambda_3 + J_i(\theta,\kappa)\sin\lambda_1\lambda_3]\} \qquad (27)\end{aligned}$$

The equivalence of the integral over a step function with steps of width 1 and a sum permits us to rewrite Eq.(7):

$$\mathcal{H}_{1\kappa}(\kappa) = \int_0^N I_{11}(\theta,\kappa)d\theta = \sum_{\theta=0}^{N-1} I_{11}(\theta,\kappa) \qquad (28)$$

This ends the calculation of $\mathcal{H}_{1\kappa}(\kappa)$ and $\hat{U}_{1\kappa}(\kappa)$ of Eq.(6). We shall return later to Eq.(28) when we need it for plotting. Here we turn to $\hat{U}_{2\kappa}(\kappa)$ of Eq.(5).

Again the index κ is written after the integration over ζ to indicate that the summation over κ in Eq.(3.3-23) is not carried out:

$$\hat{U}_{2\kappa}(\kappa) = 2\alpha\Psi_{00} \int_0^N \mathcal{R}e\left(\int_0^N \frac{\partial\Psi_0^*}{\partial\theta} \frac{\partial\hat{v}}{\partial\theta} \right)_\kappa d\theta \tag{29}$$

With the help of Eq.(3.3-23) we get:

$$\frac{\hat{U}_{2\kappa}(\kappa)}{2\alpha\Psi_{00}^2 L^2/c\Delta t} = \mathcal{H}_{2\kappa}(\kappa) = \frac{1}{\Psi_{00}} \int_0^N \mathcal{R}e\left(\int_0^N \frac{\partial\Psi_0^*}{\partial\theta} \frac{\partial\hat{v}}{\partial\theta} d\zeta \right)_\kappa d\theta \tag{30}$$

$$\begin{aligned}
\frac{1}{\Psi_0} \mathcal{R}e \int_0^N \left(\frac{\partial\Psi_0^*}{\partial\theta} \frac{\partial\hat{v}}{\partial\theta} d\zeta \right)_\kappa &= 2\lambda_1\lambda_3 \frac{(2\pi\kappa/N)\{1 - \exp[-(\lambda_2^2 - \lambda_1^2)^{1/2}N]\}}{\lambda_2^2 - \lambda_1^2 + (2\pi\kappa/N)^2} \\
&\times \{\mathcal{R}e[J_\mathrm{R}(\theta,\kappa) + iJ_\mathrm{I}(\theta,\kappa)] \sin 2\lambda_1\lambda_3\theta + \mathcal{R}e[J_\mathrm{I}(\theta,\kappa) - iJ_\mathrm{R}(\theta,\kappa)] \cos 2\lambda_1\lambda_3\theta]\} \\
&- N\lambda_1\lambda_3 \frac{I_\mathrm{T}(\kappa/N)}{\sin\beta_\kappa} \{\mathcal{R}e[J_\mathrm{R}(\theta,\kappa) + iJ_\mathrm{I}(\theta,\kappa)] \\
&\times [\lambda_1\lambda_3 \sin\beta_\kappa\theta \cos\lambda_1\lambda_3(\theta-1) + \beta_\kappa \cos\beta_\kappa\theta \sin\lambda_1\lambda_3(\theta-1)] \\
&- \mathcal{R}e[J_\mathrm{I}(\theta,\kappa) - iJ_\mathrm{R}(\theta\kappa)] \\
&\times [\lambda_1\lambda_3 \sin\beta_\kappa\theta \sin\lambda_1\lambda_3(\theta-1) - \beta_\kappa \cos\beta_\kappa\theta \cos\lambda_1\lambda_3(\theta-1)]\}
\end{aligned} \tag{31}$$

In order to integrate Eq.(31) over θ we need to evaluate numerically the integrals over the following terms with respect to θ:

$$I_{21}(\theta,\kappa) = \lambda_1\lambda_3 A_{1\kappa}(\kappa) \mathcal{R}e[J_\mathrm{R}(\theta,\kappa) + iJ_\mathrm{I}(\theta,\kappa)] \sin 2\lambda_1\lambda_3\theta \tag{32}$$

$$I_{22}(\theta,\kappa) = \lambda_1\lambda_3 A_{1\kappa}(\kappa) \mathcal{R}e[J_\mathrm{I}(\theta,\kappa) - iJ_\mathrm{R}(\theta,\kappa)] \cos 2\lambda_1\lambda_3\theta \tag{33}$$

$$\begin{aligned} I_{23}(\theta,\kappa) &= B_{1\kappa}(\kappa) \mathcal{R}e[J_\mathrm{R}(\theta,\kappa) + iJ_\mathrm{I}(\theta,\kappa)] \\ &\times [\lambda_1\lambda_3 \sin\beta_\kappa\theta \cos\lambda_1\lambda_3(\theta-1) + \beta_\kappa \cos\beta_\kappa\theta \sin\lambda_1\lambda_3(\theta-1)] \end{aligned} \tag{34}$$

$$\begin{aligned} I_{24}(\theta,\kappa) &= B_{1\kappa}(\kappa) \mathcal{R}e[J_\mathrm{I}(\theta,\kappa) - iJ_\mathrm{R}(\theta,\kappa)] \\ &\times [\lambda_1\lambda_3 \sin\beta_\kappa\theta \sin\lambda_1\lambda_3(\theta-1) - \beta_\kappa \cos\beta_\kappa\theta \cos\lambda_1\lambda_3(\theta-1)] \end{aligned} \tag{35}$$

The functions $J_\mathrm{R}(\theta,\kappa)$ and $J_\mathrm{I}(\theta,\kappa)$ are defined in Eq.(3.3-21) by the functions $J_\mathrm{r}(\theta,\kappa)$ and $J_\mathrm{i}(\theta,\kappa)$. In turn, $J_\mathrm{r}(\theta,\kappa)$ and $J_\mathrm{i}(\theta,\kappa)$ are defined in Eq.(3.3-17) by $J_1(\theta,\kappa)$ to $J_8(\theta,\kappa)$ of Eq.(6.3-75). We need mainly lengthy but straightforward substitutions to express $J_\mathrm{R}(\kappa,\theta)$ and $J_\mathrm{I}(\kappa,\theta)$ by $J_1(\theta,\kappa$ to $J_8(\theta,\kappa)$. The only problems are the derivatives $\partial J_\mathrm{r}/\partial\theta$ and $\partial J_\mathrm{i}(\theta,\kappa)/\partial\theta$ in Eq.(3.3-21). We obtain with the help of Eq.(3.3-17):

$$\frac{\partial J_r(\theta,\kappa)}{\partial\theta} = \left(\frac{\partial J_1(\theta,\kappa)}{\partial\theta} - J_2(\theta,\kappa) - \theta\frac{\partial J_2(\theta,\kappa)}{\partial\theta}\right)\cos\frac{2\pi\kappa\theta}{N}$$
$$- \frac{2\pi\kappa}{N}[J_1(\theta,\kappa) - \theta J_2(\theta,\kappa)]\sin\frac{2\pi\kappa\theta}{N}$$
$$- \left(\frac{\partial J_3(\theta,\kappa)}{\partial\theta} - J_4(\theta,\kappa) - \theta\frac{\partial J_4(\theta,\kappa)}{\partial\theta}\right)\sin\frac{2\pi\kappa\theta}{N}$$
$$- \frac{2\pi\kappa}{N}[J_3(\theta,\kappa) - \theta J_4(\theta,\kappa)]\cos\frac{2\pi\kappa\theta}{N} \quad (36)$$

$$\frac{\partial J_i(\theta,\kappa)}{\partial\theta} = \left(\frac{\partial J_5(\theta,\kappa)}{\partial\theta} - J_6(\theta,\kappa) - \theta\frac{\partial J_6(\theta,\kappa)}{\partial\theta}\right)\sin\frac{2\pi\kappa\theta}{N}$$
$$+ \frac{2\pi\kappa}{N}[J_5(\theta,\kappa) - \theta J_6(\theta,\kappa)]\cos\frac{2\pi\kappa\theta}{N}$$
$$+ \left(\frac{\partial J_7(\theta,\kappa)}{\partial\theta} - J_8(\theta,\kappa) - \theta\frac{\partial J_8(\theta,\kappa)}{\partial\theta}\right)\cos\frac{2\pi\kappa\theta}{N}$$
$$- \frac{2\pi\kappa}{N}[J_7(\theta,\kappa) - \theta J_8(\theta,\kappa)]\sin\frac{2\pi\kappa\theta}{N} \quad (37)$$

The functions $J_1(\theta,\kappa)$ to $J_8(\theta,\kappa)$ are defined in Eq.(6.3-75), but we must show what the derivatives $\partial J_1(\theta,\kappa)/\partial\theta$ to $\partial J_8(\theta,\kappa)/\partial\theta$ mean. We use once more the equivalence of the sum of a step function with the step width 1 and an integral over the step function. The functions $K_1(n,\kappa)$ to $K_8(n,\kappa)$ in Eq.(6.3-75) are step functions with a step width 1 for $n = 0, 1, \ldots$ and an unspecified value of κ. Using the definition of $J_j(\theta,\kappa)$ in Eq.(6.3-75) we may write:

$$\frac{\partial J_j(\theta,\kappa)}{\partial\theta} = \frac{\partial}{\partial\theta}\sum_{n=0}^{\theta-1} K_j(n,\kappa) = \frac{\partial}{\partial\theta}\int_0^\theta K_j(\theta_n,\kappa)d\theta_n$$
$$= \int_0^\theta \frac{\partial K_j(\theta_n,\kappa)}{\partial\theta_n}d\theta_n = K_j(\theta,\kappa), \quad j = 1,\ 2, \ldots,\ 8 \quad (38)$$

Using $A_{1\kappa}(\kappa)$ and $B_{1\kappa}(\kappa)$ of Eqs.(25) and (26) we may write $\mathcal{H}_{2\kappa}(\kappa)$ of Eq.(30) in the following form:

$$\mathcal{H}_{2\kappa}(\kappa) = \int_0^N [I_{21}(\theta,\kappa) + I_{22}(\theta,\kappa) - I_{23}(\theta,\kappa) + I_{24}(\theta,\kappa)]d\theta$$
$$= \sum_{\theta=0}^{N-1} [I_{21}(\theta,\kappa) + I_{22}(\theta,\kappa) - I_{23}(\theta,\kappa) + I_{24}(\theta,\kappa)] \quad (39)$$

We shall return to $\mathcal{H}_{2\kappa}(\kappa)$ later when we need it for plotting. Here we turn to $\hat{U}_{3\kappa}(\kappa)$ of Eq.(5). As before the index κ is written after the integration over ζ to indicate that the summation over κ in Eq.(3.3-26) is not carried out:

$$\hat{U}_{3\kappa}(\kappa) = 2\alpha\Psi_{00}\frac{L^2}{c\Delta t}\int_0^N \mathcal{R}e\left(\int_0^N \frac{\partial\Psi_0^*}{\partial\zeta}\frac{\partial\hat{v}}{\partial\zeta}d\zeta\right)_\kappa d\theta \tag{40}$$

With the help of Eq.(3.3-26) we get:

$$\frac{\hat{U}_{3\kappa}(\kappa)}{2\alpha\Psi_{00}^2 L^2/c\Delta t} = \mathcal{H}_{3\kappa}(\kappa) = \frac{1}{\Psi_{00}}\int_0^N \mathcal{R}e\left(\int_0^N \frac{\partial\Psi_0^*}{\partial\zeta}\frac{\partial\hat{v}}{\partial\zeta}d\zeta\right)_\kappa d\theta \tag{41}$$

$$\begin{aligned}
\frac{1}{\Psi_{00}}\mathcal{R}e\int_0^N\left(\frac{\partial\Psi_0^*}{\partial\zeta}\frac{\partial\hat{v}}{\partial\zeta}d\zeta\right)_\kappa &= \frac{(2\pi\kappa/N)\{1-\exp[-(\lambda_2^2-\lambda_1^2)^{1/2}N]\}}{\lambda_2^2-\lambda_1^2+(2\pi\kappa/N)^2}\\
&\quad\times 2\lambda_2^2\mathcal{R}e[J_\mathrm{i}(\theta,\kappa)-iJ_\mathrm{r}(\theta,\kappa)]\sin\lambda_1\lambda_3\theta\\
-N\lambda_1\lambda_3\frac{I_\mathrm{T}(\kappa/N)}{\sin\beta_\kappa}\left[\lambda_1^2+\left(\frac{2\pi\kappa}{N}\right)^2\right]&\mathcal{R}e\{[J_\mathrm{i}(\theta\kappa)-iJ_\mathrm{r}(\theta,\kappa)]\cos\lambda_1\lambda_3\\
&\quad-[J_\mathrm{r}(\theta,\kappa)+iJ_\mathrm{i}(\theta,\kappa)]\sin\lambda_1\lambda_3\}\sin\beta_\kappa\theta
\end{aligned} \tag{42}$$

In order to integrate Eq.(42) over θ we need to evaluate numerically the integrals over two terms:

$$I_{31}(\theta,\kappa) = A_{1\kappa}(\kappa)\mathcal{R}e\{2\lambda_2^2[J_\mathrm{i}(\theta,\kappa)-iJ_\mathrm{r}(\theta,\kappa)]\}\sin\lambda_1\lambda_3\theta \tag{43}$$

$$\begin{aligned}
I_{32}(\theta,\kappa) = B_{1\kappa}(\kappa)[\lambda_1^2+(2\pi\kappa/N)^2]\{&\mathcal{R}e[J_\mathrm{i}(\theta,\kappa)-iJ_\mathrm{r}(\theta,\kappa)]\cos\lambda_1\lambda_3\\
&-\mathcal{R}e[J_\mathrm{r}(\theta,\kappa)+iJ_\mathrm{i}(\theta,\kappa)]\sin\lambda_1\lambda_3\}\sin\beta_\kappa\theta
\end{aligned} \tag{44}$$

The functions $J_\mathrm{r}(\theta,\kappa)$ and $J_\mathrm{i}(\theta,\kappa)$ are defined in Eq.(3.3-17). No new problems are encountered as with Eqs.(32)–(35). We may write $\mathcal{H}_{3\kappa}(\kappa)$ of Eq.(41) in the following form:

$$\mathcal{H}_{3\kappa}(\kappa) = \int_0^N [I_{31}(\theta,\kappa)-I_{32}(\theta,\kappa)]d\theta = \sum_{\theta=0}^{N-1}[I_{31}(\theta,\kappa)-I_{32}(\theta,\kappa)] \tag{45}$$

In analogy to U_c in Eqs.(2.5-16) and (2.5-17) we write the energy $\hat{U}_{\alpha\mathrm{c}}$ of Eq.(4) in the following normalized form:

$$\frac{c\Delta t\hat{U}_{\alpha c}}{2\alpha L^2\Psi_{00}^2 N} = \frac{cT\hat{U}_{\alpha c}}{2\alpha L^2\Psi_{00}^2 N^2} = \frac{\mathcal{H}_\alpha}{N^2} \tag{46}$$

$$\hat{U}_{\alpha c} = \frac{2\alpha L^2\Psi_{00}^2}{cT}\mathcal{H}_\alpha = \frac{2\alpha L^2\Psi_{00}^2}{cT}\sum_{\kappa=1}^{N-1}\mathcal{H}_{\alpha\kappa}(\kappa) \tag{47}$$

$$\hat{U}_{\alpha c\kappa}(\kappa) = \frac{2\alpha L^2\Psi_{00}^2}{cT}\mathcal{H}_{\alpha\kappa}(\kappa) \tag{48}$$

$$\mathcal{H}_{\alpha\kappa}(\kappa) = \mathcal{H}_{2\kappa}(\kappa) + \mathcal{H}_{3\kappa}(\kappa) + \frac{m_0^2c^4(\Delta t)^2}{\hbar^2}\mathcal{H}_{1\kappa}(\kappa) \tag{49}$$

A comparison of Eqs.(48) and (49) with Eqs.(2.6-2) and (2.5-18) shows a close similarity.

In order to produce plots of $\mathcal{H}_{1\kappa}(\kappa)$, $\mathcal{H}_{2\kappa}(\kappa)$, and $\mathcal{H}_{3\kappa}(\kappa)$ according to Eqs.(28), (39), and (45) we must decide on a useful choice of N, λ_1, λ_2, λ_3, ϕ_{1e}, and A_{1m}. The easiest parameter to decide on is N. For $N = 100$ our computer required about 5 hours to produce a set of plots for $\mathcal{H}_{1\kappa}(\kappa)$, $\mathcal{H}_{2\kappa}(\kappa)$, and $\mathcal{H}_{3\kappa}(\kappa)$. An increase to $N = 200$ was estimated to require about 40 hours, which was too much.

For λ_2 we obtain from Eq.(2.3-2) the following approximate relation between λ_2 and λ_1:

$$\lambda_2 \doteq \lambda_1\frac{m_0c}{eA_{m0x}} = \frac{\Delta t}{\hbar/m_0c^2} \Longrightarrow\gg \lambda_1 \tag{50}$$

One may use $\lambda_1 = 0.1$ and $\lambda_2 = 10, 20, 40$.

For λ_3 we need large values due to Eq.(2.4-13). We choose $\lambda_3 = 1000, 2000, 10000$. Due to Eqs.(16) and (17) we choose $\phi_{1e} = \pm n\lambda_1\lambda_3$ and $A_{1m} = \pm n\lambda_1$ with n in the range from 1 to 100. We can choose larger or smaller values of n but we cannot choose $n = 0$.

With this background we use $\phi_{1e} = 10\lambda_1\lambda_3$, $A_{1m} = 10\lambda_1$, $\lambda_1 = 0.1$, $\lambda_2 = 10$, $\lambda_3 = 1000$ for Figs.3.4-1 to 3.4-3 that show $\mathcal{H}_{1\kappa}(\kappa)$, $\mathcal{H}_{2\kappa}(\kappa)$, and $\mathcal{H}_{3\kappa}(\kappa)$.

For the second set of plots in Figs.3.4-4 to 3.4-6 we choose $\lambda_2 = 20$ and $\lambda_3 = 2000$ but leave all other parameters unchanged.

The third set of plots in Figs.3.4-7 to 3.4-9 shows $\lambda_2 = 40$ and $\lambda_3 = 10000$. The values $\phi_{1e} = 10\lambda_1\lambda_3$, $A_{1m} = 10\lambda_1$, and $\lambda_1 = 0.1$ remain unchanged.

At this stage we cannot yet recognize common features of Fig.3.4-1 to Fig.3.4-9. A better resolution than $N = 100$ might help but it runs into the problem of computer time.

We have found here eigenvalues of a difference equation that depend in a wide range only on the structure of the difference equation but not on the numerical values of the parameters. One will suspect that this is a result of

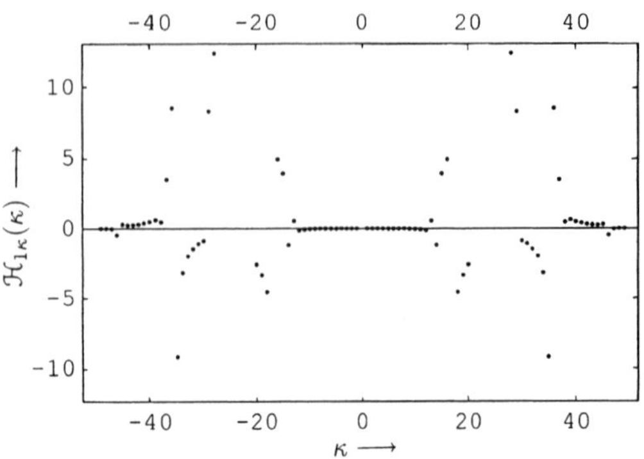

FIG.3.4-1. Plot of $\mathcal{H}_{1\kappa}(\kappa)$ according to Eq.(28) for $N = 100$, $\phi_{1e} = 10\lambda_1\lambda_3$, $A_{1m} = 10\lambda_1$, $\lambda_1 = 0.1$, $\lambda_2 = 10$, $\lambda_3 = 1000$.

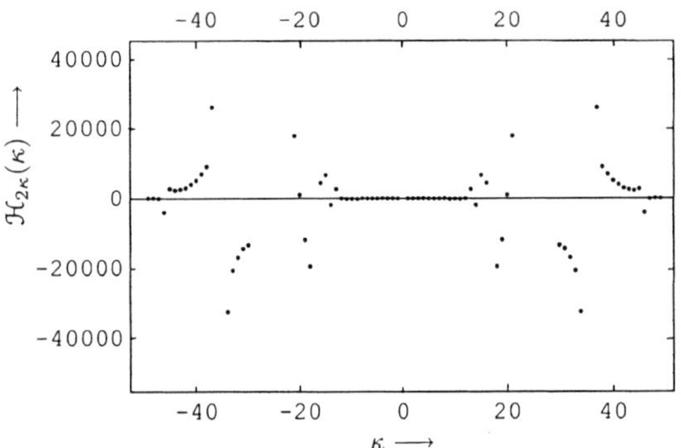

FIG.3.4-2. Plot of $\mathcal{H}_{2\kappa}(\kappa)$ according to Eq.(39) for $N = 100$, $\phi_{1e} = 10\lambda_1\lambda_3$, $A_{1m} = 10\lambda_1$, $\lambda_1 = 0.1$, $\lambda_2 = 10$, $\lambda_3 = 1000$.

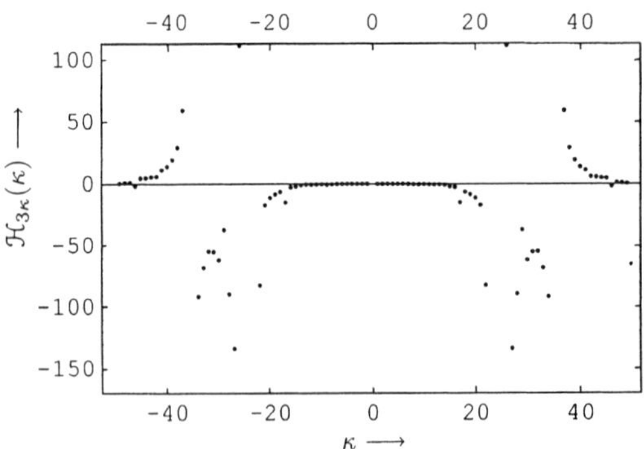

FIG.3.4-3. Plot of $\mathcal{H}_{3\kappa}(\kappa)$ according to Eq.(45) for $N = 100$, $\phi_{1e} = 10\lambda_1\lambda_3$, $A_{1m} = 10\lambda_1$, $\lambda_1 = 0.1$, $\lambda_2 = 10$, $\lambda_3 = 1000$.

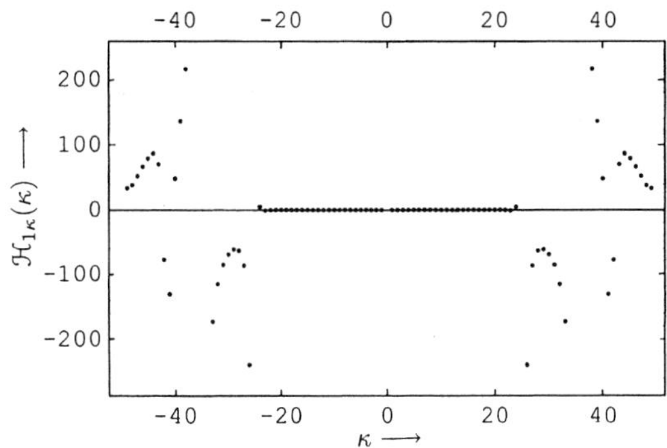

FIG.3.4-4. Plot of $\mathcal{H}_{1\kappa}(\kappa)$ according to Eq.(28) for $N = 100$, $\phi_{1e} = 10\lambda_1\lambda_3$, $A_{1m} = 10\lambda_1$, $\lambda_1 = 0.1$, $\lambda_2 = 20$, $\lambda_3 = 2000$.

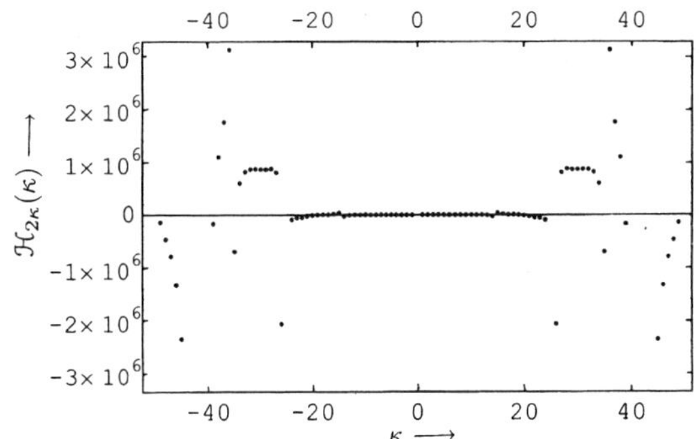

FIG.3.4-5. Plot of $\mathcal{H}_{2\kappa}(\kappa)$ according to Eq.(39) for $N = 100$, $\phi_{1e} = 10\lambda_1\lambda_3$, $A_{1m} = 10\lambda_1$, $\lambda_1 = 0.1$, $\lambda_2 = 20$, $\lambda_3 = 2000$.

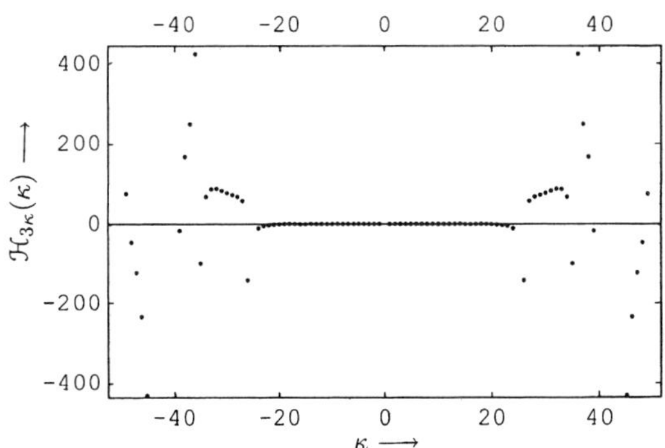

FIG.3.4-6. Plot of $\mathcal{H}_{3\kappa}(\kappa)$ according to Eq.(45) for $N = 100$, $\phi_{1e} = 10\lambda_1\lambda_3$, $A_{1m} = 10\lambda_1$, $\lambda_1 = 0.1$, $\lambda_2 = 20$, $\lambda_3 = 2000$.

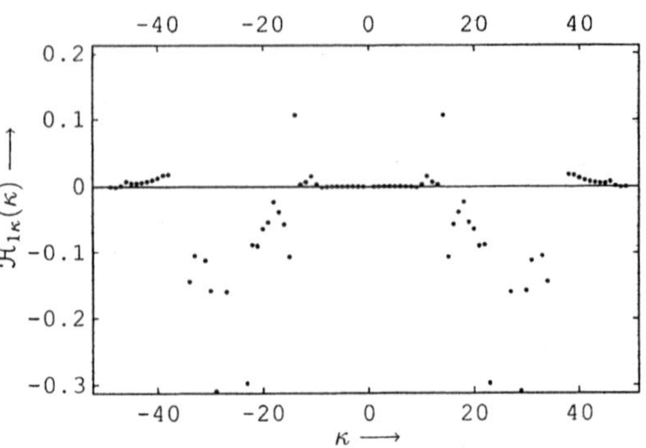

FIG.3.4-7. Plot of $\mathcal{H}_{1\kappa}(\kappa)$ according to Eq.(28) for $N = 100$, $\phi_{1e} = 10\lambda_1\lambda_3$, $A_{1m} = 10\lambda_1$, $\lambda_1 = 0.1$, $\lambda_2 = 40$, $\lambda_3 = 10000$.

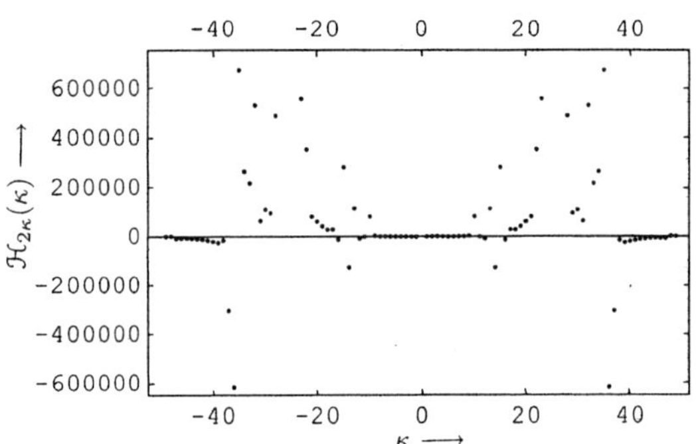

FIG.3.4-8. Plot of $\mathcal{H}_{2\kappa}(\kappa)$ according to Eq.(39) for $N = 100$, $\phi_{1e} = 10\lambda_1\lambda_3$, $A_{1m} = 10\lambda_1$, $\lambda_1 = 0.1$, $\lambda_2 = 40$, $\lambda_3 = 10000$.

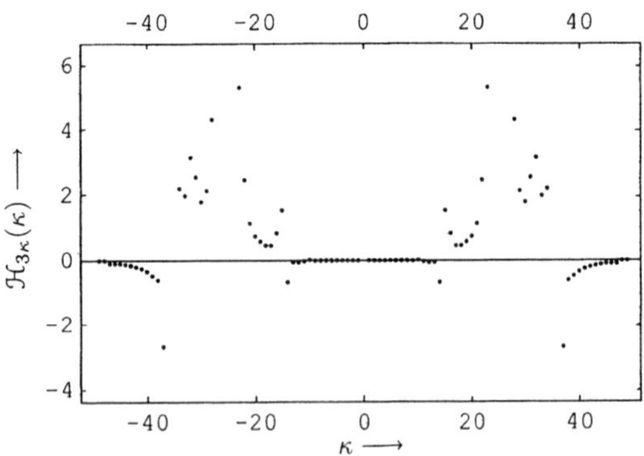

FIG.3.4-9. Plot of $\mathcal{H}_{3\kappa}(\kappa)$ according to Eq.(45) for $N = 100$, $\phi_{1e} = 10\lambda_1\lambda_3$, $A_{1m} = 10\lambda_1$, $\lambda_1 = 0.1$, $\lambda_2 = 40$, $\lambda_3 = 10000$.

replacing an infinite interval by an arbitrarily large but finite interval. For instance, sinusoidal functions that fit into a finite interval have the form $\sin 2\pi n\theta$ or $\cos 2\pi n\theta$ with $n = 1, 2, \ldots$, while there is no restriction on n in an open or half open interval. A restriction of eigenvalues is of great interest in physics, but it is too early to tell whether this particular restriction leads to results of practical interest. We shall explore the matter some more in Section 3.5 with additional plots but we refrain from drawing conclusions at this time.

3.5 Plots for Second-Order Approximation

We are presenting 18 more plots to show the effect of variations of ϕ_{1e}, A_{1m}, λ_1, λ_2, and λ_3 on $\mathcal{H}_{1\kappa}(\kappa)$ to $\mathcal{H}_{3\kappa}(\kappa)$. The parameter $N = 100$ remains unchanged from Figs.3.4-1 to 3.4-9.

Figures 3.4-1 to 3.4-3 are shown with the electric potential ϕ_{1e} reversed from $+10\lambda_1\lambda_3$ to $-10\lambda_1\lambda_3$ in Figs.3.5-1 to 3.5-3. We obtain essentially the amplitude reversed plots, but the vertical scales of Figs.3.4-2 and 3.5-2 are also changed.

In Figs.3.5-4 to 3.5-6 we see the plots of Figs.3.4-1 to 3.4-3 with the magnetic potential A_{1m} reversed from $+10\lambda_1$ to $-10\lambda_1$. We obtain essentially the same plots as in Figs.3.4-1 to 3.4-3. Again, the vertical scales of Figs.3.4-2 and 3.5-5 are not the same. A closer inspection reveals other differences between the two plots.

In Figs.3.5-7 to 3.5-9 both ϕ_{1e} and A_{1m} have a reversed amplitude compared with Figs.3.4-1 to 3.4-3. The plots of Figs.3.5-7 to 3.5-9 appear at first glance to be amplitude reversed plots of Figs.3.4-1 to 3.4-3 but a closer inspection of Fig.3.5-8 shows differences for $\pm\kappa = 15, \ldots, 20$.

In Figs.3.5-10 to 3.5-12 the parameters $\lambda_1 = 0.05$ and $\lambda_3 = 10000$ are changed compared with Figs.3.4-1 to 3.4-3. The plots are changed more drastically than by the changes of ϕ_{1e} and A_{1m}.

In Figs.3.5-13 to 3.5-15 the parameter λ_1 is reduced from 0.05 to 0.02 compared with Figs.3.5-10 to 3.5-12. The plots of Figs.3.5-13 and 3.5-15 look more like those of Figs.3.4-1 and 3.4-3 rather than those of Figs.3.5-10 and 3.5-12. If we look at the products $\lambda_1\lambda_3 = 100$ in Figs.3.4-1 to 3.4-3, $\lambda_1\lambda_3 = 500$ in Figs.3.5-10 to 3.5-12, and $\lambda_1\lambda_3 = 200$ in Figs.3.5-13 to 3.5-15 we recognize that the product $\lambda_1\lambda_3$ is important for the similarity of the illustrations, as suggested by Eq.(2.4-13).

Finally, we show in Figs.3.5-16 to 3.5-18 the plots of Figs.3.4-1 to 3.4-3 with the changes $\lambda_1 = 0.01$ and $\lambda_3 = 10000$ instead of $\lambda_1 = 0.1$ and $\lambda_3 = 1000$. The product $\lambda_1\lambda_3 = 100$ is the same in both cases and the illustrations look very similar. However, a closer inspection shows differences in Figs.3.4-2 and 3.5-17 around $\pm\kappa = 20$.

We must once more ask the reader for patience until Section 4.4 for an explanation why we show so many plots here without deriving any conclusions from them.

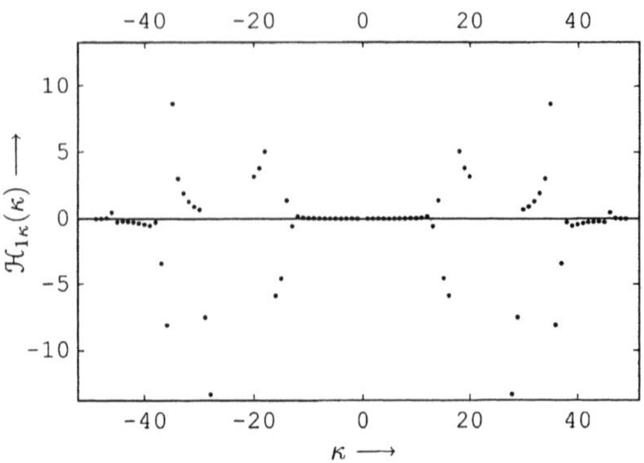

FIG.3.5-1. Plot of $\mathcal{H}_{1\kappa}(\kappa)$ according to Eq.(3.4-28) for $N = 100$, $\phi_{1\mathrm{e}} = -10\lambda_1\lambda_3$, $A_{1\mathrm{m}} = 10\lambda_1$, $\lambda_1 = 0.1$, $\lambda_2 = 10$, $\lambda_3 = 1000$.

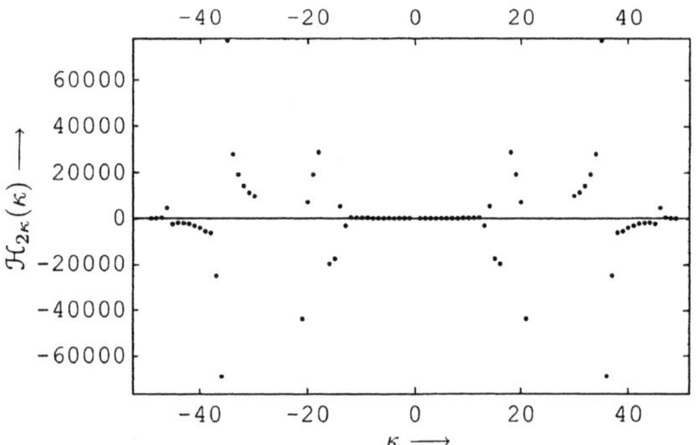

FIG.3.5-2. Plot of $\mathcal{H}_{2\kappa}(\kappa)$ according to Eq.(3.4-39) for $N = 100$, $\phi_{1\mathrm{e}} = -10\lambda_1\lambda_3$, $A_{1\mathrm{m}} = 10\lambda_1$, $\lambda_1 = 0.1$, $\lambda_2 = 10$, $\lambda_3 = 1000$.

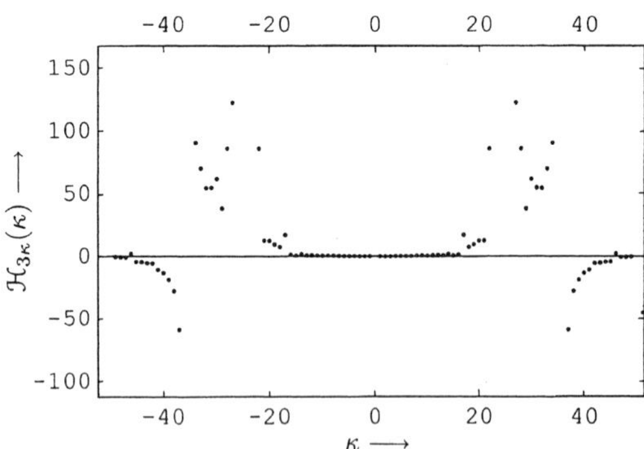

FIG.3.5-3. Plot of $\mathcal{H}_{3\kappa}(\kappa)$ according to Eq.(3.4-45) for $N = 100$, $\phi_{1\mathrm{e}} = -10\lambda_1\lambda_3$, $A_{1\mathrm{m}} = 10\lambda_1$, $\lambda_1 = 0.1$, $\lambda_2 = 10$, $\lambda_3 = 1000$.

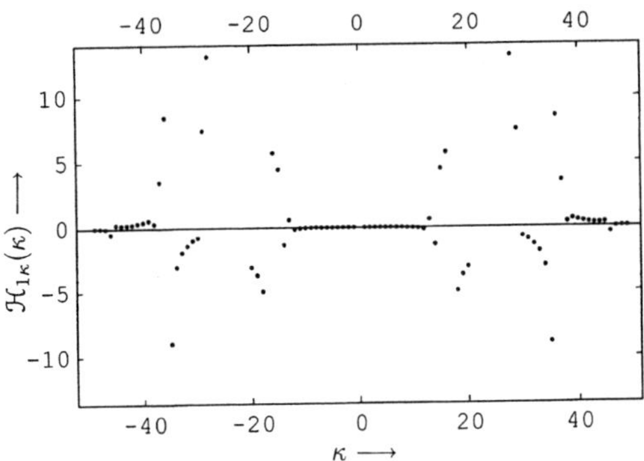

FIG.3.5-4. Plot of $\mathcal{H}_{1\kappa}(\kappa)$ according to Eq.(3.4-28) for $N = 100$, $\phi_{1e} = 10\lambda_1\lambda_3$, $A_{1m} = -10\lambda_1$, $\lambda_1 = 0.1$, $\lambda_2 = 10$, $\lambda_3 = 1000$.

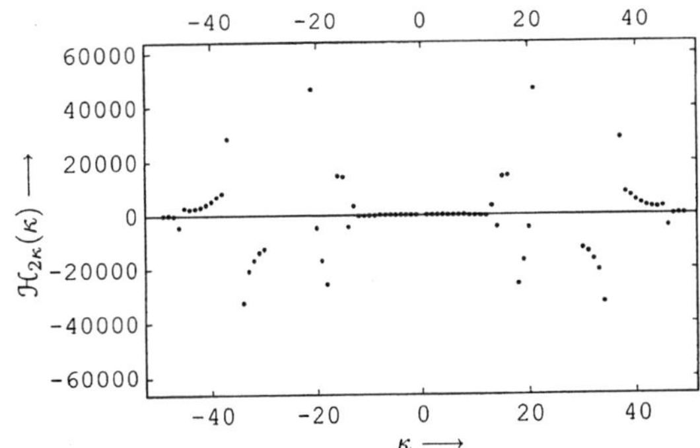

FIG.3.5-5. Plot of $\mathcal{H}_{2\kappa}(\kappa)$ according to Eq.(3.4-39) for $N = 100$, $\phi_{1e} = 10\lambda_1\lambda_3$, $A_{1m} = -10\lambda_1$, $\lambda_1 = 0.1$, $\lambda_2 = 10$, $\lambda_3 = 1000$.

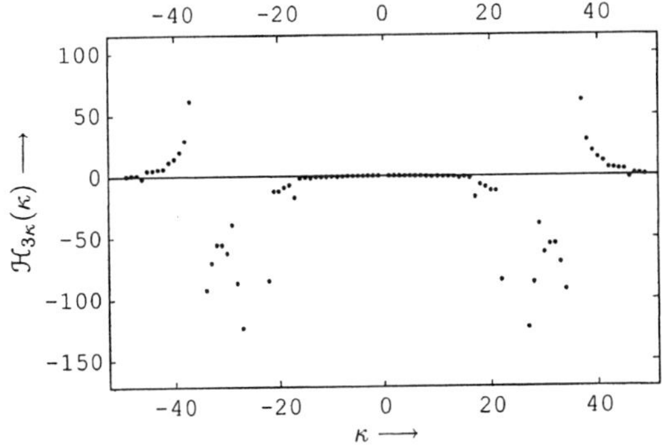

FIG.3.5-6. Plot of $\mathcal{H}_{3\kappa}(\kappa)$ according to Eq.(3.4-45) for $N = 100$, $\phi_{1e} = 10\lambda_1\lambda_3$, $A_{1m} = -10\lambda_1$, $\lambda_1 = 0.1$, $\lambda_2 = 10$, $\lambda_3 = 1000$.

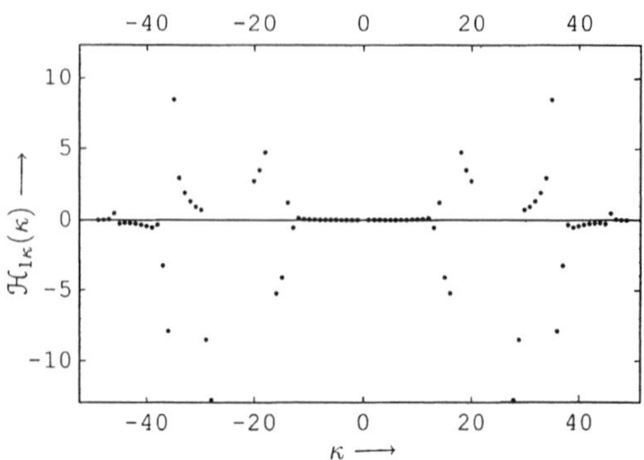

FIG.3.5-7. Plot of $\mathcal{H}_{1\kappa}(\kappa)$ according to Eq.(3.4-28) for $N = 100$, $\phi_{1e} = -10\lambda_1\lambda_3$, $A_{1m} = -10\lambda_1$, $\lambda_1 = 0.1$, $\lambda_2 = 10$, $\lambda_3 = 1000$.

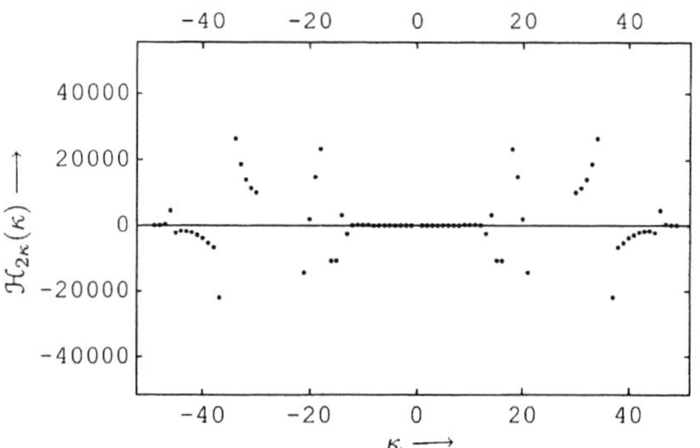

FIG.3.5-8. Plot of $\mathcal{H}_{2\kappa}(\kappa)$ according to Eq.(3.4-39) for $N = 100$, $\phi_{1e} = -10\lambda_1\lambda_3$, $A_{1m} = -10\lambda_1$, $\lambda_1 = 0.1$, $\lambda_2 = 10$, $\lambda_3 = 1000$.

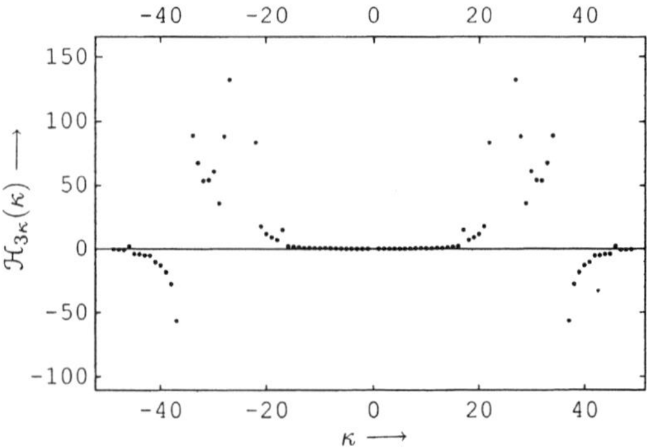

FIG.3.5-9. Plot of $\mathcal{H}_{3\kappa}(\kappa)$ according to Eq.(3.4-45) for $N = 100$, $\phi_{1e} = -10\lambda_1\lambda_3$, $A_{1m} = -10\lambda_1$, $\lambda_1 = 0.1$, $\lambda_2 = 10$, $\lambda_3 = 1000$.

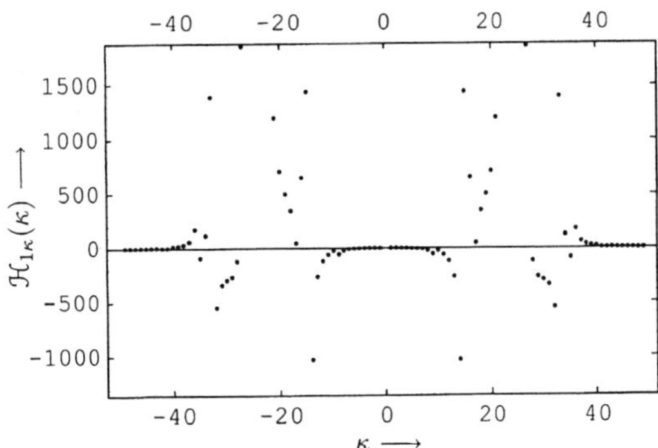

FIG.3.5-10. Plot of $\mathcal{H}_{1\kappa}(\kappa)$ according to Eq.(3.4-28) for $N = 100$, $\phi_{1e} = 10\lambda_1\lambda_3$, $A_{1m} = 10\lambda_1$, $\lambda_1 = 0.05$, $\lambda_2 = 10$, $\lambda_3 = 10000$.

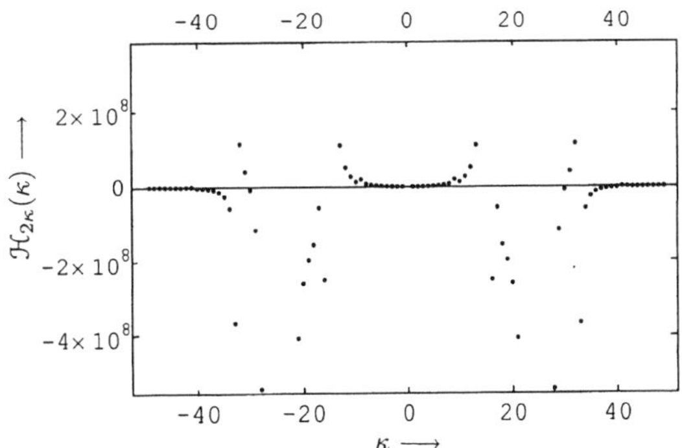

FIG.3.5-11. Plot of $\mathcal{H}_{2\kappa}(\kappa)$ according to Eq.(3.4-39) for $N = 100$, $\phi_{1e} = 10\lambda_1\lambda_3$, $A_{1m} = 10\lambda_1$, $\lambda_1 = 0.05$, $\lambda_2 = 10$, $\lambda_3 = 10000$.

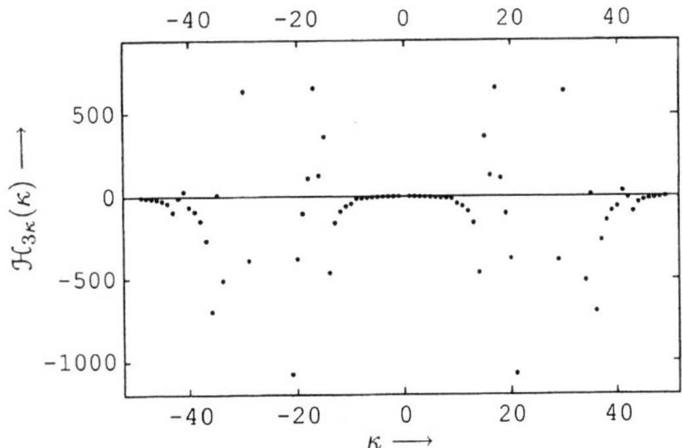

FIG.3.5-12. Plot of $\mathcal{H}_{3\kappa}(\kappa)$ according to Eq.(3.4-45) for $N = 100$, $\phi_{1e} = 10\lambda_1\lambda_3$, $A_{1m} = 10\lambda_1$, $\lambda_1 = 0.05$, $\lambda_2 = 10$, $\lambda_3 = 10000$.

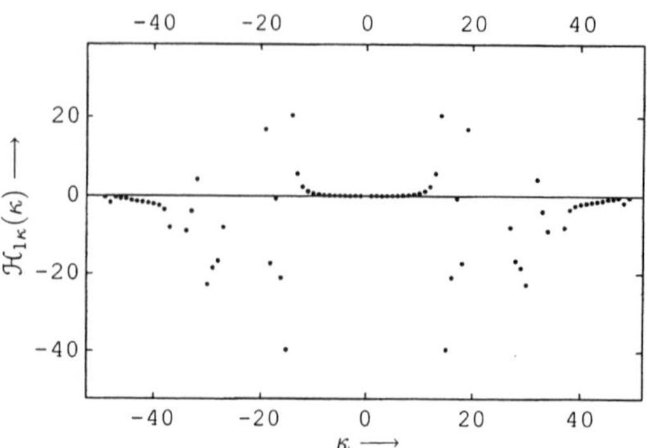

FIG.3.5-13. Plot of $\mathcal{H}_{1\kappa}(\kappa)$ according to Eq.(3.4-28) for $N = 100$, $\phi_{1e} = 10\lambda_1\lambda_3$, $A_{1m} = 10\lambda_1$, $\lambda_1 = 0.02$, $\lambda_2 = 10$, $\lambda_3 = 10000$.

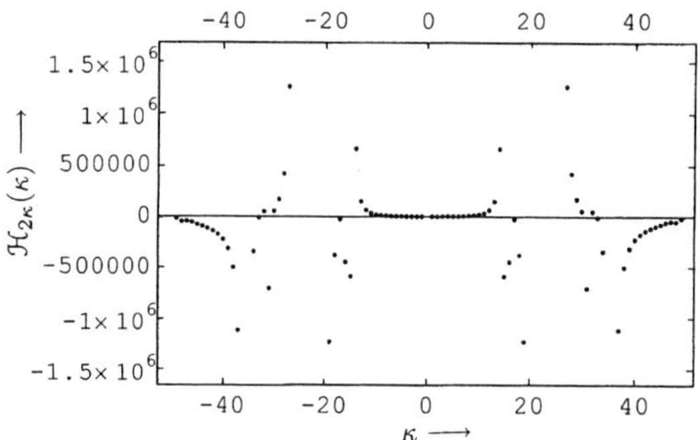

FIG.3.5-14. Plot of $\mathcal{H}_{2\kappa}(\kappa)$ according to Eq.(3.4-39) for $N = 100$, $\phi_{1e} = 10\lambda_1\lambda_3$, $A_{1m} = 10\lambda_1$, $\lambda_1 = 0.02$, $\lambda_2 = 10$, $\lambda_3 = 10000$.

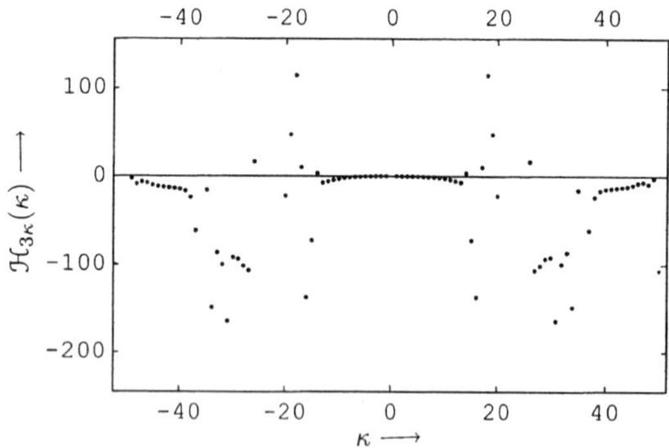

FIG.3.5-15. Plot of $\mathcal{H}_{3\kappa}(\kappa)$ according to Eq.(3.4-45) for $N = 100$, $\phi_{1e} = 10\lambda_1\lambda_3$, $A_{1m} = 10\lambda_1$, $\lambda_1 = 0.02$, $\lambda_2 = 10$, $\lambda_3 = 10000$.

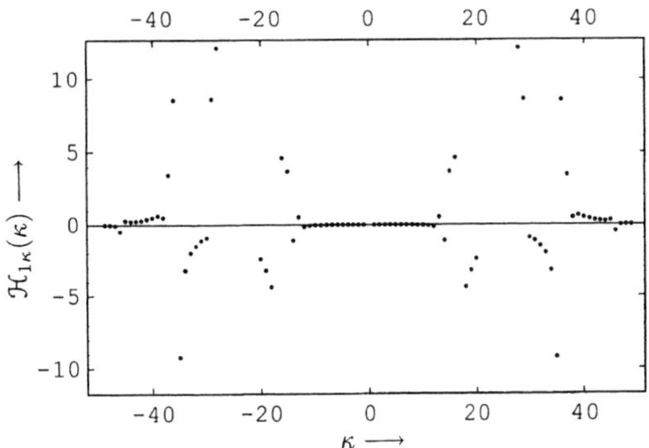

FIG.3.5-16. Plot of $\mathcal{H}_{1\kappa}(\kappa)$ according to Eq.(3.4-28) for $N = 100$, $\phi_{1e} = 10\lambda_1\lambda_3$, $A_{1m} = 10\lambda_1$, $\lambda_1 = 0.01$, $\lambda_2 = 10$, $\lambda_3 = 10000$.

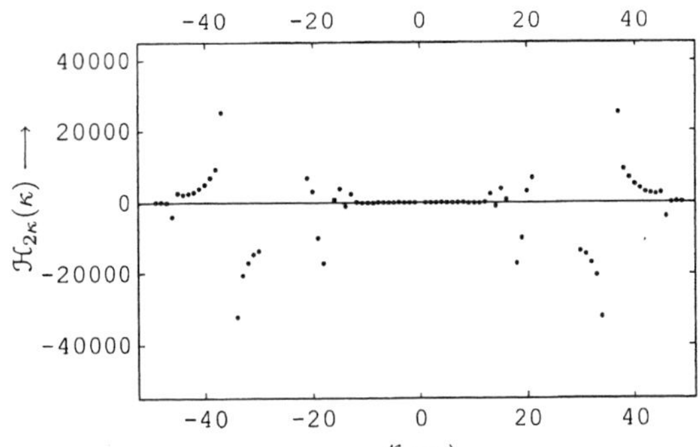

FIG.3.5-17. Plot of $\mathcal{H}_{2\kappa}(\kappa)$ according to Eq.(3.4-39) for $N = 100$, $\phi_{1e} = 10\lambda_1\lambda_3$, $A_{1m} = 10\lambda_1$, $\lambda_1 = 0.01$, $\lambda_2 = 10$, $\lambda_3 = 10000$.

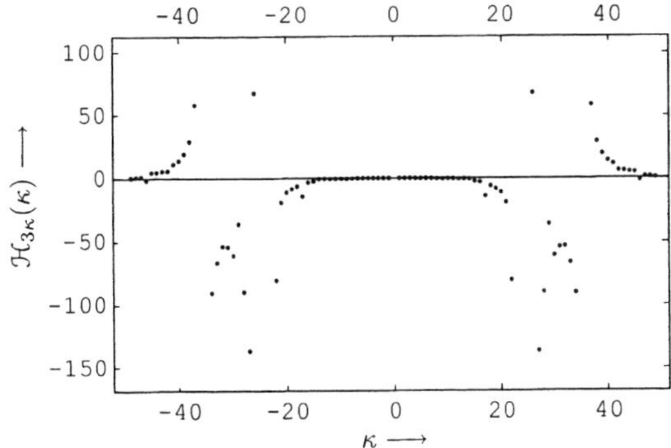

FIG.3.5-18. Plot of $\mathcal{H}_{3\kappa}(\kappa)$ according to Eq.(3.4-45) for $N = 100$, $\phi_{1e} = 10\lambda_1\lambda_3$, $A_{1m} = 10\lambda_1$, $\lambda_1 = 0.01$, $\lambda_2 = 10$, $\lambda_3 = 10000$.

For the first line of Eq.(3.1-3), written as a homogeneous equation, we obtain again Eq.(3.1-6):

$$\Big([\Psi_1(\zeta+1,\theta)-2\Psi_1(\zeta,\theta)+\Psi_1(\zeta-1,\theta)]-[\Psi_1(\zeta,\theta+1)-2\Psi_1(\zeta,\theta)+\Psi_1(\zeta,\theta-1)]$$
$$-i\lambda_1\Big\{[\Psi_1(\zeta+1,\theta)-\Psi_1(\zeta-1,\theta)]+\lambda_3[\Psi_1(\zeta,\theta+1)-\Psi_1(\zeta,\theta-1)]\Big\}$$
$$-\lambda_2^2\Psi_1(\zeta,\theta)\Big)\left(-\frac{\hbar^2}{(\Delta x)^2}\right)=0$$
$$\Psi_1=\Psi_{\mathrm{x}1x},\quad \lambda_1,\ \lambda_2,\ \lambda_3 \text{ see Eq.(2.3-2)} \tag{1}$$

For the second line of Eq.(3.1-3) we obtain with the help of Eqs.(3.1-11) and (6.2-1) for small values of $\Delta x = c\Delta t$ and $\Delta\zeta = 1$:

$$\Delta x \ll \lambda_{\mathrm{C}} = h/m_0c, \quad \Psi_0 = \Psi_{\mathrm{x}0x}$$
$$\frac{4\lambda_{\mathrm{C}}A_{\mathrm{e}}(\zeta,\theta)}{e}\left(\frac{\hbar}{i\Delta x}\frac{\tilde{\Delta}}{\tilde{\Delta}\zeta}-eA_{\mathrm{m}0x}\right)^2$$
$$\times\left[1-\frac{1}{2m_0^2c^2}\left(\frac{\hbar}{i\Delta x}\frac{\tilde{\Delta}}{\tilde{\Delta}\zeta}-eA_{\mathrm{m}0x}\right)^2\right]\Psi_0(\zeta,\theta)$$
$$\doteq -\frac{\hbar^2}{(\Delta x)^2}\frac{2\lambda_{\mathrm{C}}A_{\mathrm{e}}(\zeta,\theta)}{e}\left(\frac{\hbar}{m_0c\Delta x}\right)^2\left(\frac{\tilde{\Delta}}{\tilde{\Delta}\zeta}-\frac{ie\Delta x}{\hbar}A_{\mathrm{m}0x}\right)^4\Psi_0(\zeta,\theta)$$
$$= -\frac{\hbar^2}{(\Delta x)^2}\frac{2\lambda_{\mathrm{C}}A_{\mathrm{e}}(\zeta,\theta)}{e}\left(\frac{\hbar}{m_0c\Delta x}\right)^2\Big[[\Psi_0(\zeta+2,\theta)-4\Psi_0(\zeta+1,\theta)$$
$$+6\Psi_0(\zeta,\theta)-4\Psi_0(\zeta-1,\theta)+\Psi_0(\zeta-2,\theta)]$$
$$+2\frac{ie\Delta xA_{\mathrm{m}0x}}{\hbar}[\Psi_0(\zeta+2,\theta)-2\Psi_0(\zeta+1,\theta)+2\Psi_0(\zeta-1,\theta)-\Psi_0(\zeta-2,\theta)]$$
$$-6\left(\frac{ie\Delta xA_{\mathrm{m}0x}}{\hbar}\right)^2[\Psi_0(\zeta+1,\theta)-2\Psi_0(\zeta,\theta)+\Psi_0(\zeta-1,\theta)]$$
$$+2\left(\frac{ie\Delta xA_{\mathrm{m}0x}}{\hbar}\right)^3[\Psi_0(\zeta+1,\theta)-\Psi_0(\zeta-1,\theta)]$$
$$-\left(\frac{ie\Delta xA_{\mathrm{m}0x}}{\hbar}\right)^4\Psi_0(\zeta,\theta)\Big] \tag{2}$$

We turn to the terms in lines 3 and 4 of Eq.(3.1-3) multiplied by $L_{\mathrm{cx}11}$. They are shown for $\Delta x \ll h/m_0c$ in Eq.(3.1-18). This equation is worked out in Section 6.2 and Eq.(6.2-3) is obtained. We choose $\Delta\theta = 1$, $\Delta\zeta = 1$ in Eq.(6.2-3):

$$-i\frac{m_0(\Delta x)^2}{\alpha\hbar\Delta t}V_{\text{cx31}}(\zeta,\theta)\bigg(\frac{1}{2}[\Psi_0(\zeta,\theta{+}1)-\Psi_0(\zeta,\theta{-}1)]+\frac{ie\Delta t}{\hbar}\phi_{\text{e0}}\Psi_0(\zeta,\theta)\bigg) \qquad 18$$

$$-i\frac{m_0(\Delta x)^2}{2\alpha\hbar\Delta t}C_{\text{cx51}}\bigg(\frac{1}{2}[\Psi_0(\zeta,\theta+1)-\Psi_0(\zeta,\theta-1)]+\frac{ie\Delta t}{\hbar}\phi_{\text{e0}}\Psi_0(\zeta,\theta)\bigg) \qquad 19$$

$$-e\left(\frac{\Delta x}{\hbar}\right)^2\bigg[A_{\text{m1}x}(\zeta,\theta)\bigg(\frac{\hbar}{2i\Delta x}[\Psi_0(\zeta{+}1,\theta)-\Psi_0(\zeta{-}1,\theta)]-eA_{\text{m0}x}\Psi_0(\zeta,\theta)\bigg) \qquad 20$$

$$+\frac{\hbar}{2i\Delta x}[\Psi_0(\zeta+1,\theta)A_{\text{m1}x}(\zeta+1,\theta)-\Psi_0(\zeta-1,\theta)A_{\text{m1}x}(\zeta-1,\theta)] \qquad 21$$

$$-2eA_{\text{m0}x}A_{\text{m1}x}(\zeta,\theta)\Psi_0(\zeta,\theta)\bigg] \qquad 22$$

$$-\frac{e}{c^2}\left(\frac{\Delta x}{\hbar}\right)^2\bigg[\phi_{\text{e1}}(\zeta,\theta)\bigg(\frac{\hbar}{2i\Delta t}[\Psi_0(\zeta,\theta{+}1)-\Psi_0(\zeta,\theta{-}1)]+e\phi_{\text{e0}}\Psi_0(\zeta,\theta)\bigg) \qquad 23$$

$$+\frac{\hbar}{2i\Delta t}[\Psi_0(\zeta,\theta+1)\phi_{\text{e1}}(\zeta,\theta+1)-\Psi_0(\zeta,\theta-1)\phi_{\text{e1}}(\zeta,\theta-1)] \qquad 24$$

$$+2e\phi_{\text{e0}}\phi_{\text{e1}}(\zeta,\theta)\Psi_0(\zeta,\theta)\bigg] \qquad 25 \quad (4)$$

A comparison of this equation with Eq.(3.1-32) shows that the transition from $\Delta x \gg \lambda_{\text{C}}$ to $\Delta x \ll \lambda_{\text{C}}$ increases the number of lines in the equation from 16 to 25.

4.2 Evaluation of Eq.(4.1-4)

Before attempting to find solutions of Eq.(4.1-4) we must radically simplify it. A first step is to make as many as possible of the constant components of $\mathbf{A}_{\text{e}}$, $\mathbf{A}_{\text{m}}$, ϕ_{e}, and ϕ_{m} in Eq.(4.1-4) zero. As in Section 3.2 one must retain the constants $A_{\text{m0}x}$ and ϕ_{e0} to keep Eq.(2.3-2) valid. In addition, the magnitude $A_{\text{e}}(\zeta,\theta)$ of the electric vector potential $\mathbf{A}_{\text{e}}$ cannot be ignored now as will be seen presently. Only the following components can be chosen to be zero:

$$A_{\text{m0}y}=A_{\text{m0}z}=0, \quad \phi_{\text{m0}}=0 \qquad (1)$$

Lines 1 and 2 of Eq.(4.1-4) remain unchanged. They are equal to the left side of Eq.(2.3-2) except that Ψ_0 is replaced by Ψ_1.

In line 3 of Eq.(4.1-4) we see the factor $A_{\text{e}}(\zeta,\theta)=|\mathbf{A}_{\text{e}}(\zeta,\theta)|$ that made us discard lines 3 and 4 of Eq.(3.1-32) in the third paragraph of Section 3.2. This is now different because there is the additional term $(\hbar/m_0c\Delta x)^2$ that becomes very large for small values of Δx. Hence, we retain lines 3 to 7 of Eq.(4.1-4).

Lines 8–17 of Eq.(4.1-1) are multiplied by the factor $A_{\text{e0}x}A_{\text{m0}y}-A_{\text{e0}y}A_{\text{m0}x}$, which is zero according to Eq.(1). These lines are discarded.

Line 18 is equal to line 9 in Eq.(3.1-32) and is discarded. Line 19 equals line 10 in Eq.(3.1-32); it becomes very small for small values of Δx and $\Delta t=\Delta x/c$.

Lines 20–25 in Eq.(4.1-4) are all multiplied by Δx if we substitute $c\Delta t = \Delta x$. These lines become very small for small values of Δx unless at least one of the potentials $A_{\mathrm{m}1x}(\zeta,\theta)$ or $\phi_{\mathrm{e}1}(\zeta,\theta)$ becomes very large for some value of ζ or θ. We exclude here such large potentials in order to demonstrate the effect of lines 3 to 7 of Eq.(4.1-4).

Equation (4.1-4) is reduced more than Eq.(3.1-32) was to yield Eqs.(3.2-3) and (3.2-4). We write $\tilde{\Psi}_1$ to distinguish it from Ψ_1 in Eq. (3.2-3):

$$\begin{aligned}
&[\tilde{\Psi}_1(\zeta+1,\theta) - 2\tilde{\Psi}_1(\zeta,\theta) + \tilde{\Psi}_1(\zeta-1,\theta)] - [\tilde{\Psi}_1(\zeta,\theta+1) - 2\tilde{\Psi}_1(\zeta,\theta) + \tilde{\Psi}_1(\zeta,\theta-1)] \\
&-i\lambda_1\Big\{[\tilde{\Psi}_1(\zeta+1,\theta) - \tilde{\Psi}_1(\zeta-1,\theta)] + \lambda_3[\tilde{\Psi}_1(\zeta,\theta+1) - \tilde{\Psi}_1(\zeta,\theta-1)]\Big\} - \lambda_2^2\tilde{\Psi}_1(\zeta,\theta) \\
&= G_2(\zeta,\theta) \qquad (2)
\end{aligned}$$

$$\begin{aligned}
G_2(\zeta,\theta) = -\frac{2\lambda_{\mathrm{C}} A_{\mathrm{e}}(\zeta,\theta)}{e}\left(\frac{\hbar}{m_0 c\Delta x}\right)^2 \Bigg[&[\Psi_0(\zeta+2,\theta) - 4\Psi_0(\zeta+1,\theta) \\
&+ 6\Psi_0(\zeta,\theta) - 4\Psi_0(\zeta-1,\theta) + \Psi_0(\zeta-2,\theta)] \\
+ 2i\frac{e\Delta x A_{\mathrm{m}0x}}{\hbar}&[\Psi_0(\zeta+2,\theta) - 2\Psi_0(\zeta+1,\theta) + 2\Psi_0(\zeta-1,\theta) - \Psi_0(\zeta-2,\theta)] \\
+ 6\left(\frac{e\Delta x A_{\mathrm{m}0x}}{\hbar}\right)^2 &[\Psi_0(\zeta+1,\theta) - 2\Psi_0(\zeta,\theta) + \Psi_0(\zeta-1,\theta)] \\
-2i\left(\frac{e\Delta x A_{\mathrm{m}0x}}{\hbar}\right)^3 &[\Psi_0(\zeta+1,\theta) - \Psi_0(\zeta-1,\theta)] - \left(\frac{e\Delta x A_{\mathrm{m}0x}}{\hbar}\right)^4 \Psi_0(\zeta,\theta)\Bigg] \qquad (3)
\end{aligned}$$

Let us compare these two equations with the corresponding Eqs.(3.2-3) and (3.2-4) for $\Delta x \ll \lambda_{\mathrm{C}} = \hbar/m_0 c$. The homogeneous Eqs.(3.2-3) and (2) are equal; the only difference is in the inhomogeneous terms $G_1(\zeta,\theta)$ and $G_2(\zeta,\theta)$. A conspicuous difference is the magnitude $A_{\mathrm{e}}(\zeta,\theta)$ of the electric vector potential $\mathbf{A}_{\mathrm{e}}(\zeta,\theta)$ that is caused by magnetic (dipole) current densities according to Eq.(1.1-19). It would occur in Eq.(3.2-4) too but there we were able to make it a small contribution that was ignored. The factor $(\hbar/m_0 c\Delta x)^2$ makes this impossible now. But the same factor makes it possible to ignore now all the terms of $G_1(\zeta,\theta)$ in Eq.(3.2-4) if we exclude extreme values of the magnitudes $A_{\mathrm{m}1x}$, $A_{\mathrm{m}0x}$, $\phi_{\mathrm{e}1}$, and $\phi_{\mathrm{e}0}$ of magnetic and electric potentials.

This result implies that the electric vector potential $\mathbf{A}_{\mathrm{e}}$ introduced by the modified Maxwell equations is of great importance for differences $\Delta x \ll \lambda_{\mathrm{C}}$ but of much less importance for $\Delta x \gg \lambda_{\mathrm{C}}$. Vice versa one might say the usual Maxwell equations may or may not work for $\Delta x \gg \lambda_{\mathrm{C}}$ but they cannot work for $\Delta x \ll \lambda_{\mathrm{C}}$.

We substitute $\Delta x = cT/N$ and $\hbar/m_0 c = 2\pi\lambda_{\mathrm{C}}$ in Eq.(3). Note that $A_{\mathrm{e}}(\zeta,\theta)$ as magnitude of $\mathbf{A}_{\mathrm{e}}(\zeta,\theta)$ is never negative.

$$\frac{\lambda_C}{\Delta x} = \frac{h}{m_0 c \Delta x} = \frac{\lambda_C}{cT/N} = \frac{\lambda_C}{cT} N \tag{4}$$

$$\begin{aligned} \frac{2\lambda_C A_e(\zeta,\theta)}{e}\left(\frac{\hbar}{m_0 c \Delta x}\right)^2 &= \frac{1}{2\pi^2}\frac{\lambda_C A_e(\zeta,\theta)}{e}\left(\frac{\lambda_C}{cT}\right)^2 N^2 \\ &= p_1(\zeta,\theta)N^2 = p_N(\zeta,\theta) \end{aligned} \tag{5}$$

$$\frac{e\Delta x A_{m0x}}{\hbar} = \frac{ecT A_{m0x}}{\hbar N} = \lambda_1 \tag{6}$$

$$p_1(\zeta,\theta) = \frac{1}{2\pi^2}\frac{\lambda_C A_e(\zeta,\theta)}{e}\left(\frac{\lambda_C}{cT}\right)^2 \tag{7}$$

Equation (3) is rewritten:

$$\begin{aligned} G_2(\zeta,\theta) = &-p_N(\zeta,\theta)[(1+2i\lambda_1)\Psi_0(\zeta+2,\theta) \\ &-(4+4i\lambda_1-6\lambda_1^2+2i\lambda_1^3)\Psi_0(\zeta+1,\theta)+(6-12\lambda_1^2-\lambda_1^4)\Psi_0(\zeta,\theta) \\ &-(4-4i\lambda_1-6\lambda_1^2-2i\lambda_1^3)\Psi_0(\zeta-1,\theta)+(1-2i\lambda_1)\Psi_0(\zeta-2,\theta)] \end{aligned} \tag{8}$$

A general solution of the homogeneous Eq.(2) plus a particular solution of the inhomogeneous equation is needed. The boundary condition of Eq.(3.2-5) is used once more:

$$\begin{aligned} \tilde{\Psi}_1(0,\theta) = \tilde{\Psi}_{00}S(\theta)(1-e^{-\iota\theta}) &= 0 && \text{for } \theta < 0 \\ &= \tilde{\Psi}_{00}(1-e^{-\iota\theta}) && \text{for } \theta \geq 0 \end{aligned} \tag{9}$$

With the initial conditions of Eqs.(3.2-6) and (3.2-7) we again obtain Eq.(3.2-8) but we must write $\bar{u}(\zeta,\theta)$ rather than $\hat{u}(\zeta,\theta)$ since these functions will become different.

$$\tilde{\Psi}_1(\zeta,\theta) = \tilde{\Psi}_{00}[(1-e^{-\iota\theta})F(\zeta)+\bar{u}(\zeta,\theta)] \tag{10}$$

The function $F(\zeta)$ and the constant ι are the same as in Eqs.(3.2-9) and (3.2-10). The boundary and initial conditions for $\bar{u}(\zeta,\theta)$ follows from Eqs. (3.2-11) to (3.2-13) by the substitution of $\bar{u}$ for $\hat{u}$:

$$\bar{u}(0,\theta) = 0 \qquad \theta \geq 0 \tag{11}$$

$$\bar{u}(\zeta,0) = 0 \qquad \zeta > 0 \tag{12}$$

$$\bar{u}(\zeta,1) - \bar{u}(\zeta,0) = -(1-e^{-\iota})F(\zeta) \qquad \zeta > 0 \tag{13}$$

Substitution of $\bar{u}(\zeta,\theta)$ of Eq.(10) into Eq.(2) yields the same equation but $\tilde{\Psi}_1$ is replaced by $\bar{u}$:

$$\begin{aligned}&[\bar{u}(\zeta+1,\theta)-2\bar{u}(\zeta,\theta)+\bar{u}(\zeta-1,\theta)]-[\bar{u}(\zeta,\theta+1)+2\bar{u}(\zeta,\theta)+\bar{u}(\zeta,\theta-1)]\\ &\quad -i\lambda_1\Big\{[\bar{u}(\zeta+1,\theta)-\bar{u}(\zeta-1,\theta)]+\lambda_3[\bar{u}(\zeta,\theta+1)-\bar{u}(\zeta,\theta-1)]\Big\}-\lambda_2^2\bar{u}(\zeta,\theta)\\ &\qquad = G_2(\zeta,\theta)\end{aligned} \tag{14}$$

The solution of Eq.(14) is written in the following form:

$$\bar{u}(\zeta,\theta)=u(\zeta,\theta)+\bar{v}(\zeta,\theta) \tag{15}$$

where $u(\zeta,\theta)$ stands for the solution of the homogeneous equation and $\bar{v}(\zeta,\theta)$ for a particular solution of the inhomogeneous equation. The homogeneous solution was expressed in terms of Bernoulli's product method in Eqs.(3.2-16) to (3.2-20). We use here the coefficients A_{50} to A_{53} and A_{60} to A_{63}:

$$u_\kappa(\zeta,\theta)=\phi_\kappa(\zeta)\psi_\kappa(\theta) \tag{16}$$

$$\begin{aligned}\phi_\kappa(\zeta)&=e^{i\lambda_1\zeta}(A_{50}e^{i\varphi_\kappa\zeta}+A_{51}e^{-i\varphi_\kappa\zeta})=e^{i\lambda_1\zeta}(A_{52}\cos\varphi_\kappa\zeta+iA_{53}\sin\varphi_\kappa\zeta)\\ \varphi_\kappa&=2\pi\kappa/N,\quad \kappa=0,\ \pm1,\dots,\ \pm N/2\end{aligned} \tag{17}$$

$$\begin{aligned}\psi_\kappa(\theta)&=e^{i\lambda_1\lambda_3\theta}(A_{61}e^{i\varphi_\kappa\theta}+A_{60}e^{-i\varphi_\kappa\theta})\\ &=e^{i\lambda_1\lambda_3\theta}(A_{62}\cos\varphi_\kappa\theta+iA_{63}\sin\varphi_\kappa\theta)\end{aligned} \tag{18}$$

Consider next the inhomogeneous solution $\bar{v}(\zeta,\theta)$. We go to Eq.(3.2-21) but write $\bar{v}$, $\tilde{S}$, and $\tilde{T}$ instead of $\hat{v}$, S, and T:

$$\begin{aligned}\bar{v}(\zeta,\theta)&=\sum_{\kappa>-\kappa_0}^{<\kappa_0}\left(\tilde{S}(\theta)e^{i\lambda_1\zeta}\cos\frac{2\pi\kappa\zeta}{N}+\tilde{T}(\theta)e^{i\lambda_1\zeta}\sin\frac{2\pi\kappa\zeta}{N}\right)\\ &=\sum_{\kappa=-N/2+1}^{N/2-1}[\tilde{S}(\theta)\phi_c(\zeta)+\tilde{T}(\theta)\phi_s(\zeta)]\end{aligned} \tag{19}$$

Substitution of $\bar{v}(\zeta,\theta)$ into Eq.(2) brings:

$$\sum_{\kappa=-N/2+1}^{N/2-1}$$

$$\Big[\tilde{S}_\kappa(\theta)[\phi_c(\zeta+1)-2\phi_c(\zeta)+\phi_c(\zeta-1)]+\tilde{T}_\kappa(\theta)[\phi_s(\zeta+1)-2\phi_s(\zeta)+\phi_s(\zeta-1)]$$

$$-\phi_c(\zeta)[\tilde{S}_\kappa(\theta+1)-2\tilde{S}_\kappa(\theta)+\tilde{S}_\kappa(\theta-1)]-\phi_s(\zeta)[\tilde{T}_\kappa(\theta+1)-2\tilde{T}_\kappa(\theta)+\tilde{T}_\kappa(\theta-1)]$$
$$-i\lambda_1\Big(\tilde{S}_\kappa(\theta)[\phi_c(\zeta+1)-\phi_c(\zeta-1)]+\tilde{T}_\kappa(\theta)[\phi_s(\zeta+1)-\phi_s(\zeta-1)]$$
$$+\lambda_3\{\phi_c(\zeta)[\tilde{S}_\kappa(\theta+1)-\tilde{S}_\kappa(\theta-1)]+\phi_s(\zeta)[\tilde{T}_\kappa(\theta+1)-\tilde{T}_\kappa(\theta-1)]\}\Big)$$
$$-\lambda_2^2[\tilde{S}_\kappa(\theta)\phi_c(\zeta)+\tilde{T}_\kappa(\theta)\phi_s(\zeta)]\Big]=G_2(\zeta,\theta) \quad (20)$$

The substitutions of Eqs.(3.2-23)–(3.2-28) apply again. Equation (3.2-29) is obtained with S_κ, T_κ, and $G_1(\zeta,\theta)$ replaced by $\tilde{S}_\kappa$, $\tilde{T}_\kappa$, and $G_2(\zeta,\theta)$. We only write the equivalent of the reduced Eq.(3.2-30) derivable from Eq.(20):

$$\sum_{\kappa=-N/2+1}^{N/2-1}\Big[-(1+i\lambda_1\lambda_3)\tilde{S}_\kappa(\theta+1)+\Big(2(\cos\lambda_1+\lambda_1\sin\lambda_1)\cos\frac{2\pi\kappa}{N}-\lambda_2^2\Big)\tilde{S}_\kappa(\theta)$$
$$-(1-i\lambda_1\lambda_3)\tilde{S}_\kappa(\theta-1)+2i(\sin\lambda_1-\lambda_1\cos\lambda_1)\sin\frac{2\pi\kappa}{N}\tilde{T}_\kappa(\theta)\Big]e^{i\lambda_1\zeta}\cos\frac{2\pi\kappa\zeta}{N}$$
$$+\sum_{\kappa=-N/2+1}^{N/2-1}\Big[-(1+i\lambda_1\lambda_3)\tilde{T}_\kappa(\theta+1)+\Big(2(\cos\lambda_1+\lambda_1\sin\lambda_1)\cos\frac{2\pi\kappa}{N}-\lambda_2^2\Big)\tilde{T}_\kappa(\theta)$$
$$-(1-i\lambda_1\lambda_3)\tilde{T}_\kappa(\theta+1)-2i(\sin\lambda_1-\lambda_1\cos\lambda_1)\sin\frac{2\pi\kappa}{N}\tilde{S}_\kappa(\theta)\Big]e^{i\lambda_1\zeta}\sin\frac{2\pi\kappa\zeta}{N}$$
$$=G_2(\zeta,\theta) \quad (21)$$

The left side of Eq.(21) is essentially a Fourier series in terms of ζ. The right side can be represented as a Fourier series too:

$$G_2(\zeta,\theta)=\sum_{\kappa=-N/2+1}^{N/2-1}\Big(\tilde{G}_{s\kappa}(\theta,\kappa)e^{i\lambda_1\zeta}\sin\frac{2\pi\kappa\zeta}{N}+\tilde{G}_{c\kappa}(\theta,\kappa)e^{i\lambda_1\zeta}\cos\frac{2\pi\kappa\zeta}{N}\Big) \quad (22)$$

Multiplication with $2N^{-1}e^{-i\lambda_1\zeta}\sin(2\pi\nu\zeta/N)$ or $2N^{-1}e^{-i\lambda_1\zeta}\cos(2\pi\nu\zeta/N)$ and integration over the orthogonality interval $0\le\zeta\le N$ yields $\tilde{G}_{s\kappa}(\theta,\kappa)$ and $\tilde{G}_{c\kappa}(\theta,\kappa)$:

$$\tilde{G}_{c\kappa}(\theta,\kappa)=\frac{2}{N}\int_0^N G_2(\zeta,\theta)e^{-i\lambda_1\zeta}\cos\frac{2\pi\kappa\zeta}{N}d\zeta \quad \text{for } \nu=\kappa \quad (23)$$

$$\tilde{G}_{s\kappa}(\theta,\kappa)=\frac{2}{N}\int_0^N G_2(\zeta,\theta)e^{-i\lambda_1\zeta}\sin\frac{2\pi\kappa\zeta}{N}d\zeta \quad \text{for } \nu=\kappa \quad (24)$$

Equations (21) and (22) yield for every component κ of the two sums the following two equations:

$$\begin{aligned}&-(1+i\lambda_1\lambda_3)\tilde{S}_\kappa(\theta+1)+\left(2(\cos\lambda_1+\lambda_1\sin\lambda_1)\cos\frac{2\pi\kappa}{N}-\lambda_2^2\right)\tilde{S}_\kappa(\theta)\\&\quad-(1-i\lambda_1\lambda_3)\tilde{S}_\kappa(\theta-1)-2i(\sin\lambda_1-\lambda_1\cos\lambda_1)\sin\frac{2\pi\kappa}{N}\tilde{T}_\kappa(\theta)=\tilde{G}_{c\kappa}(\theta,\kappa)\end{aligned}\quad(25)$$

$$\begin{aligned}&-(1+i\lambda_1\lambda_3)\tilde{T}_\kappa(\theta+1)+\left(2(\cos\lambda_1+\lambda_1\sin\lambda_1)\cos\frac{2\pi\kappa}{N}-\lambda_2^2\right)\tilde{T}_\kappa(\theta)\\&\quad-(1-i\lambda_1\lambda_3)\tilde{T}_\kappa(\theta-1)-2i(\sin\lambda_1-\lambda_1\cos\lambda_1)\sin\frac{2\pi\kappa}{N}\tilde{S}_\kappa(\theta)=\tilde{G}_{s\kappa}(\theta,\kappa)\end{aligned}\quad(26)$$

We make the same substitutions and simplifications as made in Section 3.2 by Eqs.(3.2-36) to (3.2-40):

$$\lambda_1=ecTA_{\mathrm{m}0x}/N\hbar,\ \cos\lambda_1=1+O(\Delta t)^2,\ \lambda_1\sin\lambda_1=O(\Delta t)^2,\ T=N\Delta t$$

$$\lambda_2^2=(T/N\hbar)^2(m_0^2c^4-e\phi_{\mathrm{e}0}^2+e^2c^2A_{\mathrm{m}0x}^2)=O(\Delta t)^2\quad(27)$$

$$\lambda_1^2\lambda_3^2=(eT\phi_{\mathrm{e}0}/N\hbar)^2=O(\Delta t)^2\quad(28)$$

$$1+i\lambda_1\lambda_3\doteq e^{i\lambda_1\lambda_3},\ 1-i\lambda_1\lambda_3\doteq e^{-i\lambda_1\lambda_3},\quad(29)$$

$$2(\cos\lambda_1+\lambda_1\sin\lambda_1)\cos\frac{2\pi\kappa}{N}-\lambda_2^2\doteq 2\cos\frac{2\pi\kappa}{N}\quad(30)$$

$$2(\sin\lambda_1-\lambda_1\cos\lambda_1)\sin\frac{2\pi\kappa}{N}=O(\Delta t)^2\quad(31)$$

We may solve Eqs.(25) and (26) just as Eqs.(3.2-34) and (3.2-35) were solved. The equations obtained are like Eqs.(3.2-41) and (3.2-42) except that S, T, and G are replaced by $\tilde{S}$, $\tilde{T}$, and $\tilde{G}$:

$$\begin{aligned}&e^{2i\lambda_1\lambda_3}\tilde{S}_\kappa(\theta+2)-4\cos\frac{2\pi\kappa}{N}e^{i\lambda_1\lambda_3}\tilde{S}_\kappa(\theta+1)+2\left(2+\cos\frac{4\pi\kappa}{N}\right)\tilde{S}_\kappa(\theta)\\&\qquad-4\cos\frac{2\pi\kappa}{N}e^{-i\lambda_1\lambda_3}\tilde{S}_\kappa(\theta-1)+e^{-2i\lambda_1\lambda_3}\tilde{S}_\kappa(\theta-2)\\&\quad=-e^{i\lambda_1\lambda_3}\tilde{G}_{c\kappa}(\theta+1,\kappa)+2\cos\frac{2\pi\kappa}{N}\tilde{G}_{c\kappa}(\theta,\kappa)-e^{-i\lambda_1\lambda_3}\tilde{G}_{c\kappa}(\theta-1,\kappa)\end{aligned}\quad(32)$$

$$\begin{aligned}&e^{2i\lambda_1\lambda_3}\tilde{T}_\kappa(\theta+2)-4\cos\frac{2\pi\kappa}{N}e^{i\lambda_1\lambda_3}\tilde{T}_\kappa(\theta+1)+2\left(2+\cos\frac{4\pi\kappa}{N}\right)\tilde{T}_\kappa(\theta)\\&\qquad-4\cos\frac{2\pi\kappa}{N}e^{-i\lambda_1\lambda_3}\tilde{T}_\kappa(\theta-1)+e^{-2i\lambda_1\lambda_3}\tilde{T}_\kappa(\theta-2)\\&\quad=-e^{i\lambda_1\lambda_3}\tilde{G}_{s\kappa}(\theta+1,\kappa)+2\cos\frac{2\pi\kappa}{N}\tilde{G}_{s\kappa}(\theta,\kappa)-e^{-i\lambda_1\lambda_3}\tilde{G}_{s\kappa}(\theta-1,\kappa)\end{aligned}\quad(33)$$

For the solution of the homogeneous Eqs.(32) and (33) we can follow Section 3.2 from Eq.(3.2-43)

$$\tilde{S}_\kappa(\theta) = \tilde{c}_\kappa v_\kappa^\theta \qquad \text{or} \qquad \tilde{T}_\kappa(\theta) = \tilde{d}_\kappa v_\kappa^\theta \tag{34}$$

to (3.2-55) if S, T, c, and d are replaced by $\tilde{S}$, $\tilde{T}$, $\tilde{c}$, and $\tilde{d}$. For the solution of the inhomogeneous Eqs.(32) and (33) we follow Eqs.(3.2-56) and (3.2-57):

$$\tilde{S}_\kappa(\theta) = \tilde{c}_{\kappa 1}(\theta) v_{\kappa 1}^\theta + \tilde{c}_{\kappa 2}(\theta) v_{\kappa 2}^\theta + \tilde{c}_{\kappa 3}(\theta) \theta v_{\kappa 1}^\theta + \tilde{c}_{\kappa 4}(\theta) \theta v_{\kappa 2}^\theta \tag{35}$$

$$\tilde{T}_\kappa(\theta) = \tilde{d}_{\kappa 1}(\theta) v_{\kappa 1}^\theta + \tilde{d}_{\kappa 2}(\theta) v_{\kappa 2}^\theta + \tilde{d}_{\kappa 3}(\theta) \theta v_{\kappa 1}^\theta + \tilde{d}_{\kappa 4}(\theta) \theta v_{\kappa 2}^\theta \tag{36}$$

Equations (6.4-11) and (6.4-18) define $\tilde{d}_{\kappa i}(\theta)$ and $\tilde{c}_{\kappa i}(\theta)$.

Substitution of Eqs.(35) and (36) into Eq.(19) yields the particular solution of the inhomogeneous equation (10). The boundary and initial conditions of Eqs.(11) to (13) must still be satisfied. We obtain the following boundary and initial conditions for $\bar{v}(\zeta, \theta)$ from Eqs.(11) to (13) with the help of Eqs.(2.3-19) to (2.3-21):

$$\bar{v}(0, \theta) = 0 \quad \text{for } \theta \geq 0 \tag{37}$$

$$\bar{v}(\zeta, 0) = 0 \quad \text{for } \zeta > 0 \tag{38}$$

$$\bar{v}(\zeta, 1) - \hat{v}(\zeta, 0) = 0 \quad \text{for } \zeta > 0 \tag{39}$$

The boundary condition of Eq.(37) is satisfied if we discard the first term in Eq.(19):

$$\bar{v}(\zeta, \theta) = \sum_{\kappa=-N/2+1}^{N/2-1} \tilde{T}_\kappa(\theta) e^{i\lambda_1 \zeta} \sin \frac{2\pi\kappa\zeta}{N} \tag{40}$$

The initial condition of Eq.(38) is satisfied for the third and fourth term of $\tilde{T}_\kappa(\theta)$ in Eq.(36) due to the factor θ. The first and second terms have the coefficients $\tilde{d}_{\kappa i}(0)$ for $\theta = 0$:

$$\bar{v}(\zeta, 0) = \sum_{\kappa=-N/2+1}^{N/2-1} [\tilde{d}_{\kappa 1}(0) + \tilde{d}_{\kappa 2}(0)] e^{i\lambda_1 \zeta} \sin \frac{2\pi\kappa\zeta}{N} = 0 \tag{41}$$

From Eqs.(36) and (40) we get with Eqs.(3.2-48), (3.2-49), (39), and (38) the relation

$$\bar{v}(\zeta,1) = \sum_{\kappa=-N/2+1}^{N/2-1} \{[\tilde{d}_{\kappa1}(1) + \tilde{d}_{\kappa3}(1)]v_{\kappa1} + [\tilde{d}_{\kappa2}(1) + \tilde{d}_{\kappa4}(1)]v_{\kappa2}\}$$

$$\times e^{i\lambda_1\zeta} \sin\frac{2\pi\kappa\zeta}{N} = 0$$

$$= e^{-i\lambda_1\lambda_3} \sum_{\kappa=-N/2+1}^{N/2-1} \left([\tilde{d}_{\kappa1}(1) + \tilde{d}_{\kappa2}(1) + \tilde{d}_{\kappa3}(1) + \tilde{d}_{\kappa4}(1)] \cos\frac{2\pi\kappa}{N} \right.$$

$$\left. + i[\tilde{d}_{\kappa1}(1) - \tilde{d}_{\kappa2}(1) + \tilde{d}_{\kappa3}(1) - \tilde{d}_{\kappa4}(1)] \sin\frac{2\pi\kappa}{N} \right) e^{i\lambda_1\zeta} \sin\frac{2\pi\kappa\zeta}{N} = 0 \qquad (42)$$

Since Eqs.(41) and (42) equal Eqs.(3.2-62) and (3.2-63) if $\hat{v}$ and $d_{\kappa i}$ are replaced by $\bar{v}$ and $\tilde{d}_{\kappa i}$ we may skip from Eq.(3.2-63) to Eq.(3.2-72) and replace Eqs.(3.2-72) to (3.2-79) by the following equations:

$$\tilde{d}_{\kappa2} = -\tilde{d}_{\kappa1} \qquad (43)$$

$$\tilde{d}_{\kappa1}(1) = \Delta\tilde{d}_{\kappa1}(0) + \tilde{d}_{\kappa1} = \frac{\tilde{D}_{\kappa1}(1)}{D_{\kappa0}(1)} + \tilde{d}_{\kappa1} \qquad (44)$$

$$\tilde{d}_{\kappa3}(1) = \Delta\tilde{d}_{\kappa3}(0) + \tilde{d}_{\kappa3} = \frac{\tilde{D}_{\kappa3}(1)}{D_{\kappa0}(1)} + \tilde{d}_{\kappa3} \qquad (45)$$

$$\tilde{d}_{\kappa3} = -\tilde{d}_{\kappa1} - \frac{\tilde{D}_{\kappa1}(1) + \tilde{D}_{\kappa3}(1)}{D_{\kappa0}(1)} \qquad (46)$$

$$\tilde{d}_{\kappa2}(1) = \Delta\tilde{d}_{\kappa2}(0) + \tilde{d}_{\kappa2} = \frac{\tilde{D}_{\kappa2}(1)}{D_{\kappa0}(1)} + \tilde{d}_{\kappa2} \qquad (47)$$

$$\tilde{d}_{\kappa4}(1) = \Delta\tilde{d}_{\kappa4}(0) + \tilde{d}_{\kappa4} = \frac{\tilde{D}_{\kappa4}(1)}{D_{\kappa0}(1)} + \tilde{d}_{\kappa4} \qquad (48)$$

$$\tilde{d}_{\kappa4} = -\tilde{d}_{\kappa2} - \frac{\tilde{D}_{\kappa2}(1) + \tilde{D}_{\kappa4}(1)}{D_{\kappa0}(1)} \qquad (49)$$

$$\tilde{d}_{\kappa4} = \tilde{d}_{\kappa1} - \frac{\tilde{D}_{\kappa2}(1) + \tilde{D}_{\kappa4}(1)}{D_{\kappa0}(1)} \qquad (50)$$

We may now write the solution $\tilde{\Psi}_1(\zeta,\theta)$ of Eq.(2). Starting with Eqs.(10), (3.2-9), (3.2-10), and (15) we obtain the following expressions:

$$\tilde{\Psi}_1(\zeta,\theta) = \tilde{\Psi}_{00}\Big\{(1 - e^{-2i\lambda_1\lambda_3\theta})\exp[-(\lambda_2^2 - \lambda_1^2)^{1/2}\zeta]e^{i\lambda_1\zeta} + u(\zeta,\theta) + \bar{v}(\zeta,\theta)\Big\} \quad (51)$$

$$u(\zeta,\theta) = -2i\lambda_1\lambda_3 e^{i\lambda_1(\zeta+\lambda_3\theta)} \sum_{\kappa=-N/2+1}^{N/2-1} \frac{I_{\mathrm{T}}(\kappa/N)}{\sin\beta_\kappa}\sin\beta_\kappa\theta \sin\frac{2\pi\kappa\zeta}{N} \quad (52)$$

$I_{\mathrm{T}}(\kappa/N)$ see Eq.(2.4-29); λ_1, λ_2, λ_3 see Eqs.(27), (28)

$$\bar{v}(\zeta,\theta) = \sum_{\kappa=-N/2+1}^{N/2-1} \tilde{T}_\kappa(\theta)e^{i\lambda_1\zeta}\sin\frac{2\pi\kappa\zeta}{N} \quad (53)$$

$$\tilde{T}_\kappa(\theta) = [\tilde{d}_{\kappa1}(\theta) + \theta\tilde{d}_{\kappa3}(\theta)]v_{\kappa1}^{\theta} + [\tilde{d}_{\kappa2}(\theta) + \theta\tilde{d}_{\kappa4}(\theta)]v_{\kappa3}^{\theta} \quad (54)$$

$v_{\kappa1}$, $v_{\kappa2}$ see Eqs.(3.2-48), (3.2-49)

$\tilde{d}_{\kappa i}(\theta)$ see Eqs.(6.4-11), (6.4-28)–(6.4-31)

$\tilde{d}_{\kappa2}$, $\tilde{d}_{\kappa3}$, $\tilde{d}_{\kappa4}$ see Eqs.(43), (46), (50)

Comparison of Eq.(51) with Eq.(2.4-32) shows that $\tilde{\Psi}_1(\zeta,\theta)$ can also be written in the following form:

$$\tilde{\Psi}_1(\zeta,\theta) = \Psi_0(\zeta,\theta) + \tilde{\Psi}_{00}\bar{v}(\zeta,\theta)$$

$$\bar{v}(\zeta,\theta) - \sum_{\kappa=-N/2+1}^{N/2-1}\Big([\tilde{d}_{\kappa1}(\theta) + \theta\tilde{d}_{\kappa3}(\theta)]e^{i(2\pi\kappa/N-\lambda_1\lambda_3)\theta} + [\tilde{d}_{\kappa2}(\theta) + \theta\tilde{d}_{\kappa4}(\theta)]e^{-i(2\pi\kappa/N+\lambda_1\lambda_3)\theta}\Big)e^{i\lambda_1\zeta}\sin\frac{2\pi\kappa\zeta}{N} \quad (55)$$

4.3 Quantization of the Solution for $\Delta x \ll h/m_0c$

We follow Section 3.3 but are careful to put a tilde ˜ or replace a hat ˆ by a bar ¯ where appropriate. Equation (3.3-1) becomes:

$$\tilde{\Psi} = \Psi_0 + \alpha\tilde{\Psi}_1 = \Psi_{\mathrm{x}0x_j} + \alpha\tilde{\Psi}_{\mathrm{x}1x_j} \quad (1)$$

Equations (3.3-2) to (3.3-4) assume the following form:

$$\tilde{\Psi}^*\tilde{\Psi} = (\Psi_0^* + \alpha\tilde{\Psi}_1^*)(\Psi_0 + \alpha\tilde{\Psi}_1) = \Psi_0^*\Psi_0 + \alpha(\Psi_0^*\tilde{\Psi}_1 + \tilde{\Psi}_1^*\Psi_0) + O(\alpha^2) \quad (2)$$

$$\frac{\partial\tilde{\Psi}^*}{\partial\theta}\frac{\partial\tilde{\Psi}}{\partial\theta} = \frac{\partial\Psi_0^*}{\partial\theta}\frac{\partial\Psi_0}{\partial\theta} + \alpha\left(\frac{\partial\Psi_0^*}{\partial\theta}\frac{\partial\tilde{\Psi}_1}{\partial\theta} + \frac{\partial\tilde{\Psi}_1^*}{\partial\theta}\frac{\partial\Psi_0}{\partial\theta}\right) + O(\alpha^2) \tag{3}$$

$$\frac{\partial\tilde{\Psi}^*}{\partial\zeta}\frac{\partial\tilde{\Psi}}{\partial\zeta} = \frac{\partial\Psi_0^*}{\partial\zeta}\frac{\partial\Psi_0}{\partial\zeta} + \alpha\left(\frac{\partial\Psi_0^*}{\partial\zeta}\frac{\partial\tilde{\Psi}_1}{\partial\zeta} + \frac{\partial\tilde{\Psi}_1^*}{\partial\zeta}\frac{\partial\Psi_0}{\partial\zeta}\right) + O(\alpha^2) \tag{4}$$

The function $\Psi_0(\zeta,\theta)$ of Eq.(3.3-5) remains unchanged

$$\begin{aligned}\Psi_0(\zeta,\theta) &= \Psi_{00}[(1-e^{-2i\lambda_1\lambda_3\theta}F(\zeta)+u(\zeta,\theta)]\\ &= \Psi_{00}\bigg((1-e^{-2i\lambda_1\lambda_3\theta})\exp[-(\lambda_2^2-\lambda_1^2)^{1/2}\zeta]e^{i\lambda_1\zeta}\\ &\quad -2i\lambda_1\lambda_3 e^{i\Lambda_1\lambda_3}e^{i\lambda_1\zeta}e^{-i\lambda_1\lambda_3\theta}\sum_{\kappa=-N/2+1}^{N/2-1}\frac{I_{\mathrm{T}}(\kappa/N)}{\sin\beta_\kappa}\sin\beta_\kappa\theta\sin\frac{2\pi\kappa\zeta}{N}\bigg)\end{aligned} \tag{5}$$

but $\Psi_1(\zeta,\theta)$ of Eq.(3.3-6) becomes $\tilde{\Psi}_1(\zeta,\theta)$ according to Eq.(4.2-55):

$$\tilde{\Psi}_1(\zeta,\theta) = \Psi_0(\zeta,\theta) + \tilde{\Psi}_{00}\bar{v}(\zeta,\theta) \tag{6}$$

$$\begin{aligned}\bar{v}(\zeta,\theta) = \sum_{\kappa=-N/2+1}^{N/2-1}&\Big([\tilde{d}_{\kappa1}(\theta)+\theta\tilde{d}_{\kappa3}(\theta)]e^{i(2\pi\kappa/N-\lambda_1\lambda_3)\theta}\\ &+[\tilde{d}_{\kappa2}(\theta)+\theta\tilde{d}_{\kappa4}(\theta)]e^{-i(2\pi\kappa/N+\lambda_1\lambda_3)\theta}\Big)e^{i\lambda_1\zeta}\sin\frac{2\pi\kappa\zeta}{N}\end{aligned} \tag{7}$$

We may use Eq.(6) to rewrite Eqs.(2)–(4). The notation $\mathcal{R}e(\dots)$ is used for the real part of the expression in parentheses:

$$\begin{aligned}\tilde{\Psi}^*\tilde{\Psi} &= \Psi_0^*\Psi_0 + \alpha[2\Psi_0^*\Psi_0 + \tilde{\Psi}_{00}(\Psi_0^*\bar{v}+\bar{v}^*\Psi_0)]\\ &= \Psi_0^*\Psi_0 + 2\alpha[\Psi_0^*\Psi_0 + \tilde{\Psi}_{00}\mathcal{R}e(\Psi_0^*\bar{v})]\end{aligned} \tag{8}$$

$$\frac{\partial\tilde{\Psi}^*}{\partial\theta}\frac{\partial\tilde{\Psi}}{\partial\theta} = \frac{\partial\Psi_0^*}{\partial\theta}\frac{\partial\Psi_0}{\partial\theta} + 2\alpha\left[\frac{\partial\Psi_0^*}{\partial\theta}\frac{\partial\Psi_0}{\partial\theta} + \tilde{\Psi}_{00}\mathcal{R}e\left(\frac{\partial\Psi_0^*}{\partial\theta}\frac{\partial\bar{v}}{\partial\theta}\right)\right] \tag{9}$$

$$\frac{\partial\tilde{\Psi}^*}{\partial\zeta}\frac{\partial\tilde{\Psi}}{\partial\zeta} = \frac{\partial\Psi_0^*}{\partial\zeta}\frac{\partial\Psi_0}{\partial\zeta} + 2\alpha\left[\frac{\partial\Psi_0^*}{\partial\zeta}\frac{\partial\Psi_0}{\partial\zeta} + \tilde{\Psi}_{00}\mathcal{R}e\left(\frac{\partial\Psi_0^*}{\partial\zeta}\frac{\partial\bar{v}}{\partial\zeta}\right)\right] \tag{10}$$

We have again the terms $\Psi_0^*\Psi_0$, $(\partial\Psi_0^*/\partial\theta)(\partial\Psi_0/\partial\theta)$, as well as $(\partial\Psi_0^*/\partial\zeta) \times (\partial\Psi_0/\partial\zeta)$ obtained in Eqs.(2.5-4), (2.5-9) and (2.5-11). The terms $\Psi_0^*\bar{v}$, $(\partial\Psi_0^*/\partial\theta)(\partial\bar{v}/\partial\theta)$, and $(\partial\Psi_0^*/\partial\zeta)(\partial\bar{v}/\partial\zeta)$ in Eqs.(8)–(10) must be calculated with the help of Eqs.(5) and (7). First we show how $\bar{v}(\zeta,\theta)$ depends on $G_2(\zeta,\theta)$ of Eq.(4.2-2). This starts with $\tilde{H}_{\mathrm{s}\kappa}(\theta,\kappa)$ of Eq.(6.4-1) and $\tilde{G}_{\mathrm{s}\kappa}(\theta,\kappa)$ of Eq.(4.2-24):

$$\tilde{H}_{s\kappa}(\theta,\kappa) = \frac{2}{N}\int_0^N \Big(-e^{i\lambda_1\lambda_3}G_2(\zeta,\theta+1) + 2\cos\frac{2\pi\kappa}{N}G_2(\zeta,\theta) - e^{-i\lambda_1\lambda_3}G_2(\zeta,\theta-1)\Big) \times e^{-i\lambda_1\zeta}\sin\frac{2\pi\kappa\zeta}{N}d\zeta, \quad \theta = 1,\ 2,\ \ldots,\ N-2 \quad (11)$$

The four functions $\tilde{d}_{\kappa 1}(\theta)$ to $\tilde{d}_{\kappa 4}(\theta)$ in Eq.(7) may be written with the help of Eqs.(6.4-11), (6.4-10), and (6.4-12)–(6.4-15) as follows:

$$\tilde{d}_{\kappa 1}(\theta) = \sum_{n=0}^{\theta-1}\tilde{H}_{s\kappa}(n,\kappa)\{nF_1(n,\kappa) + F_3(n,\kappa) + i[nF_5(n,\kappa) + F_7(n,\kappa)] + \tilde{d}_{\kappa 1}\} \quad (12)$$

$$\tilde{d}_{\kappa 2}(\theta) = \sum_{n=0}^{\theta-1}\tilde{H}_{s\kappa}(n,\kappa)\{nF_2(n,\kappa) + F_4(n,\kappa) + i[nF_6(n,\kappa) + F_8(n,\kappa)] + \tilde{d}_{\kappa 2}\} \quad (13)$$

$$\tilde{d}_{\kappa 3}(\theta) = -\sum_{n=0}^{\theta-1}\tilde{H}_{s\kappa}(n,\kappa)[F_1(n,\kappa) + iF_5(n,\kappa) + \tilde{d}_{\kappa 3}] \quad (14)$$

$$\tilde{d}_{\kappa 4}(\theta) = -\sum_{n=0}^{\theta-1}\tilde{H}_{s\kappa}(n,\kappa)[F_2(n,\kappa) + iF_6(n,\kappa) + \tilde{d}_{\kappa 4}] \quad (15)$$

Equations (6.3-51)–(6.3-58) define the functions $F_1(n,\kappa)$ to $F_8(n,\kappa)$, while Eqs.(4.2-43), (4.2-46), and (4.2-50) define $\tilde{d}_{\kappa 2}$, $\tilde{d}_{\kappa 3}$, and $\tilde{d}_{\kappa 4}$. Using these equations we may rewrite the terms

$$[\tilde{d}_{\kappa 1}(\theta) + \theta\tilde{d}_{\kappa 3}(\theta)]e^{2\pi i\kappa\theta/N} + [\tilde{d}_{\kappa 2}(\theta) + \theta\tilde{d}_{\kappa 4}(\theta)]e^{-2\pi i\kappa\theta/N} \quad (16)$$

of $\bar{v}(\zeta,\theta)$ in Eq.(7) into the form of Eq.(6.4-48) and obtain the expression

$$\bar{v}(\zeta,\theta) = e^{i\lambda_1(\zeta-\lambda_3\theta)}\sum_{\kappa=-N/2+1}^{N/2-1}[\tilde{J}_{\mathrm{r}}(\theta,\kappa) + i\tilde{J}_{\mathrm{i}}(\theta,\kappa)]\sin\frac{2\pi\kappa\zeta}{N}$$

$$\tilde{J}_{\mathrm{r}}(\theta,\kappa) = [\tilde{J}_1(\theta,\kappa) - \theta\tilde{J}_2(\theta,\kappa)]\cos\frac{2\pi\kappa\theta}{N} - [\tilde{J}_3(\theta,\kappa) - \theta\tilde{J}_4(\theta,\kappa)]\sin\frac{2\pi\kappa\theta}{N}$$

$$\tilde{J}_{\mathrm{i}}(\theta,\kappa) = [\tilde{J}_5(\theta,\kappa) - \theta\tilde{J}_6(\theta,\kappa)]\sin\frac{2\pi\kappa\theta}{N} + [\tilde{J}_7(\theta,\kappa) - \theta\tilde{J}_8(\theta,\kappa)]\cos\frac{2\pi\kappa\theta}{N} \quad (17)$$

The sums over n shown in Eqs.(12)–(15) have been separated in Eq.(6.4-48).

We obtain $\Psi_0^*(\zeta,\theta)$ from Eq.(5) by changing the sign of i. The product $\tilde{\Psi}_{00}\Psi_0^*\bar{v}$ of Eq.(8) becomes with $\tilde{\Psi}_{00} = \Psi_{00}$:

$$\Psi_{00}\Psi_0^*(\zeta,\theta)\bar{v}(\zeta,\theta) = \Psi_{00}^2 e^{-i\lambda_1\lambda_3\theta}\Bigg((1-e^{2i\lambda_1\lambda_3\theta})\exp[-(\lambda_2^2-\lambda_1^2)^{1/2}\zeta]$$
$$+ 2i\lambda_1\lambda_3 e^{i\lambda_1\lambda_3(\theta-1)} \sum_{\kappa=-N/2+1}^{N/2-1} \frac{I_{\mathrm{T}}(\kappa/N)}{\sin\beta_\kappa}\sin\beta_\kappa\theta\sin\frac{2\pi\kappa\zeta}{N}\Bigg)$$
$$\times \sum_{\kappa=-N/2+1}^{N/2-1}[\tilde{J}_{\mathrm{r}}(\theta,\kappa)+i\tilde{J}_{\mathrm{i}}(\theta,\kappa)]\sin\frac{2\pi\kappa\zeta}{N} \qquad (18)$$

This is a product of two sums. The terms $\sin(2\pi\kappa\zeta/N)$ are important. We write ν for κ in the second sum and separate the products $\sin(2\pi\kappa\zeta/N)\sin(2\pi\nu\zeta/N)$ for $\kappa=\nu$ and $\kappa\neq\nu$:

$$\Psi_{00}\Psi_0^*(\zeta,\theta)\bar{v}(\zeta,\theta) = \Psi_{00}^2 e^{-i\lambda_1\lambda_3\theta}\Bigg[(1-e^{2i\lambda_1\lambda_3\theta})$$
$$\times \sum_{\kappa=-N/2+1}^{N/2-1}[\tilde{J}_{\mathrm{r}}(\theta,\kappa)+i\tilde{J}_{\mathrm{i}}(\theta,\kappa)]\exp[-(\lambda_2^2-\lambda_1^2)^{1/2}\zeta]\sin\frac{2\pi\kappa\zeta}{N}$$
$$+2i\lambda_1\lambda_3 e^{i\lambda_1\lambda_3(\theta-1)}\Bigg(\sum_{\kappa=-N/2+1}^{N/2-1}\frac{I_{\mathrm{T}}(\kappa/N)}{\sin\beta_\kappa}\sin(\beta_\kappa\theta)[\tilde{J}_{\mathrm{r}}(\theta,\kappa)+i\tilde{J}_{\mathrm{i}}(\theta,\kappa)]\sin^2\frac{2\pi\kappa\zeta}{N}$$
$$+\sum_{\kappa=-N/2+1}^{N/2-1,\neq\nu}\ \sum_{\nu=-N/2+1}^{N/2-1}\frac{I_{\mathrm{T}}(\kappa/N)}{\sin\beta_\kappa}\sin\beta_\kappa\theta$$
$$\times[\tilde{J}_{\mathrm{r}}(\theta,\nu)+i\tilde{J}_{\mathrm{i}}(\theta,\nu)]\sin\frac{2\pi\nu\zeta}{N}\sin\frac{2\pi\kappa\zeta}{N}\Bigg)\Bigg] \qquad (19)$$

According to Eqs.(2.5-3) and (8) the integral over ζ from 0 to N of this expression is needed:

$$\Psi_{00}\int_0^N \Psi_0^*(\zeta,\theta)\bar{v}(\zeta,\theta)d\zeta$$
$$= \Psi_{00}^2\sum_{\kappa=-N/2+1}^{N/2-1}\Bigg(2\sin(\lambda_1\lambda_3\theta)\frac{(2\pi\kappa/N)\{1-\exp[-(\lambda_2^2-\lambda_1^2)^{1/2}N]\}}{\lambda_2^2-\lambda_1^2+(2\pi\kappa/N)^2}$$
$$\times[\tilde{J}_{\mathrm{i}}(\theta,\kappa)-i\tilde{J}_{\mathrm{r}}(\theta,\kappa)]$$
$$+N\lambda_1\lambda_3\frac{I_{\mathrm{T}}(\kappa/N)}{\sin\beta_\kappa}\sin(\beta_\kappa\theta)\{\tilde{J}_{\mathrm{r}}(\theta,\kappa)\sin\lambda_1\lambda_3-\tilde{J}_{\mathrm{i}}(\theta,\kappa)\cos\lambda_1\lambda_3$$
$$+i[\tilde{J}_{\mathrm{r}}(\kappa,\theta)\cos\lambda_1\lambda_3+\tilde{J}_{\mathrm{i}}(\kappa,\theta)\sin\lambda_1\lambda_3]\}\Bigg) \qquad (20)$$

Let us turn to Eq.(9). The product $(\partial\Psi_0^*/\partial\theta)(\partial\Psi_0/\partial\theta)$ of Eq.(2.5-9) is required. Then we need $\partial\bar{v}/\partial\theta$, which we get from Eq.(17):

$$\frac{\partial\bar{v}(\zeta,\theta)}{\partial\theta} = e^{i\lambda_1\zeta} \sum_{\kappa=-N/2+1}^{N/2-1} [\tilde{J}_{\mathrm{R}}(\theta,\kappa) + i\tilde{J}_{\mathrm{I}}(\theta,\kappa)] \sin\frac{2\pi\kappa\zeta}{N}$$

$$\begin{aligned}
\tilde{J}_{\mathrm{R}}(\theta,\kappa) &= \left(\frac{\partial\tilde{J}_{\mathrm{r}}(\theta,\kappa)}{\partial\theta} + \lambda_1\lambda_3\tilde{J}_{\mathrm{i}}(\theta,\kappa)\right)\cos\lambda_1\lambda_3\theta \\
&\quad + \left(\frac{\partial\tilde{J}_{\mathrm{i}}(\theta,\kappa)}{\partial\theta} - \lambda_1\lambda_3\tilde{J}_{\mathrm{r}}(\theta,\kappa)\right)\sin\lambda_1\lambda_3\theta \\
\tilde{J}_{\mathrm{I}}(\theta,\kappa) &= -\left(\frac{\partial\tilde{J}_{\mathrm{r}}(\theta,\kappa)}{\partial\theta} + \lambda_1\lambda_3\tilde{J}_{\mathrm{i}}(\theta,\kappa)\right)\sin\lambda_1\lambda_3\theta \\
&\quad + \left(\frac{\partial\tilde{J}_{\mathrm{i}}(\theta,\kappa)}{\partial\theta} - \lambda_1\lambda_3\tilde{J}_{\mathrm{r}}(\theta,\kappa)\right)\cos\lambda_1\lambda_3\theta
\end{aligned} \tag{21}$$

The exponential $e^{i\lambda_1\zeta}$ was not written as $\cos\lambda_1\zeta + i\sin\lambda_1\zeta$ since it will be cancelled by an exponential $e^{-i\lambda_1\zeta}$ in the following equation for $(\partial\Psi_0^*/\partial\theta)(\partial\bar{v}/\partial\theta)$. The term $\partial\Psi_0^*/\partial\theta$ is obtained with the help of Eqs.(2.5-5) and (2.5-6). We again use $\tilde{\Psi}_{00} = \Psi_{00}$:

$$\begin{aligned}
\tilde{\Psi}_{00}\frac{\partial\Psi_0^*(\zeta,\theta)}{\partial\theta}\frac{\partial\bar{v}(\zeta,\theta)}{\partial\theta} &= \Psi_{00}^2\Bigg[-2i\lambda_1\lambda_3 e^{2i\lambda_1\lambda_3\theta} \\
&\times \sum_{\kappa=-N/2+1}^{N/2-1} [\tilde{J}_{\mathrm{R}}(\theta,\kappa) + i\tilde{J}_{\mathrm{I}}(\theta,\kappa)] \exp[-(\lambda_2^2-\lambda_1^2)^{1/2}\zeta] \sin\frac{2\pi\kappa\zeta}{N} \\
&\quad - 2\lambda_1\lambda_3 e^{i\lambda_1\lambda_3(\theta-1)}\Bigg(\sum_{\kappa=-N/2+1}^{N/2-1} \frac{I_{\mathrm{T}}(\kappa/N)}{\sin\beta_\kappa}(\lambda_1\lambda_3\sin\beta_\kappa\theta \\
&\qquad - i\beta_\kappa\cos\beta_\kappa\theta)[\tilde{J}_{\mathrm{R}}(\theta,\kappa) + i\tilde{J}_{\mathrm{I}}(\theta,\kappa)]\sin^2\frac{2\pi\kappa\zeta}{N} \\
&\quad + \sum_{\kappa=-N/2+1}^{N/2-1,\neq\nu} \sum_{\nu=-N/2+1}^{N/2-1} \frac{I_{\mathrm{T}}(\kappa/N)}{\sin\beta_\kappa}[\tilde{J}_{\mathrm{R}}(\theta,\nu) + i\tilde{J}_{\mathrm{I}}(\theta,\nu)] \\
&\qquad \times (\lambda_1\lambda_3\sin\beta_\kappa\theta - i\beta_\kappa\cos\beta_\kappa\theta)\sin\frac{2\pi\kappa\zeta}{N}\sin\frac{2\pi\nu\zeta}{N}\Bigg)\Bigg] \\
\tilde{J}_{\mathrm{R}}(\theta,\nu) &= \tilde{J}_{\mathrm{R}}(\theta,\kappa),\ \tilde{J}_{\mathrm{I}}(\theta,\nu) = \tilde{J}_{\mathrm{I}}(\theta,\kappa) \quad \text{for } \kappa\to\nu
\end{aligned} \tag{22}$$

Following Eq.(20) we integrate this expression over ζ from 0 to N:

$$\Psi_{00}\int_0^N \frac{\partial\Psi_0^*(\zeta,\theta)}{\partial\theta}\frac{\partial\bar{v}(\zeta,\theta)}{\partial\theta}d\zeta = \Psi_{00}^2\Bigg(2\lambda_1\lambda_3\sum_{\kappa=-N/2+1}^{N/2-1}\{[\tilde{J}_{\mathrm{R}}(\theta,\kappa)+i\tilde{J}_{\mathrm{I}}(\theta,\kappa)]\sin 2\lambda_1\lambda_3\theta$$

$$+[\tilde{J}_{\mathrm{I}}(\theta,\kappa)-i\tilde{J}_{\mathrm{R}}(\theta,\kappa)]\cos 2\lambda_1\lambda_3\theta\}\frac{(2\pi\kappa/N)\{1-\exp[-(\lambda_2^2-\lambda_1^2)^{1/2}N]\}}{\lambda_2^2-\lambda_1^2+(2\pi\kappa/N)^2}$$

$$-N\lambda_1\lambda_3\sum_{\kappa=-N/2+1}^{N/2-1}\frac{I_{\mathrm{T}}(\kappa/N)}{\sin\beta_\kappa}\{[\tilde{J}_{\mathrm{R}}(\theta,\kappa)+i\tilde{J}_{\mathrm{I}}(\theta,\kappa)]$$

$$\times\{[\lambda_1\lambda_3\sin\beta_\kappa\theta\cos\lambda_1\lambda_3(\theta-1)+\beta_\kappa\cos\beta_\kappa\theta\sin\lambda_1\lambda_3(\theta-1)]$$

$$-[\tilde{J}_{\mathrm{I}}(\theta,\kappa)-i\tilde{J}_{\mathrm{R}}(\theta,\kappa)]$$

$$\times[\lambda_1\lambda_3\sin\beta_\kappa\theta\sin\lambda_1\lambda_3(\theta-1)-\beta_\kappa\cos\beta_\kappa\theta\cos\lambda_1\lambda_3(\theta-1)]\}\Bigg) \qquad (23)$$

We turn to Eq.(10). The term $\partial\bar{v}/\partial\zeta$ needs to be derived from Eq.(17):

$$\frac{\partial\bar{v}(\zeta,\theta)}{\partial\zeta} = e^{i\lambda_1(\zeta-\lambda_3\theta)}\sum_{\kappa=-N/2+1}^{N/2-1}[\tilde{J}_{\mathrm{r}}(\theta,\kappa)+i\tilde{J}_{\mathrm{i}}(\theta,\kappa)]$$

$$\times\left(\frac{2\pi\kappa}{N}\cos\frac{2\pi\kappa\zeta}{N}+i\lambda_1\sin\frac{2\pi\kappa\zeta}{N}\right) \qquad (24)$$

We may now obtain the term $(\partial\Psi_0^*/\partial\zeta)(\partial\bar{v}/\partial\zeta)$ of Eq.(10) with the help of Eqs.(2.5-7) and (2.5-8):

$$\tilde{\Psi}_{00}\frac{\partial\Psi_0^*(\zeta,\theta)}{\partial\zeta}\frac{\partial\bar{v}(\zeta,\theta)}{\partial\zeta} = \Psi_{00}^2\Bigg\{-2[\lambda_1-i(\lambda_2^2-\lambda_1^2)^{1/2}]$$

$$\times\sum_{\kappa=-N/2+1}^{N/2-1}[\tilde{J}_r(\theta,\kappa)+i\tilde{J}_{\mathrm{i}}(\theta,\kappa)]\bigg(\frac{2\pi\kappa}{N}\exp[-(\lambda_2^2-\lambda_1^2)^{1/2}\zeta]\cos\frac{2\pi\kappa\zeta}{N}$$

$$+i\lambda_1\exp[-(\lambda_2^2-\lambda_1^2)^{1/2}\zeta]\sin\frac{2\pi\kappa\zeta}{N}\bigg)\sin\lambda_1\lambda_3\theta$$

$$+2i\lambda_1\lambda_3e^{-i\lambda_1\lambda_3}\sum_{\kappa=-N/2+1}^{N/2-1}\frac{I_{\mathrm{T}}(\kappa/N)}{\sin\beta_\kappa}[\tilde{J}_{\mathrm{r}}(\theta,\kappa)+i\tilde{J}_{\mathrm{i}}(\theta,\kappa)]\sin\beta_\kappa\theta$$

$$\times\left[\lambda_1^2\sin^2\frac{2\pi\kappa\zeta}{N}+\left(\frac{2\pi\kappa}{N}\right)^2\cos^2\frac{2\pi\kappa\zeta}{N}\right]$$

$$+ 2i\lambda_1\lambda_3 e^{-i\lambda_1\lambda_3} \sum_{\kappa=-N/2+1}^{N/2-1,\neq\nu} \sum_{\nu=-N/2+1}^{N/2-1} \frac{I_{\mathrm{T}}(\kappa/N)}{\sin\beta_\kappa}[\tilde{J}_{\mathrm{r}}(\theta,\nu) + i\tilde{J}_{\mathrm{i}}(\theta,\nu)]\sin\beta_\kappa\theta$$

$$\times\left(\lambda_1 \sin\frac{2\pi\kappa\zeta}{N} + i\frac{2\pi\kappa}{N}\cos\frac{2\pi\kappa\zeta}{N}\right)\left(\frac{2\pi\nu}{N}\cos\frac{2\pi\nu\zeta}{N} + i\lambda_1\sin\frac{2\pi\nu\zeta}{N}\right)\Bigg\} \quad (25)$$

This expression is integrated over ζ from 0 to N following Eqs.(20) and (23):

$$\tilde{\Psi}_{00}\int_0^N \frac{\partial\Psi_0^*(\zeta,\theta)}{\partial\zeta}\frac{\partial\bar{v}(\zeta,\theta)}{\partial\zeta}d\zeta = \Psi_{00}^2\Bigg\{2\lambda_2^2\sin\lambda_1\lambda_3\theta$$

$$\times \sum_{\kappa=-N/2+1}^{N/2-1}[\tilde{J}_{\mathrm{i}}(\theta,\kappa)-i\tilde{J}_{\mathrm{r}}(\theta,\kappa)]\frac{(2\pi\kappa/N)\{1-\exp[-(\lambda_2^2-\lambda_1^2)^{1/2}N]\}}{\lambda_2^2-\lambda_1^2+(2\pi\kappa/N)^2}$$

$$-N\lambda_1\lambda_3\sum_{\kappa=-N/2+1}^{N/2-1}\left[\lambda_1^2+\left(\frac{2\pi\kappa}{N}\right)^2\right]\{\tilde{J}_{\mathrm{i}}(\theta,\kappa)\cos\lambda_1\lambda_3-\tilde{J}_{\mathrm{r}}(\theta,\kappa)\sin\lambda_1\lambda_3$$

$$-i[\tilde{J}_{\mathrm{r}}(\theta,\kappa)\cos\lambda_1\lambda_3+\tilde{J}_{\mathrm{i}}(\theta,\kappa)\sin\lambda_1\lambda_3]\}\frac{I_{\mathrm{T}}(\kappa/N)}{\sin\beta_\kappa}\sin\beta_\kappa\theta\Bigg\} \quad (26)$$

We turn to the text following Eq.(2.5-11). In order to allow for the generalization of the terms of Eq.(2.5-3) by Eqs.(8)–(10) we define the energy $\bar{U}$ by the sum of the following three components

$$\bar{U} = \bar{U}_1 + \bar{U}_2 + \bar{U}_3 = \bar{U}_{\mathrm{c}} + \bar{U}_{\mathrm{v}}(\theta) \quad (27)$$

The terms $\bar{U}_1$ to $\bar{U}_3$ are obtained by the substitution of Eqs.(8)–(10) into Eq.(2.5-3). We use the notation $\bar{U}$ to $\bar{U}_{\mathrm{v}}(\theta)$ to distinguish the terms from the approximations U to $U_{\mathrm{v}}(\theta)$ in Eq.(2.5-12) and from $\mathring{U}$ to $\mathring{U}_{\mathrm{v}}(\theta)$ in Eq.(3.3-27):

$$\bar{U}_1 = \frac{L^2}{c\Delta t}\frac{m_0^2c^4(\Delta t)^2}{\hbar^2}\int_0^N\tilde{\Psi}^*\tilde{\Psi}d\zeta = \frac{L^2}{c\Delta t}\frac{m_0^2c^4(\Delta t)^2}{\hbar^2}\Bigg((1+2\alpha)\int_0^N\Psi_0^*\Psi_0 d\zeta$$

$$+2\alpha\tilde{\Psi}_{00}\int_0^N \mathcal{R}e(\Psi_0^*\bar{v})d\zeta\Bigg) \quad (28)$$

$$\bar{U}_2 = \frac{L^2}{c\Delta t}\int_0^N\frac{\partial\tilde{\Psi}^*}{\partial\theta}\frac{\partial\tilde{\Psi}}{\partial\theta}d\zeta = \frac{L^2}{c\Delta t}\Bigg[(1+2\alpha)\int_0^N\frac{\partial\Psi_0^*}{\partial\theta}\frac{\partial\Psi_0}{\partial\theta}d\zeta$$

$$+2\alpha\tilde{\Psi}_{00}\int_0^N\mathcal{R}e\left(\frac{\partial\Psi_0^*}{\partial\theta}\frac{\partial\bar{v}}{\partial\theta}\right)d\zeta\Bigg] \quad (29)$$

$$\bar{U}_3 = \frac{L^2}{c\Delta t}\int_0^N \frac{\partial\tilde{\Psi}^*}{\partial\zeta}\frac{\partial\tilde{\Psi}}{\partial\zeta}d\zeta = \frac{L^2}{c\Delta t}\left[(1+2\alpha)\int_0^N \frac{\partial\Psi_0^*}{\partial\zeta}\frac{\partial\Psi_0}{\partial\zeta}d\zeta + 2\alpha\tilde{\Psi}_{00}\int_0^N \mathcal{R}e\left(\frac{\partial\Psi_0^*}{\partial\zeta}\frac{\partial\bar{v}}{\partial\zeta}\right)d\zeta\right] \quad (30)$$

As in Eqs.(2.5-13)–(2.5-15) and (3.3-28)–(3.3-30) we want the time-invariant part

$$\bar{U}_\mathrm{c} = \bar{U}_\mathrm{c1} + \bar{U}_\mathrm{c2} + \bar{U}_\mathrm{c3} \quad (31)$$

of $\bar{U}_1$, $\bar{U}_2$, $\bar{U}_3$ and we ignore the time-variable part $\bar{U}_\mathrm{v}(\theta)$. Again we write $\bar{U}_\mathrm{c}$ to $\bar{U}_\mathrm{c3}$ to distinguish the terms from the approximations U_c to U_c3 in Eq.(2.5-16) and from $\hat{U}_\mathrm{c}$ in Eq.(3.3-31). When we derived Eqs.(2.5-13)–(2.5-15) we simply left out terms containing $\sin\beta_\kappa\theta$ or $\cos\beta_\kappa\theta$. The time variations of Eqs.(28)–(30) are not so obvious and we must write integrals over θ. With the help of Eqs.(2.5-13)–(2.5-15) we write the time-invariant part of Eqs.(28)–(30) in the following form in analogy to Eqs.(3.3-32)–(3.3-34):

$$\bar{U}_\mathrm{c1} = \int_0^N \bar{U}_1 d\theta = \frac{L^2}{c\Delta t}\frac{m_0^2c^4(\Delta t)^2}{\hbar^2}\left[(1+2\alpha)\tilde{\Psi}_{00}^2 N\lambda_1^2\lambda_3^2 \sum_{\kappa=-N/2+1}^{N/2-1}\left(\frac{I_\mathrm{T}(\kappa/N)}{\sin\beta_\kappa}\right)^2 + 2\alpha\tilde{\Psi}_{00}\int_0^N\left(\int_0^N \mathcal{R}e(\Psi_0^*\bar{v})d\zeta\right)d\theta\right] \quad (32)$$

$$\bar{U}_\mathrm{c2} = \int_0^N \bar{U}_2 d\theta = \frac{L^2}{c\Delta t}\left\{(1+2\alpha)\tilde{\Psi}_{00}^2 N\lambda_1^2\lambda_3^2 \sum_{\kappa=-N/2+1}^{N/2-1}\left(\frac{I_\mathrm{T}(\kappa/N)}{\sin\beta_\kappa}\right)^2(\lambda_1^2\lambda_3^2+\beta_\kappa^2) + 2\alpha\tilde{\Psi}_{00}\int_0^N\left[\int_0^N \mathcal{R}e\left(\frac{\partial\Psi_0^*}{\partial\theta}\frac{\partial\bar{v}}{\partial\theta}\right)d\zeta\right]d\theta\right\} \quad (33)$$

$$\bar{U}_\mathrm{c3} = \int_0^N \bar{U}_3 d\theta = \frac{L^2}{c\Delta t}\left\{(1+2\alpha)\tilde{\Psi}_{00}^2 N\lambda_1^2\lambda_3^2 \sum_{\kappa=-N/2+1}^{N/2-1}\left(\frac{I_\mathrm{T}(\kappa/N)}{\sin\beta_\kappa}\right)^2\left[\lambda_1^2+\left(\frac{2\pi\kappa}{N}\right)^2\right] + 2\alpha\tilde{\Psi}_{00}\int_0^N\left[\int_0^N \mathcal{R}e\left(\frac{\partial\Psi_0^*}{\partial\zeta}\frac{\partial\bar{v}}{\partial\zeta}\right)d\zeta\right]d\theta\right\} \quad (34)$$

We recognize on the right side of Eqs.(32)–(34) the energies U_{c1}, U_{c2}, U_{c3} of Eqs.(2.5-13)–(2.5-15) multiplied by $(1+2\alpha)\tilde{\Psi}_{00}^2/\Psi_{00}^2 \doteq 1.0146\tilde{\Psi}_{00}^2/\Psi_{00}^2$. The factor 1.0146 implies a difference of 1.5% that is barely visible in a plot. The interesting terms are the integrals multiplied by $2\alpha\tilde{\Psi}_{00}$. We shall analyze them in the following Section 4.4.

A comparison of Eqs.(31) and (2.5-16) shows that the only difference is the symbols $\bar{}$. Hence, we may use the results of Section 2.5 from Eq.(2.5-16) on. The energies E_κ of Eqs.(2.5-32), (2.5-34), and (2.5-48) are obtained. One could write $\bar{\mathsf{E}}_\kappa$ instead of E_κ and obtain the result $\bar{\mathsf{E}}_\kappa = \mathsf{E}_\kappa$, but we follow the last paragraph of Section 3.3 and do not do so.

4.4 Evaluation of the Energy $\bar{U}_c$ For Small Distances

Following Section 3.4 we consider only the terms in Eqs.(4.3-32)–(4.3-34) multiplied by $2\alpha\tilde{\Psi}_{00}$. We denote them with $\bar{U}_{\alpha1}$ to $\bar{U}_{\alpha3}$:

$$\bar{U}_{\alpha1} = 2\alpha\tilde{\Psi}_{00}\frac{L^2}{c\Delta t}\left(\frac{m_0c^2\Delta t}{\hbar}\right)^2 \int_0^N \left(\int_0^N \mathcal{R}e(\Psi_0^*\bar{v})d\zeta\right)d\theta \tag{1}$$

$$\bar{U}_{\alpha2} = 2\alpha\tilde{\Psi}_{00}\frac{L^2}{c\Delta t}\int_0^N \left[\int_0^N \mathcal{R}e\left(\frac{\partial\Psi_0^*}{\partial\theta}\frac{\partial\bar{v}}{\partial\theta}\right)d\zeta\right]d\theta \tag{2}$$

$$\bar{U}_{\alpha3} = 2\alpha\tilde{\Psi}_{00}\frac{L^2}{c\Delta t}\int_0^N \left[\int_0^N \mathcal{R}e\left(\frac{\partial\Psi_0^*}{\partial\zeta}\frac{\partial\bar{v}}{\partial\zeta}\right)d\zeta\right]d\theta \tag{3}$$

We want the sum $\bar{U}_{\alpha c}$ of these terms as function $\bar{U}_{\alpha\kappa}(\kappa)$ according to Eq.(2.5-16):

$$\bar{U}_{\alpha c} = \bar{U}_{\alpha1} + \bar{U}_{\alpha2} + \bar{U}_{\alpha3} = \sum_{\kappa=-N/2+1}^{N/2-1} \bar{U}_{\alpha\kappa}(\kappa) \tag{4}$$

$$\bar{U}_{\alpha1} = \sum_{\kappa=-N/2+1}^{N/2-1} \bar{U}_{1\kappa}(\kappa), \quad \bar{U}_{\alpha2} = \sum_{\kappa=-N/2+1}^{N/2-1} \bar{U}_{2\kappa}(\kappa), \quad \bar{U}_{\alpha3} = \sum_{\kappa=-N/2+1}^{N/2-1} \bar{U}_{3\kappa}(\kappa) \tag{5}$$

Equations (1)–(3) must be rewritten to correspond to this scheme. An index κ is written at the end of the kernel of the integrals of Eq.(1) to indicate that the summation over κ in Eq.(4.3-20) is not carried out:

$$\bar{U}_{1\kappa}(\kappa) = 2\alpha\tilde{\Psi}_{00}\frac{L^2}{c\Delta t}\left(\frac{m_0c^2\Delta t}{\hbar}\right)^2 \int_0^N \mathcal{R}e\left(\int_0^N \Psi_0^*(\zeta,\theta)\bar{v}(\zeta,\theta)\,d\zeta\right)_\kappa d\theta$$

$$\begin{aligned}\tilde{\mathcal{H}}_{1\kappa}(\kappa) &= \frac{\bar{U}_{1\kappa}(\kappa)}{2\alpha\tilde{\Psi}_{00}^2(L^2/c\Delta t)(m_0c^2\Delta t/\hbar)^2} = \frac{1}{\tilde{\Psi}_{00}}\int_0^N \mathcal{R}e\left(\int_0^N \Psi_0^*(\zeta,\theta)\bar{v}(\zeta,\theta)d\zeta\right)_\kappa d\theta \\ &= \frac{1}{\Psi_{00}}\int_0^N \mathcal{R}e\left(\int_0^N \breve{\Psi}_0^*(\zeta,\theta)\bar{v}(\zeta,\theta)d\zeta\right)_\kappa d\theta, \qquad \tilde{\Psi}_{00} = \Psi_{00}\end{aligned} \tag{6}$$

Dividing Eq.(4.3-20) by Ψ_{00}^2 and leaving out the summation sign yields the following expression for $\tilde{\mathcal{H}}_{1\kappa}(\kappa)$:

$$\begin{aligned}\tilde{\mathcal{H}}_{1\kappa}(\kappa) = 2\,&\frac{(2\pi\kappa/N)\{1-\exp[-(\lambda_2^2-\lambda_1^2)^{1/2}N]\}}{\lambda_2^2-\lambda_1^2+(2\pi\kappa/N)^2} \\ &\times\int_0^N \mathcal{R}e[\tilde{J}_{\mathrm{i}}(\theta,\kappa)-i\tilde{J}_{\mathrm{r}}(\theta,\kappa)]\sin\lambda_1\lambda_3\theta\,d\theta \\ &+N\lambda_1\lambda_3\frac{I_{\mathrm{T}}(\kappa/N)}{\sin\beta_\kappa}\int_0^N \mathcal{R}e\{\tilde{J}_{\mathrm{r}}(\theta,\kappa)\sin\lambda_1\lambda_3-\tilde{J}_{\mathrm{i}}(\theta,\kappa)\cos\lambda_1\lambda_3 \\ &+i[\tilde{J}_{\mathrm{r}}(\theta,\kappa)\cos\lambda_1\lambda_3+\tilde{J}_{\mathrm{i}}(\theta,\kappa)\sin\lambda_1\lambda_3]\}\sin\beta_\kappa\theta\,d\theta\end{aligned} \tag{7}$$

We note that $\tilde{J}_{\mathrm{r}}(\theta,\kappa)$ and $\tilde{J}_{\mathrm{i}}(\theta,\kappa)$ are not real but complex. As in Section 3.4 we avoid an analytical separation into real and imaginary components since the computer instruction $\mathcal{R}e$ will do this separation numerically.

For the integration of Eq.(7) over θ we need to evaluate integrals containing $\tilde{J}_{\mathrm{r}}(\theta,\kappa)$ and $\tilde{J}_{\mathrm{i}}(\theta,\kappa)$. For the calculation of $\tilde{J}_{\mathrm{r}}(\theta,\kappa)$ and $\tilde{J}_{\mathrm{i}}(\theta,\kappa)$ we start with Eq.(3.4-8) and observe that the variable θ_n will become either θ or n as required:

$$\begin{aligned}\Psi_0(\zeta,\theta_n) &= (1-e^{-2i\lambda_1\lambda_3\theta_n})\exp[-(\lambda_2^2-\lambda_1^2)^{1/2}\zeta]e^{i\lambda_1\zeta} \\ &-2i\lambda_1\lambda_3e^{i\lambda_1\lambda_3}e^{i\lambda_1(\zeta-\lambda_3\theta_n)}\sum_{\kappa=-N/2+1}^{N/2-1}\frac{I_{\mathrm{T}}(\kappa/N)}{\sin\beta_\kappa}\sin\beta_\kappa\theta_n\sin\frac{2\pi\kappa\zeta}{N}\end{aligned} \tag{8}$$

The function Ψ_0 of Eq.(8) is needed for $G_2(\zeta,\theta)$ in Eq.(4.2-8):

$$
\begin{aligned}
G_2(\zeta,\theta) = &-p_{\mathrm{N}}(\zeta,\theta)[(1+2i\lambda_1)\Psi_0(\zeta+2,\theta)\\
&-(4+4i\lambda_1-6\lambda_1^2+2i\lambda_1^3)\Psi_0(\zeta+1,\theta)+(6-12\lambda_1^2-\lambda_1^4)\Psi_0(\zeta,\theta)\\
&-(4-4i\lambda_1-6\lambda_1^2-2i\lambda_1^3)\Psi_0(\zeta-1,\theta)+(1-2i\lambda_1)\Psi_0(\zeta-2,\theta)]\\
&\lambda_1 = ecTA_{\mathrm{m}0x}/\hbar N,\ p_{\mathrm{N}}(\zeta,\theta)\ \text{see Eq.(4.2-5)}
\end{aligned}
\tag{9}
$$

We must define the space and time variation of $p_{\mathrm{N}}(\zeta,\theta_n)$. In analogy to Section 3.4 we assume that $A_{\mathrm{e}}(\zeta,\theta)$ and $p_{\mathrm{N}}(\zeta,\theta)$ in Eq.(4.2-5) are independent of time and vary linearly with the space variable ζ:

$$
\begin{aligned}
&\frac{2\lambda_{\mathrm{C}}A_{\mathrm{e}}(\zeta,\theta)}{e} = \tilde{A}_{\mathrm{e}}(\zeta,\theta) = \tilde{A}_{\mathrm{e}}(\zeta) = \tilde{A}_{\mathrm{e}0}\zeta/N\\
&p_{\mathrm{N}}(\zeta,\theta_n) = p_{\mathrm{N}}(\zeta) = \frac{1}{4\pi^2}\left(\frac{\lambda_{\mathrm{C}}}{cT}\right)^2 N^2\tilde{A}_{\mathrm{e}}(\zeta) = p_{\mathrm{C}}N^2\tilde{A}_{\mathrm{e}}(\zeta)\\
&p_{\mathrm{C}} = \frac{1}{4\pi^2}\left(\frac{\lambda_{\mathrm{C}}}{cT}\right)^2 = \frac{1}{4\pi^2N^2}\left(\frac{\lambda_{\mathrm{C}}}{\Delta x}\right)^2,\ \frac{h/m_0c}{\Delta x} = \frac{\lambda_{\mathrm{C}}}{\Delta x} = \frac{N\lambda_{\mathrm{C}}}{cT}
\end{aligned}
\tag{10}
$$

Equation (9) becomes:

$$
\begin{aligned}
G_2(\zeta,\theta_n) = &-p_{\mathrm{C}}N^2\tilde{A}_{\mathrm{e}}(\zeta)\{(1+2i\lambda_1)\Psi_0(\zeta+2,\theta_n)\\
&-[4-6\lambda_1^2+2i(2\lambda_1+\lambda_1^3)]\Psi_0(\zeta+1,\theta_n)+(6-12\lambda_1^2-\lambda_1^4)\Psi_0(\zeta,\theta_n)\\
&-[4-6\lambda_1^2-2i(2\lambda_1+\lambda_1^3)]\Psi_0(\zeta-1,\theta_n)+(1-2i\lambda_1)\Psi_0(\zeta-2,\theta_n)\}
\end{aligned}
\tag{11}
$$

We shall also need $G_2(\zeta,\theta_n-1)$ and $G_2(\zeta,\theta_n+1)$. In analogy to Eqs.(3.4-19) and (3.4-20) we substitute θ_n-1 or θ_n+1 for θ_n in Eq.(11); as before this notation helps with the writing of the computer program:

$$
\begin{aligned}
G_{2-}(\zeta,\theta_n) = G_2(\zeta,\theta_n-1) = &-p_{\mathrm{C}}N^2\tilde{A}_{\mathrm{e}}(\zeta)\{(1+2i\lambda_1)\Psi_0(\zeta+2,\theta_n-1)\\
&-[4-6\lambda_1^2+2i(2\lambda_1+\lambda_1^3)]\Psi_0(\zeta+1,\theta_n-1)+(6-12\lambda_1^2-\lambda_1^4)\Psi_0(\zeta,\theta_n-1)\\
&-[4-6\lambda_1^2-2i(2\lambda_1+\lambda_1^3)]\Psi_0(\zeta-1,\theta_n-1)+(1-2i\lambda_1)\Psi_0(\zeta-2,\theta_n-1)\}
\end{aligned}
\tag{12}
$$

$$
\begin{aligned}
G_{2+}(\zeta,\theta_n) = G_2(\zeta,\theta_n+1) = &-p_{\mathrm{C}}N^2\tilde{A}_{\mathrm{e}}(\zeta)\{(1+2i\lambda_1)\Psi_0(\zeta+2,\theta_n+1)\\
&-[4-6\lambda_1^2+2i(2\lambda_1+\lambda_1^3)]\Psi_0(\zeta+1,\theta_n+1)+(6-12\lambda_1^2-\lambda_1^4)\Psi_0(\zeta,\theta_n+1)\\
&-[4-6\lambda_1^2-2i(2\lambda_1+\lambda_1^3)]\Psi_0(\zeta-1,\theta_n+1)+(1-2i\lambda_1)\Psi_0(\zeta-2,\theta_n+1)\}
\end{aligned}
\tag{13}
$$

We can compute $\tilde{G}_{\mathrm{s}\kappa}(\theta_n,\kappa)$ of Eq.(4.2-24) once $G_2(\zeta,\theta_n)$ of Eq.(11) is obtained. The notation $\tilde{G}_{\mathrm{s}\kappa}(\theta_n,\kappa)$ is changed to $G_{\mathrm{s}\kappa 2}(\theta_n,\kappa)$ to make the use of Eqs.(12) and (13) more evident. We observe that Eq.(4.2-24) treats the function $G_2(\zeta,\theta)$ as a step function with step width 1. The integral may readily be replaced by a sum of $G_2(\zeta,\theta)$ taken at $\zeta = 0,\ 1,\ \ldots,\ N-1$:

$$\begin{aligned} G_{s\kappa 2}(\theta_n,\kappa) = \tilde{G}_{s\kappa}(\theta_n,\kappa) &= \frac{2}{N}\int_0^N G_2(\zeta,\theta_n)e^{-i\lambda_1\zeta}\sin\frac{2\pi\kappa\zeta}{N}d\zeta \\ &= \frac{2}{N}\sum_{\zeta=0}^{N-1} G_2(\zeta,\theta_n)e^{-i\lambda_1\zeta}\sin\frac{2\pi\kappa\zeta}{N} \end{aligned} \tag{14}$$

The functions $G_{s\kappa 2-}(\theta_n,\kappa)$ and $G_{s\kappa 2+}(\theta_n,\kappa)$ follow from Eqs.(12) and (13):

$$G_{s\kappa 2-}(\theta_n,\kappa) = \tilde{G}_{s\kappa}(\theta_n-1,\kappa) = \frac{2}{N}\sum_{\zeta=0}^{N-1} G_{2-}(\zeta,\theta_n)e^{-i\lambda_1\zeta}\sin\frac{2\pi\kappa\zeta}{N} \tag{15}$$

$$G_{s\kappa 2+}(\theta_n,\kappa) = \tilde{G}_{s\kappa}(\theta_n+1,\kappa) = \frac{2}{N}\sum_{\zeta=0}^{N-1} \tilde{G}_{2+}(\zeta,\theta_n)e^{-i\lambda_1\zeta}\sin\frac{2\pi\kappa\zeta}{N} \tag{16}$$

With Eqs.(14)–(16) we obtain the function $\tilde{H}_{s\kappa}(\theta_n,\kappa)$ from Eq.(6.4-1):

$$\begin{aligned} \tilde{H}_{s\kappa}(\theta_n,\kappa) &= -e^{i\lambda_1\lambda_3}\tilde{G}_{s\kappa}(\theta_n+1,\kappa)+2\cos\frac{2\pi\kappa}{N}\tilde{G}_{s\kappa}(\theta_n,\kappa)-e^{-i\lambda_1\lambda_3}\tilde{G}_{s\kappa}(\theta_n-1,\kappa) \\ &= -e^{i\lambda_1\lambda_3}\tilde{G}_{s\kappa 2+}(\theta_n,\kappa)+2\cos\frac{2\pi\kappa}{N}\tilde{G}_{s\kappa 2}(\theta_n,\kappa)-e^{-i\lambda_1\lambda_3}\tilde{G}_{s\kappa 2-}(\theta_n,\kappa) \end{aligned} \tag{17}$$

With $\tilde{H}_{s\kappa}(\theta_n,\kappa)$ we can compute the eight functions $\tilde{J}_1(\theta,\kappa)$ to $\tilde{J}_8(\theta,\kappa)$ of Eq.(6.4-48). These functions in turn yield $\tilde{J}_r(\theta,\kappa)$ and $\tilde{J}_i(\theta,\kappa)$ of Eq.(4.3-17), which yield finally $\mathcal{H}_{1\kappa}(\kappa)$ of Eq.(7). We use again $A_{1\kappa}(\kappa)$ and $B_{1\kappa}(\kappa)$ of Eqs.(3.4-25) and (3.4-26)

$$A_{1\kappa}(\kappa) = \frac{2(2\pi\kappa/N)\{1-\exp[-(\lambda_2^2-\lambda_1^2)^{1/2}N]\}}{\lambda_2^2-\lambda_1^2+(2\pi\kappa/N)^2} \tag{18}$$

$$B_{1\kappa}(\kappa) = N\lambda_1\lambda_3\frac{I_T(\kappa/N)}{\sin\beta_\kappa} \tag{19}$$

to write Eq.(7) in the following form:

$$\tilde{\mathcal{H}}_{1\kappa}(\kappa) = \int_0^N \tilde{I}_{11}(\theta,\kappa)d\theta + \int_0^N \tilde{I}_{12}(\theta,\kappa)d\theta$$

$$\tilde{I}_{11}(\theta,\kappa) = A_{1\kappa}(\kappa)\mathcal{R}e[\tilde{J}_{\mathrm{i}}(\theta,\kappa) - i\tilde{J}_{\mathrm{r}}(\theta,\kappa)]\sin\lambda_1\lambda_3\theta$$

$$\tilde{I}_{12}(\theta,\kappa) = B_{1\kappa}(\kappa)\sin(\beta_\kappa\theta)\,\mathcal{R}e\{[\tilde{J}_{\mathrm{r}}(\theta,\kappa)\sin\lambda_1\lambda_3 - \tilde{J}_{\mathrm{i}}(\theta,\kappa)]\cos\lambda_1\lambda_3 + i[\tilde{J}_{\mathrm{r}}(\theta,\kappa)\cos\lambda_1\lambda_3 + \tilde{J}_{\mathrm{i}}(\theta,\kappa)\sin\lambda_1\lambda_3\} \quad (20)$$

The equivalence of the integral over a step function with step width 1 and a sum permits us to write $\tilde{\mathcal{H}}_{1\kappa}(\kappa)$ in the form

$$\tilde{\mathcal{H}}_{1\kappa}(\kappa) = \sum_{\theta=0}^{N-1} \tilde{I}_{11}(\theta,\kappa) + \sum_{\theta=0}^{N-1} \tilde{I}_{12}(\theta,\kappa) \quad (21)$$

Let us turn to $\bar{U}_{2\kappa}(\kappa)$ of Eq.(5). The index κ is written again after the integration over ζ to show that the summation over κ in Eq.(4.3-23) is not carried out:

$$\bar{U}_{2\kappa}(\kappa) = 2\alpha\tilde{\Psi}_{00}\frac{L^2}{c\Delta t}\int_0^N \mathcal{R}e\left(\int_0^N \frac{\partial\Psi_0^*}{\partial\theta}\frac{\partial\bar{v}}{\partial\theta}d\zeta\right)_\kappa d\theta \quad (22)$$

We obtain from Eq.(4.3-23):

$$\frac{\bar{U}_{2\kappa}(\kappa)}{2\alpha\Psi_{00}^2L^2/c\Delta t} = \tilde{\mathcal{H}}_{2\kappa}(\kappa) = \frac{1}{\Psi_{00}}\int_0^N \mathcal{R}e\left(\int_0^N \frac{\partial\Psi_0^*}{\partial\theta}\frac{\partial\bar{v}}{\partial\theta}d\zeta\right)_\kappa d\theta \quad (23)$$

$$\begin{aligned}
\frac{1}{\Psi_0}\mathcal{R}e\int_0^N\left(\frac{\partial\Psi_0^*}{\partial\theta}\frac{\partial\bar{v}}{\partial\theta}d\zeta\right)_\kappa &= 2\lambda_1\lambda_3\frac{(2\pi\kappa/N)\{1-\exp[-(\lambda_2^2-\lambda_1^2)^{1/2}N]\}}{\lambda_2^2-\lambda_1^2+(2\pi\kappa/N)^2} \\
&\times\{\mathcal{R}e[\tilde{J}_{\mathrm{R}}(\theta,\kappa)+i\tilde{J}_{\mathrm{I}}(\theta,\kappa)]\sin 2\lambda_1\lambda_3\theta+\mathcal{R}e[\tilde{J}_{\mathrm{I}}(\theta,\kappa)-i\tilde{J}_{\mathrm{R}}(\theta,\kappa)]\cos 2\lambda_1\lambda_3\theta]\} \\
&- N\lambda_1\lambda_3\frac{I_{\mathrm{T}}(\kappa/N)}{\sin\beta_\kappa}\{\mathcal{R}e[\tilde{J}_{\mathrm{R}}(\theta,\kappa)+i\tilde{J}_{\mathrm{I}}(\theta,\kappa)] \\
&\times[\lambda_1\lambda_3\sin\beta_\kappa\theta\cos\lambda_1\lambda_3(\theta-1)+\beta_\kappa\cos\beta_\kappa\theta\sin\lambda_1\lambda_3(\theta-1)] \\
&- \mathcal{R}e[\tilde{J}_{\mathrm{I}}(\theta,\kappa)-i\tilde{J}_{\mathrm{R}}(\theta\kappa)] \\
&\times[\lambda_1\lambda_3\sin\beta_\kappa\theta\sin\lambda_1\lambda_3(\theta-1)-\beta_\kappa\cos\beta_\kappa\theta\cos\lambda_1\lambda_3(\theta-1)]\}
\end{aligned} \quad (24)$$

In order to integrate Eq.(24) over θ we need to evaluate numerically the integrals over the following terms with respect to θ:

$$\tilde{I}_{21}(\theta,\kappa) = \lambda_1\lambda_3 A_{1\kappa}(\kappa)\Re e[\tilde{J}_{\mathrm{R}}(\theta,\kappa) + i\tilde{J}_{\mathrm{I}}(\theta,\kappa)]\sin 2\lambda_1\lambda_3\theta \tag{25}$$

$$\tilde{I}_{22}(\theta,\kappa) = \lambda_1\lambda_3 A_{1\kappa}(\kappa)\Re e[\tilde{J}_{\mathrm{I}}(\theta,\kappa) - i\tilde{J}_{\mathrm{R}}(\theta,\kappa)]\cos 2\lambda_1\lambda_3\theta \tag{26}$$

$$\begin{aligned}\tilde{I}_{23}(\theta,\kappa) = {} & B_{1\kappa}(\kappa)\Re e[\tilde{J}_{\mathrm{R}}(\theta,\kappa) + i\tilde{J}_{\mathrm{I}}(\theta,\kappa)] \\ & \times [\lambda_1\lambda_3 \sin\beta_\kappa\theta\cos\lambda_1\lambda_3(\theta-1) + \beta_\kappa\cos\beta_\kappa\theta\sin\lambda_1\lambda_3(\theta-1)]\end{aligned} \tag{27}$$

$$\begin{aligned}\tilde{I}_{24}(\theta,\kappa) = {} & B_{1\kappa}\Re e[\tilde{J}_{\mathrm{I}}(\theta,\kappa) - i\tilde{J}_{\mathrm{R}}(\theta,\kappa)] \\ & \times [\lambda_1\lambda_3 \sin\beta_\kappa\theta\sin\lambda_1\lambda_3(\theta-1) - \beta_\kappa\cos\beta_\kappa\theta\cos\lambda_1\lambda_3(\theta-1)]\end{aligned} \tag{28}$$

The functions $\tilde{J}_{\mathrm{R}}(\theta,\kappa)$ and $\tilde{J}_{\mathrm{I}}(\theta,\kappa)$ are defined in Eq.(4.3-21) by the functions $\tilde{J}_{\mathrm{r}}(\theta,\kappa)$ and $\tilde{J}_{\mathrm{i}}(\theta,\kappa)$. In turn, $\tilde{J}_{\mathrm{r}}(\theta,\kappa)$ and $\tilde{J}_{\mathrm{i}}(\theta,\kappa)$ are defined in Eq.(4.3-17) by $\tilde{J}_1(\theta,\kappa)$ to $\tilde{J}_8(\theta,\kappa)$ of Eq.(6.4-48). We need mainly lengthy but straightforward substitutions to express $\tilde{J}_{\mathrm{R}}(\kappa,\theta)$ and $\tilde{J}_{\mathrm{I}}(\kappa,\theta)$ by $\tilde{J}_1(\theta,\kappa)$ to $\tilde{J}_8(\theta,\kappa)$. The only problem is the derivatives $\partial\tilde{J}_{\mathrm{r}}/\partial\theta$ and $\partial\tilde{J}_{\mathrm{i}}(\theta,\kappa)/\partial\theta$ in Eq.(4.3-21). We obtain with the help of Eq.(4.3-17):

$$\begin{aligned}\frac{\partial\tilde{J}_{\mathrm{r}}(\theta,\kappa)}{\partial\theta} = {} & \left(\frac{\partial\tilde{J}_1(\theta,\kappa)}{\partial\theta} - \tilde{J}_2(\theta,\kappa) - \theta\frac{\partial\tilde{J}_2(\theta,\kappa)}{\partial\theta}\right)\cos\frac{2\pi\kappa\theta}{N} \\ & - \frac{2\pi\kappa}{N}[\tilde{J}_1(\theta,\kappa) - \theta\tilde{J}_2(\theta,\kappa)]\sin\frac{2\pi\kappa\theta}{N} \\ & - \left(\frac{\partial\tilde{J}_3(\theta,\kappa)}{\partial\theta} - \tilde{J}_4(\theta,\kappa) - \theta\frac{\partial\tilde{J}_4(\theta,\kappa)}{\partial\theta}\right)\sin\frac{2\pi\kappa\theta}{N} \\ & - \frac{2\pi\kappa}{N}[\tilde{J}_3(\theta,\kappa) - \theta\tilde{J}_4(\theta,\kappa)]\cos\frac{2\pi\kappa\theta}{N}\end{aligned} \tag{29}$$

$$\begin{aligned}\frac{\partial\tilde{J}_{\mathrm{i}}(\theta,\kappa)}{\partial\theta} = {} & \left(\frac{\partial\tilde{J}_5(\theta,\kappa)}{\partial\theta} - \tilde{J}_6(\theta,\kappa) - \theta\frac{\partial\tilde{J}_6(\theta,\kappa)}{\partial\theta}\right)\sin\frac{2\pi\kappa\theta}{N} \\ & + \frac{2\pi\kappa}{N}[\tilde{J}_5(\theta,\kappa) - \theta\tilde{J}_6(\theta,\kappa)]\cos\frac{2\pi\kappa\theta}{N} \\ & + \left(\frac{\partial\tilde{J}_7(\theta,\kappa)}{\partial\theta} - \tilde{J}_8(\theta,\kappa) - \theta\frac{\partial\tilde{J}_8(\theta,\kappa)}{\partial\theta}\right)\cos\frac{2\pi\kappa\theta}{N} \\ & - \frac{2\pi\kappa}{N}[\tilde{J}_7(\theta,\kappa) - \theta\tilde{J}_8(\theta,\kappa)]\sin\frac{2\pi\kappa\theta}{N}\end{aligned} \tag{30}$$

The functions $\tilde{J}_1(\theta,\kappa)$ to $\tilde{J}_8(\theta,\kappa)$ are defined in Eq.(6.4-48). We must show what the derivatives $\partial\tilde{J}_1(\theta,\kappa)/\partial\theta$ to $\partial\tilde{J}_8(\theta,\kappa)/\partial\theta$ mean. To this end we use once more the equivalence of the sum of a step function with the step width 1 and an integral over the step function. The functions $\tilde{K}_1(n,\kappa)$ to $\tilde{K}_8(n,\kappa)$ in Eq.(6.4-48) are step functions with a step width 1 for $n = 0, 1, \ldots$ and an unspecified value of κ. Using the definition of $\tilde{J}_j(\theta,\kappa)$ in Eq.(6.4-48) we may write:

$$\frac{\partial \tilde{J}_j(\theta,\kappa)}{\partial\theta} = \frac{\partial}{\partial\theta}\sum_{n=0}^{\theta-1}\tilde{K}_j(n,\kappa) = \frac{\partial}{\partial\theta}\int_0^\theta \tilde{K}_j(\theta_n,\kappa)d\theta_n$$

$$= \int_0^\theta \frac{\partial\tilde{K}_j(\theta_n,\kappa)}{\partial\theta_n}d\theta_n = \tilde{K}_j(\theta,\kappa), \quad j = 1,\ 2,\ldots,\ 8 \tag{31}$$

Using $A_{1\kappa}(\kappa)$ and $B_{1\kappa}(\kappa)$ of Eqs.(18) and (19) we may write $\tilde{\mathcal{H}}_{2\kappa}$ of Eq.(23) in the following form:

$$\tilde{\mathcal{H}}_{2\kappa}(\kappa) = \int_0^N \{\tilde{I}_{21}(\theta,\kappa) + \tilde{I}_{22}(\theta,\kappa) - [\tilde{I}_{23}(\theta,\kappa) + \tilde{I}_{24}(\theta,\kappa)]\}d\theta$$

$$= \sum_{\theta=0}^{N-1}\{\tilde{I}_{21}(\theta,\kappa) + \tilde{I}_{22}(\theta,\kappa) - [\tilde{I}_{23}(\theta,\kappa) + \tilde{I}_{24}(\theta,\kappa)]\} \tag{32}$$

Let us turn to $\bar{U}_{3\kappa}(\kappa)$ of Eq.(5). Again the index κ is written after the integration over ζ to indicate that there is no summation over κ in Eq.(4.3-26):

$$\bar{U}_{3\kappa}(\kappa) = 2\alpha\tilde{\Psi}_{00}\frac{L^2}{c\Delta t}\int_0^N \mathcal{R}e\left(\int_0^N \frac{\partial\Psi_0^*}{\partial\zeta}\frac{\partial\bar{v}}{\partial\zeta}d\zeta\right)_\kappa d\theta \tag{33}$$

We obtain with the help of Eq.(4.3-26):

$$\frac{\bar{U}_{3\kappa}(\kappa)}{2\alpha\Psi_{00}^2L^2/c\Delta t} = \tilde{\mathcal{H}}_{3\kappa}(\kappa) = \frac{1}{\Psi_{00}}\int_0^N \mathcal{R}e\left(\int_0^N \frac{\partial\Psi_0^*}{\partial\zeta}\frac{\partial\bar{v}}{\partial\zeta}d\zeta\right)_\kappa d\theta \tag{34}$$

$$\begin{aligned}
\frac{1}{\Psi_{00}}\mathcal{R}e\int_0^N\left(\frac{\partial\Psi_0^*}{\partial\zeta}\frac{\partial\bar{v}}{\partial\zeta}d\zeta\right)_\kappa &= \frac{(2\pi\kappa/N)\{1-\exp[-(\lambda_2^2-\lambda_1^2)^{1/2}N]\}}{\lambda_2^2-\lambda_1^2+(2\pi\kappa/N)^2} \\
&\quad\times 2\lambda_2^2\mathcal{R}e[\tilde{J}_{\mathrm{i}}(\theta,\kappa)-i\tilde{J}_{\mathrm{r}}(\theta,\kappa)]\sin\lambda_1\lambda_3\theta \\
-N\lambda_1\lambda_3\frac{I_{\mathrm{T}}(\kappa/N)}{\sin\beta_\kappa}\left[\lambda_1^2+\left(\frac{2\pi\kappa}{N}\right)^2\right]&\mathcal{R}e\{[\tilde{J}_{\mathrm{i}}(\theta\kappa)-i\tilde{J}_{\mathrm{r}}(\theta,\kappa)]\cos\lambda_1\lambda_3 \\
&\quad-[\tilde{J}_{\mathrm{r}}(\theta,\kappa)+i\tilde{J}_{\mathrm{i}}(\theta,\kappa)]\sin\lambda_1\lambda_3\}\sin\beta_\kappa\theta
\end{aligned} \tag{35}$$

In order to integrate Eq.(35) over θ we need to evaluate numerically the integrals over two terms:

$$\tilde{I}_{31}(\theta,\kappa) = A_{1\kappa}(\kappa)\mathcal{R}e\{2\lambda_2^2[\tilde{J}_{\rm i}(\theta,\kappa) - i\tilde{J}_{\rm r}(\theta,\kappa)]\}\sin\lambda_1\lambda_3\theta \tag{36}$$

$$\tilde{I}_{32}(\theta,\kappa) = B_{1\kappa}(\kappa)[\lambda_1^2 + (2\pi\kappa/N)^2]\{\mathcal{R}e[\tilde{J}_{\rm i}(\theta,\kappa) - i\tilde{J}_{\rm r}(\theta,\kappa)]\cos\lambda_1\lambda_3 - \mathcal{R}e[\tilde{J}_{\rm r}(\theta,\kappa) + i\tilde{J}_{\rm i}(\theta,\kappa)]\sin\lambda_1\lambda_3\}\sin\beta_\kappa\theta \tag{37}$$

The functions $\tilde{J}_{\rm r}(\theta,\kappa)$ and $\tilde{J}_{\rm i}(\theta,\kappa)$ are defined in Eq.(4.3-17). We may write $\tilde{\mathcal{H}}_{3\kappa}(\kappa)$ of Eq.(34) in the following form:

$$\tilde{\mathcal{H}}_{3\kappa}(\kappa) = \int_0^N [\tilde{I}_{31}(\theta,\kappa) - \tilde{I}_{32}(\theta,\kappa)]d\theta = \sum_{\theta=0}^{N-1}[\tilde{I}_{31}(\theta,\kappa) - \tilde{I}_{32}(\theta,\kappa)] \tag{38}$$

In analogy to $U_{\rm c}$ in Eqs.(2.5-16) and (2.5-17) we write the energy $\bar{U}_{\alpha c}$ of Eq.(4) in the following normalized form:

$$\frac{c\Delta t\bar{U}_{\alpha c}}{2\alpha L^2\Psi_{00}^2 N} = \frac{cT\bar{U}_{\alpha c}}{2\alpha L^2\Psi_{00}^2 N^2} = \frac{\tilde{\mathcal{H}}_\alpha}{N^2} \tag{39}$$

$$\bar{U}_{\alpha c} = \frac{2\alpha L^2\Psi_{00}^2}{cT}\tilde{\mathcal{H}}_\alpha = \frac{2\alpha L^2\Psi_{00}^2}{cT}\sum_{\kappa=1}^{N-1}\tilde{\mathcal{H}}_{\alpha\kappa}(\kappa) \tag{40}$$

$$\bar{U}_{\alpha c\kappa}(\kappa) = \frac{2\alpha L^2\Psi_{00}^2}{cT}\tilde{\mathcal{H}}_{\alpha\kappa}(\kappa) \tag{41}$$

$$\tilde{\mathcal{H}}_{\alpha\kappa}(\kappa) = \tilde{\mathcal{H}}_{2\kappa}(\kappa) + \tilde{\mathcal{H}}_{3\kappa}(\kappa) + \frac{m_0^2c^4(\Delta t)^2}{\hbar^2}\tilde{\mathcal{H}}_{1\kappa}(\kappa) \tag{42}$$

In order to work out a computer program for $\tilde{\mathcal{H}}_{1\kappa}(\kappa)$, $\tilde{\mathcal{H}}_{2\kappa}(\kappa)$, and $\tilde{\mathcal{H}}_{3\kappa}(\kappa)$ we introduce a more compact notation for $\Psi_0(\zeta,\theta_n)$ of Eq.(8) when ζ varies from $\zeta+2$ to $\zeta-2$ and θ_n varies from θ_n+1 to θ_n-1:

$$\Psi_{0,j,m}(\zeta,\theta_n) = \Psi_0(\zeta+j,\theta_n+m), \quad j=2,1,0,-1,-2, \quad m=1,0,-1$$
$$\Psi_{0,0,0}(\zeta,\theta) = \Psi_0(\zeta,\theta_n) \tag{43}$$

Table 4.1-1 uses this notation in lines 2–6. Equations (11), (12), and (13) assume the following form:

TABLE 4.4-1

FUNCTIONS β_κ, $I_T(\kappa/N)$, ... AND THE EQUATIONS EQ.(2.4-14), EQ.(2.4-29), ... DEFINING THEM, THAT ARE REQUIRED FOR THE CALCULATION OF $\tilde{\mathcal{H}}_{1\kappa}(\kappa)$ OF EQS.(6) OR (7) AS WELL AS $\tilde{\mathcal{H}}_{2\kappa}(\kappa)$ AND $\tilde{\mathcal{H}}_{3\kappa}(\kappa)$ OF EQS.(32) AND (38). THE NUMERICAL VALUES OBTAINED FOR EQUATIONS SHOWN WITH BRACKETS, E.G. EQ.[2.4-14], SHOULD BE REMEMBERED BY THE COMPUTER TO REDUCE THE COMPUTING TIME.

β_κ	Eq.[2.4-14]	$I_T(\kappa/N)$	Eq.[2.4-29]	$\Psi_0(\zeta,\theta_n)$	Eq.[4.4-8]
$\Psi_{0,2,1}(\zeta,\theta_n)$	Eq.(4.4-43)	$\Psi_{0,1,1}(\zeta,\theta_n)$	Eq.(4.4-43)	$\Psi_{0,0,1}(\zeta,\theta_n)$	Eq.(4.4-43)
$\Psi_{0,-1,1}(\zeta,\theta_n)$	Eq.(4.4-43)	$\Psi_{0,-2,1}(\zeta,\theta_n)$	Eq.(4.4-43)	$\Psi_{0,2,0}(\zeta,\theta_n)$	Eq.(4.4-43)
$\Psi_{0,1,0}(\zeta,\theta_n)$	Eq.(4.4-43)	$\Psi_{0,0,0}(\zeta,\theta_n)$	Eq.(4.4-43)	$\Psi_{0,-1,0}(\zeta,\theta_n)$	Eq.(4.4-43)
$\Psi_{0,-2,1}(\zeta,\theta_n)$	Eq.(4.4-43)	$\Psi_{0,2,-1}(\zeta,\theta_n)$	Eq.(4.4-43)	$\Psi_{0,1,-1}(\zeta,\theta_n)$	Eq.(4.4-43)
$\Psi_{0,0,-1}(\zeta,\theta_n)$	Eq.(4.4-43)	$\Psi_{0,-1,-1}(\zeta,\theta_n)$	Eq.(4.4-43)	$\Psi_{0,-2,-1}(\zeta,\theta_n)$	Eq.(4.4-43)
$\tilde{A}_e(\zeta,\theta_n)$	Eq.(4.4-10)	$G_2(\zeta,\theta_n)$	Eq.[4.4-44]	$G_{2-}(\zeta,\theta_n)$	Eq.[4.4-45]
$G_{2+}(\zeta,\theta_n)$	Eq.[4.4-46]	$G_{s\kappa 2}(\theta_n,\kappa)$	Eq.[4.4-14]	$G_{s\kappa 2-}(\theta_n,\kappa)$	Eq.[4.4-15]
$G_{s\kappa 2+}(\theta_n,\kappa)$	Eq.[4.4-16]	$\tilde{H}_{s\kappa}(\theta_n,\kappa)$	Eq.[4.4-17]	$F_1(n,\kappa)$	Eq.(6.3-51)
$F_2(n,\kappa)$	Eq.(6.3-52)	$F_3(n,\kappa)$	Eq.(6.3-53)	$F_4(n,\kappa)$	Eq.(6.3-54)
$F_5(n,\kappa)$	Eq.(6.3-55)	$F_6(n,\kappa)$	Eq.(6.3-56)	$F_7(n,\kappa)$	Eq.(6.3-57)
$F_8(n,\kappa)$	Eq.(6.3-58)	$\tilde{d}_{3r}(\kappa)$	Eq.(6.4-46)	$\tilde{d}_{3i}(\kappa)$	Eq.(6.4-47)
$\tilde{J}_1(\theta,\kappa)$	Eq.[6.4-48]	$\tilde{J}_2(\theta,\kappa)$	Eq.[6.4-48]	$\tilde{J}_3(\theta,\kappa)$	Eq.[6.4-48]
$\tilde{J}_4(\theta,\kappa)$	Eq.[6.4-48]	$\tilde{J}_5(\theta,\kappa)$	Eq.[6.4-48]	$\tilde{J}_6(\theta,\kappa)$	Eq.[6.4-48]
$\tilde{J}_7(\theta,\kappa)$	Eq.[6.4-48]	$\tilde{J}_8(\theta,\kappa)$	Eq.[6.4-48]	$\tilde{J}_r(\theta,\kappa)$	Eq.[4.3-17]
$\tilde{J}_i(\theta,\kappa)$	Eq.[4.3-17]	$A_{1\kappa}(\kappa)$	Eq.[4.4-18]	$B_{1\kappa}(\kappa)$	Eq.[4.4-19]
$\tilde{I}_{11}(\theta,\kappa)$	Eq.(4.4-20)	$\tilde{I}_{12}(\theta,\kappa)$	Eq.(4.4-20)	$\tilde{\mathcal{H}}_{1\kappa}(\kappa)$	Eq.(4.4-21)
$\tilde{K}_1(\theta,\kappa$	Eq.[6.4-49]	$\tilde{K}_2(\theta,\kappa)$	Eq.[6.4-49]	$\tilde{K}_3(\theta,\kappa)$	Eq.[6.4-49]
$\tilde{K}_4(\theta,\kappa)$	Eq.[6.4-49]	$\tilde{K}_5(\theta,\kappa)$	Eq.[6.4-49]	$\tilde{K}_6(\theta,\kappa)$	Eq.[6.4-49]
$\tilde{K}_7(\theta,\kappa)$	Eq.[6.4-49]	$\tilde{K}_8(\theta,\kappa)$	Eq.[6.4-49]	$\partial\tilde{J}_r/\partial\theta$	Eq.[4.4-29]
$\partial\tilde{J}_i/\partial\theta$	Eq.[4.4-30]	$\tilde{J}_R(\theta,\kappa)$	Eq.[4.3-21]	$\tilde{J}_I(\theta,\kappa)$	Eq.[4.3-21]
$\tilde{I}_{21}(\theta,\kappa)$	Eq.(4.4-25)	$\tilde{I}_{22}(\theta,\kappa)$	Eq.(4.4-26)	$\tilde{I}_{23}(\theta,\kappa)$	Eq.(4.4-27)
$\tilde{I}_{24}(\theta,\kappa)$	Eq.(4.4-28)	$\tilde{\mathcal{H}}_{2\kappa}(\kappa)$	Eq.[4.4-32]	$\tilde{I}_{31}(\theta,\kappa)$	Eq.(4.4-36)
$\tilde{I}_{32}(\theta,\kappa)$	Eq.(4.4-37)	$\tilde{\mathcal{H}}_{3\kappa}(\kappa)$	Eq.[4.4-38]	p_C	Eq.(4.4-10)

$$\begin{aligned} G_{2+}(\zeta,\theta_n) = &-p_C N^2 \tilde{A}_e(\zeta)\{(1+2i\lambda_1)\Psi_{0,2,1}(\zeta,\theta_n) \\ &-[4-6\lambda_1^2+2i(2\lambda_1+\lambda_1^3)]\Psi_{0,1,1}(\zeta,\theta_n)+(6-12\lambda_1^2-\lambda_1^4)\Psi_{0,0,1}(\zeta,\theta_n) \\ &-[4-6\lambda_1^2-2i(2\lambda_1+\lambda_1^3)]\Psi_{0,-1,1}(\zeta,\theta_n)+(1-2i\lambda_1)\Psi_{0,-2,1}(\zeta,\theta_n)\} \end{aligned} \quad (46)$$

For the representation of the results of this section by plots we start with $\tilde{\mathcal{H}}_{1\kappa}(\kappa)$ of Eq.(21), $\tilde{\mathcal{H}}_{2\kappa}(\kappa)$ of Eq.(32), and $\tilde{\mathcal{H}}_{3\kappa}(\kappa)$ of Eq.(38). We choose again $N = 100$ since the computations according to Tables 3.4-1 and 4.4-1 appear about equally lengthy. Equations (2.3-2) or (3.4-50)

$$\lambda_2 \doteq \lambda_1 \frac{m_0 c}{eA_{\mathrm{m}0x}} = \frac{\Delta t}{\hbar/m_0c^2} = \frac{\Delta x}{\hbar/m_0c}, \quad \frac{h}{m_0c} = \lambda_{\mathrm{C}}, \ \frac{h}{m_0c^2} = \tau_{\mathrm{C}} \tag{47}$$

yield

$$\lambda_2 = \frac{\Delta x}{\hbar/m_0c} = 2\pi \frac{\Delta x}{\lambda_{\mathrm{C}}} \tag{48}$$

The choice $m_0c^2/ceA_{\mathrm{m}0x} > 1$ leads to the relation

$$\lambda_1 < \lambda_2, \quad A_{\mathrm{m}0x} < m_0c^2/ce, \quad A_{\mathrm{m}0x} = \frac{m_0c}{e}\frac{\lambda_1}{\lambda_2} \tag{49}$$

For $\lambda_3 = \phi_{\mathrm{e}0}/cA_{\mathrm{m}0x}$ in Eq.(2.3-2) we continue to prefer the simplest value $\lambda_3 = 1$. We choose $\tilde{A}_{\mathrm{e}0} = 10\lambda_1$, which makes $\tilde{A}_{\mathrm{e}0}$ as well as $A_{\mathrm{m}0x}$ of Eq.(49) vary proportionately to λ_1.

Consider the set of plots in Figs.4.4-1 to 4.4-3. The parameters listed are the same for all three plots. The important parameter is $\lambda_{\mathrm{C}}/\Delta x = 10^4$, which yields $\lambda_2 = 2\pi \times 10^{-4}$. We see three determined functions, which means there is no sign of randomness, except around $\kappa = \pm 20$ in Fig.4.4-3.

In Figs.4.4-4 to 4.4-6 the parameter $\lambda_{\mathrm{C}}/\Delta x = 5 \times 10^4$ produces $\lambda_2 = 4\pi \times 10^{-5}$. All three plots show the previous plots somewhat randomized. The plot of Fig.4.4-1 is still discernible in Fig.4.4-4. The randomization is much more evident in Figs.4.4-5 and 4.4-6.

We go one step further and increase $\lambda_{\mathrm{C}}/\Delta x$ to 10^5, which decreases λ_2 to $2\pi \times 10^{-5}$. The resulting plots in Figs.4.4-7 to 4.4-9 are now so randomized that the patterns of Figs.4.4-1 to 4.4-3 are completely unrecognizable.

The randomization by a resolution improvement from $\Delta x/\lambda_{\mathrm{C}} = 10^{-4}$ to 2×10^{-5} and further to 10^{-5} is what one would expect from the Compton effect. The small values $\Delta x/\lambda_{\mathrm{C}} = 10^{-4}$ to 10^{-5} are explained by Eq.(47)

$$\lambda_2/\lambda_1 = m_0c^2/ceA_{\mathrm{m}0x} \tag{50}$$

which for $\lambda_1 = 0.1\lambda_2$ implies a very large potential $A_{\mathrm{m}0x}$. We will expect a reduced binding force for smaller values of $A_{\mathrm{m}0x}$. Unfortunately, the computing time is increased by 50% if one replaces $\lambda_1 = 0.1\lambda_2$ with $\lambda_1 = 10^{-5}\lambda_2$ as is done in Figs.4.4-10 to 4.4-18. For this reason we could not investigate still smaller values of λ_1/λ_2.

Figures 4.4-10 to 4.4-18 hold for $\lambda_1 = 10^{-5}\lambda_2$. The parameter $\lambda_{\mathrm{C}}/\Delta x = 5 \times 10^3$ was chosen so that one obtains determined functions in Figs.4.4-10 to 4.4-12, except for a slight randomization in Fig.4.4-12 around $\kappa = \pm 20$. The choice $\lambda_{\mathrm{C}}/\Delta x = 1.25 \times 10^4$ brings some randomization in Figs.4.4-13 to 4.4-15. Full randomization is obtained for $\lambda_{\mathrm{C}}/\Delta x = 5 \times 10^4$ in Figs.4.4-16 to 4.4-18.

The plots of Figs.4.4-19 to 4.4-27 have $\lambda_3 = 1$ in Figs.4.4-10 to 4.4-18 replaced by $\lambda_3 = 0.1$. We obtain again the transition from determined to random functions. In addition, the vertical scales are reduced by factors between 10^{-2} and 10^{-14}.

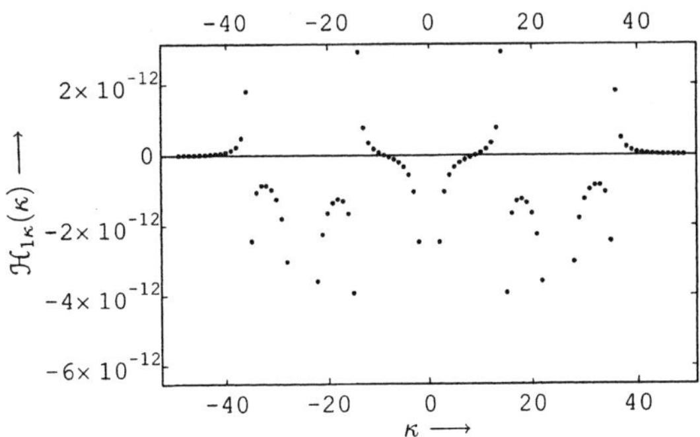

FIG.4.4-1. Plot of $\tilde{\mathcal{H}}_{1\kappa}(\kappa)$ according to Eq.(21) for $N = 100$, $\tilde{A}_{e0} = 10\lambda_1$, $\lambda_C/\Delta x = 10^4$, $\lambda_1 = 0.1\lambda_2$, $\lambda_2 = 2\pi\Delta x/\lambda_C$, $\lambda_3 = 1$.

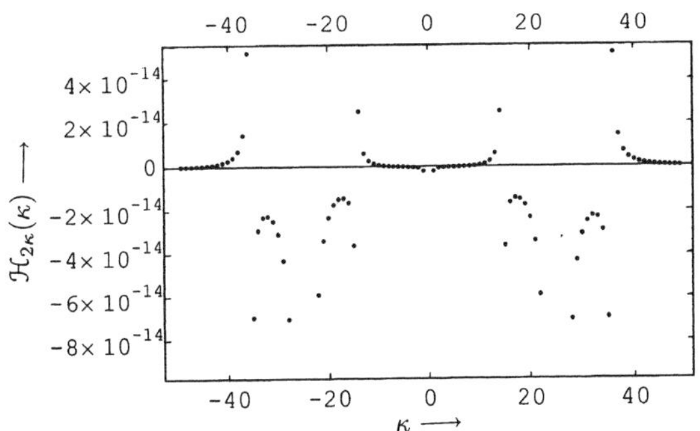

FIG.4.4-2. Plot of $\tilde{\mathcal{H}}_{2\kappa}(\kappa)$ according to Eq.(32) for $N = 100$, $\tilde{A}_{e0} = 10\lambda_1$, $\lambda_C/\Delta x = 10^4$, $\lambda_1 = 0.1\lambda_2$, $\lambda_2 = 2\pi\Delta x/\lambda_C$, $\lambda_3 = 1$.

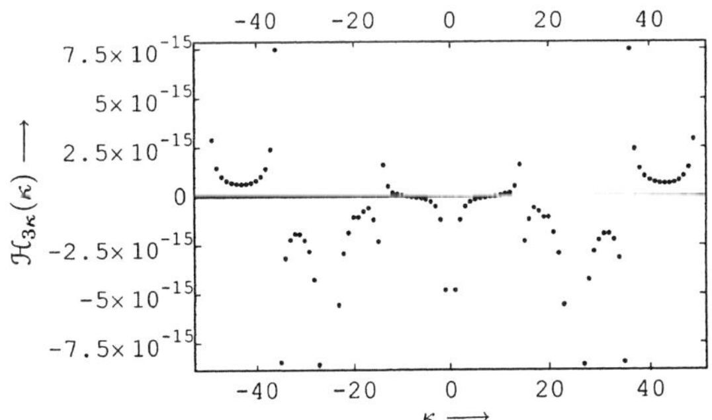

FIG.4.4-3. Plot of $\tilde{\mathcal{H}}_{3\kappa}(\kappa)$ according to Eq.(38) for $N = 100$, $\tilde{A}_{e0} = 10\lambda_1$, $\lambda_C/\Delta x = 10^4$, $\lambda_1 = 0.1\lambda_2$, $\lambda_2 = 2\pi\Delta x/\lambda_C$, $\lambda_3 = 1$.

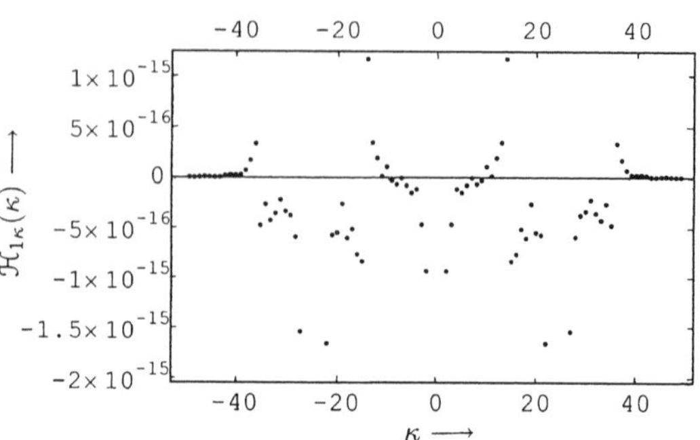

FIG.4.4-4. Plot of $\tilde{\mathcal{H}}_{1\kappa}(\kappa)$ according to Eq.(21) for $N = 100$, $\tilde{A}_{e0} = 10\lambda_1$, $\lambda_C/\Delta x = 5 \times 10^4$, $\lambda_1 = 0.1\lambda_2$, $\lambda_2 = 2\pi\Delta x/\lambda_C$, $\lambda_3 = 1$.

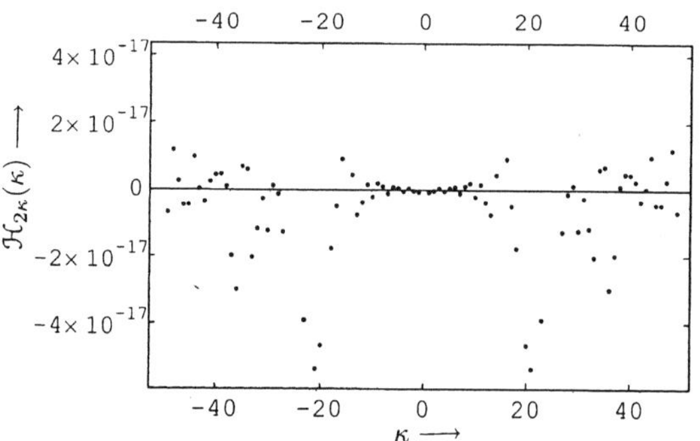

FIG.4.4-5. Plot of $\tilde{\mathcal{H}}_{2\kappa}(\kappa)$ according to Eq.(32) for $N = 100$, $\tilde{A}_{e0} = 10\lambda_1$, $\lambda_C/\Delta x = 5 \times 10^4$, $\lambda_1 = 0.1\lambda_2$, $\lambda_2 = 2\pi\Delta x/\lambda_C$, $\lambda_3 = 1$.

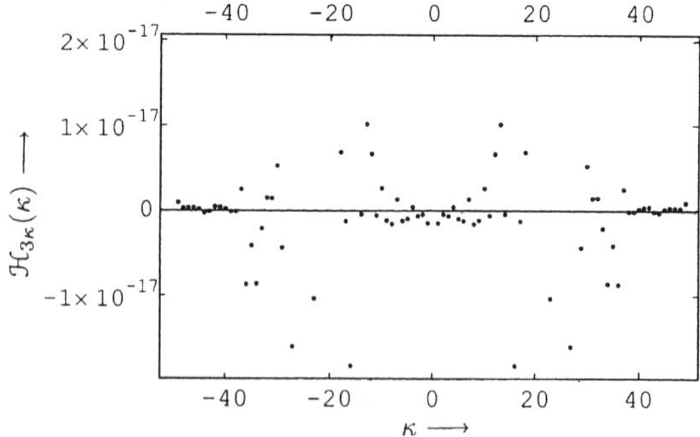

FIG.4.4-6. Plot of $\tilde{\mathcal{H}}_{3\kappa}(\kappa)$ according to Eq.(38) for $N = 100$, $\tilde{A}_{e0} = 10\lambda_1$, $\lambda_C/\Delta x = 5 \times 10^4$, $\lambda_1 = 0.1\lambda_2$, $\lambda_2 = 2\pi\Delta x/\lambda_C$, $\lambda_3 = 1$.

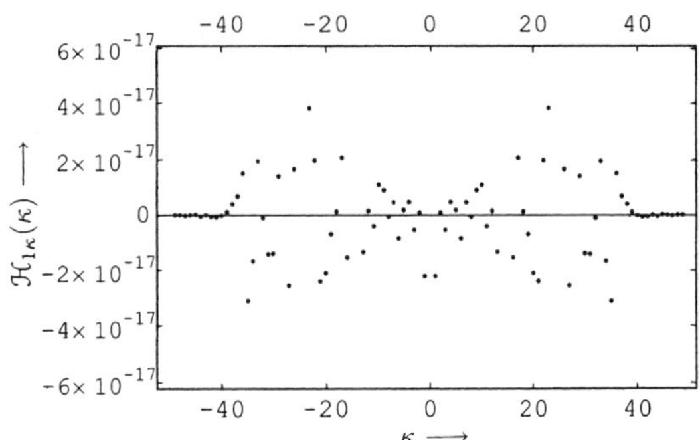

FIG.4.4-7. Plot of $\tilde{\mathcal{H}}_{1\kappa}(\kappa)$ according to Eq.(21) for $N = 100$, $\tilde{A}_{e0} = 10\lambda_1$, $\lambda_C/\Delta x = 10^5$, $\lambda_1 = 0.1\lambda_2$, $\lambda_2 = 2\pi\Delta x/\lambda_C$, $\lambda_3 = 1$.

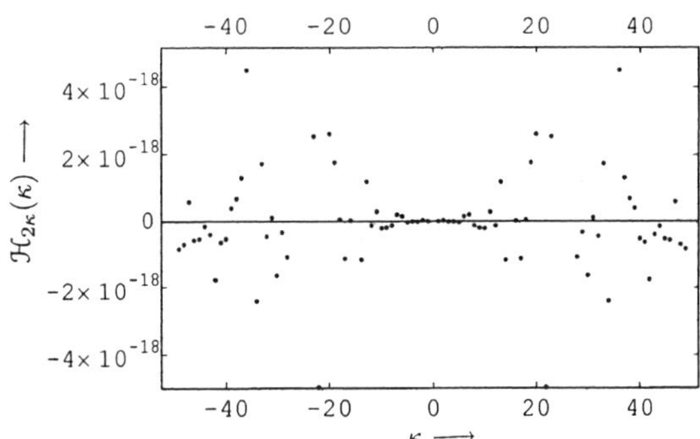

FIG.4.4-8. Plot of $\tilde{\mathcal{H}}_{2\kappa}(\kappa)$ according to Eq.(32) for $N = 100$, $\tilde{A}_{e0} = 10\lambda_1$, $\lambda_C/\Delta x = 10^5$, $\lambda_1 = 0.1\lambda_2$, $\lambda_2 = 2\pi\Delta x/\lambda_C$, $\lambda_3 = 1$.

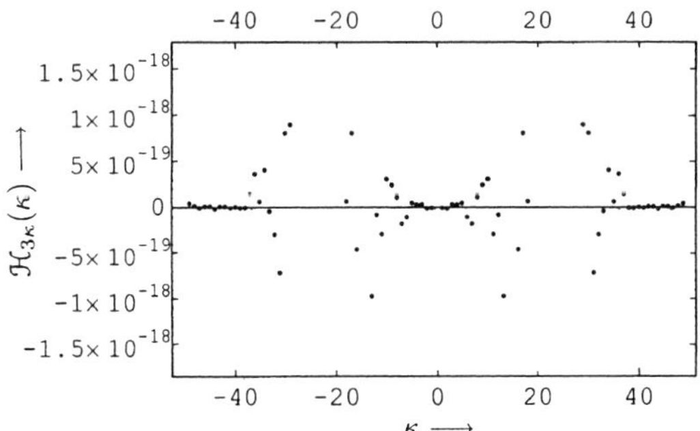

FIG.4.4-9. Plot of $\tilde{\mathcal{H}}_{3\kappa}(\kappa)$ according to Eq.(38) for $N = 100$, $\tilde{A}_{e0} = 10\lambda_1$, $\lambda_C/\Delta x = 10^5$, $\lambda_1 = 0.1\lambda_2$, $\lambda_2 = 2\pi\Delta x/\lambda_C$, $\lambda_3 = 1$.

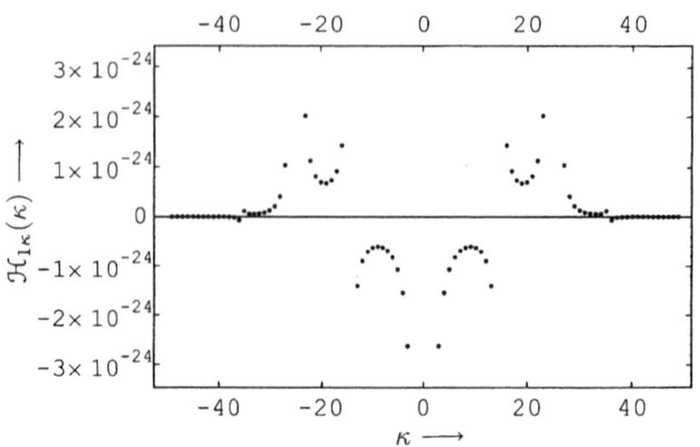

FIG.4.4-10. Plot of $\tilde{\mathcal{H}}_{1\kappa}(\kappa)$ according to Eq.(21) for $N = 100$, $\tilde{A}_{e0} = 10\lambda_1$, $\lambda_C/\Delta x = 5 \times 10^3$, $\lambda_1 = 10^{-5}\lambda_2$, $\lambda_2 = 2\pi\Delta x/\lambda_C$, $\lambda_3 = 1$.

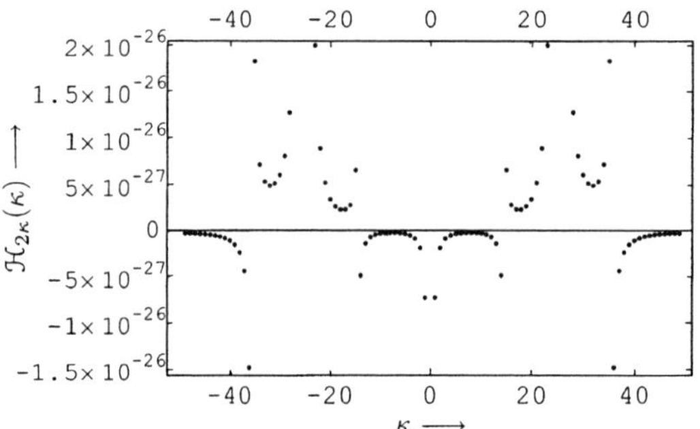

FIG.4.4-11. Plot of $\tilde{\mathcal{H}}_{2\kappa}(\kappa)$ according to Eq.(32) for $N = 100$, $\tilde{A}_{e0} = 10\lambda_1$, $\lambda_C/\Delta x = 5 \times 10^3$, $\lambda_1 = 10^{-5}\lambda_2$, $\lambda_2 = 2\pi\Delta x/\lambda_C$, $\lambda_3 = 1$.

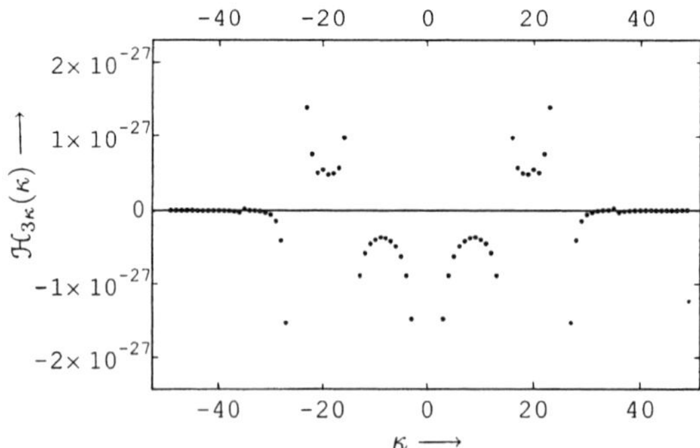

FIG.4.4-12. Plot of $\tilde{\mathcal{H}}_{3\kappa}(\kappa)$ according to Eq.(38) for $N = 100$, $\tilde{A}_{e0} = 10\lambda_1$, $\lambda_C/\Delta x = 5 \times 10^3$, $\lambda_1 = 10^{-5}\lambda_2$, $\lambda_2 = 2\pi\Delta x/\lambda_C$, $\lambda_3 = 1$.

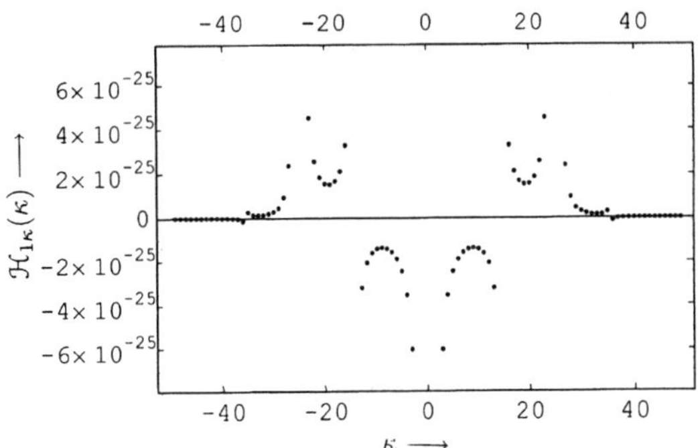

FIG.4.4-13. Plot of $\tilde{\mathcal{H}}_{1\kappa}(\kappa)$ according to Eq.(21) for $N = 100$, $\tilde{A}_{e0} = 10\lambda_1$, $\lambda_C/\Delta x = 1.25 \times 10^4$, $\lambda_1 = 10^{-5}\lambda_2$, $\lambda_2 = 2\pi\Delta x/\lambda_C$, $\lambda_3 = 1$.

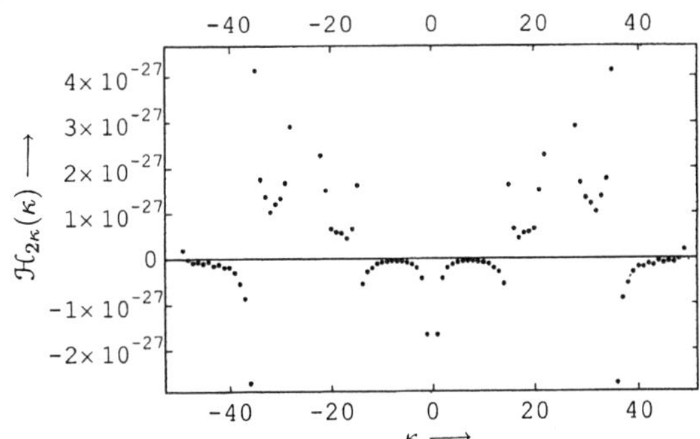

FIG.4.4-14. Plot of $\tilde{\mathcal{H}}_{2\kappa}(\kappa)$ according to Eq.(32) for $N = 100$, $\tilde{A}_{e0} = 10\lambda_1$, $\lambda_C/\Delta x = 1.25 \times 10^4$, $\lambda_1 = 10^{-5}\lambda_2$, $\lambda_2 = 2\pi\Delta x/\lambda_C$, $\lambda_3 = 1$.

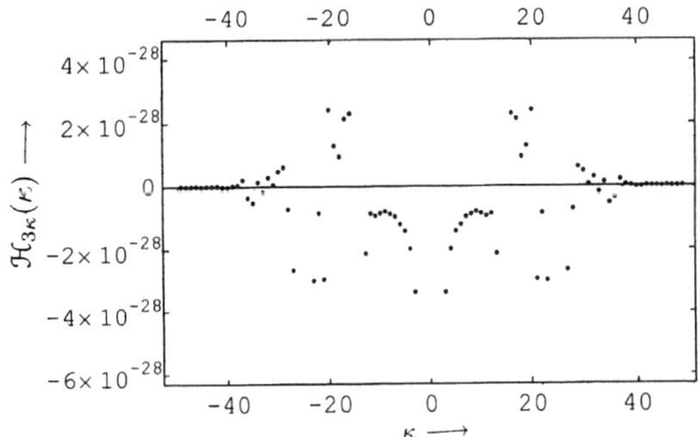

FIG.4.4-15. Plot of $\tilde{\mathcal{H}}_{3\kappa}(\kappa)$ according to Eq.(38) for $N = 100$, $\tilde{A}_{e0} = 10\lambda_1$, $\lambda_C/\Delta x = 1.25 \times 10^4$, $\lambda_1 = 10^{-5}\lambda_2$, $\lambda_2 = 2\pi\Delta x/\lambda_C$, $\lambda_3 = 1$.

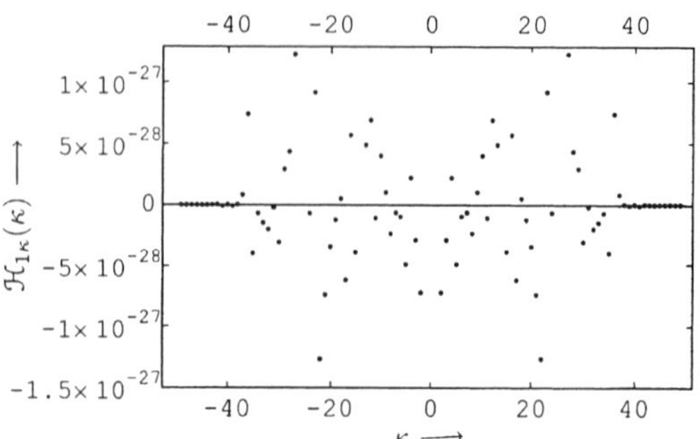

FIG.4.4-16. Plot of $\tilde{\mathcal{H}}_{1\kappa}(\kappa)$ according to Eq.(21) for $N = 100$, $\tilde{A}_{e0} = 10\lambda_1$, $\lambda_C/\Delta x = 5 \times 10^4$, $\lambda_1 = 10^{-5}\lambda_2$, $\lambda_2 = 2\pi\Delta x/\lambda_C$, $\lambda_3 = 1$.

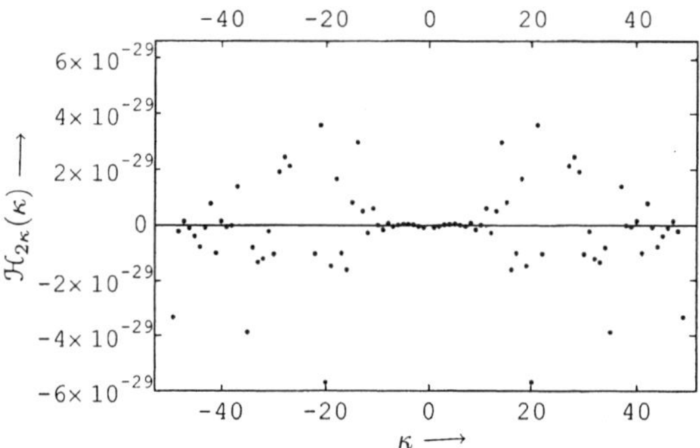

FIG.4.4-17. Plot of $\tilde{\mathcal{H}}_{2\kappa}(\kappa)$ according to Eq.(32) for $N = 100$, $\tilde{A}_{e0} = 10\lambda_1$, $\lambda_C/\Delta x = 5 \times 10^4$, $\lambda_1 = 10^{-5}\lambda_2$, $\lambda_2 = 2\pi\Delta x/\lambda_C$, $\lambda_3 = 1$.

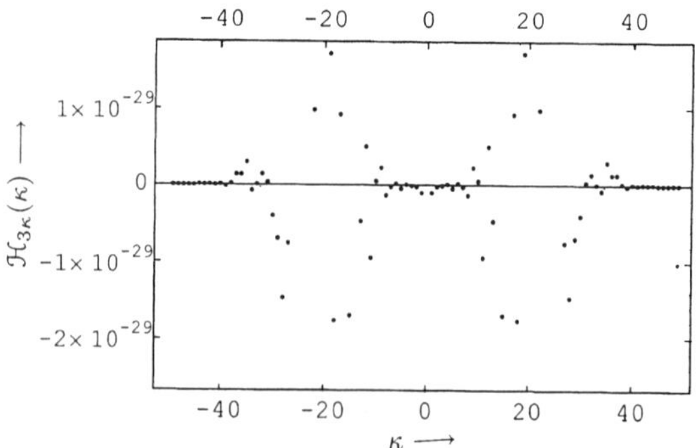

FIG.4.4-18. Plot of $\tilde{\mathcal{H}}_{3\kappa}(\kappa)$ according to Eq.(38) for $N = 100$, $\tilde{A}_{e0} = 10\lambda_1$, $\lambda_C/\Delta x = 5 \times 10^4$, $\lambda_1 = 10^{-5}\lambda_2$, $\lambda_2 = 2\pi\Delta x/\lambda_C$, $\lambda_3 = 1$.

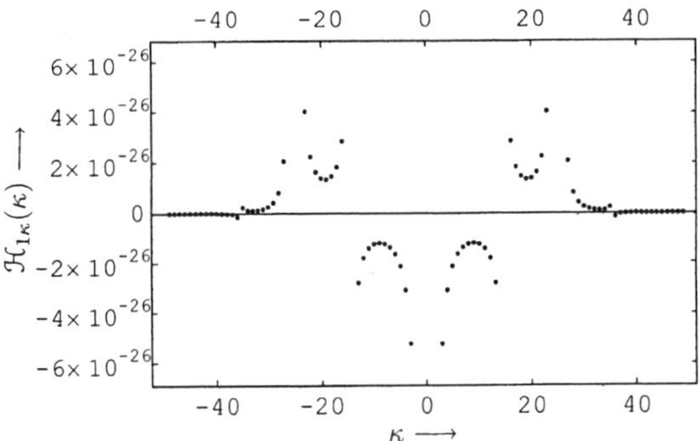

FIG.4.4-19. Plot of $\tilde{\mathcal{H}}_{1\kappa}(\kappa)$ according to Eq.(21) for $N = 100$, $\tilde{A}_{e0} = 10\lambda_1$, $\lambda_C/\Delta x = 5 \times 10^3$, $\lambda_1 = 10^{-5}\lambda_2$, $\lambda_2 = 2\pi\Delta x/\lambda_C$, $\lambda_3 = 0.1$.

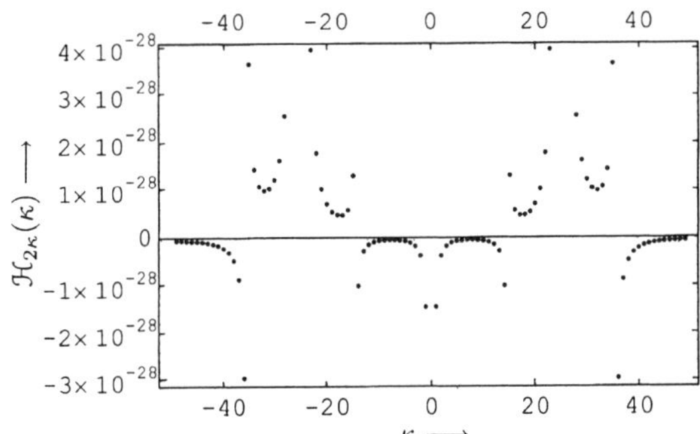

FIG.4.4-20. Plot of $\tilde{\mathcal{H}}_{2\kappa}(\kappa)$ according to Eq.(32) for $N = 100$, $\tilde{A}_{e0} = 10\lambda_1$, $\lambda_C/\Delta x = 5 \times 10^3$, $\lambda_1 = 10^{-5}\lambda_2$, $\lambda_2 = 2\pi\Delta x/\lambda_C$, $\lambda_3 = 0.1$.

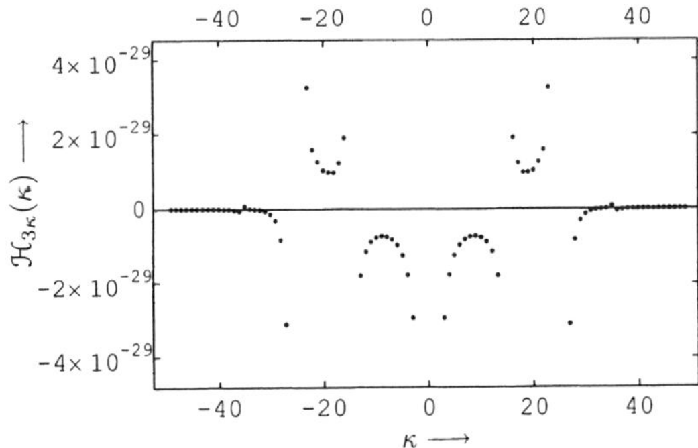

FIG.4.4-21. Plot of $\tilde{\mathcal{H}}_{3\kappa}(\kappa)$ according to Eq.(38) for $N = 100$, $\tilde{A}_{e0} = 10\lambda_1$, $\lambda_C/\Delta x = 5 \times 10^3$, $\lambda_1 = 10^{-5}\lambda_2$, $\lambda_2 = 2\pi\Delta x/\lambda_C$, $\lambda_3 = 0.1$.

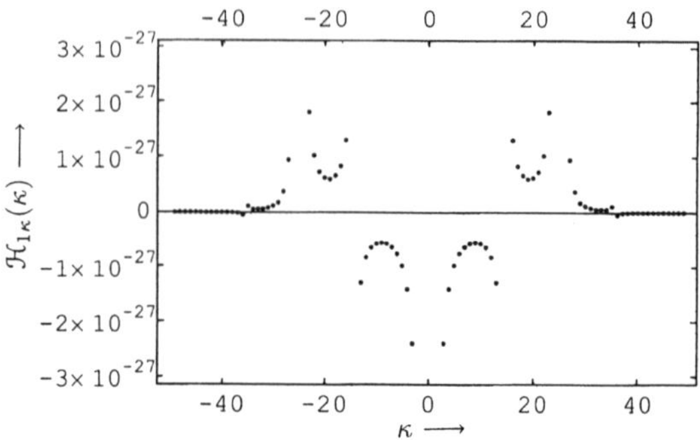

FIG.4.4-22. Plot of $\tilde{\mathcal{H}}_{1\kappa}(\kappa)$ according to Eq.(21) for $N = 100$, $\tilde{A}_{e0} = 10\lambda_1$, $\lambda_C/\Delta x = 1.25 \times 10^4$, $\lambda_1 = 10^{-5}\lambda_2$, $\lambda_2 = 2\pi\Delta x/\lambda_C$, $\lambda_3 = 0.1$.

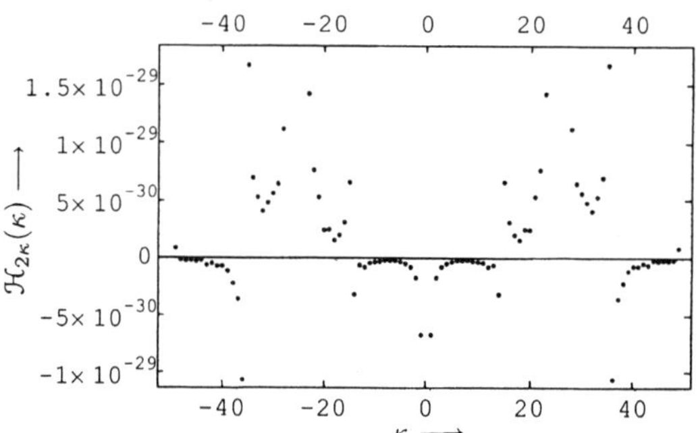

FIG.4.4-23. Plot of $\tilde{\mathcal{H}}_{2\kappa}(\kappa)$ according to Eq.(32) for $N = 100$, $\tilde{A}_{e0} = 10\lambda_1$, $\lambda_C/\Delta x = 1.25 \times 10^4$, $\lambda_1 = 10^{-5}\lambda_2$, $\lambda_2 = 2\pi\Delta x/\lambda_C$, $\lambda_3 = 0.1$.

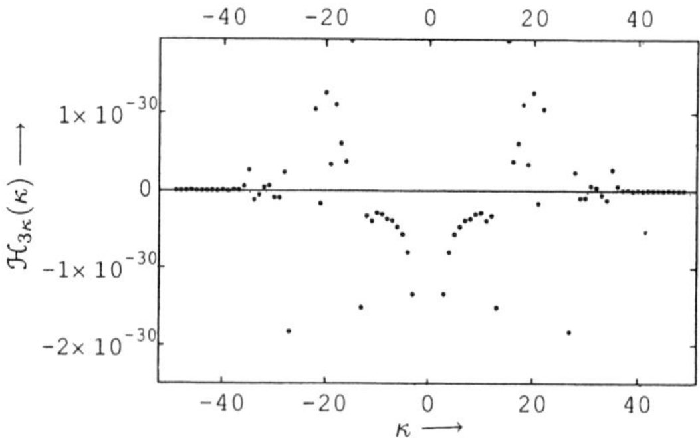

FIG.4.4-24. Plot of $\tilde{\mathcal{H}}_{3\kappa}(\kappa)$ according to Eq.(38) for $N = 100$, $\tilde{A}_{e0} = 10\lambda_1$, $\lambda_C/\Delta x = 1.25 \times 10^4$, $\lambda_1 = 10^{-5}\lambda_2$, $\lambda_2 = 2\pi\Delta x/\lambda_C$, $\lambda_3 = 0.1$.

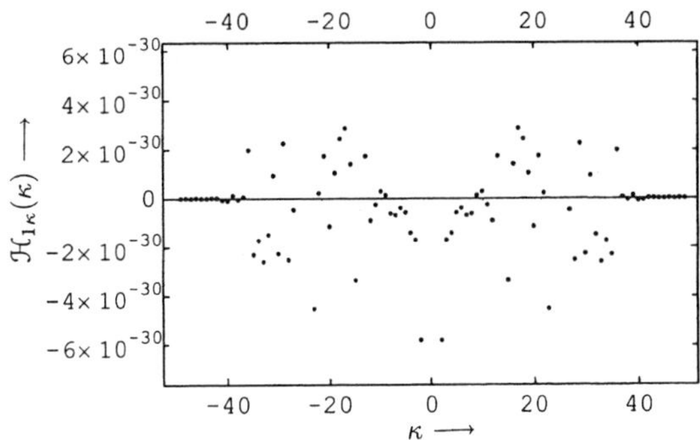

FIG.4.4-25. Plot of $\tilde{\mathcal{H}}_{1\kappa}(\kappa)$ according to Eq.(21) for $N = 100$, $\tilde{A}_{e0} = 10\lambda_1$, $\lambda_C/\Delta x = 5 \times 10^4$, $\lambda_1 = 10^{-5}\lambda_2$, $\lambda_2 = 2\pi\Delta x/\lambda_C$, $\lambda_3 = 0.1$.

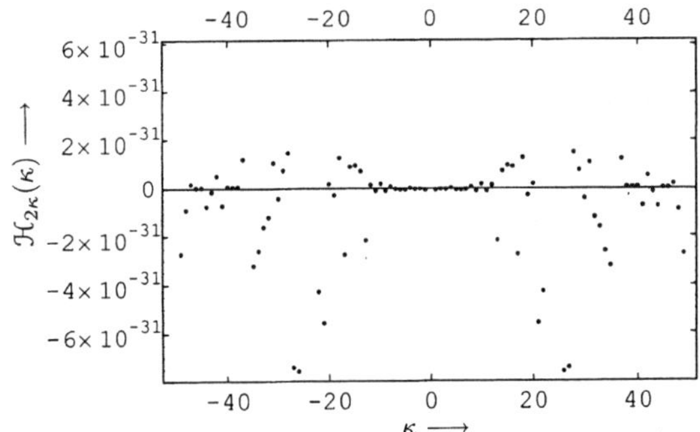

FIG.4.4-26. Plot of $\tilde{\mathcal{H}}_{2\kappa}(\kappa)$ according to Eq.(32) for $N = 100$, $\tilde{A}_{e0} = 10\lambda_1$, $\lambda_C/\Delta x = 5 \times 10^4$, $\lambda_1 = 10^{-5}\lambda_2$, $\lambda_2 = 2\pi\Delta x/\lambda_C$, $\lambda_3 = 0.1$.

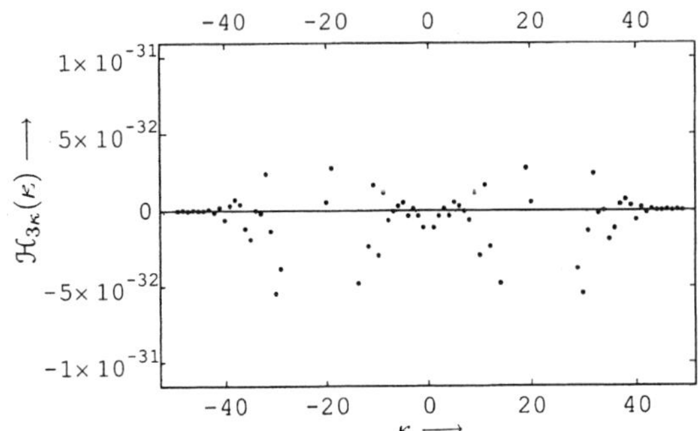

FIG.4.4-27. Plot of $\tilde{\mathcal{H}}_{3\kappa}(\kappa)$ according to Eq.(38) for $N = 100$, $\tilde{A}_{e0} = 10\lambda_1$, $\lambda_C/\Delta x = 5 \times 10^4$, $\lambda_1 = 10^{-5}\lambda_2$, $\lambda_2 = 2\pi\Delta x/\lambda_C$, $\lambda_3 = 0.1$.

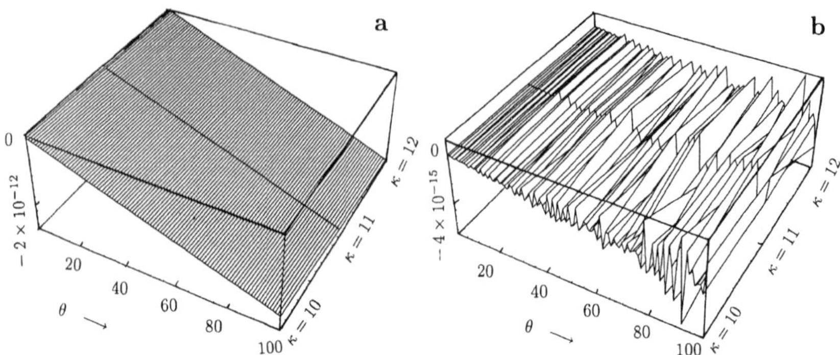

FIG.4.4-28. The real part of the function $\tilde{H}_{s\kappa}(\theta_n, \kappa)$ of Eq.(17) for Fig.4.4-1 with $\lambda_C/\Delta x = 10^4$ (a) and for Fig.4.4-7 with $\lambda_C/\Delta x = 10^5$ (b).

What causes the transition from a determined to a random function? A comparison of Tables 3.4-1 and 4.4-1 shows that the computations start in both caes with β_κ, $I_T(\kappa/N)$, and $\Psi_0(\zeta, \theta_n)$. Then they become different until $\tilde{H}_{s\kappa}(\theta_n, \kappa)$ in Table 4.4-1 is reached. From there on the computations are again equal or essentially equal. A plot of $\tilde{H}_{s\kappa}(\theta_n, \kappa)$ according to Eq.(17) is shown for $\kappa = 10$, 11, 12 and $\theta_n = 0, 1, \ldots, 100$ in Fig.4.4-28a for $N = 100$, $\tilde{A}_{e0} = 10\lambda_1$, $\lambda_C/\Delta x = 10^4$, $\lambda_1 = 0.1\lambda_2$, $\lambda_2 = 2\pi\Delta x/\lambda_C$, $\lambda_3 = 1$; these are the values of Fig.4.4-1. The plot in Fig.4.4-28b holds for the same values of the parameters, except for a change $\lambda_C/\Delta x = 10^5$; these are the values of Fig.4.4-7. It is evident that the plot of Fig.4.4-28a changes very smoothly with κ and θ. The plot in Fig.4.4-28b, on the other hand, changes radically both with κ and with θ. This makes the sums $\tilde{J}_j(\theta, \kappa)$ for $\theta_n = n$ in Eq.(6.4-48) change like random numbers.

We have shown 27 plots in Sections 3.4 and 3.5 but postponed a justification to Section 4.4. These plots did not show any randomization like Figs.4.4-1 to 4.4-27. This is the reason why we produced so many plots there.

The long calculations and the many equations in Table 4.4-1 for the computer program make it very hard to produce Figs.4.4-1 to 4.4-27 without error. It will take time and luckless PhD students to verify or correct the plots. This is the bad news. The good news is that the result of randomization requires $\tilde{\mathcal{H}}_{s\kappa}(\theta_n, \kappa)$, which is the twentysixth equation listed in Table 4.4-1. But one needs 72 equations to compute $\tilde{\mathcal{H}}_{1\kappa}(\kappa)$, $\tilde{\mathcal{H}}_{2\kappa}(\kappa)$, and $\tilde{\mathcal{H}}_{3\kappa}(\kappa)$. Hence, the result of randomness demonstrated by Fig.4.4-28 is much less error prone than Figs.4.4-1 to 4.4-27.

5 Difference Equations in Spherical Coordinates

5.1 Charged Particle in an Electromagnetic Field

According to Section 1.3 we may write the Lagrange function of an electrically charged particle in the following form using spherical coordinates:

$$\begin{aligned}\boldsymbol{\mathfrak{L}} &= \mathfrak{L}_{\mathrm{r}}\mathbf{e}_{\mathrm{r}} + \mathfrak{L}_{\theta}\mathbf{e}_{\theta} + \mathfrak{L}_{\varphi}\mathbf{e}_{\varphi} = \boldsymbol{\mathfrak{L}}_{\mathrm{M}} + \boldsymbol{\mathfrak{L}}_{\mathrm{c}} \\ &= \left(\frac{1}{2}mv^2 + e\mathbf{v}\times\mathbf{A}_{\mathrm{m}} - e\phi_{\mathrm{e}}\right)\begin{pmatrix}1&0&0\\0&1&0\\0&0&1\end{pmatrix} + \begin{pmatrix}\mathfrak{L}_{\mathrm{c}r}&0&0\\0&\mathfrak{L}_{\mathrm{c}\vartheta}&0\\0&0&\mathfrak{L}_{\mathrm{c}\varphi}\end{pmatrix}\end{aligned} \tag{1}$$

The functions $\mathfrak{L}_{\mathrm{c}r}$, $\mathfrak{L}_{\mathrm{c}\vartheta}$, $\mathfrak{L}_{\mathrm{c}\varphi}$ are shown in Eqs.(1.3-33)–(1.3-35). We shall need relations between the moments p_r, p_ϑ, p_φ and the variables r, ϑ, φ. The moments are the derivatives of the components $\mathfrak{L}_r$, $\mathfrak{L}_\vartheta$, $\mathfrak{L}_\varphi$. We get from Eqs.(1), (1.3-29), and (1.3-33)–(1.3-35):

$$p_r = \frac{\partial\mathfrak{L}_r}{\partial\dot r} = m\dot r + eA_{\mathrm{m}r} + \frac{Ze}{c}(r\dot\vartheta A_{\mathrm{e}\varphi} - r\sin\vartheta\,\dot\varphi A_{\mathrm{e}\vartheta}) \tag{2}$$

$$p_\vartheta = \frac{\partial\mathfrak{L}_\vartheta}{\partial(r\dot\vartheta)} = mr\dot\vartheta + eA_{\mathrm{m}\vartheta} + \frac{Ze}{c}(r\sin\vartheta\,\dot\varphi A_{\mathrm{e}r} - \dot r A_{\mathrm{e}\varphi}) \tag{3}$$

$$p_\varphi = \frac{\partial\mathfrak{L}_\varphi}{\partial(r\sin\vartheta\dot\varphi)} = mr\sin\vartheta\,\dot\varphi + eA_{\mathrm{m}\varphi} + \frac{Ze}{c}(\dot r A_{\mathrm{e}\vartheta} - r\dot\vartheta A_{\mathrm{e}r}) \tag{4}$$

The following relations will be used:

$$\begin{aligned}\mathbf{r} &= r\mathbf{e}_r + r\vartheta\mathbf{e}_\vartheta + r\sin\vartheta\,\varphi\mathbf{e}_\varphi \\ \mathbf{v} = \dot{\mathbf{r}} &= \dot r\mathbf{e}_r + r\dot\vartheta\mathbf{e}_\vartheta + r\sin\vartheta\,\dot\varphi\mathbf{e}_\varphi \\ \mathbf{p} &= p_r\mathbf{e}_r + p_\vartheta\mathbf{e}_\vartheta + p_\varphi\mathbf{e}_\varphi \\ \mathbf{A}_{\mathrm{m}} &= A_{\mathrm{m}r}\mathbf{e}_r + A_{\mathrm{m}\vartheta}\mathbf{e}_\vartheta + A_{\mathrm{m}\varphi}\mathbf{e}_\varphi \\ \mathbf{A}_{\mathrm{e}} &= A_{\mathrm{e}r}\mathbf{e}_r + A_{\mathrm{e}\vartheta}\mathbf{e}_\vartheta + A_{\mathrm{e}\varphi}\mathbf{e}_\varphi\end{aligned} \tag{5}$$

ISSN 1076-5670/05
DOI: 10.1016/S1076-5670(05)37005-4

We solve Eqs.(2)–(4) for $\dot{r}$, $r\dot{\vartheta}$, $r\sin\vartheta\,\dot{\varphi}$ as functions of p_r, p_ϑ, p_φ. To shorten the equations we define a common denominator D:

$$D = m\left[\left(\frac{Ze}{c}\right)^2 (A_{\mathrm{e}r}^2 + A_{\mathrm{e}\vartheta}^2 + A_{\mathrm{e}\varphi}^2) + m^2\right] = m\left[m^2 + \left(\frac{Ze}{c}\right)^2 \mathbf{A}_\mathrm{e}^2\right] \tag{6}$$

$$\begin{aligned}\dot{r} = \frac{1}{D}\bigg\{&\left(\frac{Ze}{c}\right)^2 A_{\mathrm{e}r}[A_{\mathrm{e}r}(p_r - eA_{\mathrm{m}r}) + A_{\mathrm{e}\vartheta}(p_\vartheta - eA_{\mathrm{m}\vartheta}) + A_{\mathrm{e}\varphi}(p_\varphi - eA_{\mathrm{m}\varphi})]\\ &+ \frac{Zem}{c}[A_{\mathrm{e}\vartheta}(p_\varphi - eA_{\mathrm{m}\varphi}) - A_{\mathrm{e}\varphi}(p_\vartheta - eA_{\mathrm{m}\vartheta})] + m^2(p_r - eA_{\mathrm{m}r})\bigg\}\\ = \frac{m^2}{D}\bigg[&(\mathbf{p} - e\mathbf{A}_\mathrm{m})_r + \left(\frac{Ze}{c}\right)^2 A_{\mathrm{e}r}\mathbf{A}_\mathrm{e}\cdot(\mathbf{p} - e\mathbf{A}_\mathrm{m})\\ &+ \frac{Zem}{c}[\mathbf{A}_\mathrm{e}\times(\mathbf{p} - e\mathbf{A}_\mathrm{m})]_r\bigg]\end{aligned} \tag{7}$$

$$\begin{aligned}r\dot{\vartheta} = \frac{1}{D}\bigg[&\left(\frac{Ze}{c}\right)^2 A_{\mathrm{e}\vartheta}[A_{\mathrm{e}r}(p_r - eA_{\mathrm{m}r}) + A_{\mathrm{e}\vartheta}(p_\vartheta - eA_{\mathrm{m}\vartheta}) + A_{\mathrm{e}\varphi}(p_\varphi - eA_{\mathrm{m}\varphi})]\\ &+ \frac{Zem}{c}[A_{\mathrm{e}\varphi}(p_r - eA_{\mathrm{m}r}) - A_{\mathrm{e}r}(p_\varphi - eA_{\mathrm{m}\varphi})] + m^2(p_\vartheta - eA_{\mathrm{m}\vartheta})\bigg]\\ = \frac{1}{D}\bigg[&m^2(\mathbf{p} - e\mathbf{A}_\mathrm{m})_\vartheta + \left(\frac{Ze}{c}\right)^2 A_{\mathrm{e}\vartheta}\mathbf{A}_\mathrm{e}\cdot(\mathbf{p} - e\mathbf{A}_\mathrm{m})\\ &+ \frac{Zem}{c}[\mathbf{A}_\mathrm{e}\times(\mathbf{p} - e\mathbf{A}_\mathrm{m})]_\vartheta\bigg]\end{aligned} \tag{8}$$

$$\begin{aligned}r\sin\vartheta\,\dot{\varphi} = \frac{1}{D}\bigg[&\left(\frac{Ze}{c}\right)^2 A_{\mathrm{e}\varphi}[A_{\mathrm{e}r}(p_r - eA_{\mathrm{m}r}) + A_{\mathrm{e}\vartheta}(p_\vartheta - eA_{\mathrm{m}\vartheta}) + A_{\mathrm{e}\varphi}(p_\varphi - eA_{\mathrm{m}\varphi})]\\ &+ \frac{Zem}{c}[A_{\mathrm{e}r}(p_\vartheta - eA_{\mathrm{m}\vartheta}) - A_{\mathrm{e}\vartheta}(p_r - eA_{\mathrm{m}r})] + m^2(p_\varphi - eA_{\mathrm{m}\varphi})\bigg]\\ = \frac{1}{D}\bigg[&m^2(\mathbf{p} - e\mathbf{A}_\mathrm{m})_\varphi + \left(\frac{Ze}{c}\right)^2 A_{\mathrm{e}\varphi}\mathbf{A}_\mathrm{e}\cdot(\mathbf{p} - e\mathbf{A}_\mathrm{m})\\ &+ \frac{Zem}{c}[\mathbf{A}_\mathrm{e}\times(\mathbf{p} - e\mathbf{A}_\mathrm{m})]_\varphi\bigg]\end{aligned} \tag{9}$$

Equations (1.3-33)–(1.3-35) contain the second derivatives $\ddot{r}$, $\ddot{\vartheta}$, $\ddot{\varphi}$ in addition to the first derivatives $\dot{r}$, $\dot{\vartheta}$, $\dot{\varphi}$. We differentiate Eqs.(7)–(9) with respect to t, keeping in mind that $\mathbf{A}_{\rm m}$, $\mathbf{A}_{\rm e}$, and D in Eqs.(5) and (6) are not time dependent:

$$\begin{aligned}\ddot{r} &= \frac{1}{D}\Bigg[m^2\dot{p}_r + \frac{Zem}{c}(A_{\mathrm{e}\vartheta}\dot{p}_\varphi - A_{\mathrm{e}\varphi}\dot{p}_\vartheta) \\ &\qquad + \left(\frac{Ze}{c}\right)^2 A_{\mathrm{e}r}(A_{\mathrm{e}r}\dot{p}_r + A_{\mathrm{e}\vartheta}\dot{p}_\vartheta + A_{\mathrm{e}\varphi}\dot{p}_\varphi)\Bigg] \\ &= \frac{1}{D}\left[m^2\dot{p}_r + \frac{Zem}{c}(\mathbf{A}_\mathrm{e}\times\mathbf{p})_r + \left(\frac{Ze}{c}\right)^2 A_{\mathrm{e}r}\mathbf{A}_\mathrm{e}\cdot\dot{\mathbf{p}}\right] \end{aligned} \tag{10}$$

$$\begin{aligned}r\ddot{\vartheta} &= \frac{1}{D}\Bigg[m^2\dot{p}_\vartheta + \frac{Zem}{c}(A_{\mathrm{e}\varphi}\dot{p}_r - A_{\mathrm{e}r}\dot{p}_\varphi) \\ &\qquad + \left(\frac{Ze}{c}\right)^2 A_{\mathrm{e}\vartheta}(A_{\mathrm{e}r}\dot{p}_r + A_{\mathrm{e}\vartheta}\dot{p}_\vartheta + A_{\mathrm{e}\varphi}\dot{p}_\varphi)\Bigg] \\ &= \frac{1}{D}\left[m^2\dot{p}_\vartheta + \frac{Zem}{c}(\mathbf{A}_\mathrm{e}\times\mathbf{p})_\vartheta + \left(\frac{Ze}{c}\right)^2 A_{\mathrm{e}\vartheta}\mathbf{A}_\mathrm{e}\cdot\dot{\mathbf{p}}\right] \end{aligned} \tag{11}$$

$$\begin{aligned}r\sin\vartheta\,\ddot{\varphi} &= \frac{1}{D}\Bigg[m^2\dot{p}_\varphi + \frac{Zem}{c}(A_{\mathrm{e}r}\dot{p}_\vartheta - A_{\mathrm{e}\vartheta}\dot{p}_r) \\ &\qquad + \left(\frac{Ze}{c}\right)^2 A_{\mathrm{e}\varphi}(A_{\mathrm{e}r}\dot{p}_r + A_{\mathrm{e}\vartheta}\dot{p}_\vartheta + A_{\mathrm{e}\varphi}\dot{p}_\varphi)\Bigg] \\ &= \frac{1}{D}\left[m^2\dot{p}_\varphi + \frac{Zem}{c}(\mathbf{A}_\mathrm{e}\times\mathbf{p})_\varphi + \left(\frac{Ze}{c}\right)^2 A_{\mathrm{e}\varphi}\mathbf{A}_\mathrm{e}\cdot\dot{\mathbf{p}}\right] \end{aligned} \tag{12}$$

Since the Lagrange function $\mathfrak{L}$ of Eq.(1) has three components the associated Hamilton function $\mathfrak{H}$ must have three components $\mathcal{H}_k$ too:

$$\mathcal{H}_k(p_j, x_j, t) = \sum_{j=1}^{3} p_j\dot{x}_j - \mathcal{L}_k, \quad k = r,\ \vartheta,\ \varphi \tag{13}$$

Equations (2) (4) yield the following result since the terms multiplied by Ze/c cancel:

$$\begin{aligned}\sum_{j=1}^{3} p_j\dot{x}_j &= p_r\dot{r} + p_\vartheta r\dot{\vartheta} + p_\varphi r\sin\vartheta\,\dot{\varphi} \\ &= m[\dot{r}^2 + (r\dot{\vartheta})^2 + (r\sin\vartheta\,\dot{\varphi})^2] + e(A_{\mathrm{m}r}\dot{r} + A_{\mathrm{m}\vartheta}r\dot{\vartheta} + A_{\mathrm{m}\varphi}r\sin\vartheta\,\dot{\varphi}) \\ &= m\dot{\mathbf{r}}^2 + e\mathbf{A}_\mathrm{m}\cdot\dot{\mathbf{r}}\end{aligned} \tag{14}$$

The three components $\mathcal{H}_k$ are obtained from Eqs.(1) and (1.3-33)–(1.3-35); the terms $A_{mr}\dot{r} + A_{m\vartheta}r\dot{\vartheta} + A_{m\varphi}r\sin\vartheta\,\dot{\varphi}$ cancel:

$$\mathcal{H}_r = \frac{1}{2}m(\dot{r}^2 + r^2\dot{\vartheta}^2 + r^2\sin^2\vartheta\,\dot{\varphi}^2) + e\phi_e - \mathcal{L}_{cr} = \frac{1}{2}m\dot{\mathbf{r}}^2 + e\phi_e - \mathcal{L}_{cr} \quad (15)$$

$$\mathcal{H}_\vartheta = \frac{1}{2}m(\dot{r}^2 + r^2\dot{\vartheta}^2 + r^2\sin^2\vartheta\,\dot{\varphi}^2) + e\phi_e - \mathcal{L}_{c\vartheta} = \frac{1}{2}m\dot{\mathbf{r}}^2 + e\phi_e - \mathcal{L}_{c\vartheta} \quad (16)$$

$$\mathcal{H}_\varphi = \frac{1}{2}m(\dot{r}^2 + r^2\dot{\vartheta}^2 + r^2\sin^2\vartheta\,\dot{\varphi}^2) + e\phi_e - \mathcal{L}_{c\varphi} = \frac{1}{2}m\dot{\mathbf{r}}^2 + e\phi_e - \mathcal{L}_{c\varphi} \quad (17)$$

The Hamilton function should be written in terms of the momentum $\mathbf{p}$ and the potentials ϕ_e, ϕ_m, $\mathbf{A}_e$, $\mathbf{A}_m$. This can be done by the substitution of $\dot{r}$, $r\dot{\vartheta}$, $r\sin\vartheta\,\dot{\varphi}$, $\ddot{r}$, $r\ddot{\vartheta}$, and $r\sin\vartheta\,\ddot{\varphi}$ from Eqs.(7)–(12) into Eqs.(15)–(17) as well as into Eqs.(1.3-33)–(1.3-35). Considerable effort is required. Let us first gain some understanding of the Hamilton function. To this end we consider Eqs.(2)–(4) and investigate under which conditions the terms multiplied by Ze/c will be small. Equation (2) yields the conditions

$$(Ze/c)r\dot{\vartheta}A_{e\varphi} \ll m\dot{r}, \quad (Ze/c)r\sin\vartheta\,\dot{\varphi}A_{e\vartheta} \ll m\dot{r} \quad (18)$$

which can be rewritten as follows:

$$mc^2 \gg ZecA_{e\varphi}r\dot{\vartheta}/\dot{r} \quad \text{and} \quad mc^2 \gg ZecA_{e\vartheta}r\sin\vartheta\,\dot{\varphi}/\dot{r} \quad (19)$$

In case $\dot{r}$ is close to zero while $r\dot{\vartheta}$ and $r\sin\vartheta\,\dot{\varphi}$ are not we require an alternate condition for Eq.(2):

$$A_{mr} \gg ZA_{er}r\dot{\vartheta}/c \quad \text{and} \quad A_{mr} \gg ZA_{e\vartheta}r\sin\vartheta\,\dot{\varphi}/c \quad (20)$$

Equations (19) and (20) state in essence that the energy due to the potential $\mathbf{A}_e$ should be small compared with mc^2 or the energy due to the potential $\mathbf{A}_m$. More detailed statements referring to the terms $\dot{\vartheta}/\dot{\varphi}$, $r\sin\vartheta\,\dot{\varphi}/\dot{r}$, $r\dot{\vartheta}/c$, and $r\sin\vartheta\,\dot{\varphi}/c$ are not of interest here. Relations equivalent to Eqs.(18)–(20) may be derived from Eqs.(3) and (4) too. Equations (2)–(4) may be simplified in this case:

$$p_r = m\dot{r} + eA_{mr}, \; p_\vartheta = mr\dot{\vartheta} + eA_{m\vartheta}, \; p_\varphi = mr\sin\vartheta\,\dot{\varphi} + eA_{m\varphi} \quad (21)$$

$$D = m^3 \quad (22)$$

Equations (7)–(9) become:

$$\dot{r} = \frac{1}{m}(\mathbf{p} - e\mathbf{A}_m)_r, \; r\dot{\vartheta} = \frac{1}{m}(\mathbf{p} - e\mathbf{A}_m)_\vartheta, \; r\sin\vartheta\,\dot{\varphi} = \frac{1}{m}(\mathbf{p} - e\mathbf{A}_m)_\varphi \quad (23)$$

The three components of the Hamilton function are obtained in the form

$$\mathcal{H}_r = \frac{1}{2m}(\mathbf{p} - e\mathbf{A}_\mathrm{m})^2 + e\phi_\mathrm{e} - \mathcal{L}_{cr} \tag{24}$$
$$\mathcal{H}_\vartheta = \frac{1}{2m}(\mathbf{p} - e\mathbf{A}_\mathrm{m})^2 + e\phi_\mathrm{e} - \mathcal{L}_{c\vartheta} \tag{25}$$
$$\mathcal{H}_\varphi = \frac{1}{2m}(\mathbf{p} - e\mathbf{A}_\mathrm{m})^2 + e\phi_\mathrm{e} - \mathcal{L}_{c\varphi} \tag{26}$$

These are the terms of the conventional Hamilton function of an electrically charged particle in an electromagnetic field plus correcting terms $\mathcal{L}_{cr}$, $\mathcal{L}_{c\vartheta}$, $\mathcal{L}_{c\varphi}$.

The relativistic variability of the mass m is not taken into account in Eqs.(24)–(26) due to the term $mv^2/2$ in the Lagrange function of Eq.(1). This simplification permits us to obtain the components $\mathcal{H}_r$, $\mathcal{H}_\vartheta$, $\mathcal{H}_\varphi$ of the Hamilton function and the correcting terms $\mathcal{L}_{cr}$, $\mathcal{L}_{c\vartheta}$, $\mathcal{L}_{c\varphi}$ explicitly. The relativistic variation of the mass m will be taken into account in the following Section 5.2. The Hamilton function can then be represented by means of series expansions only.

All the deviations from the conventional values in Eqs.(2)–(4) are due to the potential $\mathbf{A}_\mathrm{e}$ that is caused by magnetic monopole, dipole, or multipole current densities $\mathbf{g}_\mathrm{m}$ according to Eq.(1.1-19). The correcting terms $\mathcal{L}_{cr}$, $\mathcal{L}_{c\vartheta}$, $\mathcal{L}_{c\varphi}$ of Eqs.(1.3-33)–(1.3-35) contain mainly terms $\mathbf{A}_\mathrm{e}$ but the hypothetical magnetic charge ρ_m enters through the terms $\partial\phi_\mathrm{m}/\partial r$, $\partial\phi_\mathrm{m}/\partial\vartheta$, $\partial\phi_\mathrm{m}/\partial\varphi$ according to Eq.(1.1-22).

We turn to the evaluation of Eqs.(15)–(17) without approximation. Using Eqs.(6)–(9) we obtain $\boldsymbol{\mathcal{H}}$ in vector notation:

$$\begin{aligned}\boldsymbol{\mathcal{H}} = \frac{1}{2m}\bigg[(\mathbf{p} - e\mathbf{A}_\mathrm{m})^2 + \left(\frac{Ze}{mc}\right)^2 \{2[\mathbf{A}_\mathrm{e}\cdot(\mathbf{p} - e\mathbf{A}_\mathrm{m})]^2 + [\mathbf{A}_\mathrm{e}\times(\mathbf{p} - e\mathbf{A}_\mathrm{m})]^2\} \\ + \left(\frac{Ze}{mc}\right)^4 \mathbf{A}_\mathrm{e}^2[\mathbf{A}_\mathrm{e}\cdot(\mathbf{p} - e\mathbf{A}_\mathrm{m})]^2\bigg]\left[1 + \left(\frac{Ze}{mc}\right)^2\mathbf{A}_\mathrm{e}^2\right]^{-2} + e\phi_\mathrm{e} - \boldsymbol{\mathcal{L}}_\mathrm{c}\end{aligned} \tag{27}$$

The vector $\boldsymbol{\mathcal{L}}_\mathrm{c}$ has the three components $\mathcal{L}_{cr}$, $\mathcal{L}_{c\vartheta}$, $\mathcal{L}_{c\varphi}$ of Eq.(1). Each of them consists in turn of five components. For the first component $\mathcal{L}_{cr1}$ we obtain from Eq.(1.3-33):

$$\begin{aligned}\mathcal{L}_{cr1} = \frac{Ze}{c}\dot{r}(r\dot{\vartheta}A_{e\varphi} - r\sin\vartheta\,\dot{\varphi}A_{e\vartheta}) = \frac{Ze}{m^2c}\bigg(A_{e\varphi}(\mathbf{p} - e\mathbf{A}_\mathrm{m})_\vartheta - A_{e\vartheta}(\mathbf{p} - e\mathbf{A}_\mathrm{m})_\varphi \\ + \frac{Ze}{mc}\{A_{e\varphi}[\mathbf{A}_\mathrm{e}\times(\mathbf{p} - e\mathbf{A}_\mathrm{m})]_\vartheta - A_{e\vartheta}[\mathbf{A}_\mathrm{e}\times(\mathbf{p} - e\mathbf{A}_\mathrm{m})]_\varphi\}\bigg)\end{aligned}$$

$$\times\left[(\mathbf{p}-e\mathbf{A}_{\mathrm{m}})_r+\left(\frac{Ze}{mc}\right)^2 A_{\mathrm{e}r}\mathbf{A}_{\mathrm{e}}\cdot(\mathbf{p}-e\mathbf{A}_{\mathrm{m}})\right.$$
$$\left.+\frac{Ze}{mc}[\mathbf{A}_{\mathrm{e}}\times(\mathbf{p}-e\mathbf{A}_{\mathrm{m}})]_r\right]\left[1+\left(\frac{Ze}{mc}\right)^2\mathbf{A}_{\mathrm{e}}^2\right]^{-2} \quad (28)$$

The second component $\mathfrak{L}_{cr2}$ is defined by Eq.(1.3-33) as follows:

$$\begin{aligned}\mathfrak{L}_{cr2} &= \frac{Ze}{c}\int\left(\sin\vartheta\,\dot{\varphi}\frac{\partial\phi_{\mathrm{m}}}{\partial\vartheta}-\frac{\dot{\vartheta}}{\sin\vartheta}\frac{\partial\phi_{\mathrm{m}}}{\partial\varphi}\right)dr\\
&= \frac{Ze}{m^2c}\int\left[\frac{1}{r}\frac{\partial\phi_{\mathrm{m}}}{\partial\vartheta}(\mathbf{p}-e\mathbf{A}_{\mathrm{m}})_\varphi-\frac{1}{r\sin\vartheta}\frac{\partial\phi_{\mathrm{m}}}{\partial\varphi}(\mathbf{p}-e\mathbf{A}_{\mathrm{m}})_\vartheta\right.\\
&\quad+\frac{Ze}{mc}\left(\frac{1}{r}\frac{\partial\phi_{\mathrm{m}}}{\partial\vartheta}[\mathbf{A}_{\mathrm{e}}\times(\mathbf{p}-e\mathbf{A}_{\mathrm{m}})]_\varphi-\frac{1}{r\sin\vartheta}\frac{\partial\phi_{\mathrm{m}}}{\partial\varphi}[\mathbf{A}_{\mathrm{e}}\times(\mathbf{p}-e\mathbf{A}_{\mathrm{m}})]_\vartheta\right.\\
&\quad\left.\left.+\left(\frac{Ze}{mc}\right)^2\left(\frac{1}{r}\frac{\partial\phi_{\mathrm{m}}}{\partial\vartheta}A_{\mathrm{e}\varphi}-\frac{1}{r\sin\vartheta}\frac{\partial\phi_{\mathrm{m}}}{\partial\varphi}A_{\mathrm{e}\vartheta}\right)\mathbf{A}_{\mathrm{e}}\cdot(\mathbf{p}-e\mathbf{A}_{\mathrm{m}})\right]\\
&\qquad\times\left[1+\left(\frac{Ze}{mc}\right)^2\mathbf{A}_{\mathrm{e}}^2\right]^{-1}dr\end{aligned} \quad (29)$$

For the third component $\mathfrak{L}_{cr3}$ we obtain from Eqs.(1.3-33), (6), (11), and (12):

$$\begin{aligned}\mathfrak{L}_{cr3} &= \frac{Ze}{c}\int(r\ddot{\vartheta}A_{\mathrm{e}\varphi}-r\sin\vartheta\,\ddot{\varphi}A_{\mathrm{e}\vartheta})dr\\
&= \frac{Ze}{cD}\int\left(m^2(\mathbf{A}_{\mathrm{e}}\times\dot{\mathbf{p}})_r+\frac{Zem}{c}[(\mathbf{A}_{\mathrm{e}}\times\mathbf{p})_\vartheta-(\mathbf{A}_{\mathrm{e}}\times\mathbf{p})_\varphi]\right)dr\end{aligned} \quad (30)$$

The fourth component $\mathfrak{L}_{cr4}$ is very long. It follows from Eqs.(1.3-33), (6), (8), and (9):

$$\begin{aligned}\mathfrak{L}_{cr4} &= \frac{Ze}{c}\int\left(\dot{\vartheta}\frac{\partial}{\partial\vartheta}+\dot{\varphi}\frac{\partial}{\partial\varphi}\right)(r\dot{\vartheta}A_{\mathrm{e}\varphi}-r\sin\vartheta\,\dot{\varphi}A_{\mathrm{e}\vartheta})dr\\
&= \frac{Ze}{m^2c}\int\left\{\left[(\mathbf{p}-e\mathbf{A}_{\mathrm{m}})_\vartheta^2+2\frac{Ze}{m}(\mathbf{p}-e\mathbf{A}_{\mathrm{m}})_\vartheta[\mathbf{A}_{\mathrm{e}}\times(\mathbf{p}-e\mathbf{A}_{\mathrm{m}})]_\vartheta\right.\right.\\
&\quad+\left(\frac{Ze}{mc}\right)^2\{[\mathbf{A}_{\mathrm{e}}\times(\mathbf{p}-e\mathbf{A}_{\mathrm{m}})]_\vartheta^2+2A_{\mathrm{e}\vartheta}(\mathbf{p}-e\mathbf{A}_{\mathrm{m}})_\vartheta\mathbf{A}_{\mathrm{e}}\cdot(\mathbf{p}-e\mathbf{A}_{\mathrm{m}})\}\\
&\qquad+2\left(\frac{Ze}{mc}\right)^3A_{\mathrm{e}\vartheta}[\mathbf{A}_{\mathrm{e}}\times(\mathbf{p}-e\mathbf{A}_{\mathrm{m}})]_\vartheta\mathbf{A}_{\mathrm{e}}\cdot(\mathbf{p}-e\mathbf{A}_{\mathrm{m}})\\
&\qquad\qquad+\left(\frac{Ze}{mc}\right)^4A_{\mathrm{e}\vartheta}^2[\mathbf{A}_{\mathrm{e}}\cdot(\mathbf{p}-e\mathbf{A}_{\mathrm{m}})]^2\frac{\partial A_{\mathrm{e}\varphi}}{\partial(r\vartheta)}\end{aligned}$$

$$+ (\mathbf{p} - e\mathbf{A}_{\mathrm{m}})_\vartheta(\mathbf{p} - e\mathbf{A}_{\mathrm{m}})_\varphi + \frac{Ze}{mc}\{(\mathbf{p} - e\mathbf{A}_{\mathrm{m}})_\vartheta[\mathbf{A}_{\mathrm{e}} \times (\mathbf{p} - e\mathbf{A}_{\mathrm{m}})]_\varphi$$
$$+ (\mathbf{p} - e\mathbf{A}_{\mathrm{m}})_\varphi[\mathbf{A}_{\mathrm{e}} \times (\mathbf{p} - e\mathbf{A}_{\mathrm{m}})]_\vartheta\}$$
$$+ \left(\frac{Ze}{mc}\right)^2 \{[\mathbf{A}_{\mathrm{e}} \times (\mathbf{p} - e\mathbf{A}_{\mathrm{m}})]_\vartheta[\mathbf{A}_{\mathrm{e}} \times (\mathbf{p} - e\mathbf{A}_{\mathrm{m}})]_\varphi$$
$$+ A_{\mathrm{e}\varphi}[(\mathbf{p} - e\mathbf{A}_{\mathrm{m}})_\vartheta + (\mathbf{p} - e\mathbf{A}_{\mathrm{m}})_\varphi]\mathbf{A}_{\mathrm{e}} \cdot (\mathbf{p} - e\mathbf{A}_{\mathrm{m}})\}$$
$$+ \left(\frac{Ze}{mc}\right)^3 \{A_{\mathrm{e}\vartheta}[\mathbf{A}_{\mathrm{e}} \times (\mathbf{p} - e\mathbf{A}_{\mathrm{m}})]_\varphi + A_{\mathrm{e}\varphi}[\mathbf{A}_{\mathrm{e}} \times (\mathbf{p} - e\mathbf{A}_{\mathrm{m}})]_\vartheta\}\mathbf{A}_{\mathrm{e}} \cdot (\mathbf{p} - e\mathbf{A}_{\mathrm{m}})$$
$$+ \left(\frac{Ze}{mc}\right)^4 A_{\mathrm{e}\vartheta}A_{\mathrm{e}\varphi}[\mathbf{A}_{\mathrm{e}} \cdot (\mathbf{p} - e\mathbf{A}_{\mathrm{m}})]^2\bigg]\left(\frac{\partial A_{\mathrm{e}\varphi}}{\partial(r\vartheta)} - \frac{\partial A_{\mathrm{e}\vartheta}}{\partial(r\sin\vartheta\,\varphi)}\right)$$
$$+ \bigg[(\mathbf{p} - e\mathbf{A}_{\mathrm{m}})^2_\varphi + 2\frac{Ze}{mc}(\mathbf{p} - e\mathbf{A}_{\mathrm{m}})_\varphi[\mathbf{A}_{\mathrm{e}} \times (\mathbf{p} - e\mathbf{A}_{\mathrm{m}})]_\varphi$$
$$+ \left(\frac{Ze}{mc}\right)^2 \{[\mathbf{A}_{\mathrm{e}} \times (\mathbf{p} - e\mathbf{A}_{\mathrm{m}})]^2_\varphi + 2A_{\mathrm{e}\varphi}(\mathbf{p} - e\mathbf{A}_{\mathrm{m}})_\varphi\mathbf{A}_{\mathrm{e}} \cdot (\mathbf{p} - e\mathbf{A}_{\mathrm{m}})\}$$
$$+ 2\left(\frac{Ze}{mc}\right)^3 A_{\mathrm{e}\varphi}[\mathbf{A}_{\mathrm{e}} \times (\mathbf{p} - e\mathbf{A}_{\mathrm{m}})]_\varphi\mathbf{A}_{\mathrm{e}} \cdot (\mathbf{p} - e\mathbf{A}_{\mathrm{m}})$$
$$+ \left(\frac{Ze}{mc}\right)^4 A^2_{\mathrm{e}\varphi}[\mathbf{A}_{\mathrm{e}} \cdot (\mathbf{p} - e\mathbf{A}_{\mathrm{m}})]^2\bigg]\frac{\partial A_{\mathrm{e}\vartheta}}{\partial(r\sin\vartheta\,\varphi)}\bigg\}\left[1 + \left(\frac{Ze}{mc}\right)^2 \mathbf{A}^2_{\mathrm{e}}\right]^{-2} dr \tag{31}$$

The fifth and last component $\mathfrak{L}_{cr5}$ in Eq.(1.3-33) remains unchanged:

$$\mathfrak{L}_{cr5} = Zec\int \frac{1}{r\sin\vartheta}\left(\frac{\partial(A_{\mathrm{e}\varphi}\sin\vartheta)}{\partial\vartheta} - \frac{\partial A_{\mathrm{e}\vartheta}}{\partial\varphi}\right)dr \tag{32}$$

The components of the correcting terms $\mathfrak{L}_{c\vartheta}$ and $\mathfrak{L}_{c\varphi}$ may be obtained by analogy but we shall not write them here.

5.2 Relativistic Mass Variation

Equations (5.1-1) and (5.1-27) show a constant mass m but the Lorentz equation (1.3-1) permits a relativistically variable mass. The introduction of the rest mass m_0 yields[1] the equation of motion

$$\frac{d}{dt}\frac{m_0\mathbf{v}}{(1 - v^2/c^2)^{1/2}} = e\left(\mathbf{E} + \frac{Z}{c}\mathbf{v} \times \mathbf{H}\right) \tag{1}$$

and the conservation law of energy:

$$\frac{d}{dt}\frac{m_0c^2}{(1 - v^2/c^2)^{1/2}} = e\mathbf{E} \cdot \mathbf{v} \tag{2}$$

[1]Harmuth et al. 2001, Sec.3.3.

A four-vector can be defined with three spatial components $\mathbf{p}$

$$
\begin{aligned}
\mathbf{p} &= \frac{m_0 \mathbf{v}}{(1 - v^2/c^2)^{1/2}} \\
\mathbf{v} &= \dot{r}\mathbf{e}_r + r\dot{\vartheta}\mathbf{e}_\vartheta + r\sin\vartheta\,\dot{\varphi}\mathbf{e}_\varphi \\
\mathbf{p} &= p_r\mathbf{e}_r + p_\vartheta\mathbf{e}_\vartheta + p_\varphi\mathbf{e}_\varphi \qquad \text{for spherical coordinates}
\end{aligned} \tag{3}
$$

and the component p_4

$$
p_4 = i\frac{m_0 c}{(1 - v^2/c^2)^{1/2}} = i\frac{\mathsf{E}}{c} \tag{4}
$$

where E denotes an energy rather than the magnitude E of an electric field strength. The connection between energy E and momentum $\mathbf{p}$ is provided by the formula

$$
\mathbf{p}\cdot\mathbf{p} - \frac{\mathsf{E}^2}{c^2} = -m^2c^2 \quad \text{or} \quad \mathsf{E} = (p^2c^2 + m^2c^4)^{1/2}, \quad m = \frac{m_0}{(1 - v^2/c^2)^{1/2}} \tag{5}
$$

The relativistic generalization of the conventional part of the Lagrange function $\mathfrak{L}$ of Eq.(5.1-1) is defined by

$$
\begin{aligned}
\mathfrak{L}_{\mathrm{M}} &= -m_0c^2(1 - v^3/c^2)^{1/2} + e(-\phi_{\mathrm{e}} + \mathbf{A}_{\mathrm{m}}\cdot\mathbf{v}) \\
&= -m_0c^2\left(1 - \frac{\dot{r}^2 + r^2\dot{\vartheta}^2 + r^2\sin^2\vartheta\,\dot{\varphi}^2}{c^2}\right)^{1/2} \\
&\qquad + e(-\phi_{\mathrm{e}} + A_{\mathrm{m}r}\dot{r} + A_{\mathrm{m}\vartheta}r\dot{\vartheta} + A_{\mathrm{m}\varphi}r\sin\vartheta\,\dot{\varphi}) \\
&\doteq -m_0c^2 + \frac{1}{2}m_0v^2 + e(-\phi_{\mathrm{e}} + \mathbf{A}_{\mathrm{m}}\cdot\mathbf{v}) \quad \text{for } v^2/c^2 \ll 1
\end{aligned} \tag{6}
$$

We adopt this generalization of the part $\mathfrak{L}_{\mathrm{M}}$ of the Lagrange function $\mathfrak{L}$ of Eq.(5.1-1). The components of the correcting term $\mathfrak{L}_{\mathrm{c}}$ are left unchanged from their definitions in Eqs.(1.3-33)–(1.3-35) since the mass m does not occur there and the potentials ϕ_{m}, $\mathbf{A}_{\mathrm{e}}$ come from a relativistic theory. The relativistic generalization of the Lagrange function of Eq.(5.1-1) is

$$
\mathfrak{L} = -m_0c^2(1 - v^2/c^2)^{1/2} + e(-\phi_{\mathrm{e}} + \mathbf{A}_{\mathrm{m}}\cdot\mathbf{v}) + \mathfrak{L}_{\mathrm{c}} \tag{7}
$$

where in spherical coordinates $\mathfrak{L}_{\mathrm{c}}$ has the components $\mathfrak{L}_{\mathrm{c}r}$, $\mathfrak{L}_{\mathrm{c}\vartheta}$, $\mathfrak{L}_{\mathrm{c}\varphi}$ and $\mathfrak{L}$ has the components $\mathfrak{L}_r$, $\mathfrak{L}_\vartheta$, $\mathfrak{L}_\varphi$.

The nonrelativistic momentums p_r, p_ϑ, p_φ of Eqs.(5.1-2)–(5.1-4) are generalized to 'relativistic canonical momentums'. They assume the following form:

$$p_r = \frac{\partial \mathcal{L}_r}{\partial \dot{r}} = \frac{m_0 \dot{r}}{(1 - v^2/c^2)^{1/2}} + eA_{\mathrm{m}r} + \frac{Ze}{c}(r\dot{\vartheta}A_{\mathrm{e}\varphi} - r\sin\vartheta\,\dot{\varphi}A_{\mathrm{e}\vartheta}) \quad (8)$$

$$p_\vartheta = \frac{\partial \mathcal{L}_\vartheta}{\partial (r\dot{\vartheta})} = \frac{m_0 r\dot{\vartheta}}{(1 - v^2/c^2)^{1/2}} + eA_{\mathrm{m}\vartheta} + \frac{Ze}{c}(r\sin\vartheta\,\dot{\varphi}A_{\mathrm{e}r} - \dot{r}A_{\mathrm{e}\varphi}) \quad (9)$$

$$p_\varphi = \frac{\partial \mathcal{L}_\varphi}{\partial (r\sin\vartheta\,\dot{\varphi})} = \frac{m_0 r\sin\vartheta\,\dot{\varphi}}{(1 - v^2/c^2)^{1/2}} + eA_{\mathrm{m}\varphi} + \frac{Ze}{c}(\dot{r}A_{\mathrm{e}\vartheta} - r\dot{\vartheta}A_{\mathrm{e}r}) \quad (10)$$

For the derivation of the Hamilton function from the Lagrange function we have Eqs.(5.1-13) and (5.1-14) but we must use Eqs.(8)–(10) for p_r, p_ϑ, p_φ

$$\begin{aligned}\sum_{j=1}^{3} p_j \dot{x}_j &= \frac{m_0}{(1 - v^2/c^2)^{1/2}}(\dot{r}^2 + r^2\dot{\vartheta}^2 + r^2\sin^2\vartheta\,\dot{\varphi}^2) \\ &\qquad + e(A_{\mathrm{m}r}\dot{r} + A_{\mathrm{m}\vartheta}r\dot{\vartheta} + A_{\mathrm{m}\varphi}r\sin\vartheta\,\dot{\varphi}) \\ &= \frac{m_0}{(1 - v^2/c^2)^{1/2}}\dot{\mathbf{r}}^2 + e\mathbf{A}_{\mathrm{m}} \cdot \dot{\mathbf{r}} \end{aligned} \quad (11)$$

and the three components $\mathcal{H}_k$ of the Hamilton function are obtained in analogy to Eqs.(5.1-15)–(5.1-17):

$$\mathcal{H}_r = \frac{m_0 c^2}{\left[1 - (\dot{r}^2 + r^2\dot{\vartheta}^2 + r^2\sin^2\vartheta\,\dot{\varphi}^2)/c^2\right]^{1/2}} + e\phi_{\mathrm{e}} - \mathcal{L}_{\mathrm{c}r} \quad (12)$$

$$\mathcal{H}_\vartheta = \frac{m_0 c^2}{\left[1 - (\dot{r}^2 + r^2\dot{\vartheta}^2 + r^2\sin^2\vartheta\,\dot{\varphi}^2)/c^2\right]^{1/2}} + e\phi_{\mathrm{e}} - \mathcal{L}_{\mathrm{c}\vartheta} \quad (13)$$

$$\mathcal{H}_\varphi = \frac{m_0 c^2}{\left[1 - (\dot{r}^2 + r^2\dot{\vartheta}^2 + r^2\sin^2\vartheta\,\dot{\varphi}^2)/c^2\right]^{1/2}} + e\phi_{\mathrm{e}} - \mathcal{L}_{\mathrm{c}\varphi} \quad (14)$$

The variables $\dot{r}$, $\dot{\vartheta}$, $\dot{\varphi}$ and their derivatives $\ddot{r}$, $\ddot{\vartheta}$, $\ddot{\varphi}$ must be eliminated. The variables $\dot{r}$, $\dot{\vartheta}$, $\dot{\varphi}$ are defined by Eqs.(8)–(10). They differ by the term $(1-v^2/c^2)^{1/2}$ from Eqs.(5.1-2)–(5.1-4). As a result we no longer have a system of three linear equations. The replacement of $\dot{r}$, $\dot{\vartheta}$, $\dot{\varphi}$ becomes more difficult than in Section 5.1. As before we strive to get first an understanding by using simplifying assumptions for Eqs.(8)–(10). In Eq.(8) the terms multiplied by Ze/c will be small if the conditions

$$\frac{Ze}{c}r\dot{\vartheta}A_{\mathrm{e}\varphi} \ll \frac{m_0\dot{r}}{(1 - v^2/c^2)^{1/2}} \quad \text{and} \quad \frac{Ze}{c}r\sin\vartheta\,\dot{\varphi}A_{\mathrm{e}\vartheta} \ll \frac{m_0\dot{r}}{(1 - v^2/c^2)^{1/2}} \quad (15)$$

are satisfied. These conditions may be rewritten:

$$\frac{m_0c^2}{(1-v^2/c^2)^{1/2}} \gg Zecr\dot{\vartheta}A_{\mathrm{e}\varphi}/\dot{r} \quad \text{and} \quad \frac{m_0c^2}{(1-v^2/c^2)^{1/2}} \gg Zecr\sin\vartheta\,\dot{\varphi}A_{\mathrm{e}\vartheta}/\dot{r} \tag{16}$$

For small values of $\dot{r}$ but not $r\dot{\vartheta}$ and $r\sin\vartheta\,\dot{\varphi}$ we require alternate conditions for Eq.(8):

$$A_{\mathrm{m}r} \gg Zr\dot{\vartheta}A_{\mathrm{e}\varphi}/c \quad \text{or} \quad A_{\mathrm{m}r} \gg Zer\sin\vartheta\,\dot{\varphi}A_{\mathrm{e}\vartheta}/c \tag{17}$$

Equation (16) states in essence that the energy due to the potential $\mathbf{A}_\mathrm{e}$ should be small compared with the energy $m_0c^2/(1-v^2/c^2)^{1/2}$ whereas Eq.(17) demands that the magnitude of $\mathbf{A}_\mathrm{e}$ should be small compared with the magnitude of $\mathbf{A}_\mathrm{m}$. With these simplifying assumptions we obtain from Eqs.(8)–(10):

$$p_r = \frac{m_0\dot{r}}{(1-v^2/c^2)^{1/2}} + eA_{\mathrm{m}r} \tag{18}$$

$$p_\vartheta = \frac{m_0r\dot{\vartheta}}{(1-v^2/c^2)^{1/2}} + eA_{\mathrm{m}\vartheta} \tag{19}$$

$$p_\varphi = \frac{m_0r\sin\vartheta\,\dot{\varphi}}{(1-v^2/c^2)^{1/2}} + eA_{\mathrm{m}\varphi} \tag{20}$$

Solution of these equations for $\dot{r}$, $r\dot{\vartheta}$, and $r\sin\vartheta\,\dot{\varphi}$ yields:

$$\dot{r} = \frac{(1-v^2/c^2)^{1/2}}{m_0}(\mathbf{p}-e\mathbf{A}_\mathrm{m})_r \tag{21}$$

$$r\dot{\vartheta} = \frac{(1-v^2/c^2)^{1/2}}{m_0}(\mathbf{p}-e\mathbf{A}_\mathrm{m})_\vartheta \tag{22}$$

$$r\sin\vartheta\,\dot{\varphi} = \frac{(1-v^2/c^2)^{1/2}}{m_0}(\mathbf{p}-e\mathbf{A}_\mathrm{m})_\varphi \tag{23}$$

Squaring and summing $\dot{r}$, $r\dot{\vartheta}$, $r\sin\vartheta\,\dot{\varphi}$ yields:

$$\dot{r}^2 + (r\dot{\vartheta})^2 + (r\sin\vartheta\,\dot{\varphi})^2 = v^2 = \frac{1-v^2/c^2}{m_0^2}(\mathbf{p}-e\mathbf{A}_\mathrm{m})^2$$

$$(\mathbf{p}-e\mathbf{A}_\mathrm{m})^2 = \frac{m_0^2v^2}{1-v^2/c^2} = m_0^2c^2\left(\frac{1}{1-v^2/c^2}-1\right)$$

$$\frac{m_0c^2}{(1-v^2/c^2)^{1/2}} = \frac{m_0c^2}{\left[1-(\dot{r}^2+r^2\dot{\vartheta}^2+r^2\sin^2\vartheta\,\dot{\varphi}^2)/c^2\right]^{1/2}}$$

$$= c[(\mathbf{p}-e\mathbf{A}_\mathrm{m})^2+m_0^2c^2]^{1/2} \tag{24}$$

The last line of Eq.(24) is substituted into Eqs.(12)–(14):

$$\mathcal{H}_r = c[(\mathbf{p} - e\mathbf{A}_\mathrm{m})^2 + m_0^2c^2]^{1/2} + e\phi_\mathrm{e} - \mathcal{L}_{cr} \tag{25}$$

$$\mathcal{H}_\vartheta = c[(\mathbf{p} - e\mathbf{A}_\mathrm{m})^2 + m_0^2c^2]^{1/2} + e\phi_\mathrm{e} - \mathcal{L}_{c\vartheta} \tag{26}$$

$$\mathcal{H}_\varphi = c[(\mathbf{p} - e\mathbf{A}_\mathrm{m})^2 + m_0^2c^2]^{1/2} + e\phi_\mathrm{e} - \mathcal{L}_{c\varphi} \tag{27}$$

If we leave out the correcting terms $\mathcal{L}_{cr}$, $\mathcal{L}_{c\vartheta}$, $\mathcal{L}_{c\varphi}$ we have the conventional relativistic Hamilton function for an electrically charged particle in an EM field. The assumption we had to make to obtain Eqs.(25)–(27) was that A_e must be sufficiently small. If one wants to leave out the correcting terms $\mathcal{L}_{cr}$, $\mathcal{L}_{c\vartheta}$, $\mathcal{L}_{c\varphi}$ one gets more complicated conditions than Eq.(17) since Eqs.(1.3-33)–(1.3-35) contain ϕ_m and its derivatives in addition to A_e.

Let us turn to the solution of Eqs.(8)–(10) for $\dot{r}$, $r\dot{\vartheta}$, $r\sin\vartheta\,\dot{\varphi}$ without simplifications. The term

$$(1 - v^2/c^2)^{1/2} = [1 - (\dot{r}^2 + r^2\dot{\vartheta}^2 + r^2\sin^2\vartheta\,\dot{\varphi}^2)/c^2]^{1/2}$$

makes these equations nonlinear while the corresponding Eqs.(5.1-2)–(5.1-4) of the nonrelativistic theory were linear. There is no standard method for the solution of a system of nonlinear equations and we must find a method suitable for the case at hand. As a first step we ignore that v^2 is a function of $\dot{r}$, $r\dot{\vartheta}$, $r\sin\vartheta\,\dot{\varphi}$ and treat Eqs.(8)–(10) as a system of linear equations. We note that Eqs.(5.1-2)–(5.1-4) are transformed into Eqs.(8)–(10) by the substitution

$$m \rightarrow m_0/(1 - v^2/c^2)^{1/2}$$

and we get our 'first step of solution' of Eqs.(8)–(10) by making the same substitution in Eqs.(5.1-6)–(5.1-9). The common denominator D of Eq.(5.1-6) assumes the following form:

$$D = \frac{m_0^3}{(1 - v^2/c^2)^{3/2}}\left[1 + \alpha_\mathrm{e}^2\left(1 - \frac{v^2}{c^2}\right)\right] \tag{28}$$

$$\alpha_\mathrm{e} = \frac{ZecA_\mathrm{e}}{m_0c^2}, \quad \alpha_\mathrm{e}\left(1 - \frac{v^2}{c^2}\right)^{1/2} = \frac{ZecA_\mathrm{e}}{m_0c^2/\,(1 - v^2/c^2)^{1/2}} \tag{29}$$

The constant α_e represents the ratio of the energy due to the electric vector potential $\mathbf{A}_\mathrm{e}$ and the rest energy of the particle. The reference energy should actually be $m_0c^2/(1 - v^2/c^2)^{1/2}$ rather than m_0c^2 but we shall need v as an explicit variable. The term α_e has no physical dimension but it is a variable due to its component A_e.

The solution of $\dot{r}$, $r\dot{\vartheta}$, $r\sin\vartheta\,\dot{\varphi}$ as functions of p_r, p_ϑ, p_φ has the following form if the factor α_e of Eq.(29) is used:

$$\dot{r} = \frac{(1 - v^2/c^2)^{1/2}}{m_0}(\mathbf{p} - e\mathbf{A}_\mathrm{m})_r\left[1 + \alpha_\mathrm{e}\left(1 - \frac{v^2}{c^2}\right)^{1/2}\frac{[\mathbf{A}_\mathrm{e}\times(\mathbf{p} - e\mathbf{A}_\mathrm{m})]_r}{A_\mathrm{e}(\mathbf{p} - e\mathbf{A}_\mathrm{m})_r}\right.$$
$$\left.+ \alpha_\mathrm{e}^2\left(1 - \frac{v^2}{c^2}\right)\frac{A_{\mathrm{e}r}\mathbf{A}_\mathrm{e}\cdot(\mathbf{p} - e\mathbf{A}_\mathrm{m})}{A_\mathrm{e}^2(\mathbf{p} - e\mathbf{A}_\mathrm{m})_r}\right]\left[1 + \alpha_\mathrm{e}^2\left(1 - \frac{v^2}{c^2}\right)\right]^{-1} \tag{30}$$

$$r\dot{\vartheta} = \frac{(1 - v^2/c^2)^{1/2}}{m_0}(\mathbf{p} - e\mathbf{A}_\mathrm{m})_\vartheta\left[1 + \alpha_\mathrm{e}\left(1 - \frac{v^2}{c^2}\right)^{1/2}\frac{[\mathbf{A}_\mathrm{e}\times(\mathbf{p} - e\mathbf{A}_\mathrm{m})]_\vartheta}{A_\mathrm{e}(\mathbf{p} - e\mathbf{A}_\mathrm{m})_\vartheta}\right.$$
$$\left.+ \alpha_\mathrm{e}^2\left(1 - \frac{v^2}{c^2}\right)\frac{A_{\mathrm{e}\vartheta}\mathbf{A}_\mathrm{e}\cdot(\mathbf{p} - e\mathbf{A}_\mathrm{m})}{A_\mathrm{e}^2(\mathbf{p} - e\mathbf{A}_\mathrm{m})_\vartheta}\right]\left[1 + \alpha_\mathrm{e}^2\left(1 - \frac{v^2}{c^2}\right)\right]^{-1} \tag{31}$$

$$r\sin\vartheta\,\dot{\varphi} = \frac{(1 - v^2/c^2)^{1/2}}{m_0}(\mathbf{p} - e\mathbf{A}_\mathrm{m})_\varphi\left[1 + \alpha_\mathrm{e}\left(1 - \frac{v^2}{c^2}\right)^{1/2}\frac{[\mathbf{A}_\mathrm{e}\times(\mathbf{p} - e\mathbf{A}_\mathrm{m})]_\varphi}{A_\mathrm{e}(\mathbf{p} - e\mathbf{A}_\mathrm{m})_\varphi}\right.$$
$$\left.+ \alpha_\mathrm{e}^2\left(1 - \frac{v^2}{c^2}\right)^{1/2}\frac{A_{\mathrm{e}\varphi}\mathbf{A}_\mathrm{e}\cdot(\mathbf{p} - e\mathbf{A}_\mathrm{m})}{A_\mathrm{e}^2(\mathbf{p} - e\mathbf{A}_\mathrm{m})_\varphi}\right]\left[1 + \alpha_\mathrm{e}^2\left(1 - \frac{v^2}{c^2}\right)\right]^{-1} \tag{32}$$

Squaring and summing Eqs.(30)–(32) yields:

$$\dot{r}^2 + r^2\dot{\vartheta}^2 + r^2\sin^2\vartheta\,\dot{\varphi}^2 = v^2 = \frac{1 - v^2/c^2}{m_0^2}(\mathbf{p} - e\mathbf{A}_\mathrm{m})^2\left[1\right.$$
$$+ 2\alpha_\mathrm{e}\left(1 - \frac{v^2}{c^2}\right)^{1/2}\frac{[\mathbf{A}_\mathrm{e}\cdot(\mathbf{p} - e\mathbf{A}_\mathrm{m})]^2}{\mathbf{A}_\mathrm{e}^2(\mathbf{p} - e\mathbf{A}_\mathrm{m})^2}$$
$$+ \alpha_\mathrm{e}^2\left(1 - \frac{v^2}{c^2}\right)\frac{[\mathbf{A}_\mathrm{e}\times(\mathbf{p} - e\mathbf{A}_\mathrm{m})]^2 + 2[\mathbf{A}_\mathrm{e}\cdot(\mathbf{p} - e\mathbf{A}_\mathrm{m})]^2}{\mathbf{A}_\mathrm{e}^2(\mathbf{p} - e\mathbf{A}_\mathrm{m})^2}$$
$$+ 2\alpha_\mathrm{e}^3\left(1 - \frac{v^2}{c^2}\right)^{3/2}\frac{\mathbf{A}_\mathrm{e}\cdot[\mathbf{A}_\mathrm{e}\times(\mathbf{p} - e\mathbf{A}_\mathrm{m})]\mathbf{A}_\mathrm{e}\cdot(\mathbf{p} - e\mathbf{A}_\mathrm{m})}{A_\mathrm{e}^3(\mathbf{p} - e\mathbf{A}_\mathrm{m})^2}$$
$$\left.+ \alpha_\mathrm{e}^4\left(1 - \frac{v^2}{c^2}\right)^2\frac{[\mathbf{A}_\mathrm{e}\cdot(\mathbf{p} - e\mathbf{A}_\mathrm{m})]^2}{\mathbf{A}_\mathrm{e}^2(\mathbf{p} - e\mathbf{A}_\mathrm{m})^2}\right]\left[1 + \alpha_\mathrm{e}^2\left(1 - \frac{v^2}{c^2}\right)\right]^{-2} \tag{33}$$

To find an approximate solution of this equation for

$$\alpha_\mathrm{e}(1 - v^2/c^2)^{1/2} \ll 1 \tag{34}$$

we start by using only the first term on the right side of Eq.(33):

$$v^2 = \frac{1 - v^2/c^2}{m_0^2}(\mathbf{p} - e\mathbf{A}_\mathrm{m})^2$$
$$\left(1 - \frac{v^2}{c^2}\right)^{1/2} = \frac{m_0 c}{[m_0^2c^2 + (\mathbf{p} - e\mathbf{A}_\mathrm{m})^2]^{1/2}} \tag{35}$$

This is the same relation as shown in the first line of Eq.(24). We again get the common term of the three components $\mathcal{H}_r$, $\mathcal{H}_\vartheta$, $\mathcal{H}_\varphi$ of Eqs.(25)–(27). We call this the zero-order solution in α_e of Eq.(33).

To obtain the first-order solution in α_e of Eq.(33) we use the first two terms on the right side of Eq.(33). Since $\mathbf{p}$ will eventually be replaced by difference operators we preserve carefully the sequence of the factors:

$$v^2 = \frac{1 - v^2/c^2}{m_0^2}(\mathbf{p} - e\mathbf{A}_m)^2 \left[1 + 2\alpha_e \left(1 - \frac{v^2}{c^2}\right)^{1/2} \frac{[\mathbf{A}_e \cdot (\mathbf{p} - e\mathbf{A}_m)]^2}{\mathbf{A}_e^2(\mathbf{p} - e\mathbf{A}_m)^2}\right] \tag{36}$$

The term $(1 - v^2/c^2)^{1/2}$ multiplied by $2\alpha_e$ is replaced by the zero-order solution represented by Eq.(35). The resulting equation is solved for $1 - v^2/c^2$ and the following improvement of Eq.(35) in first order of α_e is obtained:

$$\begin{aligned} 1 - \frac{v^2}{c^2} = {} & \frac{m_0^2c^2}{m_0^2c^2 + (\mathbf{p} - e\mathbf{A}_m)^2} \\ & \times \left(1 - \frac{2\alpha_e m_0 c(\mathbf{p} - e\mathbf{A}_m)^2}{[m_0^2c^2 + (\mathbf{p} - e\mathbf{A}_m)^2]^{3/2}} \frac{[\mathbf{A}_e \cdot (\mathbf{p} - e\mathbf{A}_m)]^2}{\mathbf{A}_e^2(\mathbf{p} - e\mathbf{A}_m)^2}\right) + O(\alpha_e^2) \end{aligned} \tag{37}$$

The first-order approximation in α_e of Eq.(24) follows from Eq.(37):

$$\begin{aligned} \frac{m_0c^2}{(1 - v^2/c^2)^{1/2}} = {} & c[(\mathbf{p} - e\mathbf{A}_m)^2 + m_0^2c^2]^{1/2} \\ & \times \left(1 + \frac{\alpha_e m_0 c(\mathbf{p} - e\mathbf{A}_m)^2}{[(\mathbf{p} - e\mathbf{A}_m)^2 + m_0^2c^2]^{3/2}} \frac{[\mathbf{A}_e \cdot (\mathbf{p} - e\mathbf{A}_m)]^2}{\mathbf{A}_e^2(\mathbf{p} - e\mathbf{A}_m)^2}\right) + O(\alpha_e^2) \end{aligned} \tag{38}$$

The three components of the Hamilton function of Eqs.(12)–(14) become in this approximation:

$$\mathcal{H}_r = c[(\mathbf{p} - e\mathbf{A}_m)^2 + m_0^2c^2]^{1/2}(1 + \alpha_e Q) + e\phi_e - \mathcal{L}_{cr} \tag{39}$$

$$\mathcal{H}_\vartheta = c[(\mathbf{p} - e\mathbf{A}_m)^2 + m_0^2c^2]^{1/2}(1 + \alpha_e Q) + e\phi_e - \mathcal{L}_{c\vartheta} \tag{40}$$

$$\mathcal{H}_\varphi = c[(\mathbf{p} - e\mathbf{A}_m)^2 + m_0^2c^2]^{1/2}(1 + \alpha_e Q) + e\phi_e - \mathcal{L}_{c\varphi} \tag{41}$$

$$Q = \frac{1}{m_0^2c^2} \frac{(\mathbf{p} - e\mathbf{A}_m)^2[\mathbf{A}_e \cdot (\mathbf{p} - e\mathbf{A}_m)]^2}{[1 + (\mathbf{p} - e\mathbf{A}_m)^2/m_0^2c^2]^{3/2}\,\mathbf{A}_e^2(\mathbf{p} - e\mathbf{A}_m)^2}$$

$$\alpha_e = \frac{ZecA_e}{m_0c^2} = 2\frac{Ze^2}{2h}\frac{h}{m_0c}\frac{A_e}{e} = 2\alpha\frac{\lambda_C A_e}{e}$$

$$\alpha_e = 2.210 \times 10^5 A_e \text{ for electron, } \alpha_e = 1.204 \times 10^2 A_e \text{ for proton, } A_e[\text{As/m}]$$

$$\alpha = \frac{Ze^2}{2h} \doteq 7.297\,535 \times 10^{-9} \text{ fine structure constant, } \lambda_C = \frac{h}{m_0c} \tag{42}$$

For the evaluation of the terms $\mathfrak{L}_{cr}$, $\mathfrak{L}_{c\vartheta}$, $\mathfrak{L}_{c\varphi}$ we need $\dot{r}$, $r\dot{\vartheta}$, $r\sin\vartheta\,\dot{\varphi}$ of Eqs.(30)–(32) with the terms $1-v^2/c^2$ eliminated by means of Eq.(37). Only the zero-order approximation in $\alpha_{\rm e}$ is required:

$$\dot{r} = \frac{c(\mathbf{p}-e\mathbf{A}_{\rm m})_r}{[m_0^2c^2+(\mathbf{p}-e\mathbf{A}_{\rm m})^2]^{1/2}} + O(\alpha_{\rm e}) \tag{43}$$

$$r\dot{\vartheta} = \frac{c(\mathbf{p}-e\mathbf{A}_{\rm m})_\vartheta}{[m_0^2c^2+(\mathbf{p}-e\mathbf{A}_{\rm m})^2]^{1/2}} + O(\alpha_{\rm e}) \tag{44}$$

$$r\sin\vartheta\,\dot{\varphi} = \frac{c(\mathbf{p}-e\mathbf{A}_{\rm m})_\varphi}{[m_0^2c^2+(\mathbf{p}-e\mathbf{A}_{\rm m})^2]^{1/2}} + O(\alpha_{\rm e}) \tag{45}$$

For the first component $\mathfrak{L}_{cr1}$ of $\mathfrak{L}_{cr}$ we obtain from Eq.(1.3-33):

$$\begin{aligned}\mathfrak{L}_{cr1} = \frac{Ze}{c}\dot{r}(r\dot{\vartheta}A_{\rm e\varphi} - r\sin\vartheta\,\dot{\varphi}A_{\rm e\vartheta}) &= \frac{\alpha_{\rm e}}{A_{\rm e}m_0}\,\frac{A_{\rm e\varphi}(\mathbf{p}-e\mathbf{A}_{\rm m})_\vartheta - A_{\rm e\vartheta}(\mathbf{p}-e\mathbf{A}_{\rm m})_\varphi}{[1+(\mathbf{p}-e\mathbf{A}_{\rm m})^2/m_0^2c^2]^{1/2}}\\ &\quad\times \frac{(\mathbf{p}-e\mathbf{A}_{\rm m})_r}{[1+(\mathbf{p}-e\mathbf{A}_{\rm m})^2/m_0^2c^2]^{1/2}} + O(\alpha_{\rm e}^2)\end{aligned} \tag{46}$$

The unusual way of writing the denominators is due to the replacement of $\mathbf{p}$ by operators at a later time and the resulting noncommutability of the factors of a product. This will become important in the following sections. The second component $\mathfrak{L}_{cr2}$ is defined according to Eq.(1.3-33) as follows:

$$\begin{aligned}\mathfrak{L}_{cr2} &= \frac{Ze}{c}\int\left(\sin\vartheta\,\dot{\varphi}\frac{\partial\phi_{\rm m}}{\partial\vartheta} - \frac{\dot{\vartheta}}{\sin\vartheta}\frac{\partial\phi_{\rm m}}{\partial\varphi}\right)dr\\ &= \frac{\alpha_{\rm e}}{A_{\rm e}}\int\left(\frac{\partial\phi_{\rm m}}{\partial\vartheta}\frac{(\mathbf{p}-e\mathbf{A}_{\rm m})_\varphi}{r} - \frac{\partial\phi_{\rm m}}{\partial\varphi}\frac{(\mathbf{p}-e\mathbf{A}_{\rm m})_\vartheta}{r\sin\vartheta}\right)\\ &\qquad\times\left(1+\frac{(\mathbf{p}-e\mathbf{A}_{\rm m})^2}{m_0^2c^2}\right)^{-1/2} dr\end{aligned} \tag{47}$$

For the third component $\mathfrak{L}_{cr3}$ of Eq.(1.3-33) we require the time derivative of $\dot{\vartheta}$ and $\dot{\varphi}$ of Eqs.(44) and (45):

$$\begin{aligned}\mathfrak{L}_{cr3} &= \frac{Ze}{c}\int(r\ddot{\vartheta}A_{\rm e\varphi} - r\sin\vartheta\,\ddot{\varphi}A_{\rm e\vartheta})dr\\ &= \frac{\alpha_{\rm e}}{A_{\rm e}}\int\left[A_{\rm e\varphi}\left(\frac{\partial}{\partial t} - \frac{(\mathbf{p}-e\mathbf{A}_{\rm m})_r}{m_0 r}\right)\right.\\ &\qquad\left. - A_{\rm e\vartheta}\left(\frac{\partial}{\partial t} - \frac{1}{m_0 r}\,\frac{(\mathbf{p}-e\mathbf{A}_{\rm m})_r + \operatorname{ctg}\vartheta(\mathbf{p}-e\mathbf{A}_{\rm m})_\vartheta}{[1+(\mathbf{p}-e\mathbf{A}_{\rm m})^2/m_0^2c^2]^{1/2}}\right)\right]\end{aligned}$$

$$\times \frac{(\mathbf{p}-e\mathbf{A}_{\mathrm{m}})_{\varphi}}{[1+(\mathbf{p}-e\mathbf{A}_{\mathrm{m}})^2/m_0^2c^2]^{1/2}}dr \quad (48)$$

The fourth component $\mathfrak{L}_{cr4}$ of Eq.(1.3-33) equals:

$$\begin{aligned}\mathfrak{L}_{cr4} &= \frac{Ze}{c}\int\left(\dot{\vartheta}\frac{\partial}{\partial\vartheta}+\dot{\varphi}\frac{\partial}{\partial\varphi}\right)(r\dot{\vartheta}A_{\mathrm{e}\varphi}-r\sin\vartheta\,\dot{\varphi}A_{\mathrm{e}\vartheta})dr\\ &= \frac{\alpha_{\mathrm{e}}}{A_{\mathrm{e}}m_0}\int\left((\mathbf{p}-e\mathbf{A}_{\mathrm{m}})_{\vartheta}\frac{1}{r}\frac{\partial}{\partial\vartheta}+(\mathbf{p}-e\mathbf{A}_{\mathrm{m}})_{\varphi}\frac{1}{r\sin\vartheta}\frac{\partial}{\partial\varphi}\right)\\ &\qquad\times\frac{A_{\mathrm{e}\varphi}(\mathbf{p}-e\mathbf{A}_{\mathrm{m}})_{\vartheta}-A_{\mathrm{e}\vartheta}(\mathbf{p}-e\mathbf{A}_{\mathrm{m}})_{\varphi}}{1+(\mathbf{p}-e\mathbf{A}_{\mathrm{m}})^2/m_0^2c^2}dr \quad (49)\end{aligned}$$

The fifth and last component $\mathfrak{L}_{cr5}$ of Eq.(1.3-33) remains unchanged since there are no time-derivatives $\dot{r}$, $r\dot{\vartheta}$, $r\sin\vartheta\,\dot{\varphi}$. Note the negative sign of $\mathfrak{L}_{cr5}$:

$$\begin{aligned}\mathfrak{L}_{cr5} &= Zec\int\frac{1}{r\sin\vartheta}\left(\frac{\partial(A_{\mathrm{e}\varphi}\sin\vartheta)}{\partial\vartheta}-\frac{\partial A_{\mathrm{e}\vartheta}}{\partial\varphi}\right)dr\\ &= \frac{\alpha_{\mathrm{e}}m_0c^2}{A_{\mathrm{e}}}\int\frac{1}{r\sin\vartheta}\left(\frac{\partial(A_{\mathrm{e}\varphi}\sin\vartheta)}{\partial\vartheta}-\frac{\partial A_{\mathrm{e}\vartheta}}{\partial\varphi}\right)dr \quad (50)\end{aligned}$$

To obtain the correcting terms of $\mathfrak{L}_{c\vartheta}$ and $\mathfrak{L}_{c\varphi}$ we must change the integration differentials

$$dr \to rd\vartheta \to r\sin\vartheta\,d\varphi \quad (51)$$

Using Eqs.(42)–(45) one may derive $\mathfrak{L}_{c\vartheta 1}$ to $\mathfrak{L}_{c\vartheta 5}$ and $\mathfrak{L}_{c\varphi 1}$ to $\mathfrak{L}_{c\varphi 5}$ without any new difficulties from Eqs.(1.3-34) and (1.3-35). The derivatives $\ddot{r}$, $\ddot{\vartheta}$, and $\ddot{\varphi}$ come from Eqs.(43)–(45).

The Klein-Gordon and the Dirac equations are derived from the Hamilton functions of Eqs.(25)–(27) without the correcting terms $\mathfrak{L}_{cr}$, $\mathfrak{L}_{c\vartheta}$, $\mathfrak{L}_{c\varphi}$. The Hamilton functions of Eqs.(39)–(41) with first-order correction in α_{e} will yield first-order corrections to the Klein-Gordon and Dirac equations, while higher order solutions in α_{e} of Eq.(33) will yield higher order corrections.

The term $\mathbf{g}_{\mathrm{m}}$ in the Maxwell equation (1.1-2) has an effect on the solution even if the transition $\mathbf{g}_{\mathrm{m}} \to 0$ is made at the end of the calculation since this term produces a different differential equation[2] that yields convergent rather than divergent solutions. According to Eq.(1.1-15) the potential $\mathbf{A}_{\mathrm{e}}$ represents the magnetic current density $\mathbf{g}_{\mathrm{m}}$ here. It is prudent to expect that the transition $\mathbf{A}_{\mathrm{e}} \to 0$ at the end of the calculation may have an effect similar to $\mathbf{g}_{\mathrm{m}} \to 0$. Hence, the terms with a factor α_{e} in Eqs.(39)–(41) cannot be ignored even if one takes the limit $\mathbf{A}_{\mathrm{e}} \to 0$ and $\mathbf{g}_{\mathrm{m}} \to 0$ at the end of the calculation.

[2]See Harmuth et al. 2001, Eqs.(1.3-1) and (1.2-13).

If we want to carry the theory to second order in α_e we must use the first three terms on the right side of Eq.(33). For the term $\alpha_e^2(1-v^2/c^2)$ we must use $1-v^2/c^2$ of Eq.(35) while for $2\alpha_e(1-v^2/c^2)^{1/2}$ we must use the better approximation $1-v^2/c^2$ of Eq.(37). We may proceed in this way to the third- and fourth-order approximation in α_e. But the process does not stop there. For the fourth order approximation we use the value of $1-v^2/c^2$ of Eq.(35) for the term $\alpha_e^4(1-v^2/c^2)$ in Eq.(33), but for the fifth-order approximation we use $1-v^2/c^2$ of Eq.(37). There is no end. Every improved approximation yields new terms and perhaps new effects.

5.3 Quantization with Differential Operators

Following the matrix notation of Eq.(1.3-24) we may write Eqs.(5.2-39)–(5.2-41) in the following form:

$$\mathcal{H} = \left\{c[(\mathbf{p}-e\mathbf{A}_m)^2 + m_0^2c^2]^{1/2}(1+\alpha_e Q) + e\phi_e\right\} \times \begin{pmatrix}1&0&0\\0&1&0\\0&0&1\end{pmatrix} - \begin{pmatrix}\mathcal{L}_{cr}&0&0\\0&\mathcal{L}_{c\vartheta}&0\\0&0&\mathcal{L}_{c\varphi}\end{pmatrix} \tag{1}$$

Consider the first line of this equation:

$$\begin{aligned}\mathcal{H}_r &= c[(\mathbf{p}-e\mathbf{A}_m)^2 + m_0^2c^2]^{1/2}(1+\alpha_e Q) + e\phi_e - \mathcal{L}_{cr}\\ (\mathcal{H}_r - e\phi_e + \mathcal{L}_{cr})^2 &= c^2[(\mathbf{p}-e\mathbf{A}_m)^2 + m_0^2c^2](1+\alpha_e Q)^2\end{aligned} \tag{2}$$

Our calculation holds only in first order of α_e. Hence, we replace $(1+\alpha_e Q)^2$ by $1+2\alpha_e Q$. The term $\mathcal{L}_{cr}^2$ can be left out since its components are multiplied by α_e^2 according to Eqs.(5.2-46)–(5.2-50). The rule of Eq.(1.1-47) is used for the product $(\mathcal{H}_r - e\phi_e)\mathcal{L}_{cr}$:

$$\begin{aligned}&(\mathbf{p}-e\mathbf{A}_m)^2 - \frac{1}{c^2}(\mathcal{H}_r - e\phi_e)^2 + \alpha_e\Bigg\{2[(\mathbf{p}-e\mathbf{A}_m)^2 + m_0^2c^2]Q\\ &\qquad - \frac{1}{c^2}\left[(\mathcal{H}_r - e\phi_e)\frac{\mathcal{L}_{cr}}{\alpha_e} + \frac{\mathcal{L}_{cr}}{\alpha_e}(\mathcal{H}_r - e\phi_e)\right]\Bigg\} = -m_0^2c^2\\ &\alpha_e = ZecA_e/m_0c^2 \ll 1\end{aligned} \tag{3}$$

The operators of Eq.(3) are applied to the function

$$\Psi_r = \Psi_{r0} + \alpha_e\Psi_{r1} \tag{4}$$

and one obtains

$$\Big\{(\mathbf{p}-e\mathbf{A}_{\mathrm{m}})^2 - \frac{1}{c^2}(\mathcal{H}_r - e\phi_{\mathrm{e}})^2 + \alpha_{\mathrm{e}}\Big[2[(\mathbf{p}-e\mathbf{A}_{\mathrm{m}})^2 + m_0^2c^2]Q$$
$$-\frac{1}{c^2}\Big((\mathcal{H}_r - e\phi_{\mathrm{e}})\frac{\mathcal{L}_{cr}}{\alpha_{\mathrm{e}}} + \frac{\mathcal{L}_{cr}}{\alpha_{\mathrm{e}}}(\mathcal{H}_r - e\phi_{\mathrm{e}})\Big)\Big]\Big\}(\Psi_{r0} + \alpha_{\mathrm{e}}\Psi_{r1})$$
$$= -m_0^2c^2(\Psi_{r0} + \alpha_{\mathrm{e}}\Psi_{r1}) \quad (5)$$

All components of $\mathcal{L}_{cr}$ are multiplied by α_{e} according to Eqs.(5.2-46)–(5.2-50). Hence, we may divide Eq.(5) into one part of order $O(1)$ and a second one of order $O(\alpha_{\mathrm{e}})$:

$$\Big((\mathbf{p}-e\mathbf{A}_{\mathrm{m}})^2 - \frac{1}{c^2}(\mathcal{H}_r - e\phi_{\mathrm{e}})^2 + m_0^2c^2\Big)\Psi_{r0} = 0 \quad (6)$$

$$\Big((\mathbf{p}-e\mathbf{A}_{\mathrm{m}})^2 - \frac{1}{c^2}(\mathcal{H}_r - e\phi_{\mathrm{e}})^2 + m_0^2c^2\Big)\Psi_{r1} = -\Big[2[(\mathbf{p}-e\mathbf{A}_{\mathrm{m}})^2 + m_0^2c^2]Q$$
$$-\frac{1}{c^2}\Big((\mathcal{H}_r - e\phi_{\mathrm{e}})\frac{\mathcal{L}_{cr}}{\alpha_{\mathrm{e}}} + \frac{\mathcal{L}_{cr}}{\alpha_{\mathrm{e}}}(\mathcal{H}_r - e\phi_{\mathrm{e}})\Big)\Big]\Psi_{r0} \quad (7)$$

Equation (6) is essentially the first line of Eq.(2.1-2) while Eq.(7) is an inhomogeneous variant of that equation.

For the generalization of Eq.(6) we follow the usual path of using Cartesian coordinates for quantization and then making the transition to spherical coordinates. In analogy to the transition from Eq.(2.1-2) to Eq.(2.1-4) we obtain from Eq.(6):

$$\left[\sum_{j=1}^{3}\Big(\frac{\hbar}{i}\frac{\partial}{\partial x_j} - eA_{\mathrm{m}x_j}\Big)^2 - \frac{1}{c^2}\Big(\frac{\hbar}{i}\frac{\partial}{\partial t} + e\phi_{\mathrm{e}}\Big)^2\right]\Psi_{r0} = -m_0^2c^2\Psi_{r0}$$
$$x_j = x,\ y,\ z \quad (8)$$

With the relations

$$\Big(\frac{\hbar}{i}\frac{\partial}{\partial x_j} - eA_{\mathrm{m}x_j}\Big)^2 = -\hbar^2\frac{\partial^2}{\partial x_j^2} + 2ie\hbar A_{\mathrm{m}x_j}\frac{\partial}{\partial x_j} + ie\hbar\frac{\partial A_{\mathrm{m}x_j}}{\partial x_j} + e^2A_{\mathrm{m}x_j}^2 \quad (9)$$
$$\Big(\frac{\hbar}{i}\frac{\partial}{\partial t} + e\phi_{\mathrm{e}}\Big)^2 = -\hbar^2\frac{\partial^2}{\partial t^2} - 2ie\hbar\phi_{\mathrm{e}}\frac{\partial}{\partial t} - ie\hbar\frac{\partial\phi_{\mathrm{e}}}{\partial t} + e^2\phi_{\mathrm{e}}^2 \quad (10)$$

we may rewrite Eq.(8) into the following form:

$$\left[\sum_{j=1}^{3}\left(-\hbar^2\frac{\partial^2}{\partial x_j^2}+2ie\hbar A_{\mathrm{m}x_j}\frac{\partial}{\partial x_j}+ie\hbar\frac{\partial A_{\mathrm{m}x_j}}{\partial x_j}+e^2A_{\mathrm{m}x_j}^2\right)\right.$$
$$\left.-\frac{1}{c^2}\left(-\hbar^2\frac{\partial^2}{\partial t^2}-2ie\hbar\phi_{\mathrm{e}}\frac{\partial}{\partial t}-ie\hbar\frac{\partial\phi_{\mathrm{e}}}{\partial t}+e^2\phi_{\mathrm{e}}^2\right)\right]\Psi_{r0}=-m_0^2c^2\Psi_{r0} \quad (11)$$

In the first three terms of the first line we recognize the vector operators ∇^2, grad, and div in Cartesian coordinates. Hence, we may rewrite Eq.(11) in vector form:

$$(\hbar^2\nabla^2-2ie\hbar\mathbf{A}_{\mathrm{m}}\cdot\mathrm{grad}-ie\hbar\,\mathrm{div}\,\mathbf{A}_{\mathrm{m}}-e^2\mathbf{A}_{\mathrm{m}}^2-m_0^2c^2)\Psi_{r0}$$
$$=\frac{1}{c^2}\left(\hbar^2\frac{\partial^2}{\partial t^2}+2ie\hbar\phi_{\mathrm{e}}\frac{\partial}{\partial t}+ie\hbar\frac{\partial\phi_{\mathrm{e}}}{\partial t}-e^2\phi_{\mathrm{e}}^2\right)\Psi_{r0} \quad (12)$$

At this point we change to spherical coordinates by postulating that the vector operators ∇^2, grad, and div in Eq.(12) can be replaced by the respective operators for component notation in spherical coordinates.

The procedure used also works for the homogeneous part of Eq.(7). We only need to replace Ψ_{r0} with Ψ_{r1}. Problems are caused by the terms Q and $\mathfrak{L}_{cr}$ on the right side of Eq.(7).

The factor Q in Eq.(7) has a term $[1+(\mathbf{p}-e\mathbf{A}_{\mathrm{m}})^2/m_0^2c^2]^{-3/2}$ according to Eq.(1.1-45). After quantization the term $(\mathbf{p}-e\mathbf{A}_{\mathrm{m}})^2$ is replaced by Eq.(9) or the left side of Eq.(12) without the term $m_0^2c^2$:

$$(\mathbf{p}-e\mathbf{A}_{\mathrm{m}})^2=\hbar^2\nabla^2-2ie\hbar\mathbf{A}_{\mathrm{m}}\cdot\mathrm{grad}-ie\hbar\,\mathrm{div}\,\mathbf{A}_{\mathrm{m}}-e^2\mathbf{A}_{\mathrm{m}}^2 \quad (13)$$

If we restrict the calculation to small values of $(\mathbf{p}-e\mathbf{A}_{\mathrm{m}})^2/m_0^2c^2$ as done in Eq.(2.1-12)

$$(\mathbf{p}-e\mathbf{A}_{\mathrm{m}})^2/m_0^2c^2\ll 1 \quad (14)$$

we may write

$$[1+(\mathbf{p}-e\mathbf{A}_{\mathrm{m}})^2/m_0^2c^2]^{-3/2}\doteq 1-\frac{3}{2}(\mathbf{p}-e\mathbf{A}_{\mathrm{m}})^2/m_0^2c^2 \quad (15)$$

This defines part of the factor Q of Eq.(1.1-45) for spherical coordinates. But we still have a factor

$$Q_{02}=\frac{[\mathbf{A}_{\mathrm{e}}\cdot(\mathbf{p}-e\mathbf{A}_{\mathrm{m}})]^2}{\mathbf{A}_{\mathrm{e}}^2(\mathbf{p}-e\mathbf{A}_{\mathrm{m}})^2} \quad (16)$$

to explain. We do this in analogy to Eqs.(2.1-14)–(2.1-19), replacing the Cartesian coordinates by spherical coordinates:

$$Q = \frac{1}{m_0^2 c^2} \frac{(\mathbf{p} - e\mathbf{A}_{\mathrm{m}})^2}{[1 + (\mathbf{p} - e\mathbf{A}_{\mathrm{m}})^2/m_0^2 c^2]^{3/2}} Q_{02} \tag{17}$$

The factor Q_{02} can be made equal to 1 if we replace the vectors $\mathbf{A}_{\mathrm{e}}$ and $\mathbf{p}-e\mathbf{A}_{\mathrm{m}}$ by matrices of rank 3 whose components are vectors:

$$\mathbf{A}_{\mathrm{e}} = \begin{pmatrix} A_{\mathrm{e}r}\mathbf{e}_r & 0 & 0 \\ 0 & A_{\mathrm{e}\vartheta}\mathbf{e}_\vartheta & 0 \\ 0 & 0 & A_{\mathrm{e}\varphi}\mathbf{e}_\varphi \end{pmatrix} \tag{18}$$

$$\mathbf{p} - e\mathbf{A}_{\mathrm{m}} = \begin{pmatrix} (p_r - eA_{\mathrm{e}r})\mathbf{e}_r & 0 & 0 \\ 0 & (p_\vartheta - eA_{\mathrm{m}\vartheta})\mathbf{e}_\vartheta & 0 \\ 0 & 0 & (p_\varphi - eA_{\mathrm{m}\varphi})\mathbf{e}_\varphi \end{pmatrix} \tag{19}$$

The substitution of Eqs.(18) and (19) into Eq.(16) yields

$$Q_{02} = 1 \tag{20}$$

This is a desirably simple result but it replaces the usual Klein-Gordon equation by a matrix equation of rank 3. The form of Q in Eq.(17) is reduced to

$$\begin{aligned} Q &= \frac{1}{m_0^2 c^2} \frac{(\mathbf{p} - e\mathbf{A}_{\mathrm{m}})^2}{[1 + (\mathbf{p} - e\mathbf{A}_{\mathrm{m}})^2/m_0^2 c^2]^{3/2}} \\ &\doteq \frac{(\mathbf{p} - e\mathbf{A}_{\mathrm{m}})^2}{m_0^2 c^2} \left(1 - \frac{3}{2} \frac{(\mathbf{p} - e\mathbf{A}_{\mathrm{m}})^2}{m_0^2 c^2}\right) \end{aligned} \tag{21}$$

The term $(\mathbf{p} - e\mathbf{A}_{\mathrm{m}})^2$ is defined by Eq.(13) for spherical coordinates. There is no difference whether we multiply $(\mathbf{p} - e\mathbf{A}_{\mathrm{m}})^2/m_0^2 c^2$ on the right with $1 - \frac{3}{2}(\mathbf{p} - e\mathbf{A}_{\mathrm{m}})^2/m_0^2 c^2$ as done in Eq.(21) or on the left since $(\mathbf{p} - e\mathbf{A}_{\mathrm{m}})^2$ has only to commute with the constant 1 and itself to transform a multiplication on the right to one on the left.

We have come so far without specifying $\mathbf{p}-e\mathbf{A}_{\mathrm{m}}$ and thus the three components of the matrix of Eq.(19). Only the square $(\mathbf{p} - e\mathbf{A}_{\mathrm{m}})^2$ had to be defined, which is done by Eq.(13). This is not sufficient for the term $\mathfrak{L}_{cr}$ in Eq.(7). Equations (5.2-46)–(5.2-49) require the components

$$(\mathbf{p}-e\mathbf{A}_{\mathrm{m}})_r = p_r - eA_{\mathrm{m}r}, \quad (\mathbf{p}-e\mathbf{A}_{\mathrm{m}})_\vartheta = p_\vartheta - eA_{\mathrm{m}\vartheta}, \quad (\mathbf{p}-e\mathbf{A}_{\mathrm{m}})_\varphi = p_\varphi - eA_{\mathrm{m}\varphi}$$

The terms $A_{\mathrm{m}r}$, $A_{\mathrm{m}\vartheta}$, $A_{\mathrm{m}\varphi}$ were defined in Eq.(5.1-5) and the terms p_r, p_ϑ, p_φ in Eqs.(5.2-8)–(5.2-10). Equation (5.2-3) yields the relations

$$
\begin{aligned}
v^2 &= \dot{r}^2 + (r\dot{\vartheta})^2 + (r\sin\vartheta\,\dot{\varphi})^2 \\
(1 - v^2/c^2)^{-1/2} &\doteq 1 + \frac{1}{2}[\dot{r}^2 + (r\dot{\vartheta})^2 + (r\sin\vartheta\,\dot{\varphi})^2]
\end{aligned}
\tag{22}
$$

Equation (5.2-46) for $\mathfrak{L}_{cr1}$ may be written as follows:

$$
\begin{aligned}
\mathfrak{L}_{cr1} &\doteq \frac{2\alpha\lambda_{\rm C}}{m_0 e}\left(1 - \frac{(\mathbf{p} - e\mathbf{A}_{\rm m})^2}{m_0^2c^2}\right)[A_{{\rm e}\varphi}(\mathbf{p} - e\mathbf{A}_{\rm m})_\vartheta - A_{{\rm e}\vartheta}(\mathbf{p} - e\mathbf{A}_{\rm m})_\varphi] \\
&\qquad \times (\mathbf{p} - e\mathbf{A}_{\rm m})_r \\
&\doteq \frac{2\alpha\lambda_{\rm C}}{m_0 e}\left(1 - \frac{(\mathbf{p} - e\mathbf{A}_{\rm m})^2}{m_0^2c^2}\right) \\
&\quad \times \left[A_{{\rm e}\varphi}\left(\frac{m_0 r\dot{\vartheta}}{(1 - v^2/c^2)^{1/2}} + \frac{Ze}{c}(r\sin\vartheta\,\dot{\varphi}A_{{\rm e}r} - \dot{r}A_{{\rm e}\varphi})\right)\right. \\
&\quad \left. - A_{{\rm e}\vartheta}\left(\frac{m_0 r\sin\vartheta\,\dot{\varphi}}{(1 - v^2/c^2)^{1/2}} + \frac{Ze}{c}(\dot{r}A_{{\rm e}\vartheta} - r\dot{\vartheta}A_{{\rm e}r})\right)\right] \\
&\quad \times \left(\frac{m_0\dot{r}}{(1 - v^2/c^2)^{1/2}} + \frac{Ze}{c}(r\dot{\vartheta}A_{{\rm e}\varphi} - r\sin\vartheta\,\dot{\varphi}A_{{\rm e}\vartheta})\right)
\end{aligned}
\tag{23}
$$

We turn to $\mathfrak{L}_{cr2}$ of Eq.(5.2-47). With the help of Eqs.(13) and (22) $\mathfrak{L}_{cr2}$ is fully defined:

$$
\begin{aligned}
\mathfrak{L}_{cr2} &\doteq \frac{2\alpha\lambda_{\rm C}}{e}\int\left(\frac{\partial\phi_{\rm m}}{\partial\vartheta}\frac{p_\varphi - eA_{{\rm m}\varphi}}{r} - \frac{\partial\phi_{\rm m}}{\partial\varphi}\frac{p_\vartheta - eA_{{\rm m}\vartheta}}{r\sin\vartheta}\right)\left(1 - \frac{1}{2}\frac{(\mathbf{p} - e\mathbf{A}_{\rm m})^2}{m_0^2c^2}\right)dr \\
&\doteq \frac{2\alpha\lambda_{\rm C}}{e}\int\left[\frac{1}{r}\frac{\partial\phi_{\rm m}}{\partial\vartheta}\left(\frac{m_0 r\sin\vartheta\,\dot{\varphi}}{(1 - v^2/c^2)^{1/2}} + \frac{Ze}{c}(\dot{r}A_{{\rm e}\vartheta} - r\dot{\vartheta}A_{{\rm e}r})\right)\right. \\
&\quad \left. - \frac{1}{r\sin\vartheta}\frac{\partial\phi_{\rm m}}{\partial\varphi}\left(\frac{m_0 r\dot{\vartheta}}{(1 - v^2/c^2)^{1/2}} + \frac{Ze}{c}(r\sin\vartheta\,\dot{\varphi}A_{{\rm e}r} - \dot{r}A_{{\rm e}\varphi})\right)\right] \\
&\quad \times \left(1 - \frac{1}{2}\frac{(\mathbf{p} - e\mathbf{A}_{\rm m})^2}{m_0^2c^2}\right)dr
\end{aligned}
\tag{24}
$$

Next comes $\mathfrak{L}_{cr3}$ of Eq.(5.2-48):

$$
\begin{aligned}
\mathfrak{L}_{cr3} &\doteq \frac{2\alpha\lambda_{\rm C}}{e}\int\left\{A_{{\rm e}\varphi}\frac{\partial}{\partial t}\left[\left(1 - \frac{1}{2}\frac{(\mathbf{p} - e\mathbf{A}_{\rm m})^2}{m_0^2c^2}\right)(p_\vartheta - eA_{{\rm m}\vartheta})\right]\right. \\
&\quad \left. - A_{{\rm e}\vartheta}\frac{\partial}{\partial t}\left[\left(1 - \frac{1}{2}\frac{(\mathbf{p} - e\mathbf{A}_{\rm m})^2}{m_0^2c^2}\right)(p_\varphi - eA_{{\rm m}\varphi})\right]\right\}dr
\end{aligned}
$$

$$\begin{aligned}\doteq \frac{2\alpha\lambda_C}{e}\int\bigg\{&A_{e\varphi}\frac{\partial}{\partial t}\bigg[\bigg(1-\frac{1}{2}\frac{(\mathbf{p}-e\mathbf{A}_m)^2}{m_0^2c^2}\bigg)\\ &\times\bigg(\frac{m_0r\dot{\vartheta}}{(1-v^2/c^2)^{1/2}}+\frac{Ze}{c}(r\sin\vartheta\,\dot{\varphi}A_{er}-\dot{r}A_{e\varphi})\bigg)\bigg]\\ &-A_{e\vartheta}\frac{\partial}{\partial t}\bigg[\bigg(1-\frac{1}{2}\frac{(\mathbf{p}-e\mathbf{A}_m)^2}{m_0^2c^2}\bigg)\\ &\times\bigg(\frac{m_0r\sin\vartheta\,\dot{\varphi}}{(1-v^2/c^2)^{1/2}}+\frac{Ze}{c}(\dot{r}A_{e\vartheta}-r\dot{\vartheta}A_{er})\bigg)\bigg]\bigg\}dr \qquad (25)\end{aligned}$$

Equation (5.2-49) defines $\mathfrak{L}_{cr4}$:

$$\begin{aligned}\mathfrak{L}_{cr4}\doteq\frac{2\alpha\lambda_C}{e}\int&\bigg(1-\frac{(\mathbf{p}-e\mathbf{A}_m)^2}{m_0^2c^2}\bigg)\bigg((p_\vartheta-eA_{m\vartheta})\frac{1}{r}\frac{\partial}{\partial\vartheta}+(p_\varphi-eA_{m\varphi})\frac{1}{r\sin\vartheta}\frac{\partial}{\partial\varphi}\bigg)\\ &\times[A_{e\varphi}(p_\vartheta-eA_{m\vartheta})-A_{e\vartheta}(p_\varphi-eA_{m\varphi})]dr\\ \doteq\frac{2\alpha\lambda_C}{e}\int&\bigg(1-\frac{(\mathbf{p}-e\mathbf{A}_m)^2}{m_0^2c^2}\bigg)\\ &\times\bigg[\bigg(\frac{m_0r\dot{\vartheta}}{(1-v^2/c^2)^{1/2}}+\frac{Ze}{c}(r\sin\vartheta\,\dot{\varphi}A_{er}-\dot{r}A_{e\varphi})\bigg)\frac{1}{r}\frac{\partial}{\partial\vartheta}\\ &+\bigg(\frac{m_0r\sin\vartheta\,\dot{\varphi}}{(1-v^2/c^2)^{1/2}}+\frac{Ze}{c}(\dot{r}A_{e\vartheta}-r\dot{\vartheta}A_{er})\bigg)\frac{1}{r\sin\vartheta}\frac{\partial}{\partial\varphi}\bigg]\\ &\times\bigg[A_{e\varphi}\bigg(\frac{m_0r\dot{\vartheta}}{(1-v^2/c^2)^{1/2}}+\frac{Ze}{c}(r\sin\vartheta\,\dot{\varphi}A_{er}-\dot{r}A_{e\varphi})\bigg)\\ &-A_{e\vartheta}\bigg(\frac{m_0r\sin\vartheta\,\dot{\varphi}}{(1-v^2/c^2)^{1/2}}+\frac{Ze}{c}(\dot{r}A_{e\vartheta}-r\dot{\vartheta}A_{er})\bigg)\bigg]dr \qquad (26)\end{aligned}$$

The final term $\mathfrak{L}_{cr5}$ remains essentially unchanged. We only replace α_e by the fine structure constant α:

$$\mathfrak{L}_{cr5}=\frac{2\alpha\lambda_C m_0c^2}{e}\int\frac{1}{r\sin\vartheta}\bigg(\frac{\partial(A_{e\varphi}\sin\vartheta)}{\partial\vartheta}-\frac{\partial A_{e\vartheta}}{\partial\varphi}\bigg)dr \qquad (27)$$

5.4 Difference Operators

We replace the differential operators ∇^2, grad, and div for spherical coordinates by the respective difference operators $\tilde{\nabla}^2$, $\widetilde{\text{grad}}$, and $\widetilde{\text{div}}$. The notation $\tilde{\Delta}$ for difference operators introduced by Eqs.(1.2-1) and (1.2-7) is used:

$$\tilde{\nabla}^2\Psi = \frac{\tilde{\Delta}^2\Psi}{\tilde{\Delta}r^2} + \frac{2}{r}\frac{\tilde{\Delta}\Psi}{\tilde{\Delta}r} + \frac{1}{r^2}\frac{\tilde{\Delta}^2\Psi}{\tilde{\Delta}\vartheta^2} + \frac{\cot\vartheta}{r^2}\frac{\tilde{\Delta}\Psi}{\tilde{\Delta}\vartheta} + \frac{1}{r^2\sin^2\vartheta}\frac{\tilde{\Delta}^2\Psi}{\tilde{\Delta}\varphi^2} \tag{1}$$

$$\tilde{\text{grad}}\,\Psi = \frac{\tilde{\Delta}\Psi}{\tilde{\Delta}r}\mathbf{e}_r + \frac{1}{r}\frac{\tilde{\Delta}\Psi}{\tilde{\Delta}\vartheta}\mathbf{e}_\vartheta + \frac{1}{r\sin\vartheta}\frac{\tilde{\Delta}\Psi}{\tilde{\Delta}\varphi}\mathbf{e}_\varphi \tag{2}$$

$$\tilde{\text{div}}\,\mathbf{A} = \frac{\tilde{\Delta}A_r}{\tilde{\Delta}r} + \frac{2}{r}A_r + \frac{1}{r}\frac{\tilde{\Delta}A_\vartheta}{\tilde{\Delta}\vartheta} + \frac{\cot\vartheta}{r}A_\vartheta + \frac{1}{r\sin\vartheta}\frac{\tilde{\Delta}A_\varphi}{\tilde{\Delta}\varphi} \tag{3}$$

In Section 2.2 we replaced the variables x_j and t by normalized variables ζ_j and θ. The variables x_j and t were originally defined in the intervals $0 \le x_j \le cT$ and $0 \le t \le T$. These intervals became $0 \le \zeta_j \le N$ and $0 \le \theta \le N$ for $\zeta_j = x_j/c\Delta t$ and $\theta = t/\Delta t$. Since ϑ and φ are defined in the intervals $0 \le \vartheta \le \pi$ and $0 \le \varphi \le 2\pi$ we replace them by the normalized variables $\eta = \vartheta/\Delta\vartheta$ and $\xi = \varphi/\Delta\varphi$ to obtain the intervals $0 \le \eta \le N$ and $0 \le \xi \le N$ for the normalized angles ϑ and φ:

$$\begin{array}{llll} \theta = t/\Delta t & \rho = r/\Delta r & \eta = \vartheta/\Delta\vartheta & \xi = \varphi/\Delta\varphi \\ 0 \le \theta \le N & 0 \le \rho \le N & 0 \le \eta \le N & 0 \le \xi \le N \\ \Delta t = T/N & \Delta r = cT/N & \Delta\vartheta = \pi/N & \Delta\varphi = 2\pi/N \end{array} \tag{4}$$

Equations (1)–(3) assume the following form:

$$\tilde{\nabla}^2\Psi = \frac{1}{(\Delta r)^2}\left(\frac{\tilde{\Delta}^2\Psi}{\tilde{\Delta}\rho^2} + \frac{2}{\rho}\frac{\tilde{\Delta}\Psi}{\tilde{\Delta}\rho} + \frac{1}{\rho^2}\frac{\tilde{\Delta}^2\Psi}{\tilde{\Delta}\eta^2} + \frac{\cot(\eta\Delta\vartheta)}{\rho^2}\frac{\tilde{\Delta}\Psi}{\tilde{\Delta}\eta} + \frac{1}{\rho^2\sin^2(\eta\Delta\vartheta)}\frac{\tilde{\Delta}^2\Psi}{\tilde{\Delta}\xi^2}\right) \tag{5}$$

$$\tilde{\text{grad}}\,\Psi = \frac{1}{\Delta r}\left(\frac{\tilde{\Delta}\Psi}{\tilde{\Delta}\rho}\mathbf{e}_r + \frac{1}{\rho}\frac{\tilde{\Delta}\Psi}{\tilde{\Delta}\eta}\mathbf{e}_\vartheta + \frac{1}{\rho\sin(\eta\Delta\vartheta)}\frac{\tilde{\Delta}\Psi}{\tilde{\Delta}\xi}\mathbf{e}_\varphi\right) \tag{6}$$

$$\tilde{\text{div}}\,\mathbf{A} = \frac{1}{\Delta r}\left(\frac{\tilde{\Delta}A_r}{\tilde{\Delta}\rho} + \frac{2}{\rho}A_r + \frac{1}{\rho}\frac{\tilde{\Delta}A_\vartheta}{\tilde{\Delta}\eta} + \frac{\cot(\eta\Delta\vartheta)}{\rho}A_\vartheta + \frac{1}{\rho\sin(\eta\Delta\vartheta)}\frac{\tilde{\Delta}A_\varphi}{\tilde{\Delta}\xi}\right) \tag{7}$$

The first- and second-order difference quotient for ρ follows from Eqs.(1.2-1), (1.2-3), and (1.2-7):

$$\frac{\tilde{\Delta}\Psi}{\tilde{\Delta}\rho} = \frac{\Psi(\rho+\Delta\rho) - \Psi(\rho-\Delta\rho)}{2\Delta\rho} = \frac{1}{2}[\Psi(\rho+1) - \Psi(\rho-1)] \tag{8}$$

$$\frac{\tilde{\Delta}^2\Psi}{\tilde{\Delta}\rho^2} = \frac{\Psi(\rho+\Delta\rho) - 2\Psi(\rho) + \Psi(\rho-\Delta\rho)}{(\Delta\rho)^2} = \Psi(\rho+1) - 2\Psi(\rho) + \Psi(\rho-1) \tag{9}$$

The substitutions $\rho \to \eta \to \xi \to \theta$ yield the corresponding equations for η, ξ, and θ.

Substitution of Eqs.(8) and (9) into Eqs.(5)–(7) yields the final form of the difference operators $\tilde{\nabla}^2$, $\widetilde{\text{grad}}$, and $\widetilde{\text{div}}$ in spherical coordinates. To shorten the expressions we write $\Psi(\rho+1)$ rather than $\Psi(\rho+1,\eta,\xi,\theta)$ when ρ is varied:

$$\begin{aligned}\tilde{\nabla}^2\Psi = \frac{1}{(\Delta r)^2}\bigg(&\Psi(\rho+1) - 2\Psi(\rho) + \Psi(\rho-1) + \frac{1}{\rho}[\Psi(\rho+1) - \Psi(\rho-1)] \\ &+ \frac{1}{\rho^2}[\Psi(\eta+1) - 2\Psi(\eta) + \Psi(\eta-1)] + \frac{\cot(\eta\Delta\vartheta)}{2\rho^2}[\Psi(\eta+1) - \Psi(\eta-1)] \\ &+ \frac{1}{\rho^2\sin^2(\eta\Delta\vartheta)}[\Psi(\xi+1) - 2\Psi(\xi) + \Psi(\xi-1)]\bigg)\end{aligned} \tag{10}$$

$$\begin{aligned}\widetilde{\text{grad}}\,\Psi = \frac{1}{\Delta r}\bigg(&\frac{1}{2}[\Psi(\rho+1) - \Psi(\rho-1)]\mathbf{e}_r + \frac{1}{2\rho}[\Psi(\eta+1) - \Psi(\eta-1)]\mathbf{e}_\vartheta \\ &+ \frac{1}{2\rho\sin(\eta\Delta\vartheta)}[\Psi(\xi+1) - \Psi(\xi-1)]\mathbf{e}_\varphi\bigg)\end{aligned} \tag{11}$$

$$\begin{aligned}\widetilde{\text{div}}\,\mathbf{A} = \frac{1}{\Delta r}\bigg(&\frac{1}{2}[A_r(\rho+1) - A_r(\rho-1)] + \frac{2}{\rho}A(\rho) \\ &+ \frac{1}{2\rho}[A_\vartheta(\eta+1) - A_\vartheta(\eta-1)] + \frac{\cos(\eta\Delta\vartheta)}{\rho}A_\vartheta(\eta) \\ &+ \frac{1}{2\rho\sin(\eta\Delta\vartheta)}[A_\varphi(\xi+1) - A_\varphi(\xi-1)]\bigg)\end{aligned} \tag{12}$$

We can now write Eq.(5.3-12) as a difference equation. Since only the specialized case $\mathbf{A}_\mathrm{m} = 0$, $\phi_\mathrm{e} = \phi_\mathrm{e}(r) = \phi_\mathrm{e}(\rho)$ will eventually be used we can simplify Eq.(5.3-12) drastically:

$$c^2(\hbar\nabla^2 - m_0^2c^2)\Psi_{\mathrm{r}0} = \left(\frac{\hbar^2}{(\Delta t)^2}\frac{\partial^2}{\partial\theta^2} + \frac{2ie\phi_\mathrm{e}(\rho)\hbar}{\Delta t}\frac{\partial}{\partial\theta} - e^2\phi_\mathrm{e}^2(\rho)\right)\Psi_{\mathrm{r}0} \tag{13}$$

This equation is rewritten into a symbolic difference equation. We obtain with the help of Eqs.(5) and (8)–(10):

$$c^2(\hbar^2\tilde{\nabla}^2 - m_0^2c^2)\Psi_{\mathrm{r}0} = \left(\frac{\hbar^2}{(\Delta t)^2}\frac{\tilde{\Delta}^2}{\tilde{\Delta}\theta^2} + \frac{2ie\phi_\mathrm{e}(\rho)\hbar}{\Delta t}\frac{\tilde{\Delta}}{\tilde{\Delta}\theta} - e^2\phi_\mathrm{e}^2(\rho)\right)\Psi_{\mathrm{r}0} \tag{14}$$

The function $\Psi_{r0} = \Psi_{r0}(\boldsymbol{\rho}, \theta) = \Psi_{r0}(\rho, \eta, \xi, \theta)$ depends on the spatial variables ρ, η, ξ and the time variable θ. The operator on the left side of Eq.(14) does not contain a term dependent on θ, but the operator on the right contains θ and ρ. For the elimination of θ on the right side we follow the spirit of the differential theory augmented by some intuition. First we write Ψ_{r0} as a product of $u(\boldsymbol{\rho})$ and $v(\theta) = v^\theta$

$$\Psi_{r0}(\boldsymbol{\rho}, \theta) = u(\boldsymbol{\rho})v^\theta \tag{15}$$

then we assume the operator on the right side of Eq.(14) can be replaced by $-[\mathsf{E} - e\phi_e(\rho)]^2$, where E is a constant:

$$u(\boldsymbol{\rho})\left(\frac{\hbar^2}{(\Delta t)^2}\frac{\tilde{\Delta}^2}{\tilde{\Delta}\theta^2} + \frac{2ie\phi_e(\rho)\hbar}{\Delta t}\frac{\tilde{\Delta}}{\tilde{\Delta}\theta} - e^2\phi_e^2(\rho)\right)v^\theta = -u(\boldsymbol{\rho})v^\theta[\mathsf{E} - e\phi_e(\rho)]^2 \tag{16}$$

The explicit difference equation of v^θ becomes:

$$\frac{\hbar^2}{(\Delta t)^2}(v^{\theta+1} - 2v^\theta + v^{\theta-1}) + \frac{ie\phi_e(\rho)\hbar}{\Delta t}(v^{\theta+1} - v^{\theta-1}) - e^2\phi_e^2(\rho)v^\theta = -[\mathsf{E} - e\phi_e(\rho)]^2 v^\theta \tag{17}$$

Multiplication with $v^{-\theta}(\Delta t/\hbar)^2$ brings

$$v - 2 + v^{-1} + \frac{ie\phi_e(\rho)\Delta t}{\hbar}(v - v^{-1}) - \left(\frac{e\phi_e(\rho)\Delta t}{\hbar}\right)^2 + [\mathsf{E} - e\phi_e(\rho)]^2 = 0 \tag{18}$$

and we obtain a quadratic equation for v:

$$\left(1 + \frac{ie\phi_e(\rho)\Delta t}{\hbar}\right)v^2 - \left[2 + \left(\frac{e\phi_e(\rho)\Delta t}{\hbar}\right)^2 - [\mathsf{E} - e\phi_e(\rho)]^2\left(\frac{\Delta t}{\hbar}\right)^2\right]v + \left(1 - \frac{ie\phi_e(\rho)\Delta t}{\hbar}\right) = 0 \tag{19}$$

The solutions $v_{1,2}$ can be greatly simplified if we rewrite the terms $e\phi_e(\rho)\Delta t/\hbar$ and $\mathsf{E}\Delta t/\hbar$:

$$\frac{e\phi_e(\rho)\Delta t}{\hbar} = \frac{e\phi_e(\rho)}{m_0c^2}\frac{\Delta t}{\hbar/m_0c^2} = \frac{e\phi_e(\rho)}{m_0c^2}\frac{\Delta r}{\hbar/m_0c} \tag{20}$$

$$\frac{\mathsf{E}\Delta t}{\hbar} = \frac{\mathsf{E}}{m_0c^2}\frac{\Delta r}{\hbar/m_0c} \tag{21}$$

In the differential theory it is usual to assume $e\phi_{\mathrm{e}}(\rho)/m_0c^2 \ll 1$ and $\mathsf{E}/m_0c^2 \ll 1$. We follow these assumptions but keep the factor $\Delta r/(\hbar/m_0c)$:

$$\frac{e\phi_{\mathrm{e}}(\rho)}{m_0c^2}\frac{\Delta r}{\hbar/m_0c} \ll 1, \quad \frac{\mathsf{E}}{m_0c^2}\frac{\Delta r}{\hbar/m_0c} \ll 1 \tag{22}$$

We do not assume anything about the magnitude of $\Delta r/(\hbar/m_0c)$. The calculation will demand certain minimum values for this term. This is in line with the results of Chapters 3 and 4 where $\Delta r/(\hbar/m_0c)$ is replaced by $\Delta x/(\hbar/m_0c)$. The coarse distinction between $\Delta x/(\hbar/m_0c) \gg 1$ and $\Delta x/(\hbar/m_0c) \ll 1$ will be replaced by a much finer one. A differential theory cannot have a minimum value for dr.

The simplifications of Eq.(22) reduce the solution of Eq.(19) to a manageable form:

$$v_{1,2} \doteq \frac{1}{2}\left(1 - \frac{ie\phi_{\mathrm{e}}(\rho)\Delta t}{\hbar}\right)\left(2 \pm 2i[\mathsf{E} - e\phi_{\mathrm{e}}(\rho)]\frac{\Delta t}{\hbar}\right)$$

$$v_1 \doteq 1 - \frac{i\mathsf{E}\Delta t}{\hbar} \doteq e^{-i\mathsf{E}\Delta t/\hbar} \tag{23}$$

$$v_2 \doteq 1 + \frac{i\mathsf{E}\Delta t}{\hbar} - \frac{2ie\phi_{\mathrm{e}}(\rho)\Delta t}{\hbar} \tag{24}$$

The solution v_2 contains the variable ρ and thus contradicts the assumption made for Eq.(15), but v_1 is a usable solution that yields the result known from the differential theory:

$$v_1^{\theta} = e^{-i\mathsf{E}\Delta t\theta/\hbar} = e^{-i\mathsf{E}t/\hbar} \tag{25}$$

We return to Eq.(14) and rewrite its left side with the help of Eqs.(15) and (16):

$$c^2(\hbar^2\tilde{\nabla}^2 - m_0^2c^2)u(\boldsymbol{\rho}) = -[\mathsf{E} - e\phi_{\mathrm{e}}(\rho)]^2 u(\boldsymbol{\rho}) \tag{26}$$

as an explicit difference equation:

$$\begin{aligned}
&[u(\rho+1,\eta,\xi) - 2u(\rho,\eta,\xi) + u(\rho-1,\eta,\xi)] \\
&\qquad\qquad + \frac{1}{\rho}[u(\rho+1,\eta,\xi) - u(\rho-1,\eta,\xi)] \\
&+ \frac{1}{\rho^2}[u(\rho,\eta+1,\xi) - 2u(\rho,\eta,\xi) + u(\rho,\eta-1,\xi)] \\
&\qquad\qquad + \frac{\cot(\eta\Delta\vartheta)}{2\rho^2}[u(\rho,\eta+1,\xi) - u(\rho,\eta-1,\xi)]
\end{aligned}$$

$$+\frac{1}{\rho^2\sin^2(\eta\Delta\vartheta)}[u(\rho,\eta,\xi+1)-2u(\rho,\eta,\xi)+u(\rho,\eta,\xi-1)]$$
$$+\left[\left(\frac{\mathsf{E}-e\phi_{\mathrm{e}}(\rho)}{m_0c^2}\right)^2-1\right]\left(\frac{\Delta r}{\hbar/m_0c}\right)^2 u(\rho,\eta,\xi)=0 \quad (27)$$

We separate Eq.(27) into a function $u_\nu(\rho,\eta)$ and a function $\varphi_\nu(\xi)$ by means of Bernoulli's product method

$$u(\rho,\eta,\xi)=u_\nu(\rho,\eta)\varphi_\nu(\xi) \quad (28)$$

and obtain:

$$\Big\{[u_\nu(\rho+1,\eta)-2u_\nu(\rho,\eta)+u_\nu(\rho-1,\eta)]+\frac{1}{\rho}[u_\nu(\rho+1,\eta)-u_\nu(\rho-1,\eta)]$$
$$+\frac{1}{\rho^2}[u_\nu(\rho,\eta+1)-2u_\nu(\rho,\eta)+u_\nu(\rho,\eta-1)]+\frac{\cot(\eta\Delta\vartheta)}{2\rho^2}[u_\nu(\rho,\eta+1)-u_\nu(\rho,\eta-1)]$$
$$+\left[\left(\frac{\mathsf{E}-e\phi_{\mathrm{e}}(\rho)}{m_0c^2}\right)^2-1\right]\left(\frac{\Delta r}{\hbar/m_0c}\right)^2 u_\nu(\rho,\eta)\Big\}\varphi_\nu(\xi)$$
$$=-\frac{1}{\rho^2\sin^2(\eta\Delta\vartheta)}[\varphi_\nu(\xi+1)-2\varphi_\nu(\xi)+\varphi_\nu(\xi-1)]u_\nu(\rho,\eta) \quad (29)$$

We multiply Eq.(29) with

$$\frac{\rho^2\sin^2(\eta\Delta\vartheta)}{u_\nu(\rho,\eta)\varphi_\nu(\xi)}$$

and obtain on the left a function of the variables ρ and η, while on the right is a function of ξ. Such an equation requires that both sides are equal to a constant that we denote $-(\nu\Delta\varphi)^2=-(2\pi\nu/N)^2$:

$$\frac{\rho^2\sin^2(\eta\Delta\vartheta)}{u_\nu(\rho,\eta)}\Big\{\text{left side of Eq.(29) without }\varphi_\nu(\xi)\Big\}$$
$$=-\frac{1}{\varphi_\nu(\xi)}[\varphi_\nu(\xi+1)-2\varphi_\nu(\xi)+\varphi_\nu(\xi-1)]=-(\nu\Delta\varphi)^2=-\left(\frac{2\pi\nu}{N}\right)^2 \quad (30)$$

Equation (30) yields a relation for $\varphi_\nu(\xi)$

$$\varphi_\nu(\xi+1)-[2+(\nu\Delta\varphi)^2]\varphi_\nu(\xi)+\varphi_\nu(\xi-1)=0 \quad (31)$$

with the solution

$$\begin{aligned}
&\varphi_\nu(\xi) = v_\nu^\xi \\
&v_\nu^2 - [2 + (\nu\Delta\varphi)^2]v_\nu + 1 = 0 \\
&v_\nu = 1 + \frac{1}{2}(\nu\Delta\varphi)^2 \pm i\nu\Delta\varphi\left(1 + \frac{1}{4}(\nu\Delta\varphi)^2\right)^{1/2} \doteq 1 \pm i\nu\Delta\varphi \doteq e^{\pm i\nu\Delta\varphi} \\
&\varphi_\nu(\xi) = A_1 e^{2\pi i\nu\xi/N} + A_2 e^{-2\pi i\nu\xi/N} = A_1 e^{i\nu\varphi} + A_2 e^{-i\nu\varphi},\ \Delta\varphi = 2\pi/N \quad (32)
\end{aligned}$$

Equation (30) without the terms φ_ν can be brought into the following form:

$$\begin{aligned}
&[u_\nu(\rho+1,\eta) - 2u_\nu(\rho,\eta) + u_\nu(\rho-1,\eta)] + \frac{1}{\rho}[u_\nu(\rho+1,\eta) - u_\nu(\rho-1,\eta)] \\
&\quad + \frac{1}{\rho^2}[u_\nu(\rho,\eta+1) - 2u_\nu(\rho,\eta) + u_\nu(\rho,\eta-1)] \\
&\quad\quad + \frac{\cot(\eta\Delta\vartheta)}{2\rho^2}[u_\nu(\rho,\eta+1) - u_\nu(\rho,\eta-1)] \\
&+ \left\{\left[\left(\frac{\mathsf{E} - \phi_e(\rho)}{m_0c^2}\right)^2 - 1\right]\left(\frac{\Delta r}{\hbar/m_0c}\right)^2 + \frac{(\nu\Delta\varphi)^2}{\rho^2\sin^2(\eta\Delta\vartheta}\right\}u_\nu(\rho,\eta) = 0 \quad (33)
\end{aligned}$$

We use once more Bernoulli's product method to separate the variables ρ and η

$$u_\nu(\rho,\eta) = u_\mu(\rho)\varphi_\mu(\eta) \quad (34)$$

and substitute $u_\mu(\rho)\varphi_\mu(\eta)$ into Eq.(33):

$$\begin{aligned}
&\left\{[u_\mu(\rho+1) - 2u_\mu(\rho) + u_\mu(\rho-1)] + \frac{1}{\rho}[u_\mu(\rho+1) - u_\mu(\rho-1)]\right. \\
&\quad \left. + \left[\left(\frac{\mathsf{E} - e\phi_e(\rho)}{m_0c^2}\right)^2 - 1\right]\left(\frac{\Delta r}{\hbar/m_0c}\right)^2 u_\mu(\rho)\right\}\varphi_\mu(\eta) \\
&= -\left(\frac{1}{\rho^2}[\varphi_\mu(\eta+1) - 2\varphi_\mu(\eta) + \varphi_\mu(\eta-1)]\right. \\
&\quad \left. + \frac{\cot(\eta\Delta\vartheta)}{2\rho^2}[\varphi_\mu(\eta+1) - \varphi_\mu(\eta-1)] + \frac{(\nu\Delta\varphi)^2}{\rho^2\sin^2(\eta\Delta\vartheta)}\varphi_\mu(\eta)\right)u_\mu(\rho,\theta) \quad (35)
\end{aligned}$$

We multiply Eq.(35) by $\rho^2/u_\mu(\rho)\varphi_\mu(\eta)$ and obtain on the left a function of ρ, and on the right a function of η. Again, both sides of the equation must equal a constant that we denote $-\lambda_\vartheta$:

$$\left\{[u_\mu(\rho+1)-2u_\mu(\rho)+u_\mu(\rho-1)]+\frac{1}{\rho}[u_\mu(\rho+1)-u_\mu(\rho-1)]\right.$$
$$\left.+\left[\left(\frac{\mathsf{E}-e\phi_{\mathrm{e}}(\rho)}{m_0c^2}\right)^2-1\right]\left(\frac{\Delta r}{\hbar/m_0c}\right)^2u_\mu(\rho)\right\}\frac{\rho^2}{u_\mu(\rho)}$$
$$=-\Big([\varphi_\mu(\eta+1)-2\varphi_\mu(\eta)-\varphi_\mu(\eta-1)]+\frac{1}{2}\cot(\eta\Delta\vartheta)[\varphi_\mu(\eta+1)-\varphi_\mu(\eta-1)]$$
$$+\frac{(\nu\Delta\varphi)^2}{\sin^2(\eta\Delta\vartheta)}\varphi_\mu(\eta)\Big)\frac{1}{\varphi_\mu(\eta)}=-\lambda_\vartheta \quad (36)$$

Equation (36) yields a relation for $\varphi_\mu(\eta)$:

$$[\varphi_\mu(\eta+1)-2\varphi_\mu(\eta)+\varphi_\mu(\eta-1)]+\frac{1}{2}\cot(\eta\Delta\vartheta)[\varphi_\mu(\eta+1)-\varphi_\mu(\eta-1)]$$
$$+\left(\frac{(\nu\Delta\varphi)^2}{\sin^2(\eta\Delta\vartheta)}-\lambda_\vartheta\right)\varphi_\mu(\eta)=0 \quad (37)$$

This is the difference equation of the discrete spherical harmonic functions. Its solution is much more complicated than that of Eq.(31) for $\varphi_\nu(\xi)$. In Sections 6.6 and 6.7 we derive the eigenvalues

$$\lambda_\vartheta=-l(l+1),\quad l=0,\ 1,\ 2,\ \ldots \quad (38)$$

which are equal to those obtained in the differential theory.

Equation (36) with λ_ϑ from Eq.(38) yields a difference equation for $u_\mu(\rho)$ that equals the one of the differential theory except for the replacement of differentials by finite differences:

$$[u_\mu(\rho+1)-2u_\mu(\rho)+u_\mu(\rho-1)+\frac{1}{\rho}[u_\mu(\rho+1)-u_\mu(\rho-1)]-\frac{l(l+1)}{\rho^2}u_\mu(\rho)$$
$$=-\frac{[\mathsf{E}-e\phi_{\mathrm{e}}(\rho)]^2-m_0^2c^4}{\hbar^2c^2}(\Delta r)^2u_\mu(\rho) \quad (39)$$

A boundary condition for this difference equation follows from Eqs.(15) and (25) for $\rho=0$. There is no need for a boundary condition for $\varphi_\nu(\xi)$ of Eq.(32) or $\varphi_\mu(\eta)$ of Eq.(37):

$$\Psi_{\mathrm{r}0}(\rho,\theta)=\Psi_{00}u_\mu(\rho)e^{-i\mathsf{E}\Delta t\theta/\hbar} \quad (40)$$

For a causal solution that is zero before $\theta=0$ we have two possibilities:

$$\Psi_{r0}(0,\theta) = \Psi_{00}S(\theta)\cos\frac{\mathsf{E}\Delta t\theta}{\hbar} = 0 \qquad \text{for } \theta < 0$$
$$= \Psi_{00}\cos\frac{\mathsf{E}\Delta t\theta}{\hbar} \qquad \text{for } \theta \geq 0 \tag{41}$$

$$\Psi_{r0}(0,\theta) = \Psi_{00}S(\theta)\sin\frac{\mathsf{E}\Delta t\theta}{\hbar} = 0 \qquad \text{for } \theta < 0$$
$$= \Psi_{00}\sin\frac{\mathsf{E}\Delta t\theta}{\hbar} \qquad \text{for } \theta \geq 0 \tag{42}$$

The boundary condition of Eq.(41) starts at $\theta = 0$ like a step function while Eq.(42) yields a linear increase. More complicated boundary conditions can be represented by making $\mathsf{E} = \mathsf{E}_n$ a function of $n = 0,\ 1,\ 2,\ \ldots$ and using a Fourier series expansion.

5.5 Spatial Difference Equation

In the differential theory the equation corresponding to Eq.(5.4-39) can be written in the following form (Schiff 1949, p. 309; Messiah 1962, p. 884) if $e\phi_e(\rho)$ is replaced by $-\mathsf{Z}e^2/\rho$ and $\hat{\rho}$ is written instead of ρ. We have already defined $\rho = r/\Delta r$ in Eq.(5.4-4) and Δr has no place in a differential theory. Hence, some other normalization factor has to be used and this creates a different normalized variable $\hat{\rho}$:

$$\left(\frac{d^2}{d\hat{\rho}^2} + \frac{2}{\hat{\rho}}\frac{d}{d\hat{\rho}} + \frac{\hat{\lambda}}{\hat{\rho}} - \frac{1}{4} - \frac{l(l+1) - \hat{\gamma}^2}{\hat{\rho}^2}\right)u(\hat{\rho}) = 0$$
$$\hat{\rho} = \hat{\alpha}r,\ \gamma = \frac{\mathsf{Z}e^2}{\hbar c},\ \hat{\alpha} = 2\frac{m_0 c}{\hbar}\left(1 - \frac{\mathsf{E}^2}{m_0^2c^4}\right)^{1/2},\ \hat{\lambda} = \frac{2\mathsf{E}\hat{\gamma}}{\hbar c\hat{\alpha}},\ \mathsf{Z} = 1,\ 2,\ \ldots \tag{1}$$

The difference equation (5.4-39) can be brought into a similar form. Note that the hat ˆ is replaced by a tilde ˜:

$$[u_\mu(\rho+1) - 2u_\mu(\rho) + u_\mu(\rho-1)] + \frac{1}{\rho}[u_\mu(\rho+1) - u_\mu(\rho-1)]$$
$$+ \left(\frac{\tilde{\lambda}\tilde{\delta}}{\rho} - \frac{\tilde{\delta}^2}{4} - \frac{l(l+1) - \tilde{\gamma}^2}{\rho^2}\right)u_\mu(\rho) = 0$$

$$\Delta\tilde{r} = \tilde{\alpha}\Delta r,\quad \tilde{\delta} = \frac{\Delta\tilde{r}}{\hbar/m_0c} = \tilde{\alpha}\frac{\Delta r}{\hbar/m_0c},\quad \tilde{\gamma} = \mathsf{Z}\frac{Ze^2}{\hbar} = 4\pi\mathsf{Z}\alpha$$
$$\tilde{\alpha} = 2\left(1 - \frac{\mathsf{E}^2}{m_0^2c^4}\right)^{1/2},\quad \tilde{\lambda} = \frac{2\mathsf{E}\tilde{\gamma}}{m_0c^2\tilde{\alpha}},\quad \tilde{\lambda}\tilde{\delta} = \frac{2\Delta r\mathsf{E}\tilde{\gamma}}{\hbar c},\quad \frac{\Delta r}{\hbar/m_0c} = \frac{1}{\tilde{\alpha}}\frac{\Delta\tilde{r}}{\hbar/m_0c}$$
$$\mathsf{Z} = 1,\ 2,\ \ldots;\ Z \doteq 376.730\,[\mathrm{V/A}];\ \alpha = \frac{Ze^2}{2h}\ \text{fine structure constant} \tag{2}$$

The terms $\tilde{\delta} = \Delta\tilde{r}/(\hbar/m_0c) = \tilde{\alpha}\Delta r/(\hbar/m_0c)$, which express the length $\Delta\tilde{r} = \tilde{\alpha}\Delta r$ in relation to the Compton wavelength, are an important feature of Eq.(2). We cannot choose $\Delta\tilde{r} = \hbar/m_0c$ since we have seen in Sections 3 and 4 that $\Delta x \gg \hbar/m_0c$ and $\Delta x \ll \hbar/m_0c$ lead to completely different results. The remaining difference between the constants in Eqs.(1) and (2) is due to the use of the international system of units (SI) made conspicuous by the wave impedance Z in Eq.(2).

Equation (2) is multiplied by ρ^2 and the terms are rearranged to match the requirements of the method of solution used:

$$\rho(\rho+1)u_\mu(\rho+1)+[g\rho^2+\tilde{\lambda}\tilde{\delta}\rho-l(l+1)+\tilde{\gamma}^2]u_\mu(\rho)+\rho(\rho-1)u_\mu(\rho-1)=0$$

$$g = -\left(2+\frac{\tilde{\delta}^2}{4}\right) = -\left[2+\frac{1}{4}\left(\frac{\Delta\tilde{r}}{\hbar/m_0c}\right)^2\right] \tag{3}$$

To obtain eigenfunctions of Eq.(3) we use a method based on the Laplace transform (Nörlund 1910, 1915, 1924, 1929; Milne-Thomson 1951; Guldberg and Wallenberg 1911). It is not necessary to use this detour via differential calculus. The same solutions can be obtained by purely algebraic means, but the Laplace transform shortens the calculation.

We start with the contour integral

$$u_\mu(\rho) = \frac{1}{2\pi i}\int_\ell s^{\rho-1}w(s)ds \tag{4}$$

and take it in the complex plane from the point ℓ_1 along a contour ℓ to the point ℓ_2. Partial integration yields:

$$u_\mu(\rho) = \frac{1}{2\pi i}\left(\frac{1}{\rho}s^\rho w\Big|_{\ell_1}^{\ell_2} - \frac{1}{\rho}\int_\ell s^\rho w'(s)ds\right) \tag{5}$$

Let the points ℓ_1 and ℓ_2 be chosen so that the term

$$s^\rho w(s)\Big|_{\ell_1}^{\ell_2} \tag{6}$$

vanishes. One obtains:

$$-\rho u_\mu(\rho) = \frac{1}{2\pi i}\int_\ell s^\rho w'(s)ds \tag{7}$$

Further partial integration yields

$$\rho(\rho+1)u_\mu(\rho) = \frac{1}{2\pi i}\int_\ell s^{\rho+1}w''(s)ds \tag{8}$$

provided the expression

$$s^{\rho+1}w'(s)\Big|_{\ell_1}^{\ell_2} \tag{9}$$

vanishes.

If one makes the corresponding assumptions about the beginning and end of the contour of integration one obtains the following relations for $u_\mu(\rho+1)$ and $u_\mu(\rho-1)$:

$$(-1)^k(\rho+1)\dots(\rho+k)u_\mu(\rho+1) = \frac{1}{2\pi i}\int_\ell s^{\rho+k}w^{(k)}(s)ds$$
$$(-1)^k(\rho-1)\dots(\rho+k-2)u_\mu(\rho-1) = \frac{1}{2\pi i}\int_\ell s^{\rho+k-2}w^{(k)}(s)ds \tag{10}$$

We rewrite Eq.(3) so that the coefficients become multiples of the terms on the left side of Eqs.(7), (8), and (10):

$$\begin{aligned}&[(\rho+2)(\rho+1)-2(\rho+1)]u_\mu(\rho+1)\\ &\quad+[g\rho(\rho+1)+(-g+\tilde{\lambda}\tilde{\delta})\rho-l(l+1)+\tilde{\gamma}^2]u_\mu(\rho)\\ &\qquad+\rho(\rho-1)u_\mu(\rho-1)=0\end{aligned} \tag{11}$$

Substitution of the integral transforms for $u_\mu(\rho)$, $u_\mu(\rho+1)$, and $u_\mu(\rho-1)$ yields the following differential equation:

$$(s^2+gs+1)sw''(s)+(2s+g-\tilde{\lambda}\tilde{\delta})sw'(s)-[l(l+1)-\tilde{\gamma}^2]w(s)=0 \tag{12}$$

This is a differential equation with variable coefficients of the Fuchs-type (Fuchs 1866; Frobenius 1873). We follow the standard method for its solution. Equation (12) has a singular point at $s=\infty$. The other singular points are located at the roots of the equation

$$(s^2+gs+1)s=0 \tag{13}$$

We obtain three solutions for s:

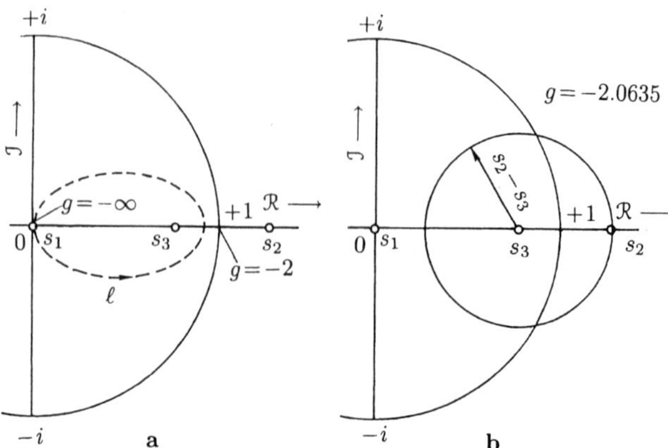

FIG.5.5-1. Location of the singular points $s_2 = s_2(g)$ and $s_3 = s_3(g)$ in the complex plane according to Eq.(14) for $-2 \geq g > -\infty$. (a) shows the integration path from 0 around s_3. (b) shows a circle of convergence around s_3 for $g = -2.0635$ that is limited by s_3 and does not reach 0.

$$s_1 = 0, \quad s_2 = -\frac{g}{2} + \left(\frac{g^2}{4} - 1\right)^{1/2}, \quad s_3 = -\frac{g}{2} - \left(\frac{g^2}{4} - 1\right)^{1/2} = \frac{1}{s_2}$$

$$s_2 + s_3 = -g, \quad s_2 - s_3 = 2\left(\frac{g^2}{4} - 1\right)^{1/2} = \tilde{\delta}\left(1 + \frac{\tilde{\delta}^2}{16}\right)^{1/2}$$

$$s_2 - s_3 = \frac{\Delta\tilde{r}}{\hbar/m_0c}\left[1 + \frac{1}{16}\left(\frac{\Delta\tilde{r}}{\hbar/m_0c}\right)^2\right]^{1/2} \doteq \frac{\Delta\tilde{r}}{\hbar/m_0c} = \frac{1}{s_3} - s_3, \quad \Delta\tilde{r} = \tilde{\alpha}\Delta r$$

$$s_3^2 + \frac{\Delta\tilde{r}}{\hbar/m_0c}s_3 - 1 = s_3^2 + \tilde{\alpha}\frac{\Delta r}{\hbar/m_0c}s_3 - 1 = 0$$

$$s_3 = -\frac{\tilde{\alpha}}{2}\frac{\Delta r}{\hbar/m_0c} \pm \left[1 + \frac{\tilde{\alpha}^2}{4}\left(\frac{\Delta r}{\hbar/m_0c}\right)^2\right]^{1/2}$$

$$\doteq 1 - \frac{\tilde{\alpha}}{2}\frac{\Delta r}{\hbar/m_0c} = 1 - \frac{\Delta r}{\hbar/m_0c}\left(1 - \frac{\mathsf{E}^2}{m_0^2c^4}\right)^{1/2}, \quad s_3 > 0 \tag{14}$$

The parameter g is always real according to Eq.(3). The loci of s_2 and s_3 for real values of g are shown in Fig.5.5-1. The value of g is slightly smaller than -2 for small values of $\tilde{\delta}$ or $\Delta\tilde{r}/(\hbar/m_0c) \ll 1$ but it is significantly smaller for large values of $\tilde{\delta}$ or $\Delta\tilde{r}/(\hbar/m_0c) \gg 1$. In either case the root $s_3 = s_3(g)$ is located on the real axis between 0 and +1. The locus of s_2 readily follows from the relation $s_2s_3 = 1$ in Eq.(14).

A formal solution of Eq.(12) is obtained by a power series expansion of $w(s)$ in either the point s_3 or s_2. Both solutions are required for a general solution. We discuss the solution in the point s_3. The power series of $w(s)$ converges for small values of $\Delta\tilde{r}$ inside a circle that passes through the point s_2 as shown by Fig.5.5-1b:

$$w(s) = w_3(s) = \sum_{\nu=0}^{N} q_3(\nu)(s - s_3)^{p_3+\nu} \tag{15}$$

The subscript 3 of $w_3(s)$, $q_3(\nu)$, and p_3 indicates that these terms apply to the solution in the point s_3.

The solution of Eq.(12) by the series of Eq.(15) is derived in Section 6.8. Here we state only some results. There is a regular solution with $p_3 = 0$. A non-regular solution with $p_3 = p$

$$\begin{aligned}
p_3 = p &= \frac{\tilde{\lambda}\tilde{\delta}}{s_3 - s_2} = \frac{\tilde{\lambda}}{s_3 - s_2}\frac{\Delta\tilde{r}}{\hbar/m_0c} = -\frac{2\mathsf{E}\tilde{\gamma}}{m_0c^2\tilde{\alpha}}\frac{1}{\tilde{\delta}(1+\tilde{\delta}^2/16)^{1/2}}\frac{\Delta\tilde{r}}{\hbar/m_0c} \\
&= -2\pi Z\alpha\frac{\mathsf{E}}{m_0c^2}\left[\left(\frac{m_0^2c^4}{m_0^2c^4-\mathsf{E}^2}\right)^2 + \left(\frac{\Delta r}{\hbar/m_0c}\right)^2\right]^{-1/2} \\
&= -2\pi Z\alpha\frac{\mathsf{E}}{m_0c^2}\left(1 - \frac{\mathsf{E}^2}{m_0^2c^4}\right)\left[1 + \left(1 - \frac{\mathsf{E}^2}{m_0^2c^4}\right)^2\left(\frac{\Delta r}{\hbar/m_0c}\right)^2\right]^{-1/2} \\
&= p_0 + O(\Delta r) \\
p_0 &= -2\pi Z\alpha\frac{\mathsf{E}}{m_0c^2}\left(1 - \frac{\mathsf{E}^2}{m_0^2c^4}\right)
\end{aligned} \tag{16}$$

exists in the neighborhood of s_3. From the results of Chapters 3 and 4 we know that Δr cannot be made arbitrarily small without risk of obtaining a meaning less result. However, the notation $O(\Delta r)$ separates only terms multiplied by Δr or powers of Δr, it is not assumed that these terms must be small compared with $|p_0|$. A recursion formula with three terms is obtained for the coefficients $q_3(\nu)$ by substituting Eq.(15) into Eq.(12):

$$\begin{aligned}
&\alpha_{3,1}(\nu)q_3(\nu+1) + \alpha_{3,0}(\nu)q_3(\nu) + \alpha_{3,-1}(\nu)q_3(\nu-1) = 0 \\
\alpha_{3,1}(\nu) &= s_3(s_3 - s_2)(p+\nu+1)(\nu+1) \\
\alpha_{3,0}(\nu) &= (2s_3 - s_2)(p+\nu)(p+\nu-1) + [3s_3 - s_2 - (s_3 - s_2)p](p+\nu) \\
&\qquad - l(l+1) + \tilde{\gamma}^2 \\
\alpha_{3,-1}(\nu) &= (p+\nu)(p+\nu-1) \\
&\qquad s_2 \neq s_3
\end{aligned} \tag{17}$$

For $\nu = 0$ we obtain a formula with two terms for $q_3(1)$:

$$\alpha_{3,1}(0)q_3(1) + \alpha_{3,0}(0)q_3(0) = 0$$

$$\alpha_{3,1}(0) = s_3(s_3 - s_2)(p+1), \quad \alpha_{3,0}(0) = s_3p(p+1) - l(l+1) + \tilde{\gamma}^2 \tag{18}$$

One may choose $q_3(0)$ but not $q_3(1)$. Since a second-order differential equation should have two choosable constants one would need the second solution in s_2 to produce a general solution.

Let the line of integration of the contour integral of Eq.(4) begin at the origin, run around s_3, and return to the origin as shown by the line ℓ in Fig.5.5-1a. The point s_2 remains outside the loop. More details about this integration may be found in the works of Guldberg and Wallenberg (1911), Nörlund (1910, 1915, 1924, 1929), and Milne-Thomson (1951, p. 485). One obtains a factorial series:

$$u_{\mu 3}(\rho) = s_3^\rho \frac{\Gamma(\rho)}{\Gamma(\rho+p+1)} \sum_{\nu=0}^{N} (-1)^\nu q_3(\nu) \frac{(p+1)\dots(p+\nu)}{(\rho+p+1)\dots(\rho+p+\nu)}$$

$$\frac{q_3(1)}{q_3(0)} = -\frac{\alpha_{3,0}(0)}{\alpha_{3,1}(0)} \tag{19}$$

The fraction $(p+1)\dots(p+\nu)/(\rho+p+1)\dots(\rho+p+\nu)$ shall always be replaced by 1 for $\nu = 0$. This convention avoids the need for writing a special term for $\nu = 0$. The upper limit N of the sum is usually replaced by ∞ since the calculus of finite differences eliminates only the infinitesimal but not the infinite. According to Eq.(5.4-4) we have N intervals $\Delta r = cT/N$ in the range $0 \le r \le cT$ or $\Delta\rho = 1$ in the range $0 \le \rho \le N$, which implies $N+1$ points 0, 1, 2, ... , N. Hence, the series in Eq.(19) can have only $N+1$ orthogonal or linearly independent functions and the upper limit $\nu = N$ must be used. This looks like the concept of divergence is eliminated, but there is no such luck.

We call the sum of Eq.(19) convergent if it approaches arbitrarily closely a limit for $N \gg 1$; it is called divergent if such a limit is not approached. This replaces the usual requirement that a limit is approached for $N \to \infty$.

In the case of Fig.5.5-1b the power series does not converge everywhere along the line ℓ of integration in Fig.5.5-1a. As a result the factorial series of Eq.(19) may diverge for all values of ρ. There are two ways to obtain a convergent series. First, one may decrease the value $g = -2(2+\tilde{\delta}^2/4) = -2.0635$ used in Fig.5.5-1b to move s_3 to the left and s_2 to the right so that the circle of convergence goes through $s_1 = 0$ rather than through s_2 as shown in Fig.5.5-2b. The distinguished value $g = -3/\sqrt{2}$ of Fig.5.5-2a is obtained. For

$$g = -\left(2 + \frac{\tilde{\delta}^2}{4}\right) = -\left[2 + \frac{1}{4}\left(\frac{\Delta\tilde{r}}{\hbar/m_0c}\right)^2\right] < -\frac{3}{\sqrt{2}}$$

$$\frac{\Delta\tilde{r}}{\hbar/m_0c} > 2\left(\frac{3}{\sqrt{2}} - 2\right)^{1/2} \doteq 0.69662 \tag{20}$$

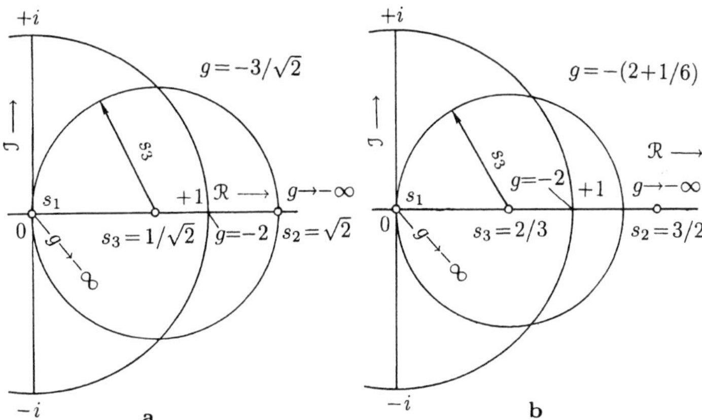

FIG.5.5-2. The convergence circle of Fig.5.5-1b around s_3 reaches both 0 and s_2 for $g = -3/\sqrt{2}$ (a). For smaller values of g the singular point s_2 becomes unimportant since the circle of convergence is limited by $s_1 = 0$ (b).

we have convergence of the power series along the integration line ℓ of Fig.5.5-1a. In essence, as long as the spatial resolution $\Delta\tilde{r}$ is larger than the Compton wavelength of the particle we have convergence.

For smaller values of $\Delta\tilde{r}$ than defined by Eq.(20) we may use a second way that requires conformal mapping to achieve convergence. The circle of convergence of the power series of Eq.(15) according to Fig.5.5-1b is mapped onto a loop that runs through the origin. Except for $s = s_3$ and $s = 0$ there must be no singularity either inside or on the loop.

The factorial series obtained in either one of these two ways do not necessarily converge in the whole interval $0 \leq \rho < N$ unless certain conditions are satisfied.

For a convergent power series we must have the following relations according to Eq.(20) and Fig.5.5-2a:

$$\tilde{\delta} = \frac{\Delta\tilde{r}}{\hbar/m_0c} = \frac{\tilde{\alpha}\Delta r}{\hbar/m_0c} = 2\Delta r\frac{m_0c}{\hbar}\left(1 - \frac{\mathsf{E}^2}{m_0^2c^4}\right)^{1/2} > 0.69662$$

$$\frac{\Delta r}{\hbar/m_0c} > \frac{0.69962}{2} = 0.34981 \quad \text{for } \mathsf{E} \ll m_0c^2 \tag{21}$$

We thus get in Eq.(16) for p the values $\alpha \doteq 7.297 \times 10^{-3}$, $\mathsf{E}/m_0c^2 \ll 1$, $[m_0^2c^4/(m_0^2c^4 - \mathsf{E}^2)]^2 \doteq 1$, and $[\Delta r/(\hbar/m_0c)]^2 > 0.122$. Hence, p is a negative number with small magnitude, $|p| \ll 1$.

We return to Eq.(19). If p could be a negative integer the factorial series would terminate. According to Eq.(16) p is negative but small compared with 1. A second possibility is that one can choose the coefficients $q_3(\nu)$ so that

$$\kappa = \frac{m_0 c}{\hbar}\left(1 - \frac{\mathsf{E}^2}{m_0^2 c^4}\right)^{1/2} = \frac{\tilde{\alpha}}{2}\frac{m_0 c}{\hbar}, \quad \mathsf{E} = m_0 c^2 \left[1 - \left(\frac{\kappa\hbar}{m_0 c}\right)^2\right]^{1/2}$$

$$s_3 = 1 - \frac{\tilde{\alpha}}{2}\frac{m_0 c}{\hbar}\Delta r = 1 - \kappa\Delta r, \; s_3 \doteq e^{-\kappa\Delta r} \tag{30}$$

and rewrite Eq.(16):

$$p = p_0 + O(\Delta r), \qquad p_0 = \tilde{\gamma}\frac{m_0 c}{\kappa\hbar}\left[1 - \left(\frac{\kappa\hbar}{m_0 c}\right)^2\right]^{1/2} \tag{31}$$

Substitution into Eq.(28) yields with the help of Eq.(23):

$$\kappa = \frac{m_0 c}{\hbar}\left(1 + \frac{p_0^2}{\tilde{\gamma}^2}\right)^{-1/2}, \quad \tilde{\gamma} = 4\pi Z\alpha, \; \alpha = \frac{Ze^2}{2h}$$

$$\kappa_1 \doteq \frac{\tilde{\gamma} m_0 c}{n\hbar}\left[1 + \frac{\tilde{\gamma}^2}{n^2}\left(\frac{n}{2l+1} - \frac{1}{2}\right)\right]$$

$$\kappa_2 \doteq \frac{\tilde{\gamma} m_0 c}{n'\hbar}\left[1 - \frac{\tilde{\gamma}^2}{n'^2}\left(\frac{n'}{2l+1} + \frac{1}{2}\right)\right]$$

$$\mathsf{E} = m_0 c^2\left(1 - \frac{\tilde{\gamma}^2}{p_0^2 + \tilde{\gamma}^2}\right)^{-1/2}$$

$$\mathsf{E}_1 \doteq m_0 c^2\left[1 - \frac{\tilde{\gamma}^2}{2n^2} - \frac{\tilde{\gamma}^4}{n^4}\left(\frac{n}{2l+1} - \frac{3}{8}\right)\right]$$

$$\mathsf{E}_2 \doteq m_0 c^2\left[1 - \frac{\tilde{\gamma}^2}{2n'^2} + \frac{\tilde{\gamma}^4}{n'^4}\left(\frac{n'}{2l+1} + \frac{3}{8}\right)\right]$$

$$n' = n - 2l - 1 \tag{32}$$

The values shown in Eq.(32) for κ and E are the same as obtained from the differential Klein-Gordon equation (1). The positive sign of the square root in Eq.(28) and thus the values κ_2 and E_2 are usually not considered in order to avoid poles in the eigenfunctions of the differential equation.

The recursion formula of Eq.(17) yields the following coefficients $q_3(\nu)$ for the terms not containing Δr

$$\frac{q_3(1)}{q_3(0)} = \frac{1}{\kappa\Delta r}\left(\frac{p_0(p_0+1) - l(l+1) + \tilde{\gamma}^2}{2(p_0+1)} + O(\Delta r)\right)$$

$$\frac{q_3(2)}{q_3(0)} = \frac{1}{(\kappa\Delta r)^2}\left(\frac{(p_0+1)(p_0+2)-l(l+1)+\tilde{\gamma}^2}{4(p_0+2)}\right.$$
$$\left.\times\frac{p_0(p_0+1)-l(l+1`)+\tilde{\gamma}^2}{2(p_0+1)} + O(\Delta r)\right) \quad (33)$$

etc. From Stirling's formula for the Gamma function one obtains the relation (Milne-Thomson 1951, pp. 254, 255)

$$\frac{\Gamma(\rho+x)}{\Gamma(\rho+y)} = \rho^{x-y}, \ \rho \gg 1$$

which yields

$$\lim_{\rho\gg 1}\frac{\Gamma(\rho-n+l+1)}{\Gamma(\rho+p_0+1)} = \rho^{-(p_0+n-l)}$$
$$\Gamma(\rho) = (\rho-1)\ldots(\rho-n+l+1)\Gamma(\rho-n+l+1) \quad (34)$$

We get from Eq.(28):

$$-(p_0+n-l) = -\frac{1}{2} \mp \left[\left(l+\frac{1}{2}\right)^2 - \tilde{\gamma}^2\right]^{1/2} = u \quad (35)$$

Equation (34) yields for $\rho \gg 1$:

$$(\rho-1)(\rho-2)\ldots(\rho-n+l+1) = \rho^{n-l-1} + O(\rho^{n-l-2}) \quad (36)$$

The factorial series of Eq.(19) becomes for $\rho \gg 1$:

$$u_{\mu 3}(\rho) = q_3(0)s_3^\rho \rho^u \rho^{n-l-1}\left(1 - \frac{q_3(1)}{q_3(0)}\frac{p_0+1}{\rho} + \ldots\right)$$
$$= q_3(0)e^{-\kappa\Delta r\rho}\rho^{u+n-l-1}\left(1 - \frac{p_0(p_0+1)-l(l+1)+\tilde{\gamma}^2}{2\kappa\Delta r\rho} + \ldots\right)$$
$$= q_{30}e^{-\kappa\Delta r\rho}(\kappa\Delta r\rho)^{u+n-l-1}\left(1 - \frac{p_0(p_0+1)-l(l+1)+\tilde{\gamma}^2}{2\kappa\Delta r\rho} + \ldots\right)$$
$$\kappa\Delta r = 2\pi\frac{\Delta r}{h/m_0c}\left(1-\frac{p_0^2}{\tilde{\gamma}^2}\right)^{-1/2}, \qquad \rho = r/\Delta r \gg 1 \quad (37)$$

Equation (37) contains a polynomial with $n-l$ terms of order $O(1)$ and a series with arbitrarily many terms of order $O(\Delta r)$, which have been left out. The polynomial is the same as that derived from the differential Klein-Gordon equation (1). It may be used as an approximation for the factorial series of Eq.(19), provided the condition $\Delta\tilde{r}/(\hbar/m_0c) > 0.69662$ of Eq.(20) is satisfied.

5.6 Quantization of the Solution $\Psi_{r0}(\boldsymbol{\rho}, \theta)$

We follow Section 2.5 and calculate the energy U of $\Psi_{r0}(\boldsymbol{\rho}, \theta)$ from the energy-impulse tensor of Eq.(2.5-1), keeping in mind that the operator ∇ is required for spherical coordinates:

$$T_{00} = \frac{1}{c^2}\frac{\partial \Psi^*_{r0}}{\partial t}\frac{\partial \Psi_{r0}}{\partial t} + \nabla\Psi^*_{r0}\cdot\nabla\Psi_{r0} + \frac{m_0^2c^2}{\hbar^2}\Psi^*_{r0}\Psi_{r0} \quad (1)$$

$$\nabla\Psi_{r0} = \operatorname{grad}\Psi_{r0} = \frac{\partial\Psi_{r0}}{\partial r}\mathbf{e}_r + \frac{1}{r}\frac{\partial\Psi_{r0}}{\partial\vartheta}\mathbf{e}_\vartheta + \frac{1}{r\sin\vartheta}\frac{\partial\Psi_{r0}}{\partial\varphi}\mathbf{e}_\varphi \quad (2)$$

We restrict the calculation to the simplest case where Ψ_{r0} is independent of ϑ and φ. The energy in a sphere with radius cT becomes:

$$U = \int_0^{2\pi}\Bigg\{\int_0^{\pi}\Bigg[\int_0^{cT}\Bigg(\frac{1}{c^2}\frac{\partial\Psi^*_{r0}(r,t)}{\partial t}\frac{\partial\Psi_{r0}(r,t)}{\partial t} + \frac{\partial\Psi^*_{r0}(r,t)}{\partial r}\frac{\partial\Psi_{r0}(r,t)}{\partial r}$$
$$+ \frac{m_0^2c^2}{\hbar^2}\Psi^*_{r0}(r,t)\Psi_{r0}(r,t)\Bigg)r^2dr\Bigg]\sin\vartheta\, d\vartheta\Bigg\}d\varphi \quad (3)$$

The dimension of U is VAs. Equation (3) is rewritten with the normalized variables $\theta = t/\Delta t$ and $\rho = r/\Delta r = r/c\Delta t$ of Eq.(5.4-4):

$$U = 2\pi c\Delta t\int_0^N\Bigg(\frac{\partial\Psi^*_{r0}(\rho,\theta)}{\partial\theta}\frac{\partial\Psi_{r0}(\rho,\theta)}{\partial\theta} + \frac{\partial\Psi^*_{r0}(\rho,\theta)}{\partial\rho}\frac{\partial\Psi_{r0}(\rho,\theta)}{\partial\rho}$$
$$+ \frac{m_0^2c^4(\Delta t)^2}{\hbar^2}\Psi^*_{r0}(\rho,\theta)\Psi_{r0}(\rho,\theta)\Bigg)\rho^2d\rho \quad (4)$$

The function Ψ_{r0} is represented by Eqs.(5.4-15) and (5.4-25) if $\boldsymbol{\rho}$ is replaced by ρ:

$$\Psi_{r0}(\rho,\theta) = \Psi_{00}u_\mu(\rho)e^{-i\mathsf{E}\Delta t\theta/\hbar} \quad (5)$$

For $u_\mu(\rho)$ we found in Eq.(5.5-19) a particular solution $u_{\mu 3}(\rho)$ for the point s_3 in Fig.5.5-2:

$$u_{\mu 3}(\rho) = s_3^\rho\frac{\Gamma(\rho)}{\Gamma(\rho+p+1)}\sum_{\nu=0}^{N}(-1)^\nu q_3(\nu)\frac{(p+1)\dots(p+\nu)}{(\rho+p+1)\dots(\rho+p+\nu)} \quad (6)$$

Here s_3 is defined by Eq.(5.5-22), $p = p_3$ by Eq.(5.5-16), and $q_3(\nu)$ by Eqs.(5.5-17) and (5.5-23). The fraction $(p+1)\dots(p+\nu)/(\rho+p+1)\dots(\rho+p+\nu)$ shall always be replaced by 1 for $\nu = 0$ to prevent having to write a special term. We obtain:

$$\Psi_{\mathrm{r0}}^*(\rho,\theta)\Psi_{\mathrm{r0}}(\rho,\theta) = \Psi_{00}^2 u_\mu^2(\rho)$$
$$= \Psi_{00}^2 s_3^{2\rho}\frac{\Gamma^2(\rho)}{\Gamma^2(\rho+p+1)}\left(\sum_{\nu=0}^{N}(-1)^\nu q_3(\nu)\frac{(p+1)\dots(p+\nu)}{(\rho+p+1)\dots(\rho+p+\nu)}\right)^2 \quad (7)$$

The differentiation of Eq.(5) with respect to θ or ρ produces the following result:

$$\frac{\partial\Psi_{\mathrm{r0}}(\rho,\theta)}{\partial\theta} = -i\Psi_{00}\frac{\mathrm{E}\Delta t}{\hbar}e^{-i\mathrm{E}\Delta t\theta/\hbar}$$
$$\times s_3^\rho\frac{\Gamma(\rho)}{\Gamma(\rho+p+1)}\sum_{\nu=0}^{N}(-1)^\nu q_3(\nu)\frac{(p+1)\dots(p+\nu)}{(\rho+p+1)\dots(\rho+p+\nu)} \quad (8)$$

$$\frac{\partial\Psi_{\mathrm{r0}}(\rho,\theta)}{\partial\rho} = \Psi_{00}e^{-i\mathrm{E}\Delta t\theta/\hbar}$$
$$\times\frac{\partial}{\partial\rho}\left(s_3^\rho\frac{\Gamma(\rho)}{\Gamma(\rho+p+1)}\sum_{\nu=0}^{N}(-1)^\nu q_3(\nu)\frac{(p+1)\dots(p+\nu)}{(\rho+p+1)\dots(\rho+p+\nu)}\right) \quad (9)$$

The first term in Eq.(4) becomes:

$$\frac{\partial\Psi_{\mathrm{r0}}^*}{\partial\theta}\frac{\partial\Psi_{\mathrm{r0}}}{\partial\theta} = \Psi_{00}^2\left(\frac{\mathrm{E}\Delta t}{\hbar}\right)^2$$
$$\times s_3^{2\rho}\frac{\Gamma^2(\rho)}{\Gamma^2(\rho+p+1)}\left(\sum_{\nu=0}^{N}(-1)^\nu q_3(\nu)\frac{(p+1)\dots(p+\nu)}{(\rho+p+1)\dots(\rho+p+\nu)}\right)^2 \quad (10)$$

For the second term in Eq.(4) we get:

$$\frac{\partial\Psi_{\mathrm{r0}}^*}{\partial\rho}\frac{\partial\Psi_{\mathrm{r0}}}{\partial\rho} = \Psi_{00}^2$$
$$\times\left[\frac{\partial}{\partial\rho}\left(s_3^\rho\frac{\Gamma(\rho)}{\Gamma(\rho+p+1)}\sum_{\nu=0}^{N}(-1)^\nu q_3(\nu)\frac{(p+1)\dots(p+\nu)}{(\rho+p+1)\dots(\rho+p+\nu)}\right)\right]^2 \quad (11)$$

Equations (10), (11), and (7) must be substituted into Eq.(4). Since θ has disappeared from Eqs.(7), (10), and (11) and ρ is being integrated over we get a constant for Eq.(4). This is different from the result of Eq.(2.5-12). The following three components of Eq.(4) are obtained:

$$U_1 = 2\pi c\Delta t \int_0^N \left(\frac{m_0c^2\Delta t}{\hbar}\right)^2 \Psi_{r0}^*\Psi_{r0}\rho^2 d\rho = 2\pi c\Delta t\Psi_{00}^2 \left(\frac{m_0c^2\Delta t}{\hbar}\right)^2$$
$$\times \int_0^N s_3^{2\rho}\frac{\Gamma^2(\rho)}{\Gamma^2(\rho+p+1)}\left(\sum_{\nu=0}^N(-1)^\nu q_3(\nu)\frac{(p+1)\dots(p+\nu)}{(\rho+p+1)\dots(\rho+p+\nu)}\right)^2 \rho^2 d\rho \quad (12)$$

$$U_2 = 2\pi c\Delta t \int_0^N \frac{\partial\Psi_{r0}^*}{\partial\theta}\frac{\partial\Psi_{r0}}{\partial\theta}\rho^2 d\rho = 2\pi c\Delta t\Psi_{00}^2 \left(\frac{E\Delta t}{\hbar}\right)^2$$
$$\times \int_0^N s_3^{2\rho}\frac{\Gamma^2(\rho)}{\Gamma^2(\rho+p+1)}\left(\sum_{\nu=0}^N(-1)^\nu q_3(\nu)\frac{(p+1)\dots(p+\nu)}{(\rho+p+1)\dots(\rho+p+\nu)}\right)^2 \rho^2 d\rho \quad (13)$$

$$U_3 = 2\pi c\Delta t \int_0^N \frac{\partial\Psi_{r0}^*}{\partial\rho}\frac{\partial\Psi_{r0}}{\partial\rho}\rho^2 d\rho = 2\pi c\Delta t\Psi_{00}^2$$
$$\times \int_0^N \left[\frac{\partial}{\partial\rho}\left(s_3^{\rho}\frac{\Gamma(\rho)}{\Gamma(\rho+p+1)}\sum_{\nu=0}^N(-1)^N q_3(\nu)\frac{(p+1)\dots(p+\nu)}{(\rho+p+1)\dots(\rho+p+\nu)}\right)\right]^2 \rho^2 d\rho \quad (14)$$

The integrals of Eqs.(12) and (13) are equal. They must be evaluated numerically. In Eq.(14) one can eliminate the differentiation $\partial/\partial\rho$ by the symmetric difference quotient $\tilde{\Delta}/\tilde{\Delta}\rho$ to facilitate computer evaluation:

$$\frac{\partial}{\partial\rho}\left(s_3^{\rho}\frac{\Gamma(\rho)}{\Gamma(\rho+p+1)}\sum_{\nu=0}^N(-1)^\nu q_3(\nu)\frac{(p+1)\dots(p+\nu)}{(\rho+p+1)\dots(\rho+p+\nu)}\right)$$
$$\to \frac{s_3^\rho}{2\Delta\rho}\sum_{\nu=0}^N(-1)^\nu q_3(\nu)(p+1)\dots(p+\nu)$$
$$\times\left(\frac{s^{\Delta\rho}\Gamma(\rho+\Delta\rho)}{\Gamma(\rho+\Delta\rho+p+1)}\frac{1}{(\rho+\Delta\rho+p+1)\dots(\rho+\Delta\rho+p+\nu)}\right.$$
$$\left.-\frac{s^{-\Delta\rho}\Gamma(\rho-\Delta\rho)}{\Gamma(\rho-\Delta\rho+p+1)}\frac{1}{(\rho-\Delta\rho+p+1)\dots(\rho-\Delta\rho+p+\nu)}\right) \quad (15)$$

If we use the approximations that led to Eq.(5.5-35) we obtain for the energies U_1 to U_3 the following expressions:

$$U_1 = 2\pi\Psi_{00}^2 \left(\frac{m_0 c}{\hbar}\right)^2 \times \int_0^{cT} \left[e^{-\kappa r} r^{u+n-l-1} \left(1 - \frac{p_0(p_0+1) - l(l+1) + \bar{\gamma}^2}{2\kappa r} + \dots\right)\right]^2 r^2 dr \quad (16)$$

$$U_2 = 2\pi\Psi_{00}^2 \left(\frac{\mathsf{E}}{\hbar c}\right)^2 \times \int_0^{cT} \left[e^{-\kappa r} r^{u+n-l-1} \left(1 - \frac{p_0(p_0+1) - l(l+1) + \bar{\gamma}^2}{2\kappa r} + \dots\right)\right]^2 r^2 dr \quad (17)$$

$$U_3 = 2\pi\Psi_{00}^2 \times \int_0^{cT} \left\{\frac{\partial}{\partial r}\left[e^{-\kappa r} r^{u+n-l-1} \left(1 - \frac{p_0(p_0+1) - l(l+1) + \bar{\gamma}^2}{2\kappa r} + \dots\right)\right]\right\}^2 r^2 dr$$

$$u = -\frac{1}{2} \mp \left[\left(l + \frac{1}{2}\right)^2 - \bar{\gamma}^2\right]^{1/2} \quad (18)$$

Equation (4) may be rewritten.

$$U = U(\kappa) = U_1 + U_2 + U_3 \quad (19)$$

The parameter κ occurs explicitly in Eqs.(16)–(18) but it is contained also in p or p_0 of Eqs.(12)–(14) or (16)–(18) according to Eqs.(5.5-16) and (5.5-31). Using U_1 to U_3 from Eqs.(12)–(14) we may write $U(\kappa)$ in the following normalized form:

$$\frac{U(\kappa)}{2\pi c\Delta t\Psi_{00}^2} = \frac{NU(\kappa)}{2\pi T\Psi_{00}^2} = N\mathcal{H}_\kappa$$
$$\mathcal{H}_\kappa = d^2(\kappa) = \frac{U(\kappa)}{2\pi T\Psi_{00}^2} \quad (20)$$

The variable $d^2(\kappa)$ is introduced to avoid having to write square roots of $\mathcal{H}_\kappa$. One may rewrite the component $\mathcal{H}_\kappa$ as follows:

$$\begin{aligned}\mathcal{H}_\kappa &= (2\pi\kappa)^2 \frac{d(\kappa)}{2\pi\kappa}(\sin 2\pi\kappa\theta - i\cos 2\pi\kappa\theta)\frac{d(\kappa)}{2\pi\kappa}(\sin 2\pi\kappa\theta + i\cos 2\pi\kappa\theta) \\ &= -2\pi i\kappa p_\kappa(\theta) q_\kappa(\theta) \end{aligned} \tag{21}$$

This is the same equation as Eq.(2.5-19). The calculation from Eqs.(2.5-20) to (2.5-48) applies again.

5.7 Convergence for Small Values of Δr

In Section 5.5 we derived Eq.(5.5-20), which shows that the solution derived for $u(\hat{\rho})$ of Eq.(5.5-1) converges for

$$\begin{aligned} \frac{\Delta\tilde{r}}{\hbar/m_0c} &= \frac{\tilde{\alpha}\Delta r}{\hbar/m_0c} = 2\left(1-\frac{\mathsf{E}^2}{m_0^2c^4}\right)^{1/2}\frac{\Delta r}{\hbar/m_0c} > 2\left(\frac{3}{\sqrt{2}}-2\right)^{1/2} \doteq 0.69662 \\ g &\le -\frac{3}{\sqrt{2}} \\ s_3 &\le \frac{1}{\sqrt{2}} \end{aligned} \tag{1}$$

The power series converges absolutely inside a somewhat smaller circle. This smaller circle may be transformed by conformal mapping into a loop around s_3 that runs through $s = 0$ in Figs.5.5-1 and 5.5-2. The point s_2 will be outside this loop. The transformed series of Eq.(5.5-15) converges absolutely and uniformly along a line of integration ℓ that is inside this loop except at the point $s = 0$ and that runs from $s = 0$ around s_3 and back to $s = 0$. One may thus integrate term by term and one obtains a factorial series that converges for $\rho > \rho_0$, where ρ_0 is the still to be determined abscissa of convergence.

The transformation

$$\zeta = \left(\frac{s}{s_3}\right)^P, \quad P = \text{real} \tag{2}$$

maps the point s_3 of the s-plane in Figs.5.5-1b and 5.7-1a into the point 1,0 of the ζ-plane, and the point s_2 into the point s_2^{2P} due to the relation $s_3 = 1/s_2$ of Eq.(5.5-14). This is shown in Fig.5.7-1. The point s_2^{2P} is on or outside the circle $|\zeta - 1| = 1$ around the point $\zeta = 1,0$ if the condition

$$\left(\frac{s_2}{s_3}\right)^P \doteq \left(\frac{e^{\kappa\Delta r}}{e^{-\kappa\Delta r}}\right)^P \ge 2, \quad P \ge \frac{\ln 2}{2\ln s_2} \doteq \frac{\ln 2}{2\kappa\Delta r} \tag{3}$$

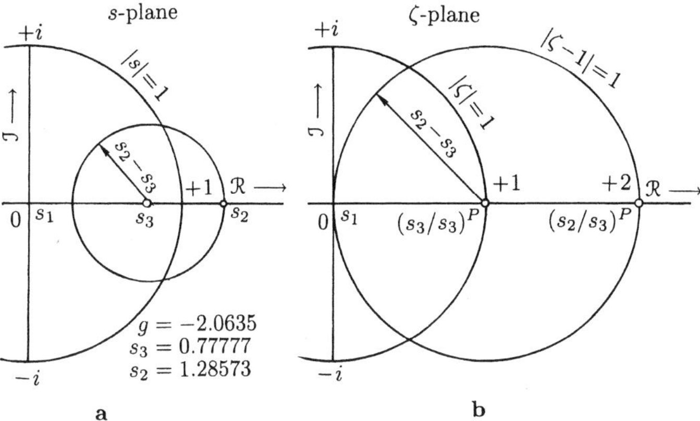

FIG.5.7-1. Transformation of Fig.5.5-1b from the s-plane (a) to the ζ-plane (b) according to Eq.(2).

is satisfied.

In Eq.(5.4-4) we introduced the relation $\rho = r/\Delta r$ to connect the normalized variable ρ with the nonnormalized distance r. We introduce in analogy a distance r_P:

$$P = \frac{r_\mathrm{P}}{\Delta r} \tag{4}$$

One recognizes from Eqs.(3) and (4) that there must be a minimum distance r_P for which one can obtain convergence. In order to determine this distance we express s_2 with the help of Eqs.(5.5-14) and (5.5-30):

$$s_2 = \frac{1}{s_3} \doteq 1 + \kappa\Delta r \doteq e^{\kappa\Delta r} \tag{5}$$

Equations (3), (4), and (5) yield:

$$r_\mathrm{P} \geq \frac{\ln 2}{\kappa} = \ln 2 \frac{\hbar}{m_0 c}\left(1 - \frac{\mathsf{E}^2}{m_0^2 c^4}\right)^{-1/2} \tag{6}$$

For small values of E we obtain in essence the result that r_P must be larger than the Compton wavelength. But for large values of E or n we can obtain much larger values of r_P according to Eq.(5.5-32).

The inverse transformation of Eq.(2)

$$s = s_3\zeta^{1/P} \tag{7}$$

maps the circle $|\zeta - 1| = 1$ of Fig.5.7-1b into a loop in the s-plane. The loop is shown in Fig.5.7-2.

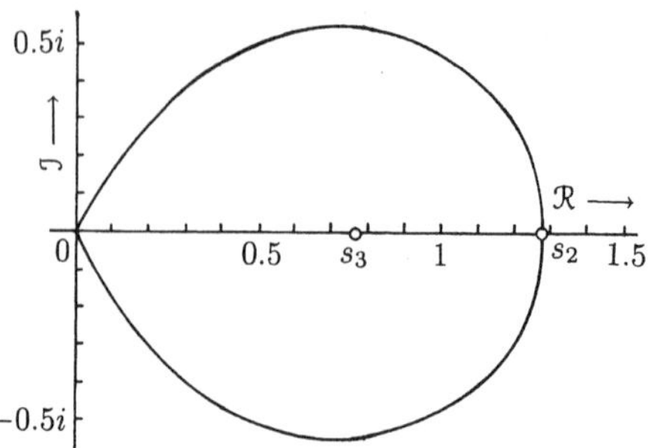

FIG.5.7-2. The circle $|\zeta - 1| = 1$ in the ζ-plane of Fig.5.7-1b mapped into a loop in the s-plane by means of Eq.(7); $s_3 = 0.77777$, $P = \ln 2/(2 \ln s_2) = 1.3791$, $s_2 = 1/s_3$.

The power series in the point s_2 may be mapped in the same manner as the one in point s_3. A complex number P must be used. This solution varies for large values of the distance r like $e^{+\kappa r}$ rather than like $e^{-\kappa r}$ as the solution in point s_3. One may infer this from $\rho = r/\Delta r$ in Eq.(5.4-4), s_3^ρ in Eq.(5.5-19), and s_2 in Eq.(5). We will not discuss the solution in point s_2 any further.

For a convergent solution according to Fig.5.7-1a we start with the differential equation (5.5-12) and write it for the point s_3 by the substitution $s = (s - s_3) + s_3$. The terms $(s - s_3)^n$ are rewritten as $s_3^n(s/s_3 - 1)^n$ and s/s_3 is replaced according to Eq.(2) by $s/s_3 = \zeta^{1/P}$. We now have a differential equation in the ζ -plane that still has to be transformed some more to make it a Fuchs-type equation. Once we have done that we may solve it with a power expansion in $\zeta - 1$ and obtain the solution

$$w(\zeta^{1/P} - 1) = \sum_{\nu=0}^{\infty} \tilde{q}(\nu)(\zeta - 1)^{\tilde{p}+\nu} \tag{8}$$

The calculation is shown in Section 6.9 from Eq.(6.9-1) to Eq.(6.9-24). The final step is to transform $w(\zeta^{1/P} - 1)$ into the s-plane by means of the substitution $\zeta = (s/s_3)^P$. This is done in Eqs.(6.9-25) to (6.9-50) and the function $w_{\mathrm{P}}(s-s_3)$ is obtained:

$$w_{\mathrm{P}}(s - s_3) = w_{\mathrm{P}}(s) = \sum_{\nu=0}^{\infty} \tilde{q}_{\mathrm{P}}(\nu)(s - s_3)^{\tilde{p}+\nu} \doteq \sum_{\nu=0}^{N} \tilde{q}_{\mathrm{P}}(\nu)(s - s_3)^{\tilde{p}+\nu} \tag{9}$$

We observe that the upper limit ∞ in Eq.(9) is replaced by a finite number N since an arbitrarily large but finite interval can be subdivided only into a finite number of arbitrarily small but finite subintervals. In other words, there is only a finite number $N + 1$ of linearly independent functions. The subscript P is used for w and q to distinguish them from w_3 and q_3 in Eq.(5.5-15), which holds for P=1 only. We note that the exponent $\tilde{p}$ in Eqs.(8) and (9) holds both in the ζ-plane and the s-plane, but the coefficients $\tilde{q}(\nu)$ and $q_{\mathrm{P}}(\nu)$ are different.

The series expansions in Section 6.9 are the bane of the Laplace transform method of solution of difference equations. A simpler and faster method is urgently needed. The operational methods of Milne-Thomson (1951, Ch. XIV) may bring this simplification. The series expansions required here are worked out in Section 6.9.

Equation (5.5-15) was transformed into a factorial series $u_{\mu 3}(\rho)$ in Eq.(5.5-19). Replacing $u_{\mu 3}(\rho)$, $q_3(\nu)$, and p there by $u_{\mathrm{P}}(\rho)$, $\tilde{q}_{\mathrm{P}}(\nu)$, and $\tilde{p}$ yields the factorial series associated with Eq.(9):

$$u_{\mathrm{P}}(\rho) = s_3^{\rho}\frac{\Gamma(\rho)}{\Gamma(\rho+\tilde{p}+1)}\sum_{\nu=0}^{N}(-1)^{\nu}\tilde{q}_{\mathrm{P}}(\nu)\frac{(\tilde{p}+1)\dots(\tilde{p}+\nu)}{(\rho+\tilde{p}+1)\dots(\rho+\tilde{p}+\nu)}$$

$$\tilde{p} = P\left(\frac{\tilde{\lambda}\tilde{\delta}}{s_3-s_2}-1\right)+1 = P(p-1)+1, \quad \text{Eq.(6.9-16)}$$

$$\frac{\tilde{q}_{\mathrm{P}}(1)}{\tilde{q}_{\mathrm{P}}(0)} = -\left(\frac{\tilde{p}[s_3P^{-1}(\tilde{p}-1+2P)-(s_3-s_2)(\tilde{p}-1)(P-1)/2P]-P[l(l+1)-\tilde{\gamma}^2]}{P^{-1}s_3(s_3-s_2)(\tilde{p}+1)} + \frac{\tilde{p}}{s_3}\frac{P-1}{2!}\right), \quad \text{Eq.(6.9-54)}$$

s_2, s_3: see Eq.(5.5-14); $\tilde{\gamma}$, $\tilde{\lambda}$, $\tilde{\delta}$: see Eq.(5.5-2); p: see Eq.(5.5-16) (10)

For $P=1$ we obtain $\tilde{p}=p$ of Eq.(5.5-16) and $\tilde{q}_P(1)/\tilde{q}_P(0) = q_3(1)/q_3(0)$ of Eqs.(5.5-19) and (5.5-18).

We want to work out a few numerical values. From Eqs.(5.5-23) and (5.5-30) we obtain the relation:

$$\Delta r = \frac{s_3(s_2-s_3)}{2\kappa} = \frac{s_3(s_2-s_3)}{2}\frac{\hbar}{m_0c}\left(1-\frac{\mathsf{E}^2}{m_0^2c^4}\right)^{-1/2} = \frac{s_3(s_2-s_3)}{4\pi}\frac{h}{m_0c}\left(1-\frac{\mathsf{E}^2}{m_0^2c^4}\right)^{-1/2} \tag{11}$$

For the limit case represented by Fig.5.5-2a with $s_3 = 1/\sqrt{2}$ and $s_2 = \sqrt{2}$ we get

$$\Delta r \geq 0.03979\frac{h}{m_0c}\left(1-\frac{\mathsf{E}^2}{m_0^2c^4}\right)^{-1/2} \tag{12}$$

From Eq.(5.5-32) we obtain:

$$\left(1-\frac{\mathsf{E}^2}{m_0^2c^4}\right)^{-1/2} = \left[1-\left(1-\frac{\tilde{\gamma}^2}{2n^2}\right)\right]^{-1/2}$$

$$\doteq \frac{\sqrt{2}n}{\tilde{\gamma}} = 15.416 \quad \text{for } n=1,\ \mathsf{Z}=1$$

$$\Delta r \geq 0.6134\frac{h}{m_0c} \tag{13}$$

Hence, the smallest value for Δr is about 61% of the Compton wavelength. If we use $s_2 = 1.2857$ and $s_3 = 0.7777$ of Fig.5.7-1a we obtain

$$\Delta r \geq 0.03144\frac{h}{m_0c}\left(1-\frac{\mathsf{E}^2}{m_0^2c^4}\right)^{-1/2} \doteq 0.48467\frac{h}{m_0c}$$

We still have to derive the numerical value of P. Equation (3) yields the relation

$$P \geq \frac{\ln 2}{2\ln s_2} \tag{14}$$

which yields $P \geq 1.3791$ for $s_2 = 1.2857$, $s_3 = 0.7777$ in Figs.5.7-1 and 5.7-2.

For plotting it is often more convenient to start with a numerical value for P. One obtains then from Eqs.(3) and (5.5-23):

$$s_2 \geq e^{\ln 2/2P} \tag{15}$$

$$s_3 \leq e^{-\ln 2/2P} \tag{16}$$

$$s_3(s_2 - s_3) = 2\kappa\Delta r = e^{-\ln 2/P} - 1 \tag{17}$$

The sign $\leq$ in Eq.(16) can readily be replaced by $=$ if the exact value of s_3 requires more decimal digits than used for plotting.

5.8 Plots for Sections 5.5 and 5.7

We start with certain constants obtained in Section 5.5 that are generally applicable. From Eqs.(5.5-2) and (5.5-32) we obtain

$$\tilde{\lambda}\tilde{\delta} = 2\mathsf{E}\tilde{\gamma}\frac{\Delta r}{\hbar c} = 2\tilde{\gamma}\left(1-\frac{\tilde{\gamma}^2}{2n^2}\right)\frac{2\pi\Delta r}{h/m_0c}, \quad \mathsf{E}_1 \doteq \mathsf{E}_2$$

$$= 4\pi\tilde{\gamma}\left(1-\frac{\tilde{\gamma}^2}{2n^2}\right)k, \quad k = \frac{\Delta r}{h/m_0c} \tag{1}$$

$$\tilde{\gamma} = 4\pi\mathsf{Z}\alpha, \quad \alpha = 7.297\,535\times 10^{-3}, \quad \mathsf{Z} = 1,\ 2,\ldots \tag{2}$$

Equation (5.5-16) yields:

$$p = -2\pi Z\alpha \frac{\mathsf{E}}{m_0c^2}\left(1-\frac{\mathsf{E}^2}{m_0^2c^4}\right)\left[1+\left(1-\frac{\mathsf{E}^2}{m_0^2c^4}\right)^2\left(\frac{\Delta r}{\hbar/m_0c}\right)^2\right]^{-1/2} \tag{3}$$

$$= -2\pi Z\alpha \frac{\mathsf{E}}{m_0c^2}\left(1-\frac{\mathsf{E}^2}{m_0^2c^4}\right)\left[1+4\pi^2\left(1-\frac{\mathsf{E}^2}{m_0^2c^4}\right)^2 k^2\right]^{-1/2} \tag{4}$$

$$p_0 = -2\pi Z\alpha \frac{\mathsf{E}}{m_0c^2}\left(1-\frac{\mathsf{E}^2}{m_0^2c^4}\right) \tag{5}$$

We see from Eqs.(5.7-12) and (5.7-13) that the choice $k = 1$ will yield convergence for any value of $P \geq 1$. But for smaller values of k we must be careful that P is chosen large enough for convergence. For E we obtain from Eq.(5.5-32)

$$\frac{\mathsf{E}}{m_0c^2} = \frac{\mathsf{E}_1}{m_0c^2} = \frac{\mathsf{E}_2}{m_0c^2} = 1 - \frac{\tilde{\gamma}^2}{2n^2} \tag{6}$$

if terms of order $\tilde{\gamma}^4$ are ignored. At this point we can choose a value $P \geq 1$. Equation (5.7-10) yields:

$$\tilde{p} = P(p-1)+1 \tag{7}$$

Further, we obtain from Eqs.(5.7-15) and (5.7-16) with $\geq$ and $\leq$ replaced by the equality sign

$$s_2 = e^{\ln 2/2P}, \quad s_3 = e^{-\ln 2/2P} \tag{8}$$

We can now calculate $\tilde{q}_{\mathrm{P}}(1)$ of Eq.(5.7-10). The free constant $\tilde{q}_{\mathrm{P}}(0)$ is chosen equal to one:

$$\tilde{q}_{\mathrm{P}}(1) = -\Bigg(\frac{\tilde{p}[s_3P^{-1}(\tilde{p}-1+2P)-(s_3-s_2)(\tilde{p}-1)(P-1)/2P]-P[l(l+1)-\tilde{\gamma}^2]}{P^{-1}s_3(s_3-s_2)(\tilde{p}+1)} + \frac{\tilde{p}}{s_3}\frac{P-1}{2}\Bigg), \quad l = 0,\ 1,\ \ldots \tag{9}$$

The first term $u_{\mathrm{P}1}(\rho)$ of the factorial series of Eq.(5.7-10) becomes

$$u_{\mathrm{P}1}(\rho) = s_3^{\rho}\frac{\Gamma(\rho)}{\Gamma(\rho+\tilde{p}+1)}, \quad \rho = 1,\ 2, \ldots,\ N \tag{10}$$

while the second term $u_{\mathrm{P}2}(\rho)$ equals:

$$u_{\mathrm{P}2}(\rho) = -u_{\mathrm{P}1}(\rho)\tilde{q}_{\mathrm{P}}(1)\frac{\tilde{p}+1}{\rho+\tilde{p}+1}, \quad \rho = 1,\ 2,\ \ldots,\ N \tag{11}$$

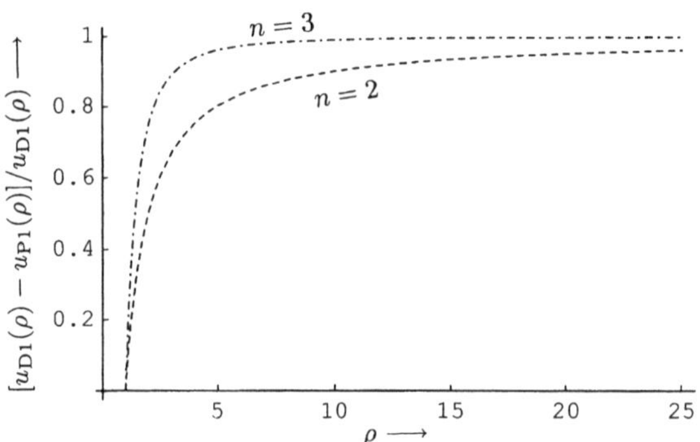

FIG.5.8-4. Relative difference $[u_{\mathrm{D1}}(\rho) - u_{\mathrm{P1}}(\rho)]/u_{\mathrm{D1}}(\rho)$ according to Eqs.(12) and (10) for $u = u_1$, $n = 2$ or $n = 3$, $l = 0$, $Z = 1$, $P = 1$, $k = 1$ of the differential and the difference result for $\rho = 1, 2, \ldots, 100$.

we obtain Fig.5.8-9 instead of Fig.5.8-5. The plots appear to be identical but if a larger scale is used, $1 \leq \rho \leq 3$ rather than $0 \leq \rho \leq 10$, one can see a deviation at $\rho = 2$ for $P = 4$. Next we try $k = 0.001$ and obtain Fig.5.8-10. The deviation of the plot for $P = 4$ at $\rho = 2$ or 3 is now clearly visible. But we also note that the vertical scale has changed, which is due to the negative value of the plot for $P = 3$ at $\rho = 1$.

Figures 5.8-11 to 5.8-13 show plots like Figs.5.8-5 to 5.8-7 but for $k = 10^{-4}$ rather than for $k = 1$. We note drastic differences between Figs.5.8-5 and 5.8-11 that become less drastic for Figs.5.8-6 and 5.8-12 and still less drastic for Figs.5.8-7 and 5.8-13. We see that the sequence of the plots for $P = 1$, 1.3791, 2, 3, 4 from top to bottom is the same in Figs.5.8-7 and 5.8-13 for $\rho \geq 12$.

While the reduction of k from 1 to 0.01 shows no noticeable difference between Figs.5.8-5 and 5.8-9 a reduction of $k = 10^{-4}$ by one-half to $k = 5 \times 10^{-5}$ in Figs.5.8-14 to 5.8-16 brings drastic changes compared with the plots of Figs.5.8-11 to 5.8-13. One such change is that the sequence of the plots for $P = 1$, 1.3791, 2, 3, 4 from top to bottom in Fig.5.8-16 is no longer restored for larger values of ρ as it was in Fig.5.8-13. We conclude from this that too small a resolution Δr affects $u_{\mathrm{P1}}(\rho)$ primarily close to the origin $\rho = 0$. But the deviation of the theory extends to larger values of ρ for smaller values of.Δr.

In order to characterize the transition from the stable plots of Figs.5.8-5 to 5.8-7 to the randomized plots of Figs.5.8-14 to 5.8-16 we show Table 5.8-1. The function $u_{\mathrm{P1}}(\rho)$ for $\rho = 1$ and either $P = 1$ or $P = 2$ retains its first five digits, 0.70713 or 0.49996, in the interval $1 \geq k \geq 0.7$. The first four digits, 0.7072 or 0.4999, are retained in the second interval shown for k, etc. The number $k = 0.6134$ is given in Eq.(5.7-13) for $P = 1$ as the transition from stable results to randomized ones. We see that this is in line with Table 5.8-1 if we do not look for a sudden change at $k = 0.6134$ but merely for a fast change.

Let us turn to the term $\nu = 1$ of $u_{\mathrm{P}}(\rho)$ in Eq.(5.7-10). It is represented by

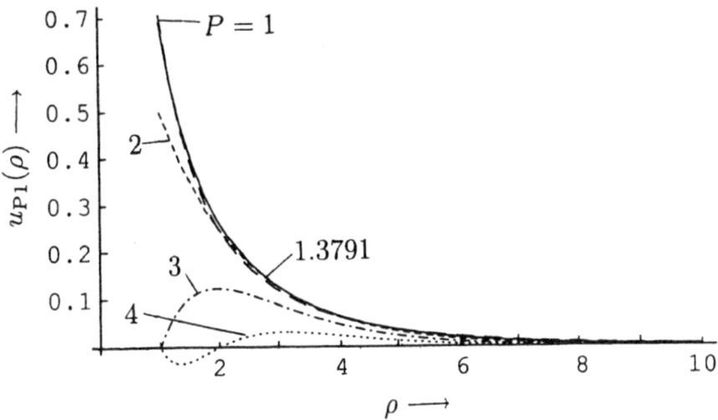

FIG.5.8-5. Plots of $u_{P1}(\rho)$ according to Eq.(10) for $n = 1$, $l = 0$, $Z = 1$, $k = 1$, and $P = 1$, 1.3791, 2, 3, 4 for $\rho = 1, 2, \ldots, 10$.

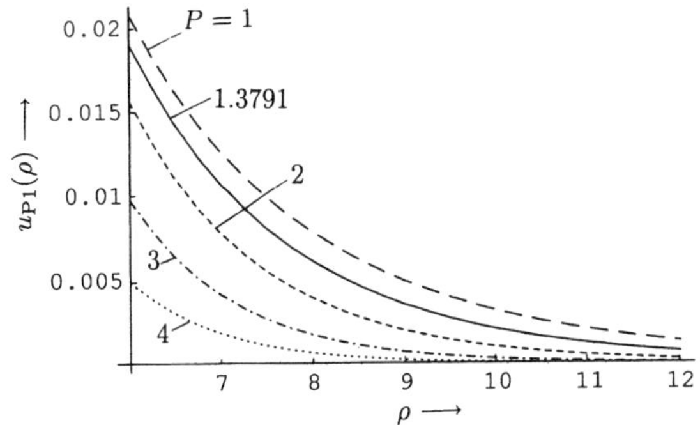

FIG.5.8-6. Plots of $u_{P1}(\rho)$ according to Eq.(10) for $n = 1$, $l = 0$, $Z = 1$, $k = 1$, and $P = 1$, 1.3791, 2, 3, 4 for $\rho = 6, 7, \ldots, 12$.

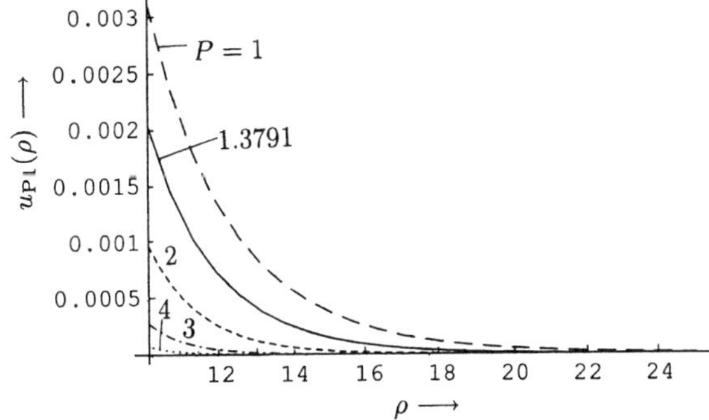

FIG.5.8-7. Plots of $u_{P1}(\rho)$ according to Eq.(10) for $n = 1$, $l = 0$, $Z = 1$, $k = 1$, and $P = 1$, 1.3791, 2, 3, 4 for $\rho = 10, 11, \ldots, 25$.

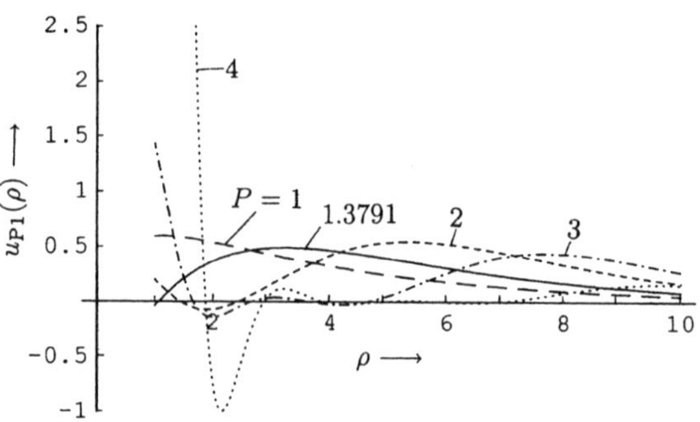

FIG.5.8-14. Plots of $u_{P1}(\rho)$ according to Eq.(10) for $n = 1$, $l = 0$, $Z = 1$, and $k = 5 \times 10^{-5}$ for various values of P; $\rho = 1, 2, \ldots, 10$.

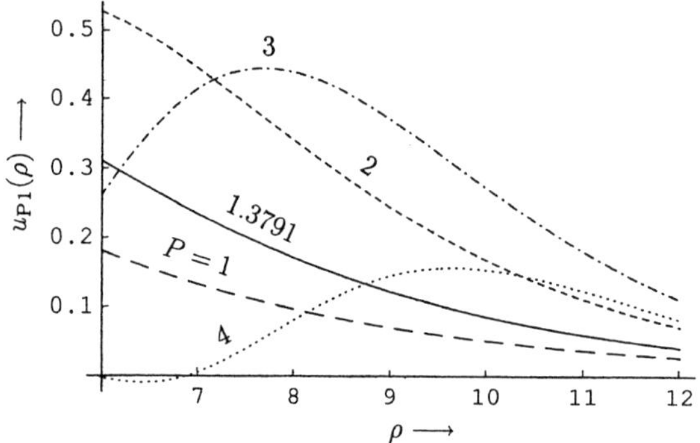

FIG.5.8-15. Plots of $u_{P1}(\rho)$ according to Eq.(10) for $n = 1$, $l = 0$, $Z = 1$, and $k = 5 \times 10^{-5}$ for various values of P; $\rho = 6, 7, \ldots, 12$.

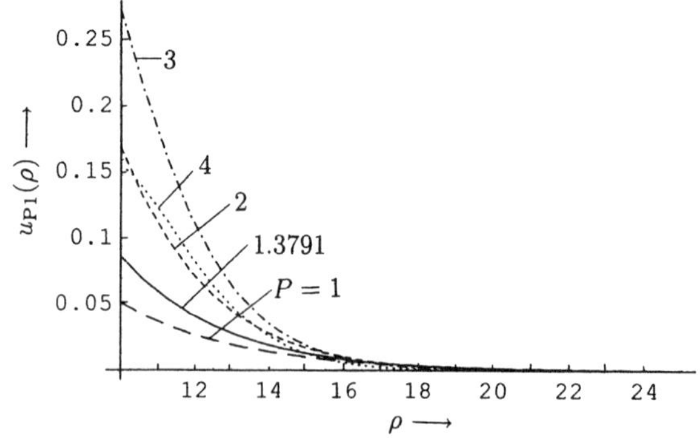

FIG.5.8-16. Plots of $u_{P1}(\rho)$ according to Eq.(10) for $n = 1$, $l = 0$, $Z = 1$, and $k = 5 \times 10^{-5}$ for various values of P; $\rho = 11, 12, \ldots, 25$.

TABLE 5.8-1

INTERVALS OF $k = \Delta r/(h/m_0c)$ FOR WHICH $u_{P1}(1)$ OF EQ.(10) HAS THE SAME FIVE, FOUR, THREE, OR TWO INITIAL DIGITS FOR $P = 1$ AND $P = 2$.

$P = 1$				
$u_{P1}(1)$	0.70713	0.7072	0.707	0.71
	$1 \geq k \geq 0.7$	$0.6 \geq k \geq 0.2$	$0.1 \geq k \geq 0.04$	$0.03 \geq k \geq 0.002$
$P = 2$				
$u_{P1}(1)$	0.49996	0.4999	0.499	0.49
	$1 \geq k \geq 0.7$	$0.6 \geq k \geq 0.3$	$0.2 \geq k \geq 0.03$	$0.02 \geq k \geq 0.003$

$u_{P2}(\rho)$ of Eq.(11). Plots for P=1, 1.3791, 2 are shown in Fig.5.8-17 for $\rho = 1$, 2, ..., 10. The plots for $P = 1$ and 1.3791 are multiplied by 100 to be able to present them together with the plot for $P = 2$. Figures 5.8-18 and 5.8-19 show the same plots for $\rho = 6$, 7, ..., 12 and $\rho = 10$, 11, ..., 25.

The two plots for $P = 3$ and $P = 4$ not shown in Fig.5.8-17 are shown in Fig.5.8-20 for $\rho = 1$, 2, ..., 10. They are as odd as the plots for $P = 3$ and $P = 4$ in Fig.5.8-5. We do not show them for $\rho = 6$, 7, ..., 12 or $\rho = 10$, 11, ..., 25.

Figure 5.8-21 shows again the plots of $u_{P2}(\rho)$ for $P = 1$, 1.3791, 2 for $\rho = 1$, 2, ..., 10, the ones for $P = 1$ and $P = 1.3791$ multiplied by 100, but $k = 1$ is replaced by $k = 0.01$. There are two noticeable small differences compared with Fig.5.8-17. The plots for $P = 1.3791$ and $P = 2$ are closer together and the plot for $P = 1$ has the value -0.3 at $\rho = 1$ rather than -0.6. We note that for $u_{P1}(\rho)$ in Figs.5.8-5, 5.8-9, and 5.8-10 we had to go to $k = 0.001$ before a difference became noticeable.

Figures 5.8-22 to 5.8-24 show plots like Figs.5.8-17 to 5.8-19 but for $k = 0.001$ rather than $k = 1$. There are significant differences between Figs.5.8-17 and 5.8-22 that remain significant for Figs.5.8-18 and 5.8-23 or Figs.5.8-19 and 5.8-24.

What we have seen in this section is primarily the impossibility of reducing the spatial resolution to less than about 10% of the Compton wavelength even for bound particles. This is in line with the results of Chapters 3 and 4, particularly Eqs.(3.1-10) and (4.1-2). Instead of the coarse distinction between $\Delta x \gg \lambda_C$ and $\Delta x \ll \lambda_C$ we have now the much finer distinction that the resolution can be about 10% of λ_C.

The conformal mapping introduced in Section 5.7 does not improve the resolution, it only permits us to study the transition from a result with non-random numbers to one with (pseudo)random numbers using a convergent series expansion. Since the apparently random numbers in Fig.5.8-8 were obtained by calculation they are not really random numbers and we must use the correct term pseudorandom numbers. The situation is the same as in the calculation of a number like π. Since the number is obtained by calculation

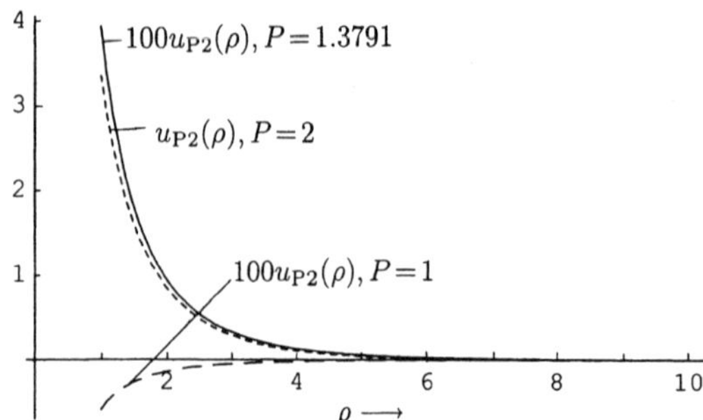

FIG.5.8-17. Plots of $u_{P2}(\rho)$ according to Eq.(11) for $n = 1$, $l = 0$, $Z = 1$, $k = 1$, and $P = 1$, 1.3791, 2 for $\rho = 1, 2, \ldots, 10$; the plots for $P = 1$ and $P = 1.3791$ are multiplied by 100.

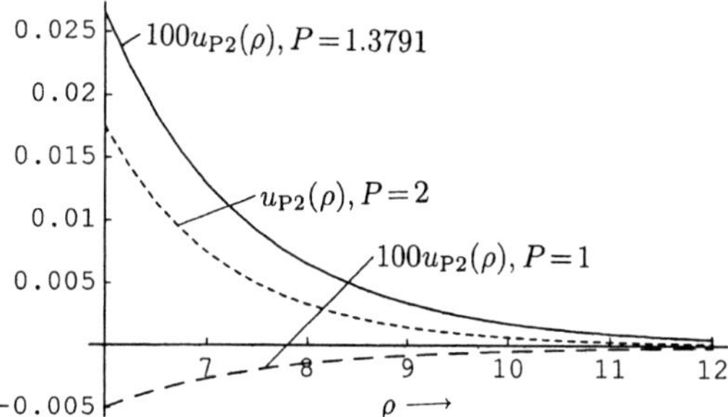

FIG.5.8-18. Plots of $u_{P2}(\rho)$ according to Eq.(11) for $n = 1$, $l = 0$, $Z = 1$, $k = 1$, and $P = 1$, 1.3791, 2 for $\rho = 6, 7, \ldots, 12$; the plots for $P = 1$ and $P = 1.3791$ are multiplied by 100.

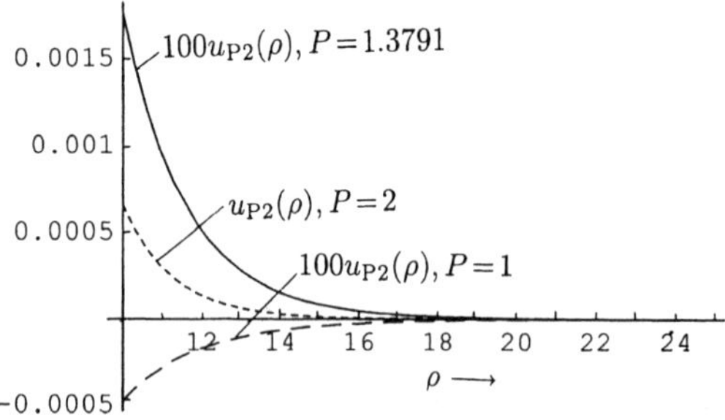

FIG.5.8-19. Plots of $u_{P2}(\rho)$ according to Eq.(11) for $n = 1$, $l = 0$, $Z = 1$, $k = 1$, and $P = 1$, 1.3791, 2 for $\rho = 10, 11, \ldots, 25$; the plots for $P = 1$ and $P = 1.3791$ are multiplied by 100.

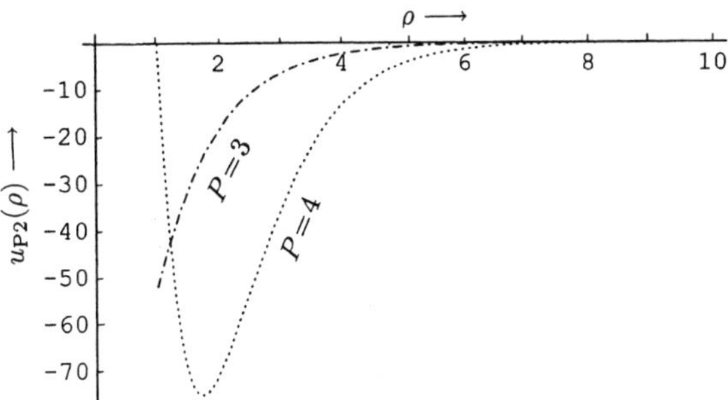

FIG.5.8-20. Plots of $u_{P2}(\rho)$ according to Eq.(11) for $n = 1$, $l = 0$, $Z = 1$, $k = 1$, and $P = 3, 4$ for $\rho = 1, 2, \dots, 10$. The magnitude of these plots is about 20 times that of the plot for $P = 2$ in Fig.5.8-17 and 2000 times that of the plots for $P = 1, 1.3791$.

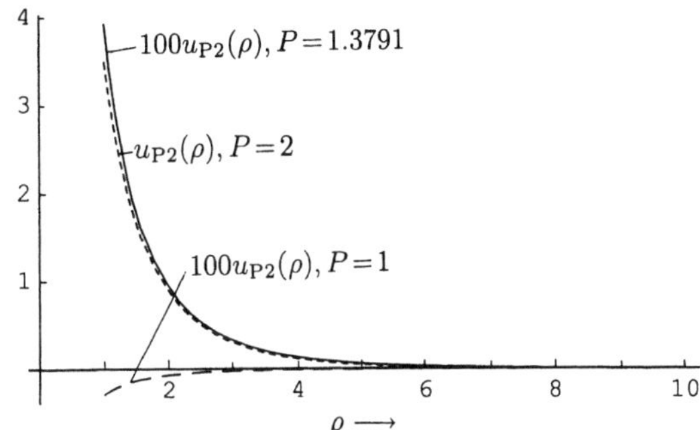

FIG.5.8-21. Plots of $u_{P2}(\rho)$ according to Eq.(11) for $n = 1$, $l = 0$, $Z = 1$, $k = 0.01$, and $P = 1, 1.3791, 2$ for $\rho = 1, 2, \dots, 10$; the plots for $P = 1$ and $P = 1.3791$ are multiplied by 100.

its digits are not a sequence of random digits, but we expect that they appear random if we apply any conceivable test of randomness. Convergence does not exclude pseudorandom numbers from our results.

From the standpoint of physics it is quite understandable that the Compton effect limits the spatial resolution even for a bound particle. This limitation of the resolution is a specific feature of a difference theory that assumes a finite resolution Δx or Δr but does not specify the magnitude of Δx or Δr. Instead, it obtains this magnitude as a result of the calculation. We shall discuss this point again in Section 5.9 from a different angle.

Figures 5.8-3 and 5.8-4 demonstrated that the results of the difference and the differential theory did not become equal for large distances ρ or $r = \rho\Delta r$. We want to show this analytically with the help of Eqs.(10) and (12). The following approximation of the Gamma function for large arguments is used (Abramovitz and Stegun 1964, p. 257):

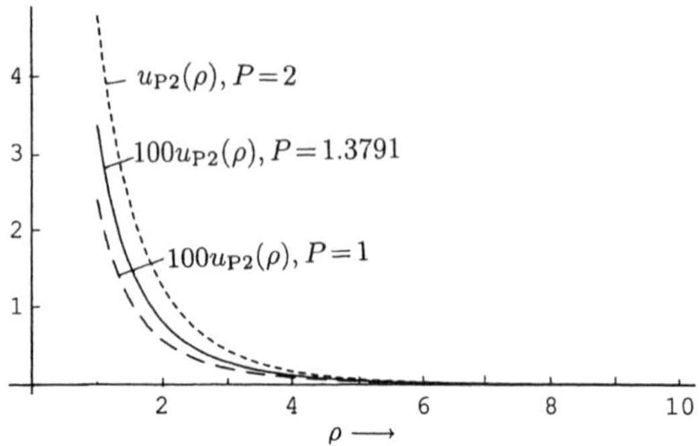

FIG.5.8-22. Plots of $u_{P2}(\rho)$ according to Eq.(11) for $n = 1$, $l = 0$, $Z = 1$, and $k = 0.001$ for various values of P; $\rho = 1, 2, \ldots, 10$.

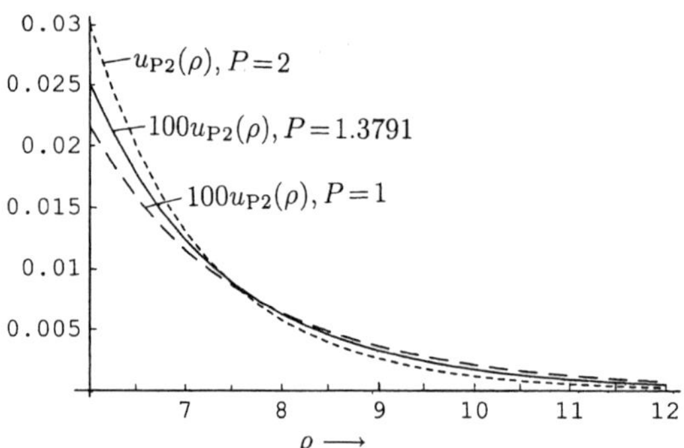

FIG.5.8-23. Plots of $u_{P2}(\rho)$ according to Eq.(11) for $n = 1$, $l = 0$, $Z = 1$, and $k = 0.001$ for various values of P; $\rho = 6, 7, \ldots, 12$.

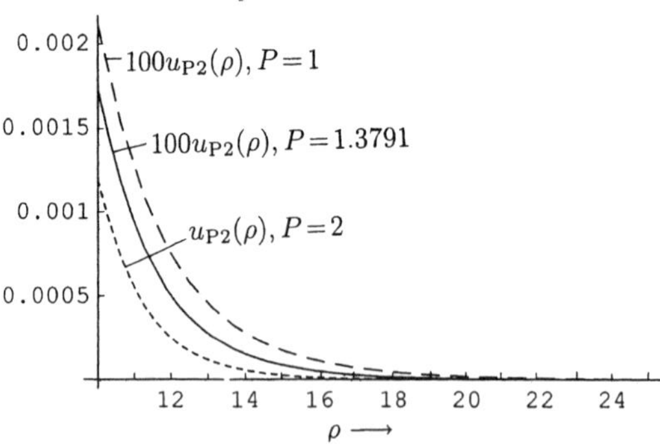

FIG.5.8-24. Plots of $u_{P2}(\rho)$ according to Eq.(11) for $n = 1$, $l = 0$, $Z = 1$, and $k = 0.001$ for various values of P; $\rho = 10, 11, \ldots, 25$.

$$\Gamma(\rho) \approx \sqrt{2\pi}e^{-\rho}e^{(\rho-1/2)\ln\rho}, \qquad \rho \gg 1 \tag{15}$$

The ratio $u_{P1}(\rho)/u_{D1}(\rho)$ becomes for large values of ρ:

$$\frac{u_{P1}(\rho)}{u_{D1}(\rho)} = \frac{s_3^{\rho}\Gamma(\rho)/\Gamma(\rho+\tilde{p}+1)}{s_3^{\rho}\rho^{u}\rho^{n-l-1}} \approx e^{\tilde{p}+1}\rho^{-(\tilde{p}+u+n-l)} \tag{16}$$

5.9 Origin of the Coulomb Field

In Section 5.8 all plots held for $\rho = r/\Delta r = 1, 2, \ldots$. Since the circle of convergence in Fig.5.7-1b goes through the origin, the convergence of the power series in the point s_3 is assured in a somewhat smaller circle, which implies that the factorial series of Eq.(5.7-10) converges for $\rho > 0$. The convergence *on* the abscissa of convergence $\rho = 0$ is a problem comparable to that of the convergence of a power series *on* the circle of convergence. It turns out that we do not have to address this problem.

For the explanation of such a fortuitous result let us state first that a result $u_P(\rho)$ of Eq.(5.7-10) holds not only in the point ρ but in the interval $\rho - 1/2 < r/\Delta r < \rho + 1/2$ since we cannot resolve an interval smaller than Δr. Hence, we have solutions for the intervals $\ldots$, $5/2 < r/\Delta r < 3/2$, $3/2 < r/\Delta r < 1/2$, $1/2 < r/\Delta r < -1/2$. In all these intervals except the last one we have a charged boson in a Coulomb field. This situation changes if the boson reaches the interval $1/2 < r/\Delta r < -1/2$ since this is where the charge is located that produces the Coulomb field. The boson cannot be distinguished from whatever carries this charge. Instead of a boson in a Coulomb field we have only a Coulomb field with a changed charge number Z. If the charge $+e$ created the Coulomb field and the boson has the charge $-e$ we get a neutral particle and no Coulomb field. Hence, there is no interest in $u_P(\rho)$ of Eq.(5.7-10) for $\rho = 0$. This is a solution of the problem of convergence for $\rho = 0$ that reminds one of the egg of Columbus.

The differential theory typically has problems with convergence at $r = 0$. We distinguish between weak poles for $r = 0$ that are acceptable and the strong poles that are not. To recognize the reason for the different behavior of a differential and a difference theory consider the function $1/r$ as it approaches $r = 0$. In the difference theory we have for $1/r$ the succession of values $\ldots$, $1/3\Delta r$, $1/2\Delta r$, $1/\Delta r$, $(1/0)$, but the last value $1/0$ was just ruled out as being of no interest. A finite value $\Delta r > 0$ yields a finite value $1/\Delta r$. In a differential theory we would have the sequence $\ldots$, $1/3dr$, $1/2dr$, $1/dr$, $(1/0)$. Again, we can ignore the last value $1/0$ since the boson would be indistinguishable from the central charge of the Coulomb field at $r = 0$. But $1/(n\,dr)$ exceeds any finite bound for any finite or denumerable number $n = 1, 2, \ldots$. Hence, the problem of divergence arises not only for $r = 0$ but for any distance $n\,dr$ from $r = 0$. We have found a mechanism that avoids or sidesteps convergence problems that works only in a difference theory and not in a differential theory.

The problem with convergence for $r \to 0$ is evidently caused by the assumption of infinitesimal distances dr. Since we cannot observe infinitesimal distances, any problem caused by their assumption is an important argument in favor of the calculus of finite differences.

5.10 Unbounded Bosons in a Coulomb Field

We turn to the solution of the differential equation (5.5-12) for which s_2 and s_3 are located on the circle $|s| = 1$ as shown in Fig.5.10-1a rather than on the real axis as shown in Fig.5.5-1a. One obtains for small values of Δr the following relations instead of Eqs.(5.5-14), (5.5-16), (5.5-23), (5.5-30), and (5.7-5):

$$s_2 \doteq 1 + i\kappa'\Delta r, \quad s_3 \doteq 1 - i\kappa'\Delta r, \quad s_2 - s_3 \doteq 2i\kappa'\Delta r$$

$$\mathsf{E} = m_0c^2\left[1 + \left(\frac{\kappa'\hbar}{m_0c}\right)^2\right]^{1/2}, \quad \kappa' = -i\kappa \doteq \frac{m_0c}{\hbar}\left(\frac{\mathsf{E}^2}{m_0^2c^4} - 1\right)^{1/2}$$

$$p = ip' = i\frac{\tilde{\lambda}\tilde{\delta}}{2\kappa'\Delta r}, \quad p_0 = ip_0' = -2\pi i Z\alpha\frac{\mathsf{E}}{m_0c^2}\left(1 - \frac{\mathsf{E}^2}{m_0^2c^4}\right) \tag{1}$$

No solution in the form of asymptotic polynomials–meaning the terms of order $O(1)$ are polynomials–can exist since p is imaginary and Eq.(5.5-26) cannot hold; its real and imaginary part do not vanish simultaneously. With the help of Stirling's relation we obtain instead of Eq.(5.5-34)

$$\lim_{\rho\to\infty}\frac{\Gamma(\rho)}{\Gamma(\rho + ip' + 1)} = \frac{1}{\rho^{1+ip'}} = \frac{K}{\rho}e^{-ip'\ln\rho} \tag{2}$$

Equation (5.5-19) yields the following relation in place of Eq.(5.5-37) for the point $s_3 = e^{-i\kappa'\Delta r}$:

$$u_{\mu 3}(\rho) = q_3'(0)e^{-ip_0'/\ln\rho}e^{i\kappa'\Delta r\rho} \times \frac{1}{\Delta r\rho}\left(1 + i\frac{ip_0'(1 + ip_0') - l(l+1) + \tilde{\gamma}^2}{2\kappa'\Delta r\rho} + \ldots\right) \cdot \tag{3}$$

The term $\exp(-ip_0'\ln\rho)$ is a phase factor that depends essentially on the value of Δr. It disappears if the product $u_{\mu 3}(\rho)u_{\mu 3}^*(\rho)$ is used. The function $\psi_{\mu 3}(\rho, \theta) = u_{\mu 3}(\rho)e^{-i\mathsf{E}\Delta t\theta/\hbar}$ according to Eqs.(5.4-15) and (5.4-25) is at large distances ρ an expanding spherical wave:

$$\psi_{\mu 3}(\rho, \theta) \approx q_0''e^{ip_0'\ln\rho}\frac{1}{\Delta r\rho}e^{i(\kappa'\Delta r\rho - \mathsf{E}\Delta t\theta/\hbar)} \tag{4}$$

The series expansion in the point s_2 yields a contracting spherical wave.

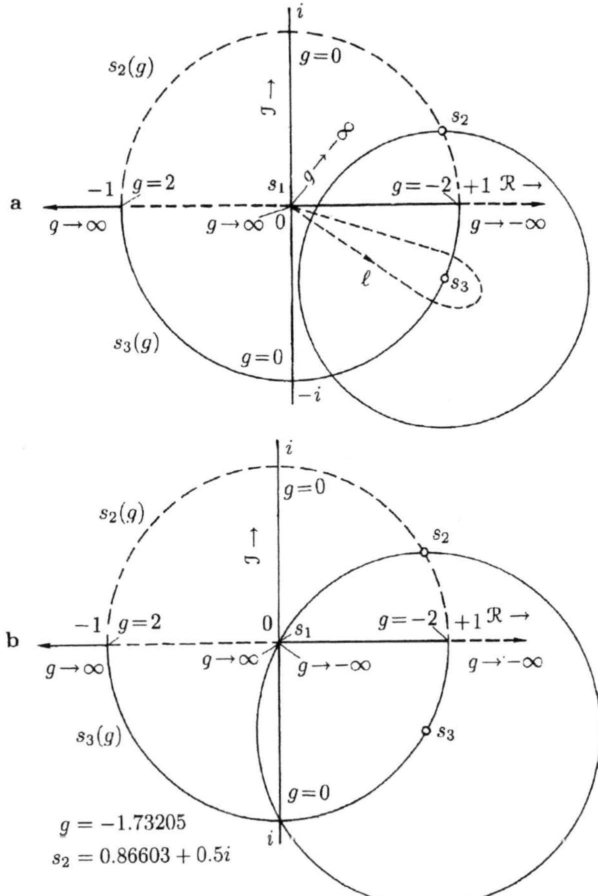

FIG.5.10-1. Location of the singular points s_2 (dashed line) and s_3 (solid line) in the complex s-plane as function of g for the Klein-Gordon equation for general values of g and the line ℓ of integration (a), and for $g = -2\cos\pi/6 = -1.73205$ (b). These illustrations also hold for the nonrelativistic Schrödinger equation and the iterated Dirac equations.

We recognize the relations $s_2 = \cos\pi/6 + i\sin\pi/6$ and $s_3 = \cos\pi/6 - i\sin\pi/6$ from Fig.5.10-1b. Equation (5.5-14) yields

$$g = -(s_2 + s_3) \geq -2\cos\pi/6 \tag{5}$$

as a condition for the circle of convergence around either s_2 or s_3 to reach $s = s_1 = 0$. For Δr we obtain from Eq.(1):

$$\Delta r = \frac{s_2 - s_3}{2i\kappa'} = \frac{1}{\kappa'}\sin\frac{\pi}{6} = \frac{\hbar}{m_0 c}\left(\frac{\mathsf{E}^2}{m_0^2 c^4} - 1\right)^{-1/2} \sin\frac{\pi}{6}$$

$$\frac{\Delta r}{h/m_0 c} = \frac{1}{\pi}\left(\frac{\mathsf{E}^2}{m_0^2 c^4} - 1\right)^{-1/2} \tag{6}$$

Hence, $\Delta r/(h/m_0 c)$ approaches infinity for an energy $\mathsf{E} \to m_0 c^2$ of a free boson but drops to smaller values for larger energies; Δr can never be zero.

As in Section 5.7 the transformation

$$\zeta = \left(\frac{s}{s_3}\right)^P, \quad P = \text{real} \tag{7}$$

maps the point s_3 of the s-plane in Fig.5.10-2a into the point 1,0 of the ζ-plane, and the point s_2 into the point s_2^{2P} due to the relation $s_2 s_2^* = s_2 s_3 = 1$ of Eq.(5.5-14). The point s_2^{2P} is on or outside the circle $|\zeta - 1| = 1$ around the point $\zeta = 1,0$ if the condition

$$\left(\frac{s_2}{s_3}\right)^P = \left(\frac{e^{i\kappa'\Delta r}}{e^{-i\kappa'\Delta r}}\right)^P \geq e^{i\pi/3}$$

$$P \geq \frac{\pi}{6}\frac{1}{\kappa'\Delta r} = \frac{\pi}{6}\frac{2i}{s_3^* - s_3} = \frac{\pi}{6}\frac{1}{|\mathfrak{Im}(s_3)|} \tag{8}$$

is satisfied [$\mathfrak{Im}(s_3)$ = imaginary part of s_3]. In Fig.5.10-2 we have

$$\kappa'\Delta r = \pi/12, \quad P = 2 \tag{9}$$

The inverse transformation of Eq.(7)

$$s = s_3\zeta^{1/P} = e^{-i\pi/12}\zeta^{1/P}, \quad P = 2 \tag{10}$$

maps the circle $|\zeta - 1| = 1$ of Fig.5.10-2b into a loop in the s-plane that is shown in Fig.5.10-3.

For a convergent solution according to Fig.5.10-2a we start with the differential equation (5.5-12). It is written for the point s_3 by the substitution $s = (s - s_3) + s_3$ as in Section 5.7. The terms $(s - s_3)^n$ are rewritten as $s_3^n(s/s_3 - 1)^n$ and s/s_3 is replaced according to Eq.(7) by $s/s_3 = \zeta^{1/P}$. We now have a differential equation in the ζ-plane. It requires some additional transformations that bring it into a form that permits a solution by a power expansion in $\zeta - 1$:

$$w(\zeta^{1/P} - 1) = \sum_{\nu=0}^{\infty} \hat{q}(\nu)(\zeta - 1)^{\hat{p}+\nu} \tag{11}$$

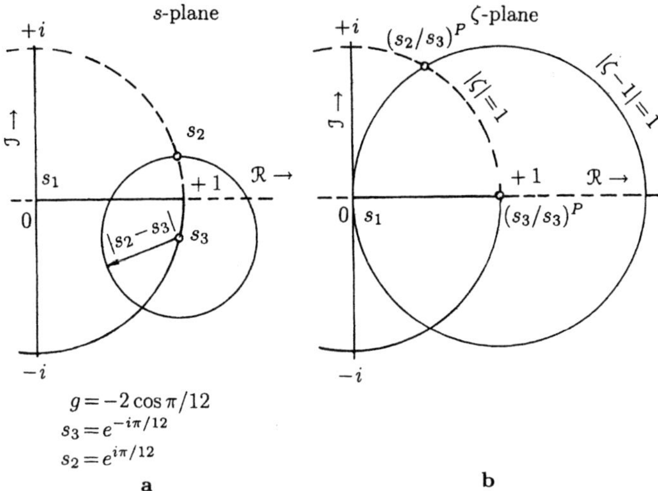

FIG.5.10-2. Transformation from the s-plane to the ζ-plane according to Eq.(7) in analogy to Fig.5.7-1; $P = 2$, $\kappa'\Delta r = \pi/12$.

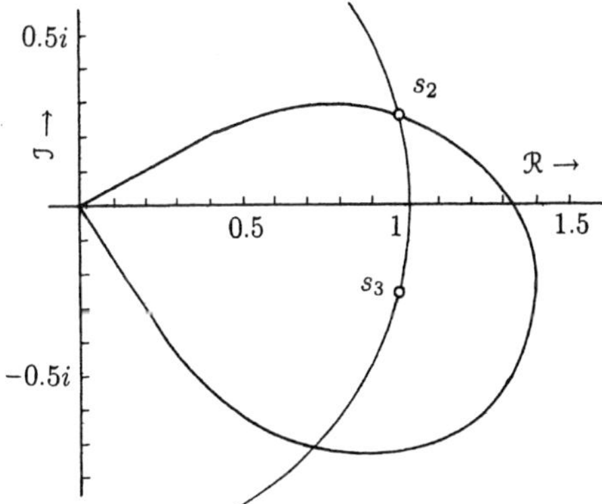

FIG.5.10-3. The circle $|\zeta - 1| = 1$ in the ζ-plane in Fig.5.10-2b is mapped into a loop in the s-plane by means of Eq.(10); $s_3 = e^{-i\pi/12} = 1/s_2$, $P = 2$, $g = -2\cos\pi/12$.

The calculation is carried out in Section 6.10. The final step is to transform $w(\zeta^{1/P} - 1)$ into the s-plane by means of the substitution $\zeta = (s/s_3)^P$. This is done in Section 6.10 too and the function $w_P(s - s_3)$ is obtained:

$$w_P(s - s_3) = w_P(s) = \sum_{\nu=0}^{\infty} \hat{q}_P(\nu)(s - s_3)^{\hat{p}+\nu} \doteq \sum_{\nu=0}^{N} \hat{q}_P(\nu)(s - s_3)^{\hat{p}+\nu} \qquad (12)$$

All this is as in Section 5.7 except that s_3 is no longer a real number but is now a complex number.

Equation (5.5-15) was transformed into a factorial series $u_{\mu 3}(\rho)$ in Eq.(5.5-19), and the same transformation produced the factorial series of Eq.(5.7-10) from the power series of Eq.(5.7-9). If we change $\tilde{q}_{\mathrm{P}}(\nu)$, $\tilde{p}$ to $\hat{q}_{\mathrm{P}}(\nu)$, $\hat{p}$ in Eq.(5.7-9) we obtain Eq.(12), and the same change transforms the factorial series of Eq.(5.7-10) into the one associated with Eq.(12):

$$u_{\mathrm{P}}(\rho) = s_3^\rho \frac{\Gamma(\rho)}{\Gamma(\rho+\hat{p}+1)} \sum_{\nu=0}^{N} (-1)^\nu \hat{q}_{\mathrm{P}}(\nu) \frac{(\hat{p}+1)\dots(\hat{p}+\nu)}{(\rho+\hat{p}+1)\dots(\rho+\hat{p}+\nu)}$$

$$\hat{p} = P\left(\frac{\tilde{\lambda}\tilde{\delta}}{s_3 - s_3^*} - 1\right) + 1 = P(ip' - 1) + 1, \quad \text{Eq.(6.10-1)}$$

$$\frac{\hat{q}_{\mathrm{P}}(1)}{\hat{q}_{\mathrm{P}}(0)} = -\left(\frac{\hat{p}[s_3 P^{-1}(\hat{p}-1+2P) - (s_3 - s_3^*)(\hat{p}-1)(P-1)/2P] - P[l(l+1) - \tilde{\gamma}^2]}{P^{-1}s_3(s_3 - s_3^*)(\hat{p}+1)} + \frac{\hat{p}}{s_3}\frac{P-1}{2!}\right), \quad \text{Eq.(6.10-14)}$$

$$s_3,\ p' \quad \text{see Eq.(1)}, \quad \tilde{\gamma},\ \tilde{\lambda},\ \tilde{\delta} \quad \text{see Eq.(5.5-2)} \tag{13}$$

As in Section 5.7 we work out a few numerical values. From Eq.(1) we obtain for the spatial resolution Δr:

$$\Delta r = \frac{s_3^* - s_3}{2i\kappa'} = -i\frac{s_3^* - s_3}{4\pi}\frac{h}{m_0 c}\left(\frac{\mathsf{E}^2}{m_0^2 c^4} - 1\right)^{-1/2} \tag{14}$$

From the limit case for convergence shown for $P = 1$ by Fig.5.10-1b we see the condition

$$s_3^* - s_3 = s_2 - s_3 \geq i \tag{15}$$

and obtain

$$\Delta r \geq \frac{1}{4\pi}\frac{h}{m_0 c}\left(\frac{\mathsf{E}^2}{m_0^2 c^4} - 1\right)^{-1/2} = 0.079577\frac{h}{m_0 c}\left(\frac{\mathsf{E}^2}{m_0^2 c^4} - 1\right)^{-1/2} \tag{16}$$

which is in line with Eq.(5.7-12). The lowest possible energy E for an unbounded boson is $m_0 c^2$, which yields $\Delta r \to \infty$. A localization of such a boson is completely impossible. For larger values of E the irresolvable spatial interval decreases to an arbitrarily small but finite one, which implies that a boson with very high energy is not much disturbed by the Compton effect.

The equivalent formula for Eq.(5.7-14) is expressed by Eq.(8). If we choose a value of P we obtain an equation for the magnitude of the imaginary part of s_3

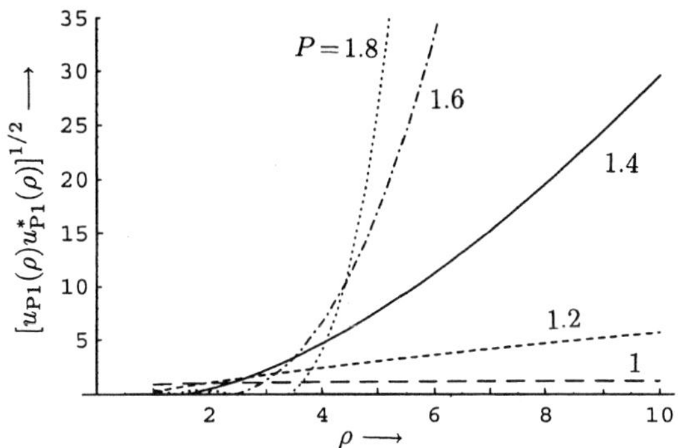

FIG.5.10-4. Plots of $[u_{P1}(\rho)u^*_{P1}(\rho)]^{1/2}$ according to Eq.(23) for $Z = 1$, $\Delta r/(h/m_0c) = k = 1$, and $P = 1$, 1.2, 1.4, 1.6, 1.8 for $\rho = 1, 2, \ldots, 10$.

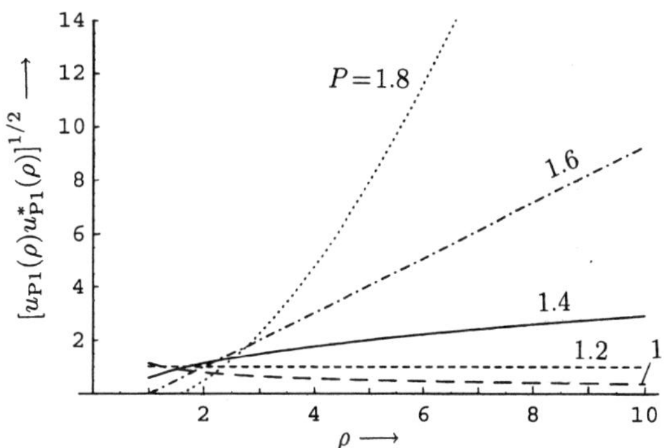

FIG.5.10-5. Plots of $[u_{P1}(\rho)u^*_{P1}(\rho)]^{1/2}$ according to Eq.(23) for $Z = 1$, $\Delta r/(h/m_0c) = k = 0.5$, and $P = 1$, 1.2, 1.4, 1.6, 1.8 for $\rho = 1, 2, \ldots, 10$.

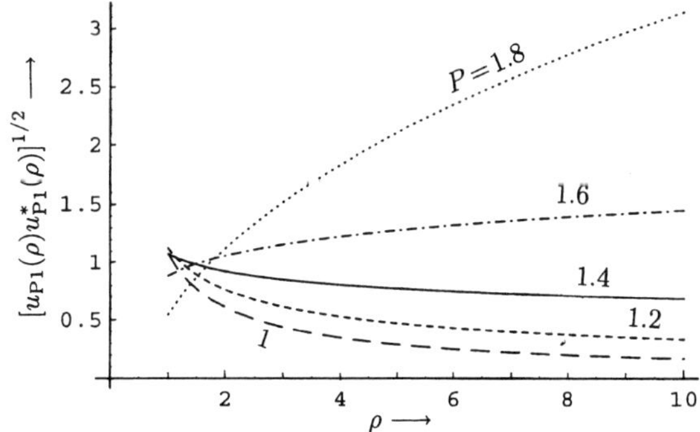

FIG.5.10-6. Plots of $[u_{P1}(\rho)u^*_{P1}(\rho)]^{1/2}$ according to Eq.(23) for $Z = 1$, $\Delta r/(h/m_0c) = k = 0.2$, and $P = 1$, 1.2, 1.4, 1.6, 1.8 for $\rho = 1, 2, \ldots, 10$.

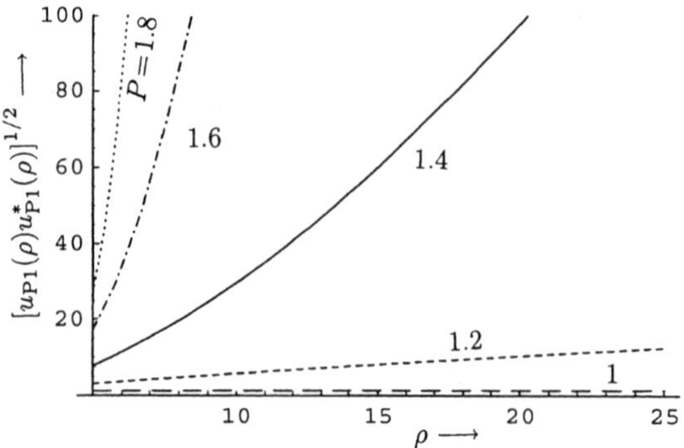

FIG.5.10-7. Plots of $[u_{P1}(\rho)u^*_{P1}(\rho)]^{1/2}$ according to Eq.(23) for $Z = 1$, $\Delta r/(h/m_0c) = k = 1$, and $P = 1$, 1.2, 1.4, 1.6, 1.8 for $\rho = 5, 6, \ldots, 25$.

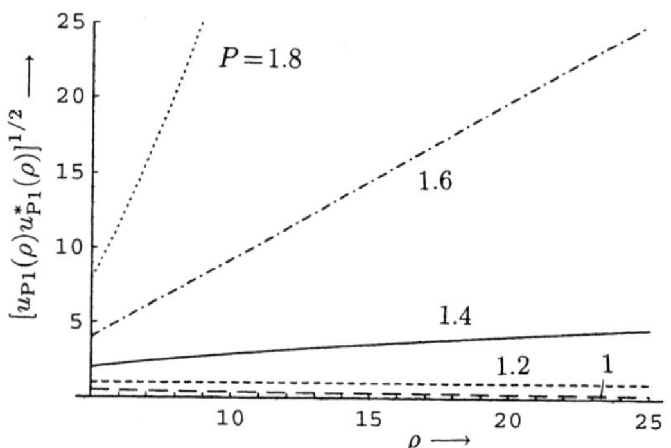

FIG.5.10-8. Plots of $[u_{P1}(\rho)u^*_{P1}(\rho)]^{1/2}$ according to Eq.(23) for $Z = 1$, $\Delta r/(h/m_0c) = k = 0.5$, and $P = 1$, 1.2, 1.4, 1.6, 1.8 for $\rho = 5, 6, \ldots, 25$.

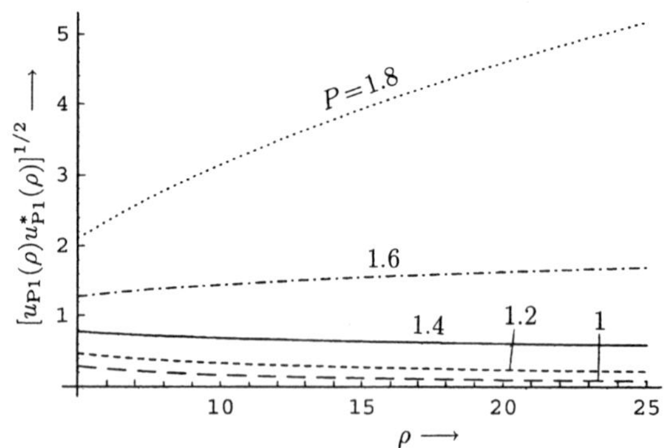

FIG.5.10-9. Plots of $[u_{P1}(\rho)u^*_{P1}(\rho)]^{1/2}$ according to Eq.(23) for $Z = 1$, $\Delta r/(h/m_0c) = k = 0.2$, and $P = 1$, 1.2, 1.4, 1.6, 1.8 for $\rho = 5, 6, \ldots, 25$.

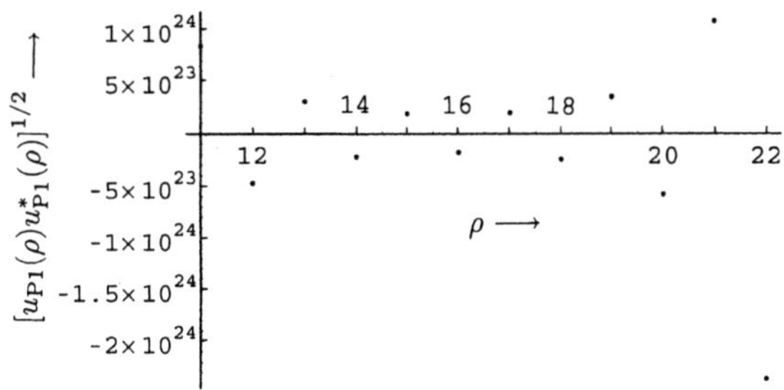

FIG.5.10-10. Plots of $[u_{\mathrm{P1}}(\rho)u^*_{\mathrm{P1}}(\rho)]^{1/2}$ according to Eq.(23) for $Z = 1$, $\Delta r/(h/m_0c) = k = 1$, $P = 5$, and $\rho = 11, 12, \dots, 22$.

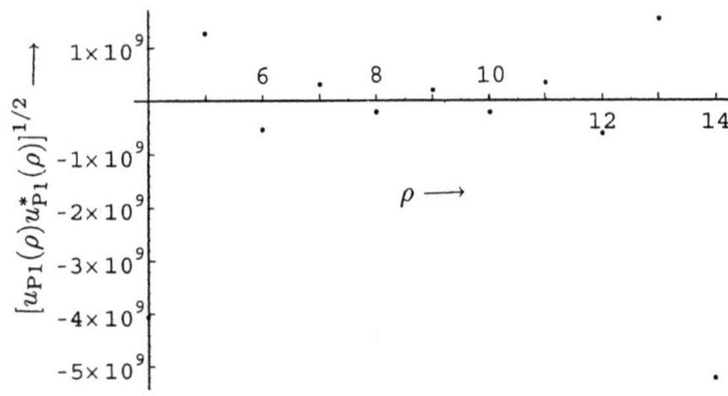

FIG.5.10-11. Plots of $[u_{\mathrm{P1}}(\rho)u^*_{\mathrm{P1}}(\rho)]^{1/2}$ according to Eq.(23) for $Z = 1$, $\Delta r/(h/m_0c) = k = 0.5$, $P = 5$, and $\rho = 4, 5, \dots, 14$.

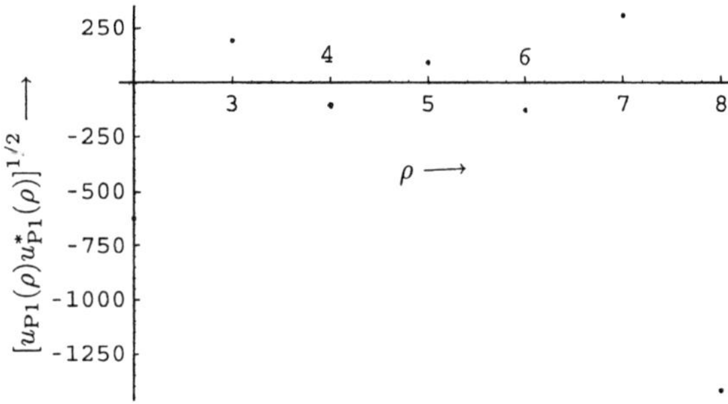

FIG.5.10-12. Plots of $[u_{\mathrm{P1}}(\rho)u^*_{\mathrm{P1}}(\rho)]^{1/2}$ according to Eq.(23) for $Z = 1$, $\Delta r/(h/m_0c) = k = 0.2$, $P = 5$, and $\rho = 2, 3, \dots, 8$.

$$|\Im m(s_3)| = \frac{\pi}{6}\frac{1}{P} \tag{17}$$

which is the equivalent of Eq.(5.7-17). The real part of s_3 is determined by the imaginary part since s_3 is always on the unit circle according to Fig.5.10-1b. We restrict at this time the location of s_3 to the interval $-2 < g \leq 0$ but we shall investigate larger values of g in the following section.

In analogy to Eq.(5.8-10) we want to plot the first term $u_{\mathrm{P1}}(\rho)$ of $u_{\mathrm{P}}(\rho)$ in Eq.(13):

$$u_{\mathrm{P1}}(\rho) = s_3^\rho \frac{\Gamma(\rho)}{\Gamma(\rho+\hat{p}+1)}, \quad \rho = 1,\ 2, \ldots,\ N \tag{18}$$

First we determine $\hat{p}$ with the help of Eq.(5.8-1):

$$\begin{aligned}
\tilde{\lambda}\tilde{\delta} &= 2\mathsf{E}\tilde{\gamma}\frac{\Delta r}{\hbar c} = 4\pi\tilde{\gamma}\frac{\mathsf{E}}{m_0c^2}\frac{\Delta r}{h/m_0c} = 4\pi\tilde{\gamma}e_{\mathrm{n}}k\\
\tilde{\gamma} &= 4\pi \mathsf{Z}\alpha,\ e_{\mathrm{n}} = \mathsf{E}/m_0c^2,\ k = \Delta r/(h/m_0c)\\
\alpha &= 7.297\,535\times 10^{-3},\ \mathsf{Z} = 1,\ 2, \ldots
\end{aligned} \tag{19}$$

From Eqs.(13) and (17) we obtain:

$$\begin{aligned}
\hat{p} &= P\left(\frac{\tilde{\lambda}\tilde{\delta}}{s_3 - s_3^*} - 1\right) + 1 = P\left(\frac{\tilde{\lambda}\tilde{\delta}}{-2|\Im m(s_3)|} - 1\right) + 1\\
&= -\frac{3}{\pi}P^2\tilde{\lambda}\tilde{\delta} - P + 1
\end{aligned} \tag{20}$$

Since $\hat{p}$ is real the only complex term in Eq.(18) is s_3. With the relations

$$s_3 = x - iy, \quad y = |\Im m(s_3)| = \frac{\pi}{6P}, \quad x^2 + y^2 = 1 \tag{21}$$

we obtain:

$$s_3 = \left[1 - (\pi/6P)^2\right]^{1/2} - i\pi/6P \tag{22}$$

In order to eliminate the complex term s_3^ρ in Eq.(18) we may plot $u_{\mathrm{P1}}(\rho)u_{\mathrm{P1}}^*(\rho)$. The square root of this expression yields a function that is easier to compare with Eq.(5.8-10):

$$\left[u_{\mathrm{P1}}(\rho)u_{\mathrm{P1}}^*(\rho)\right]^{1/2} = \frac{\Gamma(\rho)}{\Gamma(\rho+\hat{p}+1)} \tag{23}$$

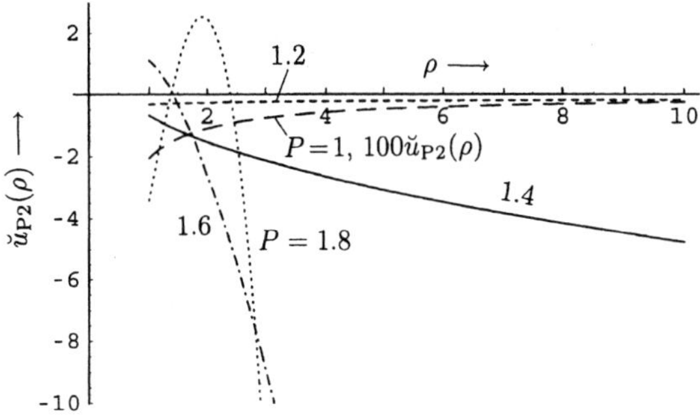

FIG.5.10-13. Plots of $\breve{u}_{\mathrm{P2}}(\rho) = [u_{\mathrm{P2}}(\rho)u^*_{\mathrm{P2}}(\rho)]^{1/2}$ according to Eq.(25) for $l = 0$, $Z = 1$, $k = \Delta r/(h/m_0c) = 1$, and $P = 1$, 1.2, 1.4, 1.6, 1.8; $\rho = 1, 2, \ldots, 10$. The plot for $P = 1$ is multiplied by 100.

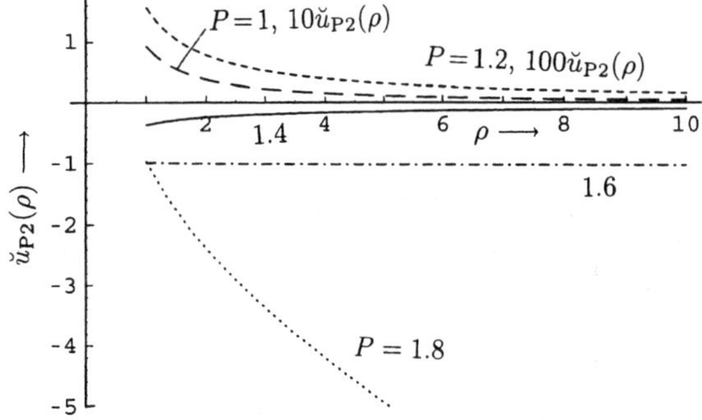

FIG.5.10-14. Plots of $\breve{u}_{\mathrm{P2}}(\rho) = [u_{\mathrm{P2}}(\rho)u^*_{\mathrm{P2}}(\rho)]^{1/2}$ according to Eq.(25) for $l = 0$, $Z = 1$, $k = \Delta r/(h/m_0c) = 0.5$, and $P = 1$, 1.2, 1.4, 1.6, 1.8; $\rho = 1, 2, \ldots, 10$. The plot for $P = 1$ is multiplied by 10, the plot for $P = 1.2$ by 100.

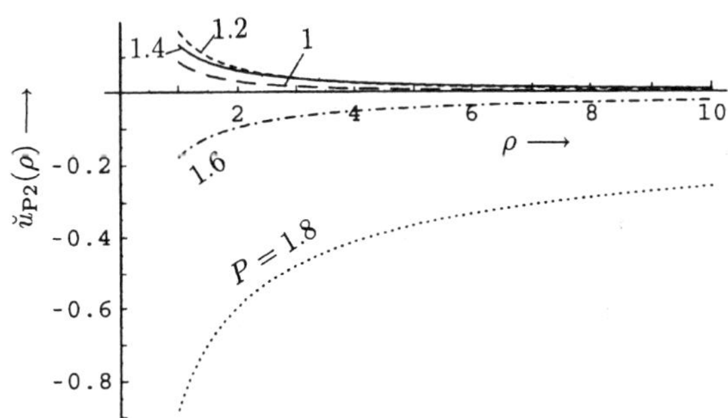

FIG.5.10-15. Plots of $\breve{u}_{\mathrm{P2}}(\rho) = [u_{\mathrm{P2}}(\rho)u^*_{\mathrm{P2}}(\rho)]^{1/2}$ according to Eq.(25) for $l = 0$, $Z = 1$, $k = \Delta r/(h/m_0c) = 0.2$, and $P = 1$, 1.2, 1.4, 1.6, 1.8; $\rho = 1, 2, \ldots, 10$.

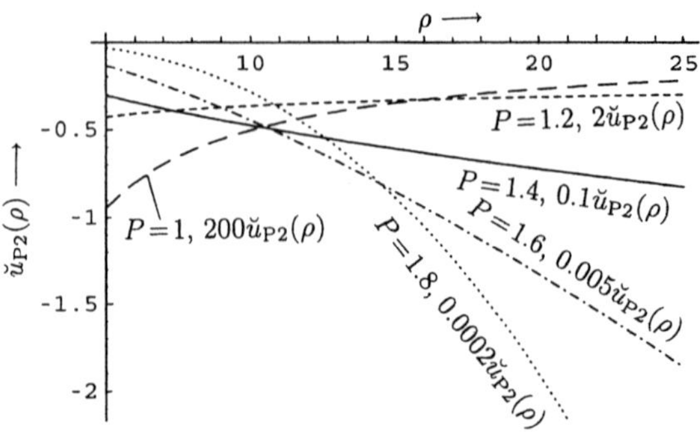

FIG.5.10-16. Plots of $\breve{u}_{P2}(\rho) = [u_{P2}(\rho)u^*_{P2}(\rho)]^{1/2}$ according to Eq.(25) for $l = 0$, $Z = 1$, $k = \Delta r/(h/m_0c) = 1$, and $P = 1$, 1.2, 1.4, 1.6, 1.8; $\rho = 5, 6, \dots, 25$. The plots for $P = 1$ to $P = 1.8$ are multiplied by 200 to 0.0002.

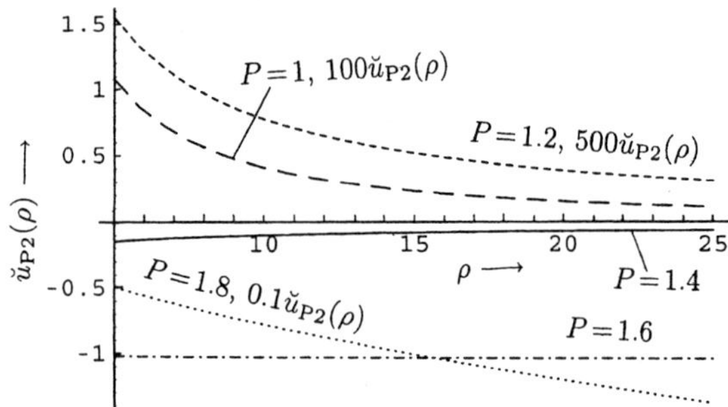

FIG.5.10-17. Plots of $\breve{u}_{P2}(\rho) = [u_{P2}(\rho)u^*_{P2}(\rho)]^{1/2}$ according to Eq.(25) for $l = 0$, $Z = 1$, $k = \Delta r/(h/m_0c) = 0.5$, and $P = 1$, 1.2, 1.4, 1.6, 1.8; $\rho = 5, 6, \dots, 25$. The plots for $P = 1$, 1.2, 1.8 are multiplied by 100, 500, 0.1.

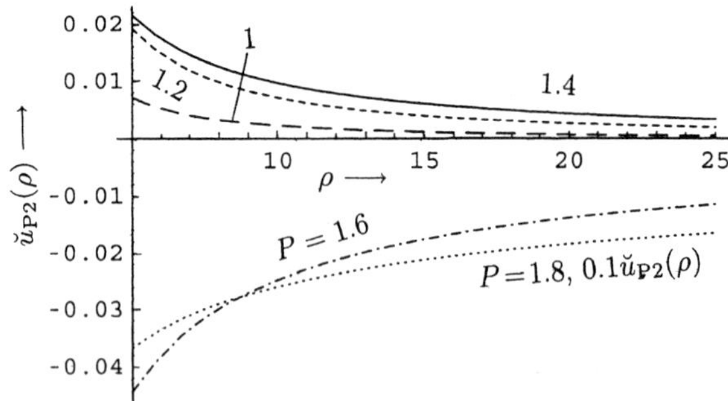

FIG.5.10-18. Plots of $\breve{u}_{P2}(\rho) = [u_{P2}(\rho)u^*_{P2}(\rho)]^{1/2}$ according to Eq.(25) for $l = 0$, $Z = 1$, $k = \Delta r/(h/m_0c) = 0.2$, and $P = 1$, 1.2, 1.4, 1.6, 1.8; $\rho = 5, 6, \dots, 25$. The plot for $P = 1$ is multiplied by 0.1.

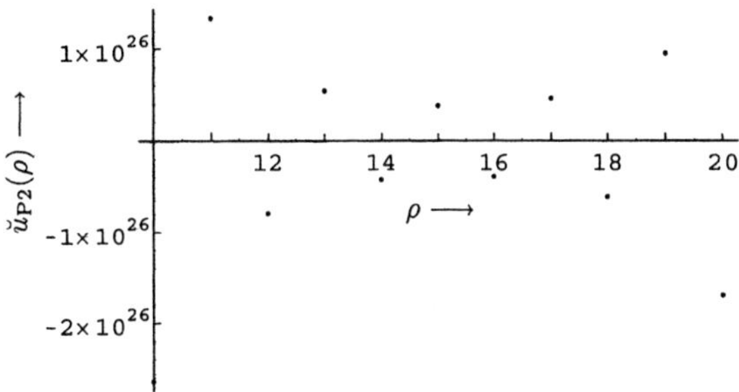

FIG.5.10-19. Plot of $\breve{u}_{P2}(\rho) = [u_{P2}(\rho)u^*_{P2}(\rho)]^{1/2}$ according to Eq.(25) for $l = 0$, $Z = 1$, $k = \Delta r/(h/m_0c) = 1$, $P = 5$, and $\rho = 10, 11, \ldots, 20$.

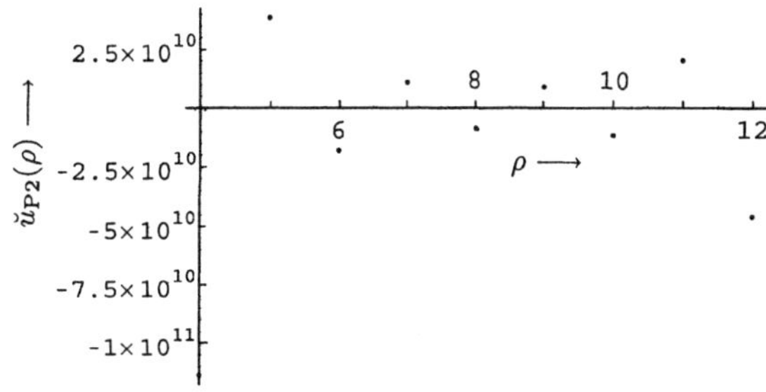

FIG.5.10-20. Plot of $\breve{u}_{P2}(\rho) = [u_{P2}(\rho)u^*_{P2}(\rho)]^{1/2}$ according to Eq.(25) for $l = 0$, $Z = 1$, $k = \Delta r/(h/m_0c) = 0.5$, $P = 5$, and $\rho = 4, 5, \ldots, 12$.

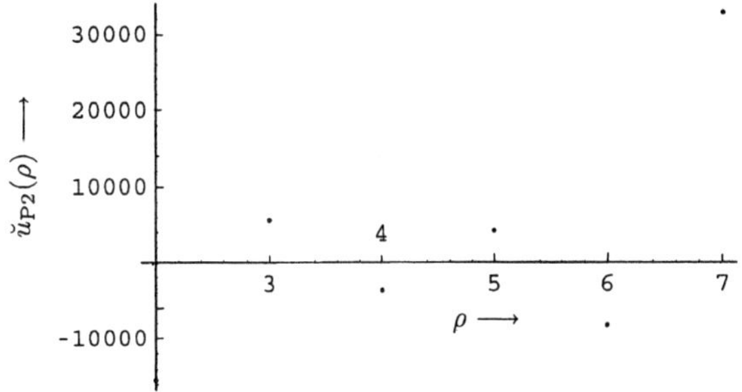

FIG.5.10-21. Plot of $\breve{u}_{P2}(\rho) = [u_{P2}(\rho)u^*_{P2}(\rho)]^{1/2}$ according to Eq.(25) for $l = 0$, $Z = 1$, $k = \Delta r/(h/m_0c) = 0.2$, $P = 5$, and $\rho = 2, 3, \ldots, 7$.

In Eq.(5.8-6) we used $\mathsf{E}/m_0c^2 = 1 - \tilde{\gamma}^2/2n^2$ with $n = 1$. To obtain comparable values we choose $\mathsf{E}/m_0c^2 = 1 + \tilde{\gamma}^2/2$ in Eq.(19) since E must now be larger than m_0c^2. As previously, we choose initially $k = \Delta r/(h/m_0c) = 1$ in Eq.(19), but reduce it to $k = 0.5$ and $k = 0.2$ rather than to 10^{-2}, 10^{-3}, and 10^{-4} as in Figs.5.8-9 to 5.8-11. This implies that the transition from non-random to (pseudo)random numbers will be much faster for unbounded bosons than for bounded ones. In line with this result we shall also use $P = 1$, 1.2, 1.4, 1.6, 1.8 instead of the larger values $P = 1, \dots, 5$ in Figs.5.8-5 to 5.8-11.

Comparing the plots of Eqs.(23) in Figs.5.10-4 to 5.10-9 with those in Figs.5.8-5 to 5.8-11 shows that large values of the plots no longer occur close to $\rho = 0$. This is to be expected since the eigenfunctions that we derived are used for probability density functions for the location of a boson and a bound boson is more likely to be close to the center of the Coulomb field than an unbound boson. The rapid change of the plots when k is reduced from 1 to smaller values is conspicuous. One would expect that the location of an unbound boson is changed more by the Compton effect than that of a bound boson.

In line with this observation are the plots of Figs.5.10-10 to 5.10-12 that show the random values obtained from Eq.(23) for $P = 5$ instead of for $P = 10$ as in Fig.5.8-8. The narrow and varying ranges of ρ in Figs.5.10-10 to 5.10-12 are due to the enormous variation of $[u_{\mathrm{P}1}(\rho)u^*_{\mathrm{P}1}(\rho)]^{1/2}$ for $P = 5$.

We turn to the second term $u_{\mathrm{P}2}$ of the factorial series of Eq.(13) and write it in the following form:

$$u_{\mathrm{P}2}(\rho) = -s_3^\rho \frac{\Gamma(\rho)}{\Gamma(\rho+\hat{p}+1)} \hat{q}_{\mathrm{P}}(1) \frac{\hat{p}+1}{\rho+\hat{p}+1}$$

$$\hat{q}_{\mathrm{P}}(1) \text{ see Eq.(13) for } \hat{q}_{\mathrm{P}}(0) = 1;\ s_3 \text{ see Eq.(22)};\ \hat{p} \text{ see Eq.(20)} \qquad (24)$$

In order to obtain for $u_{\mathrm{P}2}(\rho)$ a formula of the form of Eq.(23) we observe that $\hat{q}_{\mathrm{P}}(1)$ is a complex number. Hence, we write

$$[u_{\mathrm{P}2}(\rho)u^*_{\mathrm{P}2}(\rho)]^{1/2} = \breve{u}_{\mathrm{P}2}(\rho) = -[\hat{q}_{\mathrm{P}}(1)\hat{q}^*_{\mathrm{P}}(1)]^{1/2} \frac{\Gamma(\rho)}{\Gamma(\rho+\hat{p}+1)} \frac{\hat{p}+1}{\rho+\hat{p}+1} \qquad (25)$$

Plots of Eq.(25) are shown for the range $\rho = 1, 2, \dots, 10$ in Figs.5.10-13 to 5.10-15. The values of P run from 1 to 1.8 as in Figs.5.10-4 to 5.10-6. The value of $k = \Delta r/(h/m_0c)$ varies from 1 in Fig.5.10-13 to 0.5 in Fig.5.10-14 and 0.2 in Fig.5.10-15. It is evident that the plots for $P = 1.6$ and 1.8 are quite different from those for $P = 1$, 1.2, 1.4 and erratic. This behavior is also recognizable in Figs.5.10-16 to 5.10-18, which hold for the range $\rho = 5, 6, \dots,$ 25.

For $P = 5$ and $k = 1$, 0.5, 0.2 we obtain the random plots of points of Figs.5.10-19 to 5.10-21. They are very similar to the plots of Figs.5.10-10 to 5.10-12.

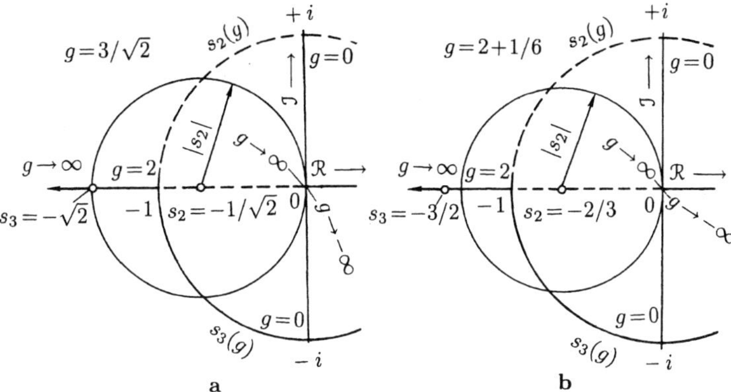

FIG.5.11-1. Plot of the left half-plane of Fig.5.10-1 in analogy to Fig.5.5-2. The values $g=-3/\sqrt{2}$ and $g=-(2+1/6)$ are replaced by $g=3/\sqrt{2}$ and $g=2+1/6$. The points s_3 and s_2 of Fig.5.5-2 become $s_2=-s_3$ and $s_3=-s_2$ in this figure. (a) holds for the limit when the circle of convergence around s_2 goes through s_3 and $s=0$, while (b) holds for smaller absolute values of s_2.

5.11 ANTIPARTICLES

In Fig.5.10-1 we have shown the loci of $s_2(g)$ and $s_3(g)$ both for positive and negative real values, while only the half for positive real values was shown in Figs.5.5-1, 5.5-2, and 5.7-1. We extend here our investigation to negative real values of the s-plane. Figure 5.11-1 is the equivalent of Fig.5.5-2 for negative real values. We see that g is replaced by $-g$, s_3 by $-s_2$, and s_2 by $-s_3$.

We substitute $s=(s-s_2)+s_2$ into Eq.(5.5-12) in order to solve the differential equation with a power series in the point s_2. The relation

$$s_2^2+gs_2+1=0$$

according to Eq.(5.5-13) is needed to obtain:

$$\begin{aligned}&[(s-s_2)^2+(2s_2-s_3)(s-s_2)^2+s_2(s_2-s_3)(s-s_2)]w''(s-s_2)\\&\quad+[2(s-s_3)^2+(4s_2+g-\tilde{\lambda}\tilde{\delta})(s-s_2)+(2s+g-\tilde{\lambda}\tilde{\delta})s_2]w'(s-s_2)\\&\quad-[l(l+1)-\tilde{\gamma}^2]w(s-s_2)=0\end{aligned} \quad (1)$$

For the solution of this differential equation we need a power series according to Eq.(5.5-15):

$$w(s-s_2)=\sum_{\nu=0}^{N}q_3'(\nu)(s-s_2)^{p_3'+\nu} \quad (2)$$

When we compare Eqs.(1) and (2) with Eqs.(6.8-1) and (6.8-2), which hold for Fig.5.5-2, we recognize that the substitutions

$$s_2 \to s_3,\ s_3 \to s_2,\ p_3 \to p_3',\ q_3(\nu) \to q_3'(\nu) \tag{3}$$

transform Eqs.(6.8-1) and (6.8-2) into Eqs.(1) and (2), which hold for Fig.5.11-1. Hence, we can take the results of Section 6.8 and make the substitutions of Eq.(3) to obtain the corresponding results for Eqs.(1) and (2). Let us start with Eq.(6.8-3):

$$\begin{aligned} p_3 &= p = \frac{\tilde{\lambda}\tilde{\delta}}{s_3 - s_2}, && s_3 - s_2 < 0 \text{ according to Fig.5.5-2} \\ p_3' &= p' = \frac{\tilde{\lambda}\tilde{\delta}}{s_2 - s_3} = -\frac{\tilde{\lambda}\tilde{\delta}}{s_3 - s_2} = -p_3, && s_3 - s_2 < 0 \text{ according to Fig.5.11-1} \end{aligned} \tag{4}$$

For the further evaluation of p_3 and p_3' we turn to Eq.(5.5-16):

$$\begin{aligned} p' &= +2\pi Z\alpha \frac{\mathsf{E}}{m_0c^2}\left(1 - \frac{\mathsf{E}^2}{m_0^2c^4}\right)\left[1 + \left(1 - \frac{\mathsf{E}^2}{m_0^2c^4}\right)^2 \left(\frac{\Delta r}{\hbar/m_0c}\right)^2\right]^{-1/2} \\ &= -2\pi Z\alpha \frac{\mathsf{E}}{-m_0c^2}\left(1 - \frac{\mathsf{E}^2}{m_0^2c^4}\right)\left[1 + \left(1 - \frac{\mathsf{E}^2}{m_0^2c^4}\right)^2 \left(\frac{\Delta r}{\hbar/m_0c}\right)^2\right]^{-1/2} \\ &= p_0' + O(\Delta r) \\ p_0' &= -2\pi Z\alpha \frac{\mathsf{E}}{-m_0c^2}\left(1 - \frac{\mathsf{E}^2}{m_0^2c^4}\right) \end{aligned} \tag{5}$$

The change of sign for $p_3' = p'$ in Eq.(4) permits us to write $-m_0$ instead of m_0 in Eq.(5). We postpone a discussion and turn to the rewriting of Eqs.(6.8-5) and (6.8-8). Since we wrote $p_3 = p$ and $p_3' = p'$ in Eq.(4) we must be careful to replace p_3 in Eqs.(6.8-5) and (6.8-8) by p' rather than p_3':

$$\begin{aligned} &\alpha_{3,1}'(0)q_3'(1) + \alpha_{3,0}'(0)q_3'(0) = 0 \\ &\alpha_{3,1}'(0) = s_2(s_2 - s_3)(p' + 1) \\ &\alpha_{3,0}'(0) = s_2p'(p' + 1) - l(l+1) + \tilde{\gamma}^2 \end{aligned} \tag{6}$$

Finally, we rewrite Eq.(6.8-8):

$$\begin{aligned} &\alpha_{3,1}'(\nu)q_3'(\nu+1) + \alpha_{3,0}'(\nu)q_3'(\nu) + \alpha_{3,-1}'(\nu)q_3'(\nu \dot{-} 1) = 0 \\ &\alpha_{3,1}'(\nu) = s_2(s_2 - s_3)(p' + \nu + 1)(\nu + 1) \\ &\alpha_{3,0}'(\nu) = (2s_2 - s_3)(p' + \nu)^2 + [s_2 - p'(s_2 - s_3)](p' + \nu) \\ &\alpha_{3,-1}'(\nu) = (p' + \nu)(p' + \nu - 1) \end{aligned} \tag{7}$$

From Eq.(5.5-23) we obtain with the substitutions of Eq.(3) for κ':

$$\begin{aligned}
\kappa &= \frac{s_3(s_3-s_2)}{2\Delta r} \doteq \frac{m_0c}{\hbar}\left(1-\frac{\mathsf{E}^2}{m_0^2c^4}\right)^{1/2} \\
\kappa' &= \frac{s_2(s_2-s_3)}{2\Delta r} \doteq -\frac{s_2(s_3-s_2)}{2\Delta r} \doteq \frac{-m_0c}{\hbar}\left(1-\frac{\mathsf{E}^2}{m_0^2c^4}\right)^{1/2}
\end{aligned} \tag{8}$$

With κ' we obtain from Eq.(5.5-32):

$$\begin{aligned}
\kappa_1' &\doteq \frac{-m_0c\tilde{\gamma}}{n\hbar}\left[1+\frac{\tilde{\gamma}^2}{n^2}\left(\frac{n}{2l+1}-\frac{1}{2}\right)\right] \\
\kappa_2' &\doteq \frac{-m_0c\tilde{\gamma}}{n'\hbar}\left[1-\frac{\tilde{\gamma}^2}{n'^2}\left(\frac{n'}{2l+1}+\frac{1}{2}\right)\right] \\
\mathsf{E}_1' &\doteq -m_0c^2\left[1-\frac{\tilde{\gamma}^2}{2n^2}-\frac{\tilde{\gamma}^4}{n^4}\left(\frac{n}{2l+1}-\frac{3}{8}\right)\right] \\
\mathsf{E} &\doteq -m_0c^2\left[1-\frac{\tilde{\gamma}^2}{2n'^2}+\frac{\tilde{\gamma}^4}{n'^4}\left(\frac{n'}{2l+1}+\frac{3}{8}\right)\right]
\end{aligned} \tag{9}$$

From Eq.(2) we get in analogy to the transition from Eq.(5.5-15) to (5.5-19) a factorial series:

$$\begin{aligned}
u_3(\rho) &= s_2^\rho\frac{\Gamma(\rho)}{\Gamma(\rho+p'+1)}\sum_{\nu=0}^{N}(-1)^\nu q_3'(\nu)\frac{(p'+1)\dots(p'+\nu)}{(\rho+p'+1)\dots(\rho+p'+\nu)} \\
&\qquad \frac{q_3'(1)}{q_3'(0)} = -\frac{\alpha_{3,0}'(0)}{\alpha_{3,1}'(0)}
\end{aligned} \tag{10}$$

The first term of Eq.(10) cannot be plotted like Eq.(5.8-10) since s_2 is negative. However, we can plot according to Eq.(5.10-23):

$$\begin{aligned}
[u_{\mathrm{P}1}(\rho)u_{\mathrm{P}1}^*(\rho)]^{1/2} &= \left[\left(s_2^\rho\frac{\Gamma(\rho)}{\Gamma(\rho+p'+1)}\right)\left(s_2^\rho\frac{\Gamma(\rho)}{\Gamma(\rho+p'+1)}\right)^*\right]^{1/2} \\
&= |s_2|^\rho\frac{\Gamma(\rho)}{\Gamma(\rho+p'+1)}, \quad \rho = 1,\ 2,\dots,\ N
\end{aligned} \tag{11}$$

We essentially obtain Eq.(5.8-10) except that $\tilde{p}$ is replaced by p'. This difference disappears if we substitute $\mathsf{E}_1'/(-m_0c^2) \doteq \mathsf{E}'/(-m_0c^2)$ or $\mathsf{E}_2'/(-m_0c^2) \doteq \mathsf{E}'/(-m_0c^2)$ from Eq.(9):

$$\frac{\mathsf{E}}{-m_0c^2} = \frac{\mathsf{E}_1'}{-m_0c^2} = \frac{\mathsf{E}_2'}{-m_0c^2} = 1 - \frac{\tilde{\gamma}^2}{2n^2} + O(\tilde{\gamma}^4) \tag{12}$$

This substitution produces for p' of Eq.(5) the same result as the substitution of Eq.(5.8-6) produces for p in Eq.(5.5-16):

$$p' = p = -2\pi \mathsf{Z}\alpha \left(1 - \frac{\tilde{\gamma}^2}{2n^2}\right)\left[1 - \left(1 - \frac{\tilde{\gamma}^2}{2n^2}\right)^2\right]$$
$$\times \left\{1 + \left[1 - \left(1 - \frac{\tilde{\gamma}^2}{2n^2}\right)^2\right]^2 (2\pi k)^2\right\}^{-1/2}$$
$$\tilde{\gamma} = 4\pi \mathsf{Z}\alpha,\ \mathsf{Z} = 1,\ 2\ldots,\ \alpha = Ze^2/2h = 7.297\ 535 \times 10^{-3}$$
$$n = 1,\ 2, \ldots,\ k = \Delta r/(h/m_0c) \tag{13}$$

Equations (5.8-10) for a particle and (11) for its antiparticle are equal.

Consider the second term of the factorial series of Eq.(10). With the help of Eq.(6) we obtain:

$$\frac{q_3'(1)}{q_3'(0)} = -\frac{\alpha_{3,0}'(0)}{\alpha_{3,1}'(0)} = \frac{p'(p'+1) - [l(l+1) - \tilde{\gamma}^2]/s_2}{(s_3 - s_2)(p'+1)} \tag{14}$$

From Eq.(6.8-5) we obtain the result:

$$\frac{q_3(1)}{q_3(0)} = -\frac{\alpha_{3,0}(0)}{\alpha_{3,1}(0)} = \frac{p(p+1) - [l(l+1) - \tilde{\gamma}^2]/s_3}{(s_2 - s_3)(p+1)} \tag{15}$$

Equations (14) and (15) are different. Even though we again have $p' = p$ we recognize in Fig.5.5-2a the values

$$s_3 = 1/\sqrt{2},\ s_2 - s_3 = \sqrt{2}/2,\ g < -2 \tag{16}$$

which must be used with Eq.(15). But Fig.5.11-1a shows the different values

$$s_2 = -1/\sqrt{2},\ s_3 - s_2 = -\sqrt{2}/2,\ g > +2 \tag{17}$$

Equations (17) and (16) bring Eqs.(14) and (15) into the following form:

$$\frac{q_3(1)}{q_3(0)} = -\frac{q_3'(1)}{q_3'(0)} = \sqrt{2}p - 2[l(l+1) - \tilde{\gamma}^2] \tag{18}$$

The points $g = 0$ in Figs.5.10-1 and 5.11-1 are of interest since a particle changes to an antiparticle or viceversa at these points. We obtain for $g = 0$ from Eq.(5.5-3):

$$g = -\left[2 + \frac{1}{4}\left(\frac{\Delta\tilde{r}}{\hbar/m_0c}\right)^2\right] = 0$$

$$\left(\frac{\Delta\tilde{r}}{\hbar/m_0c}\right)^2 = 4\left(1 - \frac{\mathsf{E}^2}{m_0^2c^4}\right)\left(\frac{\Delta r}{\hbar/m_0c}\right)^2 = -8$$

$$\frac{\mathsf{E}}{m_0c^2} = \pm\left[1 + 2\left(\frac{\Delta r}{\hbar/m_0c}\right)^2\right]^{1/2} \tag{19}$$

The energy E at $g = 0$ becomes strictly an effect of the resolution Δr of the observation and says nothing about the observed particle. If one approaches the points $g = 0$ in Fig.5.10-1 from the particle side $g < 0$ one obtains

$$\frac{\mathsf{E}}{m_0c^2} = +\left[1 + 2\left(\frac{\Delta r}{\hbar/m_0c}\right)^2\right]^{1/2} \tag{20}$$

but if one approaches $g = 0$ from the antiparticle side $g > 0$ one obtains

$$\frac{\mathsf{E}}{-m_0c^2} = +\left[1 + 2\left(\frac{\Delta r}{\hbar/m_0c}\right)^2\right]^{1/2} \tag{21}$$

Hence, the finite but otherwise undetermined resolution Δr permits a transition between particle and antiparticle without a quantum jump. This is a result characteristic for a difference theory using Δr rather than dr.

We note that $[\Delta r/(\hbar/(m_0c)]^2$ in Eq.(21) is not changed if we replace it with $\{\Delta r/[\hbar/(-m_0c)]\}^2$.

6 Appendix

6.1 Difference Operators of Higher Order

The first- and second-order difference operators have been introduced by Eqs.(1.2-1), (1.2-4), (1.2-5), and (1.2-7). For difference operators of higher order we must revisit the definitions. Difficulties are created by the use of symmetric difference operators in Eqs.(1.2-1) and (1.2-7). The nonsymmetric difference operators of Eqs.(1.2-4) and (1.2-5) are easier to extend; this extension is used by mathematicians (Nörlund 1924; Milne-Thomson 1951; Spiegel 1994). There are a number of reasons to prefer the symmetric difference operators to the nonsymmetric ones in physics (Harmuth 1989, Sec.8.1 and 8.2).

Starting with Eq.(1.2-1) we redefine the symmetric difference operator of first order:

$$\frac{\tilde{\Delta} A(\theta)}{\tilde{\Delta}\theta} = \frac{A(\theta+\Delta\theta/2) - A(\theta-\Delta\theta/2)}{\Delta\theta} \tag{1}$$

The second-, third- and fourth-order difference operators become:

$$\begin{aligned}\frac{\tilde{\Delta}^2 A(\theta)}{\tilde{\Delta}\theta^2} &= \frac{\tilde{\Delta}}{\tilde{\Delta}\theta}\left(\frac{\tilde{\Delta} A(\theta)}{\tilde{\Delta}\theta}\right) = \frac{\tilde{\Delta}}{\tilde{\Delta}\theta}\frac{A(\theta+\Delta\theta/2) - A(\theta-\Delta\theta/2)}{\Delta\theta}\\ &= \frac{A(\theta+\Delta\theta) - 2A(\theta) + A(\theta-\Delta\theta)}{(\Delta\theta)^2}\end{aligned} \tag{2}$$

$$\begin{aligned}\frac{\tilde{\Delta}^3 A(\theta)}{\tilde{\Delta}\theta^3} &= \frac{\tilde{\Delta}}{\tilde{\Delta}\theta}\left(\frac{\tilde{\Delta}^2 A(\theta)}{\tilde{\Delta}\theta^2}\right)\\ &= \frac{A(\theta+3\Delta\theta/2) - 3A(\theta+\Delta\theta/2) + 3A(\theta-\Delta\theta/2) - A(\theta-3\Delta\theta/2)}{(\Delta\theta)^3}\end{aligned} \tag{3}$$

$$\begin{aligned}\frac{\tilde{\Delta}^4 A(\theta)}{\tilde{\Delta}\theta^4} &= \frac{\tilde{\Delta}}{\tilde{\Delta}\theta}\left(\frac{\tilde{\Delta}^3 A(\theta)}{\tilde{\Delta}\theta^3}\right)\\ &= \frac{A(\theta+2\Delta\theta) - 4A(\theta+\Delta\theta) + 6A(\theta) - 4A(\theta-\Delta\theta) + A(\theta-2\Delta\theta)}{(\Delta\theta)^4}\end{aligned} \tag{4}$$

ISSN 1076-5670/05
DOI: 10.1016/S1076-5670(05)37006-6

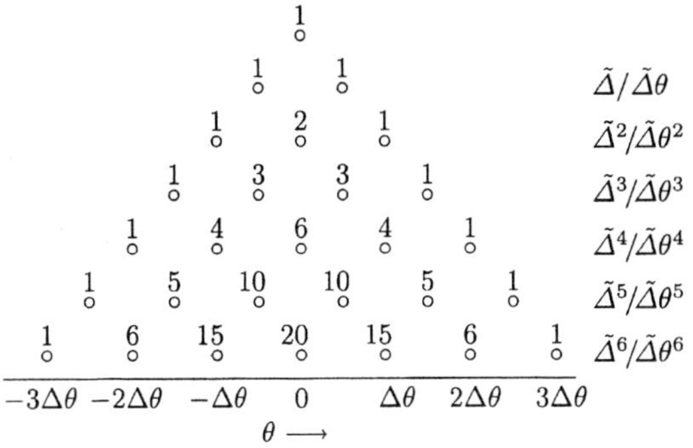

			1				
		1		1			$\tilde{\Delta}/\tilde{\Delta}\theta$
		1	2	1			$\tilde{\Delta}^2/\tilde{\Delta}\theta^2$
	1	3		3	1		$\tilde{\Delta}^3/\tilde{\Delta}\theta^3$
	1	4	6	4	1		$\tilde{\Delta}^4/\tilde{\Delta}\theta^4$
1	5	10		10	5	1	$\tilde{\Delta}^5/\tilde{\Delta}\theta^5$
1	6	15	20	15	6	1	$\tilde{\Delta}^6/\tilde{\Delta}\theta^6$
$-3\Delta\theta$	$-2\Delta\theta$	$-\Delta\theta$	0	$\Delta\theta$	$2\Delta\theta$	$3\Delta\theta$	

$\theta \longrightarrow$

FIG.6.1-1. Arguments and factors of symmetric difference operators from order 1 to 6 using odd and even multiples of $\Delta\theta/2$ according to Eqs.(1)–(4).

TABLE 6.1-1

DIFFERENCE OPERATORS FROM FIRST- TO SIXTH-ORDER ACCORDING TO EQS.(1) TO (4).

$$[A(\theta+\frac{1}{2\Delta\theta}) \qquad -A(\theta-\frac{1}{2\Delta\theta})]\frac{1}{\Delta\theta}$$

$$[A(\theta+\Delta\theta)-2A(\theta)+A(\theta-\Delta\theta)]\frac{1}{(\Delta\theta)^2}$$

$$[A(\theta+\frac{3\Delta\theta}{2})-3A(\theta+\frac{\Delta\theta}{2}) \qquad +3A(\theta-\frac{\Delta\theta}{2})-A(\theta-\frac{3\Delta\theta}{2})]\frac{1}{(\Delta\theta)^3}$$

$$[A(\theta+2\Delta\theta)-4A(\theta+\Delta\theta)+6A(\theta)-4A(\theta-\Delta\theta)+A(\theta-2\Delta\theta)]\frac{1}{(2\Delta\theta)^4}$$

$$[A(\theta+\frac{5\Delta\theta}{2})-5A(\theta+\frac{3\Delta\theta}{2})+10A(\theta+\frac{\Delta\theta}{2}) \qquad -10A(\theta-\frac{\Delta\theta}{2})+5A(\theta-\frac{3\Delta\theta}{2})-A(\theta-\frac{5\Delta\theta}{2})]\frac{1}{(\Delta\theta)^5}$$

$$[A(\theta+3\Delta\theta)-6A(\theta+2\Delta\theta)+15A(\theta-\Delta\theta)-20A(\theta)+15A(\theta-\Delta\theta)-6A(\theta-2\Delta\theta)+A(\theta-3\Delta\theta)]\frac{1}{(\Delta\theta)^6}$$

The operators of odd order, $\tilde{\Delta}/\tilde{\Delta}\theta$ and $\tilde{\Delta}^3/\tilde{\Delta}\theta^3$, contain odd multiples of the difference $\Delta\theta/2$ while the operators of even order, $\tilde{\Delta}^2/\tilde{\Delta}\theta^2$ and $\tilde{\Delta}^4/\tilde{\Delta}\theta^4$, contain even multiples of the difference $\Delta\theta/2$. Figure 6.1-1 shows what causes this complication. In Eq.(1.2-1) we had sidestepped it by using the arguments $\theta \pm \Delta\theta$ and the denominator $2\Delta\theta$. The first six difference operators listed in Table 6.1-1 make the connection clear.

An alternative way to generalize the symmetric difference operators of Eqs.(1.2-1) and (1.2-7) is as follows:

$$\frac{\tilde{\Delta}A(\theta)}{\tilde{\Delta}\theta} = \frac{A(\theta+\Delta\theta)-A(\theta-\Delta\theta)}{2\Delta\theta} \tag{5}$$

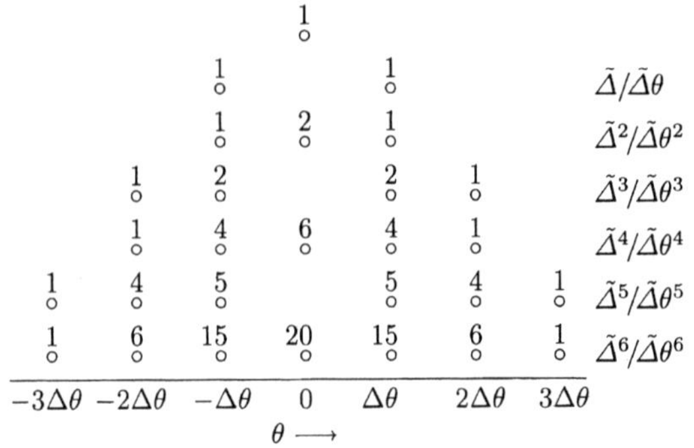

			1				
		1		1			$\tilde{\Delta}/\tilde{\Delta}\theta$
		1	2	1			$\tilde{\Delta}^2/\tilde{\Delta}\theta^2$
	1	2		2	1		$\tilde{\Delta}^3/\tilde{\Delta}\theta^3$
	1	4	6	4	1		$\tilde{\Delta}^4/\tilde{\Delta}\theta^4$
1	4	5		5	4	1	$\tilde{\Delta}^5/\tilde{\Delta}\theta^5$
1	6	15	20	15	6	1	$\tilde{\Delta}^6/\tilde{\Delta}\theta^6$
$-3\Delta\theta$	$-2\Delta\theta$	$-\Delta\theta$	0	$\Delta\theta$	$2\Delta\theta$	$3\Delta\theta$	

$\theta \longrightarrow$

FIG.6.1-2. Arguments and factors of symmetric difference operators from order 1 to 6 using integer multiples of $\Delta\theta$ according to Eqs.(5)–(8).

TABLE 6.1-2

DIFFERENCE OPERATORS FROM FIRST- TO SIXTH-ORDER ACCORDING TO EQS.(5) TO (8).

$$[A(\theta+\Delta\theta) \qquad -A(\theta-\Delta\theta)]\frac{1}{2\Delta\theta}$$

$$[A(\theta+\Delta\theta)-2A(\theta)+A(\theta-\Delta\theta)]\frac{1}{(\Delta\theta)^2}$$

$$[A(\theta+2\Delta\theta)-2A(\theta+\Delta\theta) \qquad +2A(\theta-\Delta\theta)-A(\theta-2\Delta\theta)]\frac{1}{2(\Delta\theta)^3}$$

$$[A(\theta+2\Delta\theta)-4A(\theta+\Delta\theta)+6A(\theta)-4A(\theta-\Delta\theta)+A(\theta-2\Delta\theta)]\frac{1}{(\Delta\theta)^4}$$

$$[A(\theta+3\Delta\theta)-4A(\theta+2\Delta\theta)+5A(\theta+\Delta\theta) \qquad -5A(\theta-\Delta\theta)+4A(\theta-2\Delta\theta)-A(\theta-3\Delta\theta)]\frac{1}{2(\Delta\theta)^5}$$

$$[A(\theta+3\Delta\theta)-6A(\theta+2\Delta\theta)+15A(\theta-\Delta\theta)-20A(\theta)+15A(\theta-\Delta\theta)-6A(\theta-2\Delta\theta)+A(\theta-3\Delta\theta)]\frac{1}{(\Delta\theta)^6}$$

$$\frac{\tilde{\Delta}^2 A(\theta)}{\tilde{\Delta}\theta^2} = \frac{A(\theta+\Delta\theta)-2A(\theta)+A(\theta+\Delta\theta)}{(\Delta\theta)^2} \tag{6}$$

$$\frac{\tilde{\Delta}^3 A(\theta)}{\tilde{\Delta}\theta^3} = \frac{\tilde{\Delta}}{\tilde{\Delta}\theta}\left(\frac{\tilde{\Delta}^2 A(\theta)}{\tilde{\Delta}\theta^2}\right) = \frac{1}{2(\Delta\theta)^3}\Big\{[A(\theta+2\Delta\theta)-2A(\theta+\Delta\theta)+A(\theta)]$$
$$-[A(\theta)-2A(\theta-\Delta\theta)+A(\theta-2\Delta\theta)]\Big\}$$
$$= \frac{A(\theta+2\Delta\theta)-2A(\theta+\Delta\theta)+2A(\theta-\Delta\theta)-A(\theta-2\Delta\theta)}{2(\Delta\theta)^3} \tag{7}$$

$$\frac{\tilde{\Delta}^4 A(\theta)}{\tilde{\Delta}\theta^4} = \frac{\tilde{\Delta}^2}{\tilde{\Delta}\theta^2}\left(\frac{\tilde{\Delta}^2 A(\theta)}{\tilde{\Delta}\theta^2}\right) = \frac{1}{(\Delta\theta)^4}\Big\{[A(\theta+2\Delta\theta) - 2A(\theta+\Delta\theta) + A(\theta)]$$
$$- 2[A(\theta+\Delta\theta) - 2A(\theta) + A(\theta-\Delta\theta)] + [A(\theta) - 2A(\theta-\Delta\theta) + A(\theta-2\Delta\theta)]\Big\}$$
$$= \frac{A(\theta+2\Delta\theta) - 4A(\theta+\Delta\theta) + 6A(\theta) - 4A(\theta-\Delta\theta) + A(\theta-2\Delta\theta)}{(\Delta\theta)^4} \tag{8}$$

There are no odd and even multiples of $\Delta\theta/2$ as in Eqs.(1)–(4) but only integer multiples of $\Delta\theta$. Figure 6.1-2 shows the scheme of amplitudes and arguments. For difference operators of even order, $\tilde{\Delta}^{2n}/\tilde{\Delta}\theta^{2n}$, we obtain the same binomial numbers as in Fig.6.1-1, but for difference operators of odd order, $\tilde{\Delta}^{2n+1}/\tilde{\Delta}\theta^{2n+1}$, we obtain different numbers. In analogy to Table 6.1-1 we list the difference operators up to order 6 in Table 6.1-2.

In order to decide whether to use the operators of Table 6.1-1 or of Table 6.1-2 consider a finite set of numbers $f(\theta)$, $f(\theta+\Delta\theta)$, ... , $f(\theta+m\Delta\theta)$. These numbers may have been produced by calculation to be checked by measurements or they may have been obtained by measurements to check a theory that should yield the same numbers by calculation. The operators of Table 6.1-2 correspond to this condition better than the operators of Table 6.1-1. One could interpolate the set of numbers by writing

$$f[\theta + (2n+1)\Delta\theta/2] = \frac{1}{2}\Big\{f(\theta+n\Delta\theta) + f[\theta+(n+1)\Delta\theta]\Big\} \tag{9}$$

to obtain values for integer multiples of $\Delta\theta/2$. This interpolation would not add any information if the numbers $f(\theta)$, $f(\theta+\Delta\theta)$, ... $f(\theta+m\Delta\theta)$ represent physical measurements, it would only create the illusion of an improved resolution.

As an example let us write Eq.(3) with the terms $A(\theta+3\Delta\theta/2)$ to $A(\theta-3\Delta\theta/2)$ interpolated from $A(\theta+2\Delta\theta)$, $A(\theta+\Delta\theta)$, ... , $A(\theta-2\Delta\theta)$:

$$\begin{aligned}
A(\theta+3\Delta\theta/2) &= \frac{1}{2}[A(\theta+2\Delta\theta) + A(\theta+\Delta\theta)] \\
A(\theta+\Delta\theta/2) &= \frac{1}{2}[A(\theta+\Delta\theta) + A(\theta)] \\
A(\theta-\Delta\theta/2) &= \frac{1}{2}[A(\theta) + A(\theta-\Delta\theta)] \\
A(\theta-3\Delta\theta/2) &= \frac{1}{2}[A(\theta-\Delta\theta) + A(\theta-2\Delta\theta)]
\end{aligned} \tag{10}$$

If we substitute Eq.(10) into Eq.(3) we obtain Eq.(7).

This becomes quite different if $f(\theta)$, ... , $f(\theta+m\Delta\theta)$ represent the values of a defined mathematical function, say the Gamma function $\Gamma(\theta)$. The value of

$\Gamma(2.5)$ is not the average of $\Gamma(2)$ and $\Gamma(3)$. Hence, the information contained in the set $\Gamma(\theta)$, $\Gamma(\theta+\Delta\theta)$, ... , $\Gamma(\theta+m\Delta\theta)$ is increased by specifying the additional values $\Gamma(\theta+0.5\Delta\theta)$, ... , $\Gamma[\theta+(m-1/2)\Delta\theta]$. Unlimited information can be obtained from many defined mathematical functions by sufficiently fine interpolation. This is completely different from the finite set of numbers that can be produced by or checked by physical measurements. No mathematical manipulations can increase the information contained in them.

The transition $\Delta\theta \to d\theta$ yields the same result for the difference operators of Tables 6.1-1 and 6.1-2, but the inverse transition $d\theta \to \Delta\theta$ is not unique and we must decide which set of difference operators to use. For a finite set of numbers $f(\theta)$, $f(\theta+\Delta\theta)$, ... , $f(\theta+m\Delta\theta)$ that can be obtained by or for observation in physics the operators according to Fig.6.1-2 or Table 6.1-2 are simpler than the ones of Fig.6.1-1 or Table 6.1-1 plus substitutions according to Eq.(10); we shall use them. This is a good example of how concepts of pure mathematics must be carefully analyzed before using them in physics.

6.2 Extension of Section 3.1 for $\Delta x \ll \lambda_C$

With the help of Table 6.1-2 we may rewrite Eq.(3.1-11), which holds for small values of Δx, in explicit form:

$$
\begin{aligned}
&-\frac{2\lambda_C A_e(\zeta,\theta)\hbar^2}{e(\Delta x)^2}\left(\frac{\hbar}{m_0 c\Delta x}\right)^2\left(\frac{\tilde{\Delta}}{\tilde{\Delta}\zeta}-\frac{ie\Delta x A_{m0x}}{\hbar}\right)^4\Psi_0(\zeta,\theta)\\
&=-\frac{2\lambda_C A_e(\zeta,\theta)\hbar^2}{e(\Delta x)^2}\left(\frac{\hbar}{m_0 c\Delta x}\right)^2\left[\frac{\tilde{\Delta}^4}{\tilde{\Delta}\zeta^4}-4\frac{ie\Delta x A_{m0x}}{\hbar}\frac{\tilde{\Delta}^3}{\tilde{\Delta}\zeta^3}+6\left(\frac{ie\Delta x A_{m0x}}{\hbar}\right)^2\frac{\tilde{\Delta}^2}{\tilde{\Delta}\zeta^2}\right.\\
&\qquad\left.-4\left(\frac{ie\Delta x A_{m0x}}{\hbar}\right)^3\frac{\tilde{\Delta}}{\tilde{\Delta}\zeta}+\left(\frac{ie\Delta x A_{m0x}}{\hbar}\right)^4\right]\Psi_0(\zeta,\theta)\\
&=-\frac{2\lambda_C A_e(\zeta,\theta)\hbar^2}{e(\Delta x)^2}\left(\frac{\hbar}{m_0 c\Delta x}\right)^2\left[\frac{1}{(\Delta\zeta)^4}[\Psi_0(\zeta+2\Delta\zeta,\theta)-4\Psi_0(\zeta+\Delta\zeta,\theta)\right.\\
&\qquad+6\Psi_0(\zeta,\theta)-4\Psi_0(\zeta-\Delta\zeta,\theta)+\Psi_0(\zeta-2\Delta\zeta,\theta)]\\
&+4\frac{ie\Delta x A_{m0x}}{\hbar}\frac{1}{2(\Delta\zeta)^3}[\Psi_0(\zeta+2\Delta\zeta,\theta)-2\Psi_0(\zeta+\Delta\zeta,\theta)+2\Psi_0(\zeta-\Delta\zeta,\theta)-\Psi_0(\zeta-2\Delta\zeta,\theta)]\\
&\qquad-6\left(\frac{ie\Delta x A_{m0x}}{\hbar}\right)^2\frac{1}{(\Delta\zeta)^2}[\Psi_0(\zeta+\Delta\zeta,\theta)-2\Psi_0(\zeta,\theta)+\Psi_0(\zeta-\Delta\zeta,\theta)]\\
&\qquad+4\left(\frac{ie\Delta x A_{m0x}}{\hbar}\right)^3\frac{1}{2\Delta\zeta}[\Psi_0(\zeta+\Delta\zeta,\theta)-\Psi_0(\zeta-\Delta\zeta,\theta)]\\
&\qquad\left.-\left(\frac{ie\Delta x A_{m0x}}{\hbar}\right)^4\Psi_0(\zeta,\theta)\right]
\end{aligned}
\tag{1}
$$

Equation (3.1-18) is another equation that holds for small values of Δx and requires Table 6.1-2 for writing it in explicit form. First we write the term with $\tilde{\Delta}/\tilde{\Delta}\zeta$:

$$\left(\frac{\tilde{\Delta}}{\tilde{\Delta}\zeta} - \frac{ie\Delta x A_{\mathrm{m}0x}}{\hbar}\right)^3 \Psi_0(\zeta,\theta) = \left[\frac{\tilde{\Delta}^3}{\tilde{\Delta}\zeta^3} - 3\frac{ie\Delta x A_{\mathrm{m}0x}}{\hbar}\frac{\tilde{\Delta}^2}{\tilde{\Delta}\zeta^2} + 3\left(\frac{ie\Delta x A_{\mathrm{m}0x}}{\hbar}\right)^2 \frac{\tilde{\Delta}}{\tilde{\Delta}\zeta} - \left(\frac{ie\Delta x A_{\mathrm{m}0x}}{\hbar}\right)^3\right]\Psi_0(\zeta,\theta)$$

$$\begin{aligned}
&= \frac{1}{2(\Delta\zeta)^3}[\Psi_0(\zeta+2\Delta\zeta,\theta) - 2\Psi_0(\zeta+\Delta\zeta,\theta) + 2\Psi_0(\zeta-\Delta\zeta,\theta) - \Psi_0(\zeta-2\Delta\zeta,\theta)] \\
&\quad - 3\frac{ie\Delta x A_{\mathrm{m}0x}}{\hbar}\frac{1}{(\Delta\zeta)^2}[\Psi_0(\zeta+\Delta\zeta,\theta) - 2\Psi_0(\zeta,\theta) + \Psi_0(\zeta-\Delta\zeta,\theta)] \\
&\quad + 3\left(\frac{ie\Delta x A_{\mathrm{m}0x}}{\hbar}\right)^2 \frac{1}{2\Delta\zeta}[\Psi_0(\zeta+\Delta\zeta,\theta) - \Psi_0(\zeta-\Delta\zeta,\theta)] \\
&\quad - \left(\frac{ie\Delta x A_{\mathrm{m}0x}}{\hbar}\right)^3 \Psi_0(\zeta,\theta) \qquad (2)
\end{aligned}$$

The whole Eq.(3.1-18) becomes:

$$\frac{2Ze^2}{\alpha c}\left(\frac{\hbar}{m_0 c\Delta x}\right)^4 (A_{\mathrm{e}0z}A_{\mathrm{m}0y} - A_{\mathrm{e}0y}A_{\mathrm{m}0z})\left(\frac{\tilde{\Delta}}{\tilde{\Delta}\theta} + \frac{ie\Delta t\phi_{\mathrm{e}0}}{\hbar}\right) \times \left(\frac{\tilde{\Delta}}{\tilde{\Delta}\zeta} - \frac{ie\Delta x A_{\mathrm{m}0x}}{\hbar}\right)^3 \Psi_0(\zeta,\theta)$$

$$\begin{aligned}
&= \frac{2Ze^2}{\alpha c}\left(\frac{\hbar}{m_0 c\Delta x}\right)^4 (A_{\mathrm{e}0z}A_{\mathrm{m}0y} - A_{\mathrm{e}0y}A_{\mathrm{m}0z})\left\{\frac{1}{2\Delta\theta}\left[\frac{1}{2(\Delta\zeta)^3}\left\{[\Psi_0(\zeta+2\Delta\zeta,\theta+\Delta\theta)\right.\right.\right. \\
&\quad - \Psi_0(\zeta+2\Delta\zeta,\theta-\Delta\theta)] - 2[\Psi_0(\zeta+\Delta\zeta,\theta+\Delta\theta) - \Psi_0(\zeta+\Delta\zeta,\theta-\Delta\theta)] \\
&\quad + 2[\Psi_0(\zeta-\Delta\zeta,\theta+\Delta\theta) - \Psi_0(\zeta-\Delta\zeta,\theta-\Delta\theta)] - [\Psi_0(\zeta-2\Delta\zeta,\theta+\Delta\theta) \\
&\quad \left. - \Psi_0(\zeta-2\Delta\zeta,\theta-\Delta\theta)]\right\} \\
&\quad - 3\frac{ie\Delta x A_{\mathrm{m}0x}}{\hbar}\frac{1}{(\Delta\zeta)^2}\left\{[\Psi_0(\zeta+\Delta\zeta,\theta+\Delta\theta) - \Psi_0(\zeta+\Delta\zeta,\theta-\Delta\theta)]\right. \\
&\quad - 2[\Psi_0(\zeta,\theta+\Delta\theta) - \Psi_0(\zeta,\theta-\Delta\theta)] + [\Psi_0(\zeta-\Delta\zeta,\theta+\Delta\theta) \\
&\quad \left. - \Psi_0(\zeta-\Delta\zeta,\theta-\Delta\theta)]\right\} \\
&\quad + 3\left(\frac{ie\Delta x A_{\mathrm{m}0y}}{\hbar}\right)^2 \frac{1}{2\Delta\zeta}\left\{[\Psi_0(\zeta+\Delta\zeta,\theta+\Delta\theta) - \Psi_0(\zeta+\Delta\zeta,\theta-\Delta\theta)]\right.
\end{aligned}$$

$$
\begin{aligned}
&\qquad - [\Psi_0(\zeta - \Delta\zeta, \theta + \Delta\theta) - \Psi_0(\zeta - \Delta\zeta, \theta - \Delta\theta)]\Big\} \\
&\qquad\qquad - \left(\frac{ie\Delta x A_{\mathrm{m}0x}}{\hbar}\right)^3 [\Psi_0(\zeta, \theta + \Delta\theta) - \Psi_0(\zeta, \theta - \Delta\theta)]\Bigg] \\
&+ \frac{ie\Delta t\phi_{\mathrm{e}0}}{\hbar}\Bigg[\frac{1}{2(\Delta\zeta)^3}[\Psi_0(\zeta{+}2\Delta\zeta,\theta) - 2\Psi_0(\zeta{+}\Delta\zeta,\theta) + 2\Psi_0(\zeta{-}\Delta\zeta,\theta) - \Psi_0(\zeta{-}2\Delta\zeta,\theta)] \\
&\qquad - 3\frac{ie\Delta x A_{\mathrm{m}0x}}{\hbar}\frac{1}{(\Delta\zeta)^2}[\Psi_0(\zeta + \Delta\zeta, \theta) - 2\Psi_0(\zeta, \theta) + \Psi_0(\zeta - \Delta\zeta, \theta)] \\
&\qquad + 3\left(\frac{ie\Delta x A_{\mathrm{m}0x}}{\hbar}\right)^2 \frac{1}{2\Delta\zeta}[\Psi_0(\zeta + \Delta\zeta, \theta) - \Psi_0(\zeta - \Delta\zeta, \theta)] \\
&\qquad\qquad - \left(\frac{ie\Delta x A_{\mathrm{m}0x}}{\hbar}\right)^3 \Psi_0(\zeta, \theta)\Bigg]\Bigg\} \qquad (3)
\end{aligned}
$$

6.3 Solution of Inhomogeneous Difference Equations

We have to solve the inhomogeneous difference equations (3.2-41) and (3.2-42). The equations are equal except for the inhomogeneous term. We solve Eq.(3.2-42) for $T_\kappa(\theta)$ and indicate only at the end the changes required for Eq.(3.2-41) and $S_\kappa(\theta)$. This is done because $S_\kappa(\theta)$ will be eliminated by a boundary condition while the intermediate steps of the solution of $T_\kappa(\theta)$ will be frequently referred to. A shorter notation is used. Equation (3.2-42) is brought into the following form[1]:

$$
\begin{aligned}
&p_2 s(\theta + 2) + p_1 s(\theta + 1) + p_0 s(\theta) + p_{-1} s(\theta - 1) + p_{-2} s(\theta - 2) = H_{\mathrm{s}\kappa}(\theta, \kappa) \\
&p_2 = p_{-2} = e^{2i\lambda_1\lambda_3}, \; p_1 = p_{-1} = -4\cos\frac{2\pi\kappa}{N} e^{i\lambda_1\lambda_3}, \; p_0 = 2\left(2 + \cos\frac{2\pi\kappa}{N}\right) \\
&H_{\mathrm{s}\kappa}(\theta, \kappa) = -e^{i\lambda_1\lambda_3} G_{\mathrm{s}\kappa}(\theta{+}1, \kappa) + 2\cos\frac{2\pi\kappa}{N} G_{\mathrm{s}\kappa}(\theta, \kappa) - e^{-i\lambda_1\lambda_3} G_{\mathrm{s}\kappa}(\theta{-}1, \kappa) \qquad (1)
\end{aligned}
$$

The general solution of the homogeneous equation

$$
p_2 s(\theta + 2) + p_1 s(\theta + 1) + p_0 s(\theta) + p_{-1}(\theta - 1) + p_{-2} s(\theta - 2) = 0 \qquad (2)
$$

is given by the four functions in Eq.(3.2-54):

$$
\begin{aligned}
s_1(\theta) &= v^\theta_{\kappa 1} = e^{i(2\pi\kappa/N - \lambda_1\lambda_3)\theta}, &\quad s_3(\theta) &= \theta v^\theta_{\kappa 1} = \theta e^{i(2\pi\kappa/N - \lambda_1\lambda_3)\theta} \\
s_2(\theta) &= v^\theta_{\kappa 2} = e^{-i(2\pi\kappa/N + \lambda_1\lambda_3)\theta}, &\quad s_4(\theta) &= \theta v^\theta_{\kappa 2} = \theta e^{-i(2\pi\kappa/N + \lambda_1\lambda_3)\theta} \qquad (3)
\end{aligned}
$$

[1] See Nörlund 1924, p. 396; 1929, p. 22, 125; Milne-Thomson 1951, p. 374.

We use the method of variation of the constant to find a particular solution $s(\theta)$ of Eq.(1) with the help of these four solutions of the homogeneous Eq.(2):

$$v(\theta-2) = d_{\kappa 1}(\theta)s_1(\theta-2)+d_{\kappa 2}(\theta)s_2(\theta-2)+d_{\kappa 3}(\theta)s_3(\theta-2)+d_{\kappa 4}(\theta)s_4(\theta-2) \quad (4)$$

Since we have four arbitrary functions $d_{\kappa 1}(\theta)$ to $d_{\kappa 4}(\theta)$ we can choose three more conditions in addition to Eq.(4). Using intuition we make the following choice that will be justified by its success:

$$\begin{aligned}
v(\theta-1) &= d_{\kappa 1}(\theta)s_1(\theta-1) + d_{\kappa 2}(\theta)s_2(\theta-1) \\
&\quad + d_{\kappa 3}(\theta)s_3(\theta-1) + d_{\kappa 4}(\theta)s_4(\theta-1) &(5)\\
v(\theta) &= d_{\kappa 1}(\theta)s_1(\theta) + d_{\kappa 2}(\theta)s_2(\theta) + d_{\kappa 3}(\theta)s_3(\theta) + d_{\kappa 4}(\theta)s_4(\theta) &(6)\\
v(\theta+1) &= d_{\kappa 1}(\theta)s_1(\theta+1) + d_{\kappa 2}(\theta)s_2(\theta+1) \\
&\quad + d_{\kappa 3}(\theta)s_3(\theta+1) + d_{\kappa 4}(\theta)s_4(\theta+1) &(7)
\end{aligned}$$

If we increase θ of $d_{\kappa 1}(\theta)$, $d_{\kappa 2}(\theta)$, $d_{\kappa 3}(\theta)$, and $d_{\kappa_4}(\theta)$ by 1 in Eqs.(4), (5), and (6) we obtain

$$\begin{aligned}
v(\theta-1) &= d_{\kappa 1}(\theta+1)s_1(\theta-1) + d_{\kappa 2}(\theta+1)s_2(\theta-1) \\
&\quad + d_{\kappa 3}(\theta+1)s_3(\theta-1) + d_{\kappa 4}(\theta+1)s_4(\theta-1) &(8)\\
v(\theta) &= d_{\kappa 1}(\theta+1)s_1(\theta) + d_{\kappa 2}(\theta+1)s_2(\theta) \\
&\quad + d_{\kappa 3}(\theta+1)s_3(\theta) + d_{\kappa 4}(\theta+1)s_4(\theta) &(9)\\
v(\theta+1) &= d_{\kappa 1}(\theta+1)s_1(\theta+1) + d_{\kappa 2}(\theta+1)s_2(\theta+1) \\
&\quad + d_{\kappa 3}(0+1)s_3(0+1) + d_{\kappa 4+1}(0)s_4(0+1) &(10)
\end{aligned}$$

Subtraction of Eqs.(5), (6), and (7) from Eqs.(8), (9), and (10) yields with the notation

$$\Delta d_{\kappa i}(\theta) = d_{\kappa i}(\theta+1) - d_{\kappa i}(\theta), \quad i = 1,\ 2,\ 3,\ 4 \quad (11)$$

the result

$$\begin{aligned}
&s_1(\theta-1)\Delta d_{\kappa 1}(\theta) + s_2(\theta-1)\Delta d_{\kappa 2}(\theta) \\
&\quad + s_3(\theta-1)\Delta d_{\kappa 3}(\theta) + s_4(\theta-1)\Delta d_{\kappa 4}(\theta) = 0 \quad (12)
\end{aligned}$$

$$s_1(\theta)\Delta d_{\kappa 1}(\theta) + s_2(\theta)\Delta d_{\kappa 2}(\theta) + s_3(\theta)\Delta d_{\kappa 3}(\theta) + s_4(\theta)\Delta d_{\kappa 4}(\theta) = 0 \quad (13)$$

$$\begin{aligned}
&s_1(\theta+1)\Delta d_{\kappa 1}(\theta) + s_2(\theta+1)\Delta d_{\kappa 2}(\theta) \\
&\quad + s_3(\theta+1)\Delta d_{\kappa 3}(\theta) + s_4(\theta+1)\Delta d_{\kappa 4}(\theta) = 0 \quad (14)
\end{aligned}$$

For $v(\theta+2)$ we write with the help of Eqs.(7) and (11):

$$\begin{aligned} v(\theta+2) &= d_{\kappa 1}(\theta+1)s_1(\theta+2) + d_{\kappa 2}(\theta+1)s_2(\theta+2) \\ &\qquad + d_{\kappa 3}(\theta+1)s_3(\theta+2) + d_{\kappa 4}(\theta+1)s_4(\theta+2) \\ &= d_{\kappa 1}(\theta)s_1(\theta+2) + d_{\kappa 2}(\theta)s_2(\theta+2) + d_{\kappa 3}(\theta)s_3(\theta+2) + d_{\kappa 4}(\theta)s_4(\theta+2) \\ &\qquad + s_1(\theta+2)\Delta d_{\kappa 1}(\theta) + s_2(\theta+2)\Delta d_{\kappa 2}(\theta) \\ &\qquad + s_3(\theta+2)\Delta d_{\kappa 3}(\theta) + s_4(\theta+2)\Delta d_{\kappa 4}(\theta) \end{aligned} \tag{15}$$

We substitute the complete solution $v(\theta-2)$ to $v(\theta+2)$ of Eqs.(4), (5), (6), (7), and (15) for $s(\theta-2)$ to $s(\theta+2)$ into Eq.(1). Since $s_1(\theta)$, $s_2(\theta)$, $s_3(\theta)$, and $s_4(\theta)$ are solutions of the homogeneous Eq.(2) we get:

$$\begin{aligned} s_1(\theta+2)\Delta d_{\kappa 1}(\theta) + s_2(\theta+2)\Delta d_{\kappa 2}(\theta) + s_3(\theta+2)\Delta d_{\kappa 3}(\theta) + s_4(\theta+2)\Delta d_{\kappa 4}(\theta) \\ = \frac{H_{s\kappa}(\theta,\kappa)}{p_2} \end{aligned} \tag{16}$$

Equations (12), (13), (14), and (16) contain the four unknown functions $\Delta d_{\kappa 1}(\theta)$ to $\Delta d_{\kappa 4}(\theta)$ and the known functions s_1, s_2, s_3, s_4, $H_{s\kappa}(\theta)$, p_2. Hence, we can obtain $\Delta d_{\kappa 1}(\theta)$ to $\Delta d_{\kappa 4}(\theta)$ from these four equations and then the coefficients $d_{\kappa 1}(\theta)$ to $d_{\kappa 4}(\theta)$ by means of a summation according to Eq.(11) or (1.2-23).

The solution of Eqs.(12), (13), (14), and (16) by Cramer's rule for $\Delta d_{\kappa 1}(\theta)$, $\Delta d_{\kappa 2}(\theta)$, $\Delta d_{\kappa 3}(\theta)$, $\Delta d_{\kappa 4}(\theta)$ calls for five determinants:

$$D_{\kappa 0}(\theta) = \begin{vmatrix} s_1(\theta-1) & s_2(\theta-1) & s_3(\theta-1) & s_4(\theta-1) \\ s_1(\theta) & s_2(\theta) & s_3(\theta) & s_4(\theta) \\ s_1(\theta+1) & s_2(\theta+1) & s_3(\theta+1) & s_4(\theta+1) \\ s_1(\theta+2) & s_2(\theta+2) & s_3(\theta+2) & s_4(\theta+2) \end{vmatrix}$$
$$\theta = 1,\ 2,\ \ldots,\ N-2 \tag{17}$$

$$D_{\kappa 1}(\theta) = \begin{vmatrix} 0 & s_2(\theta-1) & s_3(\theta-1) & s_4(\theta-1) \\ 0 & s_2(\theta) & s_3(\theta) & s_4(\theta) \\ 0 & s_2(\theta+1) & s_3(\theta+1) & s_4(\theta+1) \\ H_{s\kappa}(\theta,\kappa)/p_2 & s_2(\theta+2) & s_3(\theta+2) & s_4(\theta+2) \end{vmatrix} \tag{18}$$

$$D_{\kappa 3}(\theta) = \begin{vmatrix} s_1(\theta-1) & s_2(\theta-1) & 0 & s_4(\theta-1) \\ s_1(\theta) & s_2(\theta) & 0 & s_4(\theta) \\ s_1(\theta+1) & s_2(\theta+1) & 0 & s_4(\theta+1) \\ s_1(\theta+2) & s_2(\theta+2) & H_{s\kappa}(\theta,\kappa)/p_2 & s_4(\theta+2) \end{vmatrix} \tag{20}$$

$$D_{\kappa 4}(\theta) = \begin{vmatrix} s_1(\theta-1) & s_2(\theta-1) & s_3(\theta-1) & 0 \\ s_1(\theta) & s_2(\theta) & s_3(\theta) & 0 \\ s_1(\theta+1) & s_2(\theta+1) & s_3(\theta+1) & 0 \\ s_1(\theta+2) & s_2(\theta+2) & s_3(\theta+2) & H_{s\kappa}(\theta,\kappa)/p_2 \end{vmatrix} \tag{21}$$

We note that the variable θ has the values $\theta = 0,\ 1,\ \ldots,\ N$ according to Eq.(2.2-6). Hence, the values of s_i, $i = 1,\ 2,\ 3,\ 4$, are defined for this range in Eq.(3), while $s_i(-1)$, $s_i(N+1)$ and $s_i(N+2)$ are not defined. As a result the determinants $D_{\kappa i}(\theta)$ are defined only for $\theta = 1,\ 2,\ \ldots, N-2$.

The functions $s_1(\theta)$ to $s_4(\theta)$ are defined in Eq.(3) while $H_{s\kappa}(\theta)$ and p_2 are defined in Eq.(1). We obtain for $\Delta d_{\kappa 1}(\theta)$ to $\Delta d_{\kappa 4}(\theta)$:

$$\Delta d_{\kappa 1}(\theta) = \frac{D_{\kappa 1}(\theta)}{D_{\kappa 0}(\theta)},\ \Delta d_{\kappa 2}(\theta) = \frac{D_{\kappa 2}(\theta)}{D_{\kappa 0}(\theta)},\ \Delta d_{\kappa 3}(\theta) = \frac{D_{\kappa 3}(\theta)}{D_{\kappa 0}(\theta)},\ \Delta d_{\kappa 4}(\theta) = \frac{D_{\kappa 4}(\theta)}{D_{\kappa 0}(\theta)}$$
$$\theta = 1,\ 2,\ \ldots,\ N-2 \tag{22}$$

The functions $d_{\kappa i}(\theta)$ required in Eqs.(4)–(7) follow from Eq.(22) by summation. A summation constant $d_{\kappa i}$ is required that corresponds to the integration constant of differential calculus:

$$d_{\kappa i}(\theta) = \sum_{n=0}^{\theta-1} \Delta d_{\kappa i}(n) + d_{\kappa \mathrm{i}},\quad i = 1,\ 2,\ 3,\ 4;\ \theta = 1,\ 2,\ \ldots, N-2 \tag{23}$$

The correctness of this equation becomes evident if we substitute Eq.(11)

$$\begin{aligned} d_{\kappa i}(\theta) &= [d_{\kappa i}(0+1) - d_{\kappa i}(0)] + [d_{\kappa i}(2) - d_{\kappa i}(1)] + \ldots \\ &\qquad + \{d_{\kappa i}[(\theta-1)+1] + d_{\kappa i}(\theta-1)\} + d_{\kappa i} \\ &= d_{\kappa i}(\theta) - d_{\kappa i}(0) + d_{\kappa \mathrm{i}},\quad i = 1, 2, 3, 4;\ \theta = 1, 2, \ldots, N-2 \end{aligned} \tag{24}$$

and choose $d_{\kappa \mathrm{i}} = d_{\kappa i}(0)$. The constant $d_{\kappa i}(0)$ is as arbitrary as $d_{\kappa i}$. A constant only adds a solution of the homogeneous Eq.(2) to $v(\theta-2)$ in Eq.(4) or to $v(\theta)$ in Eq.(6). We note that the simplicity of the proof of Eq.(23) by Eq.(24) compared with the proof of Eq.(1.2-23) is due to the elimination of the infinitely large as well as the infinitesimally small.

The determinant $D_{\kappa 0}(\theta)$ in Eq.(17) contains only the solutions of the homogeneous equation according to Eq.(3) while $D_{\kappa 1}(\theta)$ to $D_{\kappa 4}(\theta)$ contain the inhomogeneous term, which means the determinants have to be recalculated every time the inhomogeneous term is changed. This situation is remedied by an expansion of $D_{\kappa 1}(\theta)$ to $D_{\kappa 4}(\theta)$:

$$\begin{aligned} D_{\kappa 1}(\theta) &= -\frac{H_{s\kappa}(\theta,\kappa)}{p_2}\hat{D}_{\kappa 1}(\theta) \\ \hat{D}_{\kappa 1}(\theta) &= \begin{vmatrix} s_2(\theta-1) & s_3(\theta-1) & s_4(\theta-1) \\ s_2(\theta) & s_3(\theta) & s_4(\theta) \\ s_2(\theta+1) & s_3(\theta+1) & s_4(\theta+1) \end{vmatrix} \end{aligned} \tag{25}$$

$$\begin{aligned} D_{\kappa 2}(\theta) &= \frac{H_{s\kappa}(\theta,\kappa)}{p_2}\hat{D}_{\kappa 2}(\theta) \\ \hat{D}_{\kappa 2}(\theta) &= \begin{vmatrix} s_1(\theta-1) & s_3(\theta-1) & s_4(\theta-1) \\ s_1(\theta) & s_3(\theta) & s_4(\theta) \\ s_1(\theta+1) & s_3(\theta+1) & s_4(\theta+1) \end{vmatrix} \end{aligned} \tag{26}$$

$$\begin{aligned} D_{\kappa 3}(\theta) &= -\frac{H_{s\kappa}(\theta,\kappa)}{p_2}\hat{D}_{\kappa 3}(\theta) \\ \hat{D}_{\kappa 3}(\theta) &= \begin{vmatrix} s_1(\theta-1) & s_2(\theta-1) & s_4(\theta-1) \\ s_1(\theta) & s_2(\theta) & s_4(\theta) \\ s_1(\theta+1) & s_2(\theta+1) & s_4(\theta+1) \end{vmatrix} \end{aligned} \tag{27}$$

$$\begin{aligned} D_{\kappa 4}(\theta) &= \frac{H_{s\kappa}(\theta,\kappa)}{p_2}\hat{D}_{\kappa 4}(\theta) \\ \hat{D}_{\kappa 4}(\theta) &= \begin{vmatrix} s_1(\theta-1) & s_2(\theta-1) & s_3(\theta-1) \\ s_1(\theta) & s_2(\theta) & s_3(\theta) \\ s_1(\theta+1) & s_2(\theta+1) & s_3(\theta+1) \end{vmatrix} \end{aligned} \tag{28}$$

Substitution of the functions $d_{\kappa i}(\theta)$ of Eq.(23) into Eq.(6) yields a solution of the inhomogeneous Eq.(3.2-3) according to Eq.(3.2-43).

In order to obtain a solution of the inhomogeneous Eq.(3.2-41) we must replace according to Eq.(3.2-43) the functions $d_{\kappa i}(\theta)$ in Eq.(4) by $c_{\kappa i}(\theta)$, which is a strictly notational change. In addition we must replace $H_{s\kappa}(\theta,\kappa)$ in Eq.(1) by $H_{c\kappa}(\theta,\kappa)$ according to Eq.(3.2-41):

$$H_{c\kappa}(\theta,\kappa) = -e^{i\lambda_1\lambda_3}G_{c\kappa}(\theta+1,\kappa) + 2\cos\frac{2\pi\kappa}{N}G_{c\kappa}(\theta,\kappa) - e^{-i\lambda_1\lambda_3}G_{c\kappa}(\theta-1,\kappa) \tag{29}$$

where $G_{c\kappa}(\theta,\kappa)$ is defined by Eq.(3.2-32). This change affects Eqs.(25)–(28). Equation (6) becomes:

$$v(\theta) = c_{\kappa 1}(\theta)s_1(\theta) + c_{\kappa 2}(\theta)s_2(\theta) + c_{\kappa 3}(\theta)s_3(\theta) + c_{\kappa 4}(\theta)s_4(\theta)$$
$$d_{\kappa i}(\theta) \to c_{\kappa i}(\theta),\ \ H_{s\kappa}(\theta) \to H_{c\kappa}(\theta) \tag{30}$$

For the evaluation of the determinants $\hat{D}_{\kappa 1}(\theta)$ we substitute Eq.(3) into Eq.(25):

$$\hat{D}_{\kappa 1}(\theta) = \begin{vmatrix} e^{-i(2\pi\kappa/N+\lambda_1\lambda_3)(\theta-1)} & (\theta-1)e^{i(2\pi\kappa/N-\lambda_1\lambda_3)(\theta-1)} & (\theta-1)e^{-i(2\pi\kappa/N+\lambda_1\lambda_3)(\theta-1)} \\ e^{-i(2\pi\kappa/N+\lambda_1\lambda_3)\theta} & \theta e^{i(2\pi\kappa/N-\lambda_1\lambda_3)\theta} & \theta e^{-i(2\pi\kappa/N+\lambda_1\lambda_3)\theta} \\ e^{-i(2\pi\kappa/N+\lambda_1\lambda_3)(\theta+1)} & (\theta+1)e^{i(2\pi\kappa/N-\lambda_1\lambda_3)(\theta+1)} & (\theta+1)e^{-i(2\pi\kappa/N+\lambda_1\lambda_3)(\theta+1)} \end{vmatrix} \tag{31}$$

The three factors $e^{-i(2\pi\kappa/N+\lambda_1\lambda_3)\theta}$, $e^{i(2\pi\kappa/N-\lambda_1\lambda_3)\theta}$, and $e^{-i(2\pi\kappa/N+\lambda_1\lambda_3)\theta}$ can be pulled in front of the determinant. Furthermore, we multiply the first column with θ and subtract the product from the third column:

$$\hat{D}_{\kappa 1}(\theta) = e^{-i(2\pi\kappa/N+3\lambda_1\lambda_3)\theta} \times \begin{vmatrix} e^{i(2\pi\kappa/N+\lambda_1\lambda_3)} & (\theta-1)e^{-i(2\pi\kappa/N-\lambda_1\lambda_3)} & -e^{i(2\pi\kappa/N+\lambda_1\lambda_3)} \\ 1 & \theta & 0 \\ e^{-i(2\pi\kappa/N+\lambda_1\lambda_3)} & (\theta+1)e^{i(2\pi\kappa/N-\lambda_1\lambda_3)} & e^{-i(2\pi\kappa/N+\lambda_1\lambda_3)} \end{vmatrix} \tag{32}$$

Development of the third column yields:

$$\begin{aligned} \hat{D}_{\kappa 1}(\theta) =& e^{-i(2\pi\kappa/N+3\lambda_1\lambda_3)\theta} \\ & \times \{-e^{i(2\pi\kappa/N+\lambda_1\lambda_3)}[(\theta+1)e^{i(2\pi\kappa/N-\lambda_1\lambda_3)} - \theta e^{-i(2\pi\kappa/N+\lambda_1\lambda_3)}] \\ & + e^{-i(2\pi\kappa/N+\lambda_1\lambda_3)}[\theta e^{i(2\pi\kappa/N+\lambda_1\lambda_3)} - (\theta-1)e^{-i(2\pi\kappa/N-\lambda_1\lambda_3)}]\} \\ =& 2e^{-2i\pi\kappa\theta/N}e^{-3i\lambda_1\lambda_3\theta}\left(2\theta\sin^2\frac{2\pi\kappa}{N} - i\sin\frac{4\pi\kappa}{N}\right) \end{aligned} \tag{33}$$

Following Eqs.(31)–(33) we obtain $\hat{D}_{\kappa 2}(\theta)$, $\hat{D}_{\kappa 3}(\theta)$, and $\hat{D}_{\kappa 4}(\theta)$ from Eqs.(26) to (28):

$$\hat{D}_{\kappa 2}(\theta) = -2e^{2i\pi\kappa\theta/N}e^{-3i\lambda_1\lambda_3\theta}\left(2\theta\sin^2\frac{2\pi\kappa}{N} + i\sin\frac{4\pi\kappa}{N}\right) \tag{34}$$

$$\hat{D}_{\kappa 3}(\theta) = -4e^{-2i\pi\kappa\theta/N} e^{-3i\lambda_1\lambda_3\theta} \sin^2 \frac{2\pi\kappa}{N} \tag{35}$$

$$\hat{D}_{\kappa 4}(\theta) = +4e^{2i\pi\kappa\theta/N} e^{-3i\lambda_1\lambda_3\theta} \sin^2 \frac{2\pi\kappa}{N} \tag{36}$$

We obtain further $D_{\kappa 1}(\theta)$ to $D_{\kappa 4}(\theta)$ from Eqs.(25) to (28) with p_2 taken from Eq.(1):

$$D_{\kappa 1}(\theta) = -2H_{s\kappa}(\theta, \kappa) e^{-2i\pi\kappa\theta/N} e^{-i\lambda_1\lambda_3(3\theta+2)} \left(2\theta \sin^2 \frac{2\pi\kappa}{N} - i \sin \frac{4\pi\kappa}{N}\right) \tag{37}$$

$$D_{\kappa 2}(\theta) = -2H_{s\kappa}(\theta, \kappa) e^{2i\pi\kappa\theta/N} e^{-i\lambda_1\lambda_3(3\theta+2)} \left(2\theta \sin^2 \frac{2\pi\kappa}{N} + i \sin \frac{4\pi\kappa}{N}\right) \tag{38}$$

$$D_{\kappa 3}(\theta) = 4H_{s\kappa}(\theta, \kappa) e^{-2i\pi\kappa\theta/N} e^{-i\lambda_1\lambda_3(3\theta+2)} \sin^2 \frac{2\pi\kappa}{N} \tag{39}$$

$$D_{\kappa 4}(\theta) = 4H_{s\kappa}(\theta, \kappa) e^{2i\pi\kappa\theta/N} e^{-i\lambda_1\lambda_3(3\theta+2)} \sin^2 \frac{2\pi\kappa}{N} \tag{40}$$

Our next task is the evaluation of $D_{\kappa 0}(\theta)$ of Eq.(17). A comparison with Eqs.(25)–(28) suggests developing the determinant by its fourth row:

$$D_{\kappa 0}(\theta) = -s_1(\theta+2)\hat{D}_{\kappa 1}(\theta) + s_2(\theta+2)\hat{D}_{\kappa 2}(\theta) - s_3(\theta+2)\hat{D}_{\kappa 3}(\theta) + s_4(\theta+2)\hat{D}_{\kappa 4}(\theta) \tag{41}$$

Substitution of $s_1(\theta+2)$ to $s_4(\theta+2)$ from Eq.(3) and $\hat{D}_{\kappa 1}(\theta)$ to $\hat{D}_{\kappa 4}(\theta)$ from Eqs.(33) to (36) yields:

$$D_{\kappa 0}(\theta) = -4e^{-2i\lambda_1\lambda_3(2\theta+1)} \sin^2 \frac{4\pi\kappa}{N} \left(1 + 4\cos \frac{4\pi\kappa}{N}\right) \tag{42}$$

The functions $\Delta d_{\kappa 1}(\theta)$ to $\Delta d_{\kappa 4}(\theta)$ of Eq.(22) follow from Eqs.(37) to (40) plus Eq.(42):

$$\Delta d_{\kappa 1}(\theta) = \frac{1}{2} H_{s\kappa}(\theta, \kappa) e^{-2i\pi\kappa\theta/N} e^{i\lambda_1\lambda_3\theta} \frac{2\theta \sin^2(2\pi\kappa/N) - i \sin(4\pi\kappa/N)}{\sin^2(4\pi\kappa/N)[1 + 4\cos(4\pi\kappa/N)]} \tag{43}$$

$$\Delta d_{\kappa 2}(\theta) = \frac{1}{2} H_{s\kappa}(\theta, \kappa) e^{+2i\pi\kappa\theta/N} e^{i\lambda_1\lambda_3\theta} \frac{2\theta \sin^2(2\pi\kappa/N) + i \sin(4\pi\kappa/N)}{\sin^2(4\pi\kappa/N)[1 + 4\cos(4\pi\kappa/N)]} \tag{44}$$

$$\Delta d_{\kappa 3}(\theta) = -H_{s\kappa}(\theta, \kappa) e^{-2i\pi\kappa\theta/N} e^{i\lambda_1\lambda_3\theta} \frac{\sin^2(2\pi\kappa/N)}{\sin^2(4\pi\kappa/N)[1 + 4\cos(4\pi\kappa/N)]} \tag{45}$$

$$\Delta d_{\kappa 4}(\theta) = -H_{s\kappa}(\theta, \kappa) e^{+2i\pi\kappa\theta/N} e^{i\lambda_1\lambda_3\theta} \frac{\sin^2(2\pi\kappa/N)}{\sin^2(4\pi\kappa/N)[1 + 4\cos(4\pi\kappa/N)]} \tag{46}$$

We replace θ by n in $H_{s\kappa}(\theta,\kappa)$ of Eq.(1) and $G_{s\kappa}(\theta,\kappa)$ of Eq.(3.2-33). The functions $d_{\kappa 1}(\theta)$ to $d_{\kappa 4}(\theta)$ of Eq.(23) may then be written explicitly:

$$d_{\kappa 1}(\theta)=\frac{1}{2}\sum_{n=0}^{\theta-1}H_{s\kappa}(n,\kappa)e^{-2i\pi\kappa n/N}e^{i\lambda_1\lambda_3 n}\frac{2n\sin^2(2\pi\kappa/N)-i\sin(4\pi\kappa/N)}{\sin^2(4\pi\kappa/N)[1+4\cos(4\pi\kappa/N)]}+d_{\kappa 1} \tag{47}$$

$$d_{\kappa 2}(\theta)=\frac{1}{2}\sum_{n=0}^{\theta-1}H_{s\kappa}(n,\kappa)e^{+2i\pi\kappa n/N}e^{i\lambda_1\lambda_3 n}\frac{2n\sin^2(2\pi\kappa/N)+i\sin(4\pi\kappa/N)}{\sin^2(4\pi\kappa/N)[1+4\cos(4\pi\kappa/N)]}+d_{\kappa 2} \tag{48}$$

$$d_{\kappa 3}(\theta)=-\sum_{n=0}^{\theta-1}H_{s\kappa}(n,\kappa)e^{-2i\pi\kappa n/N}e^{i\lambda_1\lambda_3 n}\frac{\sin^2(2\pi\kappa/N)}{\sin^2(4\pi\kappa/N)[1+4\cos(4\pi\kappa/N)]}+d_{\kappa 3} \tag{49}$$

$$d_{\kappa 4}(\theta)=-\sum_{n=0}^{\theta-1}H_{s\kappa}(n,\kappa)e^{+2i\pi\kappa n/N}e^{i\lambda_1\lambda_3 n}\frac{\sin^2(2\pi\kappa/N)}{\sin^2(4\pi\kappa/N)[1+4\cos(4\pi\kappa/N)]}+d_{\kappa 4} \tag{50}$$

In order to separate Eqs.(47)–(50) into real and imaginary parts we define eight functions $F_1(n,\kappa)$ to $F_8(n,\kappa)$:

$$F_1(n,\kappa)=\frac{\sin^2(2\pi\kappa/N)\cos n(\lambda_1\lambda_3-2\pi\kappa/N)}{\sin^2(4\pi\kappa/N)[1+4\cos(4\pi\kappa/N)]} \tag{51}$$

$$F_2(n,\kappa)=\frac{\sin^2(2\pi\kappa/N)\cos n(\lambda_1\lambda_3+2\pi\kappa/N)}{\sin^2(4\pi\kappa/N)[1+4\cos(4\pi\kappa/N)]} \tag{52}$$

$$F_3(n,\kappa)=\frac{\sin(4\pi\kappa/N)\sin n(\lambda_1\lambda_3-2\pi\kappa/N)}{2\sin^2(4\pi\kappa/N)[1+4\cos(4\pi\kappa/N)]} \tag{53}$$

$$F_4(n,\kappa)=-\frac{\sin(4\pi\kappa/N)\sin n(\lambda_1\lambda_3+2\pi\kappa/N)}{2\sin^2(4\pi\kappa/N)[1+4\cos(4\pi\kappa/N)]} \tag{54}$$

$$F_5(n,\kappa)=\frac{\sin^2(2\pi\kappa/N)\sin n(\lambda_1\lambda_3-2\pi\kappa/N)}{\sin^2(4\pi\kappa/N)[1+4\cos(4\pi\kappa/N)]} \tag{55}$$

$$F_6(n,\kappa)=\frac{\sin^2(2\pi\kappa/N)\sin n(\lambda_1\lambda_3+2\pi\kappa/N)}{\sin^2(4\pi\kappa/N)[1+4\cos(4\pi\kappa/N)]} \tag{56}$$

$$F_7(n,\kappa)=-\frac{\sin(4\pi\kappa/N)\cos n(\lambda_1\lambda_3-2\pi\kappa/N)}{2\sin^2(4\pi\kappa/N)[1+4\cos(4\pi\kappa/N)]} \tag{57}$$

$$F_8(n,\kappa)=\frac{\sin(4\pi\kappa/N)\cos n(\lambda_1\lambda_3+2\pi\kappa/N)}{2\sin^2(4\pi\kappa/N)[1+4\cos(4\pi\kappa/N)]} \tag{58}$$

We may rewrite Eqs.(47)–(50) into the following form:

$$d_{\kappa 1}(\theta)=\sum_{n=0}^{\theta-1} H_{s\kappa}(n,\kappa)\{nF_1(n,\kappa)+F_3(n,\kappa)+i[nF_5(n,\kappa)+F_7(n,\kappa)]+d_{\kappa 1}\} \quad (59)$$

$$d_{\kappa 2}(\theta)=\sum_{n=0}^{\theta-1} H_{s\kappa}(n,\kappa)\{nF_2(n,\kappa)+F_4(n,\kappa)+i[nF_6(n,\kappa)+F_8(n,\kappa)]+d_{\kappa 2}\} \quad (60)$$

$$d_{\kappa 3}(\theta)=-\sum_{n=0}^{\theta-1} H_{s\kappa}(n,\kappa)\{F_1(n,\kappa)+iF_5(n,\kappa)+d_{\kappa 3}\} \quad (61)$$

$$d_{\kappa 4}(\theta)=-\sum_{n=0}^{\theta-1} H_{s\kappa}(n,\kappa)\{F_2(n,\kappa)+iF_6(n,\kappa)+d_{\kappa 4}\} \quad (62)$$

The terms $d_{\kappa 1}$ to $d_{\kappa 4}$ are summation constants equivalent to integration constants of the differential theory. According to Eqs.(3.2-72), (3.2-75), and (3.2-79) only $d_{\kappa 1}$ is choosable. We choose $d_{\kappa 1}$ equal to zero and obtain the relations:

$$\begin{aligned} d_{\kappa 1} &= 0, \quad d_{\kappa 3} = -[D_{\kappa 1}(1)+D_{\kappa 3}(1)]/D_{\kappa 0}(1) = d_{3\mathrm{r}}(\kappa)+id_{3\mathrm{i}}(\kappa) \\ d_{\kappa 2} &= 0, \quad d_{\kappa 4} = -[D_{\kappa 2}(1)+D_{\kappa 4}(1)]/D_{\kappa 0}(1) = d_{4\mathrm{r}}(\kappa)+id_{4\mathrm{i}}(\kappa) \end{aligned} \quad (63)$$

Certain terms required in Eq.(3.3-7) may now be written in the form

$$\begin{aligned} d_{\kappa 1}(\theta)+\theta d_{\kappa 3}(\theta) &= \sum_{n=0}^{\theta-1} H_{s\kappa}(n,\kappa)[F_{10}(n,\kappa)+iF_{11}(n,\kappa)] \\ &\qquad -\theta\sum_{n=0}^{\theta-1} H_{s\kappa}(n,\kappa)[F_{12}(n,\kappa)+iF_{13}(n,\kappa)] \\ F_{10}(n,\kappa) &= nF_1(n,\kappa)+F_3(n,\kappa) \\ F_{11}(n,\kappa) &= nF_5(n,\kappa)+F_7(n,\kappa) \\ F_{12}(n,\kappa) &= F_1(n,\kappa)+d_{3\mathrm{r}}(\kappa) \\ F_{13}(n,\kappa) &= F_5(n,\kappa)+d_{3\mathrm{i}}(\kappa) \end{aligned} \quad (64)$$

$$\begin{aligned} d_{\kappa 2}(\theta)+\theta d_{\kappa 4}(\theta) &= \sum_{n=0}^{\theta-1} H_{s\kappa}(n,\kappa)[F_{20}(n,\kappa)+iF_{21}(n,\kappa)] \\ &\qquad -\theta\sum_{n=0}^{\theta-1} H_{s\kappa}(n,\kappa)[F_{22}(n,\kappa)+iF_{23}(n,\kappa)] \end{aligned}$$

$$\begin{aligned}
F_{20}(n,\kappa) &= nF_2(n,\kappa) + F_4(n,\kappa)\\
F_{21}(n,\kappa) &= nF_6(n,\kappa) + F_8(n,\kappa)\\
F_{22}(n,\kappa) &= F_2(n,\kappa) + d_{4\mathrm{r}}(\kappa)\\
F_{23}(n,\kappa) &= F_6(n,\kappa) + d_{4\mathrm{i}}(\kappa)
\end{aligned} \tag{65}$$

We still have to derive $d_{3\mathrm{r}}(\kappa)$, $d_{3\mathrm{i}}(\kappa)$, $d_{4\mathrm{r}}(\kappa)$, and $d_{4\mathrm{i}}(\kappa)$ explicitly. We obtain from Eqs.(37)–(40) and (42):

$$D_{\kappa 1}(1) = -2H_{\mathrm{s}\kappa}(1,\kappa)e^{-2i\pi\kappa/N}e^{-2i\lambda_1\lambda_3}\left(2\sin^2\frac{2\pi\kappa}{N} + i\sin\frac{4\pi\kappa}{N}\right) \tag{66}$$

$$D_{\kappa 2}(1) = -2H_{\mathrm{s}\kappa}(1,\kappa)e^{+2i\pi\kappa/N}e^{-2i\lambda_1\lambda_3}\left(2\sin^2\frac{2\pi\kappa}{N} + i\sin\frac{4\pi\kappa}{N}\right) \tag{67}$$

$$D_{\kappa 3}(1) = 4H_{\mathrm{s}\kappa}(1,\kappa)e^{-2i\pi\kappa/N}e^{-2i\lambda_1\lambda_3}\sin^2\frac{2\pi\kappa}{N} \tag{68}$$

$$D_{\kappa 4}(1) = 4H_{\mathrm{s}\kappa}(1,\kappa)e^{+2i\pi\kappa/N}e^{-2i\lambda_1\lambda_3}\sin^2\frac{2\pi\kappa}{N} \tag{69}$$

$$D_{\kappa 0}(1) = -4e^{-2i\lambda_1\lambda_3}\sin^2\frac{2\pi\kappa}{N}\left(1 + 4\cos\frac{4\pi\kappa}{N}\right) \tag{70}$$

Substitution into Eq.(63) brings

$$-\frac{D_{\kappa 1}(1) + D_{\kappa 3}(1)}{D_{\kappa 0}(1)} = -\frac{H_{\mathrm{s}\kappa}(1,\kappa)[\sin(2\pi\kappa/N) + i\cos(2\pi\kappa/N)]}{2\sin(4\pi\kappa/N)[1 + 4\cos(4\pi\kappa/N)]} \tag{71}$$

$$-\frac{D_{\kappa 2}(1) + D_{\kappa 4}(1)}{D_{\kappa 0}(1)} = +\frac{H_{\mathrm{s}\kappa}(1,\kappa)[\sin(2\pi\kappa/N) - i\cos(2\pi\kappa/N)]}{2\sin(4\pi\kappa/N)[1 + 4\cos(4\pi\kappa/N)]} \tag{72}$$

and we get finally $d_{3\mathrm{r}}(\kappa)$, $d_{3\mathrm{i}}(\kappa)$, $d_{4\mathrm{r}}(\kappa)$, and $d_{4\mathrm{i}}(\kappa)$:

$$d_{3\mathrm{r}}(\kappa) = -d_{4\mathrm{r}}(\kappa) = -\frac{H_{\mathrm{s}\kappa}(1,\kappa)\sin(2\pi\kappa/N)}{2\sin(4\pi\kappa/N)[1 + 4\cos(4\pi\kappa/N)]} \tag{73}$$

$$d_{3\mathrm{i}}(\kappa) = +d_{4\mathrm{i}}(\kappa) = -\frac{H_{\mathrm{s}\kappa}(1,\kappa)\cos(2\pi\kappa/N)}{2\sin(4\pi\kappa/N)[1 + 4\cos(4\pi\kappa/N)]} \tag{74}$$

Our next task is to separate the kernel of the sum of Eq.(3.3-7) into a real and an imaginary part. This is very tedious but not mathematically difficult:

$$\begin{aligned}
&[d_{\kappa 1}(\theta) + \theta d_{\kappa 3}(\theta)]e^{2\pi i\kappa\theta/N} + [d_{\kappa 2}(\theta) + \theta d_{\kappa 4}(\theta)]e^{-2\pi i\kappa\theta/N}\\
&\quad = [J_1(\theta,\kappa) - \theta J_2(\theta,\kappa)]\cos\frac{2\pi\kappa\theta}{N} - [J_3(\theta,\kappa) - \theta J_4(\theta,\kappa)]\sin\frac{2\pi\kappa\theta}{N}\\
&\qquad + i\left([J_5(\theta,\kappa) - \theta J_6(\theta,\kappa)]\sin\frac{2\pi\kappa\theta}{N} + [J_7(\theta,\kappa) - \theta J_8(\theta,\kappa)]\cos\frac{2\pi\kappa\theta}{N}\right)
\end{aligned}$$

$$J_j(\theta,\kappa) = \sum_{n=0}^{\theta-1} K_j(n,\kappa), \qquad j = 1,\ 2,\ \ldots,\ 8$$

$$\begin{aligned}
K_1(n,\kappa) &= H_{s\kappa}(n,\kappa)\{n[F_1(n,\kappa) + F_2(n,\kappa)] + F_3(n,\kappa) + F_4(n,\kappa)\} \\
K_2(n,\kappa) &= H_{s\kappa}n,\kappa)[F_1(n,\kappa) + F_2(n,\kappa)] \\
K_3(n,\kappa) &= H_{s\kappa}(n,\kappa)\{n[F_5(n,\kappa) - F_6(n,\kappa)] + F_7(n,\kappa) - F_8(n,\kappa)\} \\
K_4(n,\kappa) &= H_{s\kappa}(n,\kappa)[F_5(n,\kappa) - F_6(n,\kappa)] \\
K_5(n,\kappa) &= H_{s\kappa}(n,\kappa)\{n[F_1(n,\kappa) - F_2(n,\kappa)] + F_3(n,\kappa) - F_4(n,\kappa)\} \\
K_6(n,\kappa) &= H_{s\kappa}(n,\kappa)[F_5(n,\kappa) - F_6(n,\kappa) + 2d_{3\mathrm{r}}(\kappa)] \\
K_7(n,\kappa) &= H_{s\kappa}(n,\kappa)\{n[F_5(n,\kappa) + F_6(n,\kappa)] + F_7(n,\kappa) + F_8(n,\kappa)\} \\
K_8(n,\kappa) &= H_{s\kappa}(n,\kappa)[F_5(n,\kappa) + F_6(n,\kappa) + 2d_{3\mathrm{i}}(\kappa)] \qquad (75)
\end{aligned}$$

The variable n of $K_j(n,\kappa)$ sometimes needs to be replaced by the variable θ. The following transformation will do that:

$$K_j(\theta,\kappa) = \sum_{n=\theta-1}^{\theta-1} K_j(n,\kappa) = \frac{\partial J_j(\theta,\kappa)}{\partial\theta}, \quad j = 1,\ 2,\ldots,\ 8 \qquad (76)$$

6.4 Calculations for Section 4.2

The inhomogeneous difference equations (4.2-32) and (4.2-33) have to be solved. Except for the inhomogeneous terms the equations are equal. We solve Eq.(4.2-33) for $\tilde{T}_\kappa(\theta)$ and indicate only at the end the changes required for Eq.(4.2-32) and $\tilde{S}_\kappa(\theta)$. Following Eq.(6.3-1) we introduce a shorter notation:

$$\begin{aligned}
&p_2 s(\theta+2) + p_1 s(\theta+1) + p_0 s(\theta) + p_{-1}s(\theta-1) + p_{-2}s(\theta-2) = \tilde{H}_{s\kappa}(\theta,\kappa) \\
&p_2 = p_{-2} = e^{2i\lambda_1\lambda_3},\ p_1 = p_{-1} = -4\cos\frac{2\pi\kappa}{N}e^{i\lambda_1\lambda_3},\ p_0 = 2\left(2+\cos\frac{2\pi\kappa}{N}\right) \\
&\tilde{H}_{s\kappa}(\theta,\kappa) = -e^{i\lambda_1\lambda_3}\tilde{G}_{s\kappa}(\theta+1,\kappa) + 2\cos\frac{2\pi\kappa}{N}\tilde{G}_{s\kappa}(\theta,\kappa) - e^{-i\lambda_1\lambda_3}\tilde{G}_{s\kappa}(\theta-1,\kappa) \quad (1)
\end{aligned}$$

Except for the change $G \to \tilde{G}$, $H \to \tilde{H}$ this is the same equation as Eq.(6.3-1). Hence, Eqs.(6.3-2) and (6.3-3) apply unchanged, but Eq.(6.3-4) must be rewritten for $d \to \tilde{d}$ and $v \to \tilde{v}$:

$$\tilde{v}(\theta-2) = \tilde{d}_{\kappa1}(\theta)s_1(\theta-2) + \tilde{d}_{\kappa2}(\theta)s_2(\theta-2) + \tilde{d}_{\kappa3}(\theta)s_3(\theta-2) + \tilde{d}_{\kappa4}(\theta)s_4(\theta-2) \quad (2)$$

Equations (6.3-5) to (6.3-10) apply again with a tilde over v and d. Equation (6.3-11) must be rewritten to define $\Delta\tilde{d}_{\kappa i}(\theta)$:

$$\Delta\tilde{d}_{\kappa i}(\theta) = \tilde{d}_{\kappa i}(\theta+1) - \tilde{d}_{\kappa i}(\theta), \quad i = 1,\ 2,\ 3,\ 4 \tag{3}$$

Equations (6.3-12) to (6.3-15) remain applicable if a tilde is written over every v and d. Equation (6.3-16) is rewritten because of the changed inhomogeneous term:

$$s_1(\theta+2)\Delta\tilde{d}_{\kappa 1}(\theta) + s_2(\theta+2)\Delta\tilde{d}_{\kappa 2}(\theta) + s_3(\theta+2)\Delta\tilde{d}_{\kappa 3}(\theta) + s_4(\theta+2)\Delta\tilde{d}_{\kappa 4}(\theta) = \frac{\tilde{H}_{s\kappa}(\theta,\kappa)}{p_2} \tag{4}$$

Following the text after Eq.(6.3-16) we need five determinants for $\Delta\tilde{d}_{\kappa 1}(\theta)$, $\Delta\tilde{d}_{\kappa 2}(\theta)$, $\Delta\tilde{d}_{\kappa 3}(\theta)$, $\Delta\tilde{d}_{\kappa 4}(\theta)$. We denote them $\tilde{D}_{\kappa i}(\theta)$ with $i = 0,\ 1,\ 2,\ 3,\ 4$. They follow from Eqs.(6.3-17) to (6.3-21) for $H \to \tilde{H}$:

$$\tilde{D}_{\kappa 0}(\theta) = D_{\kappa 0}(\theta), \quad \theta = 1,\ 2,\ \ldots,\ N-2; \quad \text{see Eq.(6.3-17)} \tag{5}$$

$$\tilde{D}_{\kappa 1}(\theta) = \begin{vmatrix} 0 & s_2(\theta-1) & s_3(\theta-1) & s_4(\theta-1) \\ 0 & s_2(\theta) & s_3(\theta) & s_4(\theta) \\ 0 & s_2(\theta+1) & s_3(\theta+1) & s_4(\theta+1) \\ \tilde{H}_{s\kappa}(\theta,\kappa)/p_2 & s_2(\theta+2) & s_3(\theta+2) & s_4(\theta+2) \end{vmatrix} \tag{6}$$

$$\tilde{D}_{\kappa 2}(\theta) = \begin{vmatrix} s_1(\theta-1) & 0 & s_3(\theta-1) & s_4(\theta-1) \\ s_1(\theta) & 0 & s_3(\theta) & s_4(\theta) \\ s_1(\theta+1) & 0 & s_3(\theta+1) & s_4(\theta+1) \\ s_1(\theta+2) & \tilde{H}_{s\kappa}(\theta,\kappa)/p_2 & s_3(\theta+2) & s_4(\theta+2) \end{vmatrix} \tag{7}$$

$$\tilde{D}_{\kappa 3}(\theta) = \begin{vmatrix} s_1(\theta-1) & s_2(\theta-1) & 0 & s_4(\theta-1) \\ s_1(\theta) & s_2(\theta) & 0 & s_4(\theta) \\ s_1(\theta+1) & s_2(\theta+1) & 0 & s_4(\theta+1) \\ s_1(\theta+2) & s_2(\theta+2) & \tilde{H}_{s\kappa}(\theta,\kappa)/p_2 & s_4(\theta+2) \end{vmatrix} \tag{8}$$

$$\tilde{D}_{\kappa 4}(\theta) = \begin{vmatrix} s_1(\theta-1) & s_2(\theta-1) & s_3(\theta-1) & 0 \\ s_1(\theta) & s_2(\theta) & s_3(\theta) & 0 \\ s_1(\theta+1) & s_2(\theta+1) & s_3(\theta+1) & 0 \\ s_1(\theta+2) & s_2(\theta+2) & s_3(\theta+2) & \tilde{H}_{s\kappa}(\theta,\kappa)/p_2 \end{vmatrix} \tag{9}$$

The functions $s_1(\theta)$ to $s_4(\theta)$ are defined in Eq.(6.3-3) while $\tilde{H}_{s\kappa}(\theta)$ and p_2 are defined in Eq.(1). We obtain for $\Delta\tilde{d}_{\kappa 1}(\theta)$ to $\Delta\tilde{d}_{\kappa 4}(\theta)$:

$$\Delta\tilde{d}_{\kappa 1}(\theta)=\frac{\tilde{D}_{\kappa 1}(\theta)}{D_{\kappa 0}(\theta)},\ \Delta\tilde{d}_{\kappa 2}(\theta)=\frac{\tilde{D}_{\kappa 2}(\theta)}{D_{\kappa 0}(\theta)},\ \Delta\tilde{d}_{\kappa 3}(\theta)=\frac{\tilde{D}_{\kappa 3}(\theta)}{D_{\kappa 0}(\theta)},\ \Delta\tilde{d}_{\kappa 4}(\theta)=\frac{\tilde{D}_{\kappa 4}(\theta)}{D_{\kappa 0}(\theta)}$$
$$\theta = 1,\ 2,\ \ldots,\ N-2 \tag{10}$$

The functions $\tilde{d}_{\kappa i}(\theta)$ required in Eq.(2) follow from Eq.(10) by summation. A summation constant $\tilde{d}_{\kappa \mathrm{i}}$ is required that corresponds to the integration constant of differential calculus:

$$\tilde{d}_{\kappa i}(\theta) = \sum_{n=0}^{\theta-1} \Delta\tilde{d}_{\kappa i}(n) + \tilde{d}_{\kappa \mathrm{i}}, \quad i = 1,\ 2,\ 3,\ 4;\ \theta = 1,\ 2,\ \ldots, N-2 \tag{11}$$

With the help of the determinants $\hat{D}_{\kappa 1}(\theta)$ to $\hat{D}_{\kappa 4}(\theta)$ of Eqs.(6.3-25) to (6.3-28) we may write the determinants $\tilde{D}_{\kappa 1}(\theta)$ to $\tilde{D}_{\kappa 4}(\theta)$ of Eqs.(6) to (9) in the following form:

$$\tilde{D}_{\kappa 1}(\theta) = -\frac{\tilde{H}_{\mathrm{s}\kappa}(\theta,\kappa)}{p_2}\hat{D}_{\kappa 1}(\theta) \tag{12}$$
$$\tilde{D}_{\kappa 2}(\theta) = +\frac{\tilde{H}_{\mathrm{s}\kappa}(\theta,\kappa)}{p_2}\hat{D}_{\kappa 2}(\theta) \tag{13}$$
$$\tilde{D}_{\kappa 3}(\theta) = -\frac{\tilde{H}_{\mathrm{s}\kappa}(\theta,\kappa)}{p_2}\hat{D}_{\kappa 3}(\theta) \tag{14}$$
$$\tilde{D}_{\kappa 4}(\theta) = +\frac{\tilde{H}_{\mathrm{s}\kappa}(\theta,\kappa)}{p_2}\hat{D}_{\kappa 4}(\theta) \tag{15}$$

$\hat{D}_{\kappa 1}(\theta),\ \hat{D}_{\kappa 2}(\theta),\ \hat{D}_{\kappa 3},\ \hat{D}_{\kappa 4}(\theta)$ see Eqs.(6.3-25)–(6.3-28)

Substitution of the functions $\tilde{d}_{\kappa i}(\theta)$ of Eq.(11) into the following equation derived for $v \to \tilde{v}$, $d_{\kappa i} \to \tilde{d}_{\kappa i}$ from Eq.(6.3-6)

$$\tilde{v}(\theta) = \tilde{d}_{\kappa 1}(\theta)s_1(\theta) + \tilde{d}_{\kappa 2}(\theta)s_2(\theta) + \tilde{d}_{\kappa 3}(\theta)s_3(\theta) + \tilde{d}_{\kappa 4}(\theta)s_4(\theta) \tag{16}$$

yields a solution of the inhomogeneous Eq.(4.2-2) according to Eq.(4.2-34).

For the solution of the inhomogeneous Eq.(4.2-32) we must replace according to Eq.(4.2-34) the functions $\tilde{d}_{\kappa i}(\theta)$ by $\tilde{c}_{\kappa i}(\theta)$. Furthermore, we must replace $\tilde{H}_{\mathrm{s}\kappa}(\theta,\kappa)$ in Eq.(1) by $\tilde{H}_{\mathrm{c}\kappa}(\theta,\kappa)$ according to Eq.(4.2-32)

$$\tilde{H}_{\mathrm{c}\kappa}(\theta,\kappa) = -e^{i\lambda_1\lambda_3}\tilde{G}_{\mathrm{c}\kappa}(\theta+1,\kappa) + 2\cos\frac{2\pi\kappa}{N}\tilde{G}_{\mathrm{c}\kappa}(\theta,\kappa) - e^{-i\lambda_1\lambda_3}\tilde{G}_{\mathrm{c}\kappa}(\theta-1,\kappa) \tag{17}$$

where $\tilde{G}_{c\kappa}(\theta,\kappa)$ is defined by Eq.(4.2-23). Equation (16) becomes:

$$\tilde{v}(\theta) = \tilde{c}_{\kappa 1}(\theta)s_1(\theta) + \tilde{c}_{\kappa 2}(\theta)s_2(\theta) + \tilde{c}_{\kappa 3}(\theta)s_3(\theta) + \tilde{c}_{\kappa 4}(\theta)s_4(\theta)$$
$$\tilde{d}_{\kappa i}(\theta) \to \tilde{c}_{\kappa i}(\theta),\ \tilde{H}_{s\kappa}(\theta) \to \tilde{H}_{c\kappa}(\theta) \tag{18}$$

Equations (6.3-31) to (6.3-36) apply again and $\tilde{D}_{\kappa 1}(\theta)$ to $\tilde{D}_{\kappa 4}(\theta)$ may be written according to Eqs.(6.3-37) to (6.3-40) as follows:

$$\tilde{D}_{\kappa 1}(\theta) = -2\tilde{H}_{s\kappa}(\theta,\kappa)e^{-2i\pi\kappa\theta/N}e^{-i\lambda_1\lambda_3(3\theta+2)}\left(2\theta\sin^2\frac{2\pi\kappa}{N} - i\sin\frac{4\pi\kappa}{N}\right) \tag{19}$$

$$\tilde{D}_{\kappa 2}(\theta) = -2\tilde{H}_{s\kappa}(\theta,\kappa)e^{2i\pi\kappa\theta/N}e^{-i\lambda_1\lambda_3(3\theta+2)}\left(2\theta\sin^2\frac{2\pi\kappa}{N} + i\sin\frac{4\pi\kappa}{N}\right) \tag{20}$$

$$\tilde{D}_{\kappa 3}(\theta) = +4\tilde{H}_{s\kappa}(\theta,\kappa)e^{-2i\pi\kappa\theta/N}e^{-i\lambda_1\lambda_3(3\theta+2)}\sin^2\frac{2\pi\kappa}{N} \tag{21}$$

$$\tilde{D}_{\kappa 4}(\theta) = +4\tilde{H}_{s\kappa}(\theta,\kappa)e^{2i\pi\kappa\theta/N}e^{-i\lambda_1\lambda_3(3\theta+2)}\sin^2\frac{2\pi\kappa}{N} \tag{22}$$

For $\tilde{D}_{\kappa 0}(\theta) = D_{\kappa 0}(\theta)$ we retain Eq.(6.3-42):

$$D_{\kappa 0}(\theta) = -4e^{-2i\lambda_1\lambda_3(2\theta+1)}\sin^2\frac{4\pi\kappa}{N}\left(1+4\cos\frac{4\pi\kappa}{N}\right) \tag{23}$$

The functions $\Delta\tilde{d}_{\kappa 1}(\theta)$ to $\Delta\tilde{d}_{\kappa 4}(\theta)$ of Eq.(10) follow from Eqs.(19) to (23):

$$\Delta\tilde{d}_{\kappa 1}(\theta) = +\frac{1}{2}\tilde{H}_{s\kappa}(\theta,\kappa)e^{-2i\pi\kappa\theta/N}e^{i\lambda_1\lambda_3\theta}\frac{2\theta\sin^2(2\pi\kappa/N) - i\sin(4\pi\kappa/N)}{\sin^2(4\pi\kappa/N)[1+4\cos(4\pi\kappa/N)]} \tag{24}$$

$$\Delta\tilde{d}_{\kappa 2}(\theta) = +\frac{1}{2}\tilde{H}_{s\kappa}(\theta,\kappa)e^{+2i\pi\kappa\theta/N}e^{i\lambda_1\lambda_3\theta}\frac{2\theta\sin^2(2\pi\kappa/N) + i\sin(4\pi\kappa/N)}{\sin^2(4\pi\kappa/N)[1+4\cos(4\pi\kappa/N)]} \tag{25}$$

$$\Delta\tilde{d}_{\kappa 3}(\theta) = -\tilde{H}_{s\kappa}(\theta,\kappa)e^{-2i\pi\kappa\theta/N}e^{i\lambda_1\lambda_3\theta}\frac{\sin^2(2\pi\kappa/N)}{\sin^2(4\pi\kappa/N)[1+4\cos(4\pi\kappa/N)]} \tag{26}$$

$$\Delta\tilde{d}_{\kappa 4}(\theta) = -\tilde{H}_{s\kappa}(\theta,\kappa)e^{+2i\pi\kappa\theta/N}e^{i\lambda_1\lambda_3\theta}\frac{\sin^2(2\pi\kappa/N)}{\sin^2(4\pi\kappa/N)[1+4\cos(4\pi\kappa/N)]} \tag{27}$$

We replace θ by n in $\tilde{H}_{s\kappa}(\theta,\kappa)$ of Eq.(1) and $\tilde{G}_{s\kappa}(\theta,\kappa)$ of Eq.(4.2-24). One may then write the functions $\tilde{d}_{\kappa 1}(\theta)$ to $\tilde{d}_{\kappa 4}(\theta)$ of Eq.(11) explicitly:

$$\tilde{d}_{\kappa 1}(\theta) = \frac{1}{2}\sum_{n=0}^{\theta-1}\tilde{H}_{s\kappa}(n,\kappa)e^{-2i\pi\kappa n/N}e^{i\lambda_1\lambda_3 n}\frac{2n\sin^2(2\pi\kappa/N) - i\sin(4\pi\kappa/N)}{\sin^2(4\pi\kappa/N)[1+4\cos(4\pi\kappa/N)]} + \tilde{d}_{\kappa 1} \tag{28}$$

$$\tilde{d}_{\kappa 2}(\theta)=\frac{1}{2}\sum_{n=0}^{\theta-1}\tilde{H}_{s\kappa}(n,\kappa)e^{+2i\pi\kappa n/N}e^{i\lambda_1\lambda_3 n}\frac{2n\sin^2(2\pi\kappa/N)+i\sin(4\pi\kappa/N)}{\sin^2(4\pi\kappa/N)[1+4\cos(4\pi\kappa/N)]}+\tilde{d}_{\kappa 2} \tag{29}$$

$$\tilde{d}_{\kappa 3}(\theta)=-\sum_{n=0}^{\theta-1}\tilde{H}_{s\kappa}(n,\kappa)e^{-2i\pi\kappa n/N}e^{i\lambda_1\lambda_3 n}\frac{\sin^2(2\pi\kappa/N)}{\sin^2(4\pi\kappa/N)[1+4\cos(4\pi\kappa/N)]}+\tilde{d}_{\kappa 3} \tag{30}$$

$$\tilde{d}_{\kappa 4}(\theta)=-\sum_{n=0}^{\theta-1}\tilde{H}_{s\kappa}(n,\kappa)e^{+2i\pi\kappa n/N}e^{i\lambda_1\lambda_3 n}\frac{\sin^2(2\pi\kappa/N)}{\sin^2(4\pi\kappa/N)[1+4\cos(4\pi\kappa/N)]}+\tilde{d}_{\kappa 4} \tag{31}$$

Equations (28)–(31) differ from Eqs.(6.3-47)–(6.3-50) only by the substitutions $\tilde{H}$ and $\tilde{d}$ for H and d. Hence, we can use Eqs.(6.3-51)–(6.3-58) to rewrite Eqs.(28)–(31) into the form of Eqs.(6.3-59)–(6.3-62):

$$\tilde{d}_{\kappa 1}(\theta)=\sum_{n=0}^{\theta-1}\tilde{H}_{s\kappa}(n,\kappa)\{nF_1(n,\kappa)+F_3(n,\kappa)+i[nF_5(n,\kappa)+F_7(n,\kappa)]+\tilde{d}_{\kappa 1}\} \tag{32}$$

$$\tilde{d}_{\kappa 2}(\theta)=\sum_{n=0}^{\theta-1}\tilde{H}_{s\kappa}(n,\kappa)\{nF_2(n,\kappa)+F_4(n,\kappa)+i[nF_6(n,\kappa)+F_8(n,\kappa)]+\tilde{d}_{\kappa 2}\} \tag{33}$$

$$\tilde{d}_{\kappa 3}(\theta)=-\sum_{n=0}^{\theta-1}\tilde{H}_{s\kappa}(n,\kappa)\{F_1(n,\kappa)+iF_5(n,\kappa)+\tilde{d}_{\kappa 3}\} \tag{34}$$

$$\tilde{d}_{\kappa 4}(\theta)=-\sum_{n=0}^{\theta-1}\tilde{H}_{s\kappa}(n,\kappa)\{F_2(n,\kappa)+iF_5(n,\kappa)+\tilde{d}_{\kappa 4}\} \tag{35}$$

The terms $\tilde{d}_{\kappa 1}$ to $\tilde{d}_{\kappa 4}$ are summation constants equivalent to integration constants of the differential theory. According to Eqs.(4.2-43), (4.2-46), and (4.2-50) only $\tilde{d}_{\kappa 1}$ is choosable. We choose $\tilde{d}_{\kappa 1}$ equal to zero and obtain the relations

$$\begin{aligned}\tilde{d}_{\kappa 1}&=0,\quad \tilde{d}_{\kappa 3}=-[\tilde{D}_{\kappa 1}(1)+\tilde{D}_{\kappa 3}(1)]/D_{\kappa 0}(1)=\tilde{d}_{3\mathrm{r}}(\kappa)+i\tilde{d}_{3\mathrm{i}}(\kappa)\\ \tilde{d}_{\kappa 2}&=0,\quad \tilde{d}_{\kappa 4}=-[\tilde{D}_{\kappa 2}(1)+\tilde{D}_{\kappa 4}(1)]/D_{\kappa 0}(1)=\tilde{d}_{4\mathrm{r}}(\kappa)+i\tilde{d}_{4\mathrm{i}}(\kappa)\end{aligned} \tag{36}$$

Certain terms required in Eq.(4.3-7) may be written in the following form:

$$\begin{aligned}\tilde{d}_{\kappa 1}(\theta)+\theta\tilde{d}_{\kappa 3}(\theta)&=\sum_{n=0}^{\theta-1}\tilde{H}_{s\kappa}(n,\kappa)[F_{10}(n,\kappa)+iF_{11}(n,\kappa)]\\ &\quad-\theta\sum_{n=0}^{\theta-1}\tilde{H}_{s\kappa}(n,\kappa)[F_{12}(n,\kappa)+iF_{13}(n,\kappa)]\end{aligned}$$

$$F_{10}(n,\kappa),\ F_{11}(n,\kappa),\ F_{12}(n,\kappa),\ F_{13}(n,\kappa)\text{ see Eq.(6.3-64)} \tag{37}$$

$$\tilde{d}_{\kappa 2}(\theta)+\theta\tilde{d}_{\kappa 4}(\theta)=\sum_{n=0}^{\theta-1}\tilde{H}_{s\kappa}(n,\kappa)[F_{20}(n,\kappa)+iF_{21}(n,\kappa)]$$
$$-\theta\sum_{n=0}^{\theta-1}\tilde{H}_{s\kappa}(n,\kappa)[F_{22}(n,\kappa)+iF_{23}(n,\kappa)]$$
$$F_{20}(n,\kappa),\ F_{21}(n,\kappa),\ F_{22}(n,\kappa),\ F_{23}(n,\kappa)\text{ see Eq.(6.3-65)} \tag{38}$$

We still have to derive $\tilde{d}_{3r}(\kappa)$, $\tilde{d}_{3i}(\kappa)$, $\tilde{d}_{4r}(\kappa)$, and $\tilde{d}_{4i}(\kappa)$ explicitly. We obtain from Eqs.(19)–(23):

$$\tilde{D}_{\kappa 1}(1)=-2\tilde{H}_{s\kappa}(1,\kappa)e^{-2i\pi\kappa/N}e^{-2i\lambda_1\lambda_3}\left(2\sin^2\frac{2\pi\kappa}{N}+i\sin\frac{4\pi\kappa}{N}\right) \tag{39}$$
$$\tilde{D}_{\kappa 2}(1)=-2\tilde{H}_{s\kappa}(1,\kappa)e^{+2i\pi\kappa/N}e^{-2i\lambda_1\lambda_3}\left(2\sin^2\frac{2\pi\kappa}{N}+i\sin\frac{4\pi\kappa}{N}\right) \tag{40}$$
$$\tilde{D}_{\kappa 3}(1)=4\tilde{H}_{s\kappa}(1,\kappa)e^{-2i\pi\kappa/N}e^{-2i\lambda_1\lambda_3}\sin^2\frac{2\pi\kappa}{N} \tag{41}$$
$$\tilde{D}_{\kappa 4}(1)=4\tilde{H}_{s\kappa}(1,\kappa)e^{+2i\pi\kappa/N}e^{-2i\lambda_1\lambda_3}\sin^2\frac{2\pi\kappa}{N} \tag{42}$$
$$D_{\kappa 0}(1)=-4e^{-2i\lambda_1\lambda_3}\sin^2\frac{2\pi\kappa}{N}\left(1+4\cos\frac{4\pi\kappa}{N}\right) \tag{43}$$

Substitution into Eqs.(36) brings

$$-\frac{\tilde{D}_{\kappa 1}(1)+\tilde{D}_{\kappa 3}(1)}{D_{\kappa 0}(1)}=-\frac{\tilde{H}_{s\kappa}(1,\kappa)[\sin(2\pi\kappa/N)+i\cos(2\pi\kappa/N)]}{2\sin(4\pi\kappa/N)[1+4\cos(4\pi\kappa/N)]} \tag{44}$$
$$-\frac{\tilde{D}_{\kappa 2}(1)+\tilde{D}_{\kappa 4}(1)}{D_{\kappa 0}(1)}=+\frac{\tilde{H}_{s\kappa}(1,\kappa)[\sin(2\pi\kappa/N)-i\cos(2\pi\kappa/N)]}{2\sin(4\pi\kappa/N)[1+4\cos(4\pi\kappa/N)]} \tag{45}$$

and we obtain $\tilde{d}_{3r}(\kappa)$, $\tilde{d}_{3i}(\kappa)$, $\tilde{d}_{4r}(\kappa)$, $\tilde{d}_{4i}(\kappa)$ separated:

$$\tilde{d}_{3r}(\kappa)=-\tilde{d}_{4r}(\kappa)=-\frac{\tilde{H}_{s\kappa}(1,\kappa)\sin(2\pi\kappa/N)}{2\sin(4\pi\kappa/N)[1+4\cos(4\pi\kappa/N)]} \tag{46}$$
$$\tilde{d}_{3i}(\kappa)=+\tilde{d}_{4i}(\kappa)=-\frac{\tilde{H}_{s\kappa}(1,\kappa)\cos(2\pi\kappa/N)}{2\sin(4\pi\kappa/N)[1+4\cos(4\pi\kappa/N)]} \tag{47}$$

We still want to separate the kernel of Eq.(4.3-7) into a real and an imaginary part. To this end we copy Eq.(6.3-75) but replace d, J, K with $\tilde{d}$, $\tilde{J}$, $\tilde{K}$:

$$
\begin{aligned}
&[\tilde{d}_{\kappa 1}(\theta) + \theta \tilde{d}_{\kappa 3}(\theta)]e^{2\pi i\kappa\theta/N} + [\tilde{d}_{\kappa 2}(\theta) + \theta \tilde{d}_{\kappa 4}(\theta)]e^{-2\pi i\kappa\theta/N} \\
&\quad = [\tilde{J}_1(\theta,\kappa) - \theta \tilde{J}_2(\theta,\kappa)]\cos\frac{2\pi\kappa\theta}{N} - [\tilde{J}_3(\theta,\kappa) - \theta \tilde{J}_4(\theta,\kappa)]\sin\frac{2\pi\kappa\theta}{N} \\
&\qquad + i\left([\tilde{J}_5(\theta,\kappa) - \theta \tilde{J}_6(\theta,\kappa)]\sin\frac{2\pi\kappa\theta}{N} + [\tilde{J}_7(\theta,\kappa) - \theta \tilde{J}_8(\theta,\kappa)]\cos\frac{2\pi\kappa\theta}{N}\right)
\end{aligned}
$$

$$
\tilde{J}_j(\theta,\kappa) = \sum_{n=0}^{\theta-1} \tilde{K}_j(n,\kappa), \qquad j = 1,\ 2,\ldots,\ 8
$$

$$
\begin{aligned}
\tilde{K}_1(n,\kappa) &= \tilde{H}_{s\kappa}(n,\kappa)\{n[F_1(n,\kappa) + F_2(n,\kappa)] + F_3(n,\kappa) + F_4(n,\kappa)\} \\
\tilde{K}_2(n,\kappa) &= \tilde{H}_{s\kappa}(n,\kappa)[F_1(n,\kappa) + F_2(n,\kappa)] \\
\tilde{K}_3(n,\kappa) &= \tilde{H}_{s\kappa}(n,\kappa)\{n[F_5(n,\kappa) - F_6(n,\kappa)] + F_7(n,\kappa) - F_8(n,\kappa)\} \\
\tilde{K}_4(n,\kappa) &= \tilde{H}_{s\kappa}(n,\kappa)[F_5(n,\kappa) - F_6(n,\kappa)] \\
\tilde{K}_5(n,\kappa) &= \tilde{H}_{s\kappa}(n,\kappa)\{n[F_1(n,\kappa) - F_2(n,\kappa)] + F_3(n,\kappa) - F_4(n,\kappa)\} \\
\tilde{K}_6(n,\kappa) &= \tilde{H}_{s\kappa}(n,\kappa)[F_5(n,\kappa) - F_6(n,\kappa) + 2\tilde{d}_{3\mathrm{r}}(\kappa)] \\
\tilde{K}_7(n,\kappa) &= \tilde{H}_{s\kappa}(n,\kappa)\{n[F_5(n,\kappa) + F_6(n,\kappa)] + F_7(n,\kappa) + F_8(n,\kappa)\} \\
\tilde{K}_8(n,\kappa) &= \tilde{H}_{s\kappa}(n,\kappa)[F_5(n,\kappa) + F_6(n,\kappa) + 2\tilde{d}_{3\mathrm{i}}(\kappa)]
\end{aligned} \tag{48}
$$

The variable n of $\tilde{K}_j(n,\kappa)$ sometimes needs to be replaced by the variable θ. The following transformation will do that in a computer program:

$$
\tilde{K}_j(\theta,\kappa) = \sum_{n=\theta-1}^{\theta-1} \tilde{K}_j(n,\kappa) = \frac{\partial \tilde{J}_j(\theta,\kappa)}{\partial\theta}, \qquad j = 1,\ 2,\ldots,\ 8 \tag{49}
$$

6.5 Formulas for Spherical Coordinates

We want to rewrite Eq.(1.3-13) in component form like Eq.(1.3-20) for spherical coordinates:

$$
\frac{Ze}{c}\frac{d}{dt}(\mathbf{v}\times\mathbf{A}_{\mathrm{e}}) = \frac{Ze}{c}\left(-\mathbf{v}\times\operatorname{grad}\phi_{\mathrm{m}} + \frac{\partial\mathbf{v}}{\partial t}\times\mathbf{A}_{\mathrm{e}} + \mathbf{v}\cdot[\nabla(\mathbf{v}\times\mathbf{A}_{\mathrm{e}})] - c^2\operatorname{curl}\mathbf{A}_{\mathrm{e}}\right) \tag{1}
$$

We start with the evaluation of the first, second, and fourth term on the right side of Eq.(1):

$$
\operatorname{grad}\phi_{\mathrm{m}} = \frac{\partial\phi_{\mathrm{m}}}{\partial r}\mathbf{e}_r + \frac{1}{r}\frac{\partial\phi_{\mathrm{m}}}{\partial\vartheta}\mathbf{e}_\vartheta + \frac{1}{r\sin\vartheta}\frac{\partial\phi_{\mathrm{m}}}{\partial\varphi}\mathbf{e}_\varphi \tag{2}
$$

The velocity $\mathbf{v}$ of the point $r\mathbf{e}_r + r\vartheta\mathbf{e}_\vartheta + r\sin\vartheta\,\varphi\mathbf{e}_\varphi$ is needed for the three directions $\mathbf{e}_r$, $\mathbf{e}_\vartheta$, $\mathbf{e}_\varphi$:

$$\begin{aligned}\mathbf{v} &= \frac{\partial r}{\partial r}\frac{\partial r}{\partial t}\mathbf{e}_r + \frac{\partial(r\vartheta)}{\partial\vartheta}\frac{\partial\vartheta}{\partial t}\mathbf{e}_\vartheta + \frac{\partial(r\sin\vartheta\,\varphi)}{\partial\varphi}\frac{\partial\varphi}{\partial t}\mathbf{e}_\varphi \\ &= \dot{r}\mathbf{e}_r + r\dot{\vartheta}\mathbf{e}_\vartheta + r\sin\vartheta\,\dot{\varphi}\mathbf{e}_\varphi \end{aligned} \tag{3}$$

$$\begin{aligned}\mathbf{v}\times\operatorname{grad}\phi_\mathrm{m} &= \left(\frac{\dot{\vartheta}}{\sin\vartheta}\frac{\partial\phi_\mathrm{m}}{\partial\varphi} - \sin\vartheta\,\dot{\varphi}\frac{\partial\phi_\mathrm{m}}{\partial\vartheta}\right)\mathbf{e}_r \\ &+ \left(r\sin\vartheta\,\dot{\varphi}\frac{\partial\phi_\mathrm{m}}{\partial r} - \frac{\dot{r}}{r\sin\vartheta}\frac{\partial\phi_\mathrm{m}}{\partial\varphi}\right)\mathbf{e}_\vartheta + \left(\frac{\dot{r}}{r}\frac{\partial\phi_\mathrm{m}}{\partial\vartheta} - r\dot{\vartheta}\frac{\partial\phi_\mathrm{m}}{\partial r}\right)\mathbf{e}_\varphi \end{aligned} \tag{4}$$

The second term on the right side of Eq.(1) is evaluated:

$$\begin{aligned}\frac{\partial\mathbf{v}}{\partial t} &= \frac{\partial\dot{r}}{\partial\dot{r}}\frac{\partial\dot{r}}{\partial t}\mathbf{e}_r + \frac{\partial(r\dot{\vartheta})}{\partial\dot{\vartheta}}\frac{\partial\dot{\vartheta}}{\partial t}\mathbf{e}_\vartheta + \frac{\partial(r\sin\vartheta\,\dot{\varphi})}{\partial\dot{\varphi}}\frac{\partial\dot{\varphi}}{\partial t}\mathbf{e}_\varphi \\ &= \ddot{r}\mathbf{e}_r + r\ddot{\vartheta}\mathbf{e}_\vartheta + r\sin\vartheta\,\ddot{\varphi}\mathbf{e}_\varphi \end{aligned} \tag{5}$$

$$\mathbf{A}_\mathrm{e} = A_{\mathrm{e}r}\mathbf{e}_r + A_{\mathrm{e}\vartheta}\mathbf{e}_\vartheta + A_{\mathrm{e}\varphi}\mathbf{e}_\varphi \tag{6}$$

$$\begin{aligned}\frac{\partial\mathbf{v}}{\partial t}\times\mathbf{A}_\mathrm{e} &= (r\ddot{\vartheta}A_{\mathrm{e}\varphi} - r\sin\vartheta\,\ddot{\varphi}A_{\mathrm{e}\vartheta})\mathbf{e}_r + (r\sin\vartheta\,\ddot{\varphi}A_{\mathrm{e}r} - \ddot{r}A_{\mathrm{e}\varphi})\mathbf{e}_\vartheta \\ &+ (\ddot{r}A_{\mathrm{e}\vartheta} - r\ddot{\vartheta}A_{\mathrm{e}r})\mathbf{e}_\varphi \end{aligned} \tag{7}$$

The fourth term in Eq.(1) becomes:

$$\begin{aligned}c^2\operatorname{curl}\mathbf{A}_\mathrm{e} &= \frac{c^2}{r\sin\vartheta}\left(\frac{\partial(A_{\mathrm{e}\varphi}\sin\vartheta)}{\partial\vartheta} - \frac{\partial A_{\mathrm{e}\vartheta}}{\partial\varphi}\right)\mathbf{e}_r \\ &+ \frac{c^2}{r}\left(\frac{1}{\sin\vartheta}\frac{\partial A_{\mathrm{e}r}}{\partial\varphi} - \frac{\partial(rA_{\mathrm{e}\varphi})}{\partial r}\right)\mathbf{e}_\vartheta + \frac{c^2}{r}\left(\frac{\partial(rA_{\mathrm{e}\vartheta})}{\partial r} - \frac{\partial A_{\mathrm{e}r}}{\partial\vartheta}\right)\mathbf{e}_\varphi \end{aligned} \tag{8}$$

We turn to the dyadic $\nabla(\mathbf{v}\times\mathbf{A}_\mathrm{e})$ in Eq.(1). Some textbooks show the dyadic $\nabla\mathbf{F}$ in component form but it is too complicated for spherical coordinates. However, we find the equation

$$\frac{d\mathbf{A}}{dt} = \frac{\partial\mathbf{A}}{\partial t} + \mathbf{v}\cdot(\nabla\mathbf{A}) \tag{9}$$

in the literature[1] and we may thus write:

$$\begin{aligned}\mathbf{v}\cdot[\nabla(\mathbf{v}\times\mathbf{A}_\mathrm{e})] &= \frac{d(\mathbf{v}\times\mathbf{A}_\mathrm{e})}{dt} - \frac{\partial(\mathbf{v}\times\mathbf{A}_\mathrm{e})}{\partial t}\\ &= \frac{\partial(\mathbf{v}\times\mathbf{A}_\mathrm{e})}{\partial r}\dot{r} + \frac{\partial(\mathbf{v}\times\mathbf{A}_\mathrm{e})}{\partial\vartheta}\dot{\vartheta} + \frac{\partial(\mathbf{v}\times\mathbf{A}_\mathrm{e})}{\partial\varphi}\dot{\varphi}\end{aligned} \tag{10}$$

We obtain with the help of Eqs. (3) and (6):

$$\begin{aligned}\mathbf{v}\times\mathbf{A}_\mathrm{e} &= (r\dot{\vartheta}A_{\mathrm{e}\varphi} - r\sin\vartheta\,\dot{\varphi}A_{\mathrm{e}\vartheta})\mathbf{e}_r\\ &\quad + (r\sin\vartheta\,\dot{\varphi}A_{\mathrm{e}r} - \dot{r}A_{\mathrm{e}\varphi})\mathbf{e}_\vartheta + (\dot{r}A_{\mathrm{e}\vartheta} - r\dot{\vartheta}A_{\mathrm{e}r})\mathbf{e}_\varphi\end{aligned} \tag{11}$$

$$\begin{aligned}\frac{\partial(\mathbf{v}\times\mathbf{A}_\mathrm{e})}{\partial r}\dot{r} &= \dot{r}\frac{\partial}{\partial r}[(r\dot{\vartheta}A_{\mathrm{e}\varphi} - r\sin\vartheta\,\dot{\varphi}A_{\mathrm{e}\vartheta})\mathbf{e}_r\\ &\quad + (r\sin\vartheta\,\dot{\varphi}A_{\mathrm{e}r} - \dot{r}A_{\mathrm{e}\varphi})\mathbf{e}_\vartheta + (\dot{r}A_{\mathrm{e}\vartheta} - r\dot{\vartheta}A_{\mathrm{e}r})\mathbf{e}_\varphi]\end{aligned} \tag{12}$$

$$\begin{aligned}\frac{\partial(\mathbf{v}\times A_\mathrm{e})}{\partial\vartheta}\dot{\vartheta} &= \dot{\vartheta}\frac{\partial}{\partial\vartheta}[(r\dot{\vartheta}A_{\mathrm{e}\varphi} - r\sin\vartheta\,\dot{\varphi}A_{\mathrm{e}\vartheta})\mathbf{e}_r\\ &\quad + (r\sin\vartheta\,\dot{\varphi}A_{\mathrm{e}r} - \dot{r}A_{\mathrm{e}\varphi})\mathbf{e}_\vartheta + (\dot{r}A_{\mathrm{e}\vartheta} - r\dot{\vartheta}A_{\mathrm{e}r})\mathbf{e}_\varphi]\end{aligned} \tag{13}$$

$$\begin{aligned}\frac{\partial(\mathbf{v}\times\mathbf{A}_\mathrm{e})}{\partial\varphi}\dot{\varphi} &= \dot{\varphi}\frac{\partial}{\partial\varphi}[(r\dot{\vartheta}A_{\mathrm{e}\varphi} - r\sin\vartheta\,\dot{\varphi}A_{\mathrm{e}\vartheta})\mathbf{e}_r\\ &\quad + (r\sin\vartheta\,\dot{\varphi}A_{\mathrm{e}r} - \dot{r}A_{\mathrm{e}\varphi})\mathbf{e}_\vartheta + (\dot{r}A_{\mathrm{e}\vartheta} - r\dot{\vartheta}A_{\mathrm{e}r})\mathbf{e}_\varphi]\end{aligned} \tag{14}$$

Equation (10) becomes:

$$\begin{aligned}\mathbf{v}\cdot[\nabla(\mathbf{v}\times\mathbf{A}_\mathrm{e})] &= \left(\dot{r}\frac{\partial}{\partial r} + \dot{\vartheta}\frac{\partial}{\partial\vartheta} + \dot{\varphi}\frac{\partial}{\partial\varphi}\right)[(r\dot{\vartheta}A_{\mathrm{e}\varphi} - r\sin\vartheta\,\dot{\varphi}A_{\mathrm{e}\vartheta})\mathbf{e}_r\\ &\quad + (r\sin\vartheta\,\dot{\varphi}A_{\mathrm{e}r} - \dot{r}A_{\mathrm{e}\varphi})\mathbf{e}_\vartheta + (\dot{r}A_{\mathrm{e}\vartheta} - r\dot{\vartheta}A_{\mathrm{e}r})\mathbf{e}_\varphi]\end{aligned} \tag{15}$$

Equation (1) is brought into the following component form with the help of Eqs.(4), (7), (15), and (8):

$$\begin{aligned}\frac{d}{dt}(\mathbf{v}\times\mathbf{A}_\mathrm{e}) &= \frac{d}{dt}[(r\dot{\vartheta}A_{\mathrm{e}\varphi} - r\sin\vartheta\,\dot{\varphi}A_{\mathrm{e}\vartheta})\mathbf{e}_r + (r\sin\vartheta\,\dot{\varphi}A_{\mathrm{e}r} - \dot{r}A_{\mathrm{e}\varphi})\mathbf{e}_\vartheta\\ &\quad + (\dot{r}A_{\mathrm{e}\vartheta} - r\dot{\vartheta}A_{\mathrm{e}r})\mathbf{e}_\varphi]\end{aligned}$$

[1] Morse and Feshbach 1953, Part I, p. 295.

$$= \Bigg[-\left(\frac{\dot{\vartheta}}{\sin\vartheta} \frac{\partial \phi_\mathrm{m}}{\partial \varphi} - \sin\vartheta\, \dot{\varphi} \frac{\partial \phi_\mathrm{m}}{\partial \vartheta} \right) + r\ddot{\vartheta} A_{\mathrm{e}\varphi} - r\sin\vartheta\, \ddot{\varphi} A_{\mathrm{e}\vartheta}$$

$$+\left(\dot{r}\frac{\partial}{\partial r} + \dot{\vartheta}\frac{\partial}{\partial \vartheta} + \dot{\varphi}\frac{\partial}{\partial \varphi} \right)(r\dot{\vartheta} A_{\mathrm{e}\varphi} - r\sin\vartheta\, \dot{\varphi} A_{\mathrm{e}\vartheta}) - \frac{c^2}{r\sin\vartheta}\left(\frac{\partial (A_{\mathrm{e}\varphi}\sin\vartheta)}{\partial\vartheta} - \frac{\partial A_{\mathrm{e}\vartheta}}{\partial\varphi} \right) \Bigg] \mathbf{e}_r$$

$$+ \Bigg[-\left(r\sin\vartheta\, \dot{\varphi}\frac{\partial \phi_\mathrm{m}}{\partial r} - \frac{\dot{r}}{r\sin\vartheta}\frac{\partial \phi_\mathrm{m}}{\partial \varphi} \right) + r\sin\vartheta\, \ddot{\varphi} A_{\mathrm{e}r} - \ddot{r} A_{\mathrm{e}\varphi}$$

$$+\left(\dot{r}\frac{\partial}{\partial r} + \dot{\vartheta}\frac{\partial}{\partial \vartheta} + \dot{\varphi}\frac{\partial}{\partial \varphi} \right)(r\sin\vartheta\, \dot{\varphi} A_{\mathrm{e}r} - \dot{r} A_{\mathrm{e}\varphi}) - \frac{c^2}{r}\left(\frac{1}{\sin\vartheta}\frac{\partial A_{\mathrm{e}r}}{\partial\varphi} - \frac{\partial (r A_{\mathrm{e}\varphi})}{\partial r} \right) \Bigg] \mathbf{e}_\vartheta$$

$$+ \Bigg[-\left(\frac{\dot{r}}{r}\frac{\partial \phi_\mathrm{m}}{\partial \vartheta} - r\dot{\vartheta}\frac{\partial \phi_\mathrm{m}}{\partial r} \right) + \ddot{r} A_{\mathrm{e}\vartheta} - r\ddot{\vartheta} A_{\mathrm{e}r}$$

$$+\left(\dot{r}\frac{\partial}{\partial r} + \dot{\vartheta}\frac{\partial}{\partial \vartheta} + \dot{\varphi}\frac{\partial}{\partial \varphi} \right)(\dot{r} A_{\mathrm{e}\vartheta} - r\dot{\vartheta} A_{\mathrm{e}r}) - \frac{c^2}{r}\left(\frac{\partial (r A_{\mathrm{e}\vartheta})}{\partial r} - \frac{\partial A_{\mathrm{e}r}}{\partial\vartheta} \right) \Bigg] \mathbf{e}_\varphi \qquad (16)$$

The Lagrange functions $\mathfrak{L}_{\mathrm{c}r}$, $\mathfrak{L}_{\mathrm{c}\vartheta}$, $\mathfrak{L}_{\mathrm{c}\varphi}$ for Eq.(1.3-13) may be inferred from the factor of $\mathbf{e}_r$ using the similarity with Eq.(1.3-20) and the way Eq.(1.3-21) was inferred there:

$$\mathfrak{L}_{\mathrm{c}r} = \frac{Ze}{c} \int \Bigg[\sin\vartheta\, \dot{\varphi}\frac{\partial \phi_\mathrm{m}}{\partial \vartheta} - \frac{\dot{\vartheta}}{\sin\vartheta}\frac{\partial \phi_\mathrm{m}}{\partial \varphi} + r\ddot{\vartheta} A_{\mathrm{e}\varphi} - r\sin\vartheta\, \ddot{\varphi} A_{\mathrm{e}\vartheta}$$

$$+ \left(\dot{r}\frac{\partial}{\partial r} + \dot{\vartheta}\frac{\partial}{\partial \vartheta} + \dot{\varphi}\frac{\partial}{\partial \varphi} \right)(r\dot{\vartheta} A_{\mathrm{e}\varphi} - r\sin\vartheta\, \dot{\varphi} A_{\mathrm{e}\vartheta})$$

$$- \frac{c^2}{r\sin\vartheta}\left(\frac{\partial (A_{\mathrm{e}\varphi}\sin\vartheta)}{\partial\vartheta} - \frac{\partial A_{\mathrm{e}\vartheta}}{\partial\varphi} \right) \Bigg] dr \qquad (17)$$

The term multiplied with $\dot{r}\partial/\partial r$ may be pulled in front of the integral and differentiated with respect to $\dot{r}$. The first term on the right side of Eq.(16) is obtained:

$$\frac{\partial}{\partial \dot{r}}[\dot{r}(r\dot{\vartheta} A_{\mathrm{e}\varphi} - r\sin\vartheta\, \dot{\varphi} A_{\mathrm{e}\vartheta})] = r\dot{\vartheta} A_{\mathrm{e}\varphi} - r\sin\vartheta\, \dot{\varphi} A_{\mathrm{e}\vartheta} \qquad (18)$$

The functions $\mathfrak{L}_{\mathrm{c}\vartheta}$ and $\mathfrak{L}_{\mathrm{c}\varphi}$ follow from the factors of $\mathbf{e}_\vartheta$ and $\mathbf{e}_\varphi$ in Eq.(16) in analogy to Eq.(17):

$$\mathfrak{L}_{\mathrm{c}\vartheta} = \frac{Ze}{c} \int \Bigg[\frac{\dot{r}}{r\sin\vartheta}\frac{\partial \phi_\mathrm{m}}{\partial \varphi} - r\sin\vartheta\, \dot{\varphi}\frac{\partial \phi_\mathrm{m}}{\partial r} + r\sin\vartheta\, \ddot{\varphi} A_{\mathrm{e}r} - \ddot{r} A_{\mathrm{e}\varphi}$$

$$+ \left(\dot{r}\frac{\partial}{\partial r} + \dot{\vartheta}\frac{\partial}{\partial \vartheta} + \dot{\varphi}\frac{\partial}{\partial \varphi} \right)(r\sin\vartheta\, \dot{\varphi} A_{\mathrm{e}r} - \dot{r} A_{\mathrm{e}\varphi})$$

$$- \frac{c^2}{r}\left(\frac{1}{\sin\vartheta}\frac{\partial A_{\mathrm{e}r}}{\partial\varphi} - \frac{\partial (r A_{\mathrm{e}\varphi})}{\partial r} \right) \Bigg] r\, d\vartheta \qquad (19)$$

$$\mathfrak{L}_{c\varphi} = \frac{Ze}{c} \int \Big[r\dot{\vartheta}\frac{\partial \phi_m}{\partial r} - \frac{\dot{r}}{r}\frac{\partial \phi_m}{\partial \vartheta} + \ddot{r}A_{e\vartheta} - r\ddot{\vartheta}A_{er} + \left(\dot{r}\frac{\partial}{\partial r} + \dot{\vartheta}\frac{\partial}{\partial \vartheta} + \dot{\varphi}\frac{\partial}{\partial \varphi}\right)(\dot{r}A_{e\vartheta} - r\dot{\vartheta}A_{er}) - \frac{c^2}{r}\left(\frac{\partial(rA_{e\vartheta})}{\partial r} - \frac{\partial A_{er}}{\partial \vartheta}\right)\Big] r\sin\vartheta\, d\varphi \quad (20)$$

In Eq.(17) one can pull the factor $\dot{r}(r\dot{\vartheta}A_{e\varphi} - r\sin\vartheta\,\dot{\varphi}A_{e\vartheta})$ in front of the integral, in Eq.(19) the factor $\dot{\vartheta}(r\sin\vartheta\,\dot{\varphi}A_{er} - \dot{r}A_{e\varphi})$, and in Eq.(20) the factor $\dot{\varphi}(\dot{r}A_{e\vartheta} - r\dot{\vartheta}A_{er})$. The terms $\mathfrak{L}_{cr}$, $\mathfrak{L}_{c\vartheta}$, and $\mathfrak{L}_{c\varphi}$ are shown in this form in Eqs.(1.3-33)–(1.3-35).

6.6 Difference Equations Solved by Polynomials

We investigate when a second order difference equation with variable coefficients is satisfied by polynomials. Consider the equation

$$P_1(X)v(X+1) + P_0(X)v(X) + P_{-1}(X)v(X-1) = 0 \quad (1)$$

The factors $P_1(X)$ and $P_{-1}(X)$ shall be polynomials of the same degree, while $P_0(X)$ is a polynomial of the same or lower degree:

$$\begin{aligned} P_1(X) &= d_{10}X^j + d_{11}X^{j-1} + \cdots + d_{1j} \\ P_0(X) &= d_{00}X^j + d_{01}X^{j-1} + \cdots + d_{0j} \\ P_{-1}(X) &= d_{-10}X^j + d_{-11}X^{j-1} + \cdots + d_{-1j} \\ & d_{10},\ d_{-10} \neq 0 \end{aligned} \quad (2)$$

In differential calculus we like to write polynomials in terms of powers of X since these powers are easy to differentiate. In the calculus of finite differences one simplifies calculations by replacing powers X^j by products of the form $X(X+1)(X+2)\ldots(X+j-1)$. Hence, we write $P_1(X)$, $P_0(X)$, and $P_{-1}(X)$ in the following form:

$$\begin{aligned} P_1(X) &= c_{10}(X+1)(X+2)\ldots(X+j) + c_{11}(X+1)\ldots(X+j-1) + \cdots + c_{1j} \\ P_0(X) &= c_{00}X(X+1)\ldots(X+j-1) + c_{01}X\ldots(X+j-2) + \cdots + c_{0j} \\ P_{-1}(X) &= c_{-10}(X-1)X\ldots(X+j-2) + c_{-11}(X-1)\ldots(X+j-3) + \cdots + c_{-1j} \end{aligned} \quad (3)$$

The coefficients $c_{..}$ of Eq.(3) can be expressed in terms of the coefficients $d_{..}$ of Eq.(2):

$$
\begin{aligned}
c_{10} &= d_{10}, \; c_{00} = d_{00}, \; c_{-10} = d_{-10} \\
c_{11} &= d_{11} - c_{10} \sum_{\nu=1}^{j} \nu = d_{11} - \frac{1}{2} j(j+1) d_{10} \\
c_{01} &= d_{00} - \frac{1}{2}(j-1) j d_{00} \\
c_{-11} &= d_{-11} - \frac{1}{2} j(j-3) d_{-10} \\
c_{12} &= d_{12} - c_{11} \sum_{\nu=1}^{j-1} \nu - c_{10} \sum_{\nu=1}^{j-1} \sum_{\mu=\nu+1}^{j} \nu\mu \\
&= d_{12} - \frac{1}{2}(j-1) j d_{11} - \left(\sum_{\nu=1}^{j-1} \sum_{\mu=\nu+1}^{j} \nu\mu - \frac{1}{4} j^2 (j^2 - 1) \right) d_{10} \\
c_{02} &= d_{02} - \frac{1}{2}(j-2)(j-1) d_{00} - \left(\sum_{\nu=1}^{j-2} \sum_{\mu=\nu+1}^{j-1} \nu\mu - \frac{1}{4} j(j-1)^2 (j-2) \right) d_{00} \\
c_{-12} &= d_{-12} - \frac{1}{2}(j-1)(j-4) d_{-11} \\
&\quad - \left(\sum_{\nu=1}^{j-3} \sum_{\mu=\nu+1}^{j-2} \nu\mu - \frac{1}{2}(j-1)(j-2) - \frac{1}{4} j(j-1)(j-3)(j-4) \right) d_{-10}
\end{aligned} \tag{4}
$$

Coefficients of higher order than c_{12}, c_{02}, c_{-12} will not be needed. The double sums in Eq.(4) may be simplified:

$$
\begin{aligned}
\sum_{\nu=1}^{j-3} \sum_{\mu=\nu+1}^{j-2} \nu\mu &= \frac{1}{2} \sum_{\nu=1}^{j-3} \nu(\nu + j - 1)(j - \nu - 2) = K \\
\sum_{\nu=1}^{j-2} \sum_{\mu=\nu+1}^{j-1} \nu\mu &= K + \frac{1}{2}(j-1)^2 (j-2) \\
\sum_{\nu=1}^{j-1} \sum_{\mu=\nu+1}^{j} \nu\mu &= K + \frac{1}{2}(j-1)^2 (j-2) + \frac{1}{2} j^2 (j-1)
\end{aligned} \tag{5}
$$

The integral transformation

$$
v(X) = \frac{1}{2\pi i} \int_{\ell} s^{X-1} y(s) \, ds
$$

transforms the difference equation (1) into a differential equation, if the line of integration ℓ is chosen so that the terms

$$s^X y\Big|_{\ell_1}^{\ell_2}, \quad s^{X+1}y'\Big|_{\ell_1}^{\ell_2} = s^{X+1}y^{(1)}\Big|_{\ell_1}^{\ell_2}, \quad \ldots$$

vanish:

$$\begin{aligned}(c_{10}s^2 + c_{00}s + c_{-10})s^j y^{(j)} - (c_{11}s^2 + c_{01}s + c_{-11})s^{j-1}y^{(j-1)}\\ +(c_{12}s^2+c_{02}s+c_{-12})s^{j-2}y^{(j-2)} - \cdots + (-1)^j(c_{1j}s^2 + c_{0j}s+c_{-1j})y = 0 \quad (6)\end{aligned}$$

The characteristic equation

$$(c_{10}s^2 + c_{00}s + c_{-10})s^j = 0 \tag{7}$$

has the simple roots:

$$\begin{aligned} s_2 &= \frac{1}{2}\frac{1}{d_{10}}\left(-d_{00} + (d_{00}^2 - 4d_{-10}d_{10})^{1/2}\right)\\ s_3 &= \frac{1}{2}\frac{1}{d_{10}}\left(-d_{00} - (d_{00}^2 - 4d_{-10}d_{10})^{1/2}\right)\\ s_2 s_3 &= \frac{d_{-10}}{d_{10}} \end{aligned} \tag{8}$$

In addition, there is the j-fold root $s_1 = 0$.

We make the substitution $z = s - s_2$ in the differential equation (6) and obtain:

$$\begin{aligned} c_{10}y^{(j)} - \frac{c_{11}(z+s_2)^2 + c_{01}(z+s_2) + c_{-11}}{z(z+s_2)(z+s_2-s_3)}y^{(j-1)}\\ + \frac{c_{12}(z+s_2)^2 + c_{02}(z+s_2) + c_{-12}}{z^2(z+s_2)(z+s_2-s_3)}y^{(j-2)} - \cdots = 0 \quad (9)\end{aligned}$$

Equation (9) is solved by means of a power series:

$$y = \sum_{\nu=0}^{\infty} q_\nu z^{p+\nu} \tag{10}$$

The following determining equation for the initial power p is obtained if s_2 is not equal to s_1 or s_3:

$$p(p-1)\cdots(p-j+2)\left((p-j+1)c_{10} - \frac{c_{11}s_2^2 + c_{01}s_2 + c_{-11}}{s_2(s_2-s_3)}\right) = 0 \tag{11}$$

Equation (11) has the trivial roots $p = 0,\ 1, \ldots,\ j-2$ and the one nontrivial root:

$$p = \frac{1}{d_{10}s_2(s_2 - s_3)}(d_{11}s_2^2 + d_{01}s_2 + d_{-11}) - 1 \tag{12}$$

We interrupt this course of calculations and consider a series expansion in negative powers of X:

$$\begin{aligned} P_1'(X) &= d_{10} + d_{11}X^{-1} + \cdots + d_{1j}X^{-j} \\ P_0'(X) &= d_{00} + d_{01}X^{-1} + \cdots + d_{0j}X^{-j} \\ P_{-1}'(X) &= d_{-10} + d_{-11}X^{-1} + \cdots + d_{-1j}X^{-j} \end{aligned} \tag{13}$$

Multiplication by X^j yields the polynomials of Eq.(2). The initial power p is independent of the degree of approximation of the series expansion of Eq.(13), since the terms containing j do not show up in Eq.(12). An interchange of s_2 and s_3 leaves the form of Eq.(12) unchanged. Hence, the same comment also applies to the solution in the point s_3, if s_3 is not equal to s_1 or s_2.

Let us return to the characteristic equation Eq.(7). In the case of a double root,

$$s_2 = s_3 = -\frac{d_{00}}{2d_{10}}$$

one may solve the difference equation (1) only, if s_2 is also a root of the equation

$$c_{11}s^2 + c_{01}s + c_{-11} = 0 \tag{14}$$

The roots of Eq.(14) shall be denoted s_4 and s_5, and it shall hold $s_2 = s_3 = s_4$. One obtains:

$$s_4 = -\frac{d_{00}}{2d_{10}}, \qquad s_5 = -\frac{2d_{10}}{d_{00}}\,\frac{2d_{-11} - j(j-3)d_{-10}}{2d_{11} - j(j+1)d_{10}} \tag{15}$$

The determining equation for the initial power p now has the following form:

$$\begin{aligned} &\left[(p-j+2)(p-j+1)c_{10} - (p-j+1)c_{11}\left(1 - \frac{s_5}{s_2}\right)\right. \\ &\qquad\qquad \left. + s_2^{-2}(c_{12}s_2^2 + c_{02}s_2 + c_{-12})\right] p \cdots (p-j-3) = 0 \end{aligned} \tag{16}$$

The nontrivial values of p are the roots of the equation:

$$p^2 + \left(3 - \frac{d_{11}}{d_{10}} + \frac{4d_{10}d_{-11}}{d_{00}^2}\right) p$$
$$+ 2\left(1 + \frac{d_{01}}{d_{00}} - \frac{d_{11}}{d_{10}} - \frac{d_{02}}{d_{00}}\right) + \frac{d_{12}}{d_{10}} + \frac{4d_{10}d_{-12}}{d_{00}^2} = 0 \quad (17)$$

Again, p is independent of j.

The recursion formula for the coefficients q_ν of the power series of Eq.(10) is obtained by the substitution of Eq.(10) into Eq.(6):

$$\alpha_{0,\nu}q_0 + \cdots + \alpha_{\nu,\nu}q_\nu + \alpha_{\nu+1,\nu}q_{\nu+1} = 0 \quad (18)$$

One obtains after lengthy calculations with the help of Eqs.(4), (5), (8), and (12):

$$\alpha_{\nu+1,\nu} = d_{10}s_2(s_2 - s_3)(\nu + 1)(p + \nu + 1)s_2^{j-1}(p + \nu)\cdots(p + \nu - j + 3)$$
$$\alpha_{\nu,\nu} = \Big[d_{12}s_2^2 + d_{02}s_2 + d_{-12} - (p + \nu + 1)(d_{11}s_2^2 - d_{-11})$$
$$+ (p + \nu + 1)(p + \nu + 2)d_{10}s_2^2 - (p + 1)(d_{10}s_2^2 - d_{-10})$$
$$+ j(d_{10}s_2^2 - d_{-10})\nu\left(\nu + p - \frac{1}{2}j + \frac{1}{2}\right)\Big] s_2^{j-2}(p + \nu)\cdots(p + \nu - j + 3) \quad (19)$$

The recursion formula of Eq.(18) may be written as a system of linear equations. The coefficient $\alpha_{\nu+1,\nu}$ vanishes for $p = -(\nu + 1)$ and a system of p equations with $p - 1$ variables is obtained. The necessary and sufficient condition for the existence of such a system is that the determinant of the coefficients vanishes. This condition is simplified if s_2 or $s_2 - s_3$ is small; one obtains in first approximation:

$$\prod_{\nu=0}^{-p-1} \alpha_{\nu,\nu} = 0 \quad (20)$$

The condition $s_2 - s_3 \ll 1$ yields from this product:

$$d_{12}s_2^2 + d_{02}s_2 + d_{-12} - (p + \nu + 1)(d_{11}s_2^2 - d_{-11})$$
$$+ (p + \nu + 1)(p + \nu + 2)d_{10}s_2^2 = 0$$
$$\nu = 0,\ 1, \ldots,\ -(p + 1) \quad (21)$$

6.7 Discrete Spherical Harmonics

We start with the difference equation (5.4-37) but rewrite it with the help of Eqs.(5.4-8) and (5.4-9) in operator form in order to connect with previously published more detailed results[1].

$$\frac{\tilde{\Delta}^2\varphi_\mu(\eta)}{\tilde{\Delta}\eta^2} + \cot(\eta\Delta\vartheta)\frac{\tilde{\Delta}\varphi_\mu(\eta)}{\tilde{\Delta}\eta} + \left(\frac{(\nu\Delta\varphi)^2}{\sin^2(\eta\Delta\vartheta)} - \lambda_\vartheta\right)\varphi_\mu(\eta) = 0 \tag{1}$$

The nonnormalized, standard form of this equation is obtained for $\eta = \vartheta/\Delta\vartheta$ of Eq.(5.4-4):

$$\frac{\tilde{\Delta}^2\varphi_\mu(\vartheta)}{\tilde{\Delta}\vartheta^2} + \cot\vartheta\frac{\tilde{\Delta}\varphi_\mu(\vartheta)}{\tilde{\Delta}\vartheta} + \left(\frac{(\nu\Delta\varphi)^2}{\sin^2\vartheta} - \lambda_\vartheta\right)\varphi_\mu(\vartheta) = 0 \tag{2}$$

We substitute

$$\cos\vartheta = x, \quad \varphi_\mu(\vartheta) = \varphi_\mu(x) \tag{3}$$

and transform the difference quotients $\tilde{\Delta}/\tilde{\Delta}\vartheta$ and $\tilde{\Delta}^2/\tilde{\Delta}\vartheta^2$ into difference quotients $\tilde{\Delta}/\tilde{\Delta}x$ and $\tilde{\Delta}^2/\tilde{\Delta}x^2$:

$$\frac{\tilde{\Delta}\varphi_\mu(\vartheta)}{\tilde{\Delta}\vartheta} = \frac{\tilde{\Delta}\varphi_\mu(x)}{\tilde{\Delta}x}\frac{\tilde{\Delta}x}{\tilde{\Delta}\vartheta} = -\sin\vartheta\frac{\sin\Delta\vartheta}{\Delta\vartheta}\frac{\tilde{\Delta}\varphi_\mu(x)}{\tilde{\Delta}x}$$

$$\begin{aligned}
\frac{\tilde{\Delta}^2\varphi_\mu(\vartheta)}{\tilde{\Delta}\vartheta^2} &= \frac{1}{2}\left[\frac{\tilde{\Delta}_{\mathrm{l}}}{\tilde{\Delta}\vartheta}\left(\frac{\tilde{\Delta}_{\mathrm{r}}\varphi_\mu(\vartheta)}{\tilde{\Delta}\vartheta}\right) + \frac{\tilde{\Delta}_{\mathrm{r}}}{\tilde{\Delta}\vartheta}\left(\frac{\tilde{\Delta}_{\mathrm{l}}\varphi_\mu(\vartheta)}{\tilde{\Delta}\vartheta}\right)\right] \\
&= \frac{1}{2}\left[\frac{\tilde{\Delta}_l}{\tilde{\Delta}\vartheta}\left(\frac{\varphi_\mu(\vartheta+\Delta\vartheta) - \varphi_\mu(\vartheta)}{\Delta\vartheta}\right) + \frac{\tilde{\Delta}_r}{\tilde{\Delta}\vartheta}\left(\frac{\varphi_\mu(\vartheta) - \varphi_\mu(\vartheta-\Delta\vartheta)}{\Delta\vartheta}\right)\right] \\
&= \frac{1}{2}\Bigg[\frac{[\varphi_\mu(\vartheta+\Delta\vartheta) - \varphi_\mu(\vartheta)] - [\varphi_\mu(\vartheta) - \varphi_\mu(\vartheta-\Delta\vartheta)]}{(\Delta\vartheta)^2} \\
&\qquad + \frac{[\varphi_\mu(\vartheta+\Delta\vartheta) - \varphi(\vartheta)] - [\varphi_\mu(\vartheta) - \varphi_\mu(\vartheta-\Delta\vartheta)]}{(\Delta\vartheta)^2}\Bigg] \\
&= \left[\sin^2\vartheta\left(\frac{\sin\Delta\vartheta}{\Delta\vartheta}\right)^2 + \cos^2\vartheta\left(\frac{1-\cos\Delta\vartheta}{\Delta\vartheta}\right)^2\right]\frac{\tilde{\Delta}^2\varphi_\mu(x)}{\tilde{\Delta}x^2} \\
&\quad - 2\cos\vartheta\frac{1-\cos\vartheta}{(\Delta\vartheta)^2}\frac{\tilde{\Delta}\varphi_\mu(x)}{\tilde{\Delta}x}
\end{aligned} \tag{4}$$

[1] Harmuth 1989, Eq.(9.1-6) and Section 12.7. A missing sign + in Eq.(9.1-6) of the English edition is corrected here.

Equation (2) assumes the following form:

$$\left\{\left[(1-x^2)\left(\frac{\sin\Delta\vartheta}{\Delta\vartheta}\right)^2+x^2\left(\frac{1-\cos\Delta\vartheta}{\Delta\vartheta}\right)^2\right]\frac{\tilde{\Delta}^2}{\tilde{\Delta}x^2}\right.$$
$$\left.-x\left[2\frac{1-\cos\Delta\vartheta}{(\Delta\vartheta)^2}+\frac{\sin\Delta\vartheta}{\Delta\vartheta}\right]\frac{\tilde{\Delta}}{\tilde{\Delta}x}-\lambda_\vartheta+\frac{(\nu\Delta\varphi)^2}{1-x^2}\right\}\varphi_\mu(x)=0 \quad (5)$$

From the relations

$$\tilde{\Delta}x=\frac{1}{2}[(x+\Delta x)-(x-\Delta x)]=\Delta x$$
$$=\frac{1}{2}[\cos(\vartheta+\Delta\vartheta)-\cos(\vartheta-\Delta\vartheta)]=-\sin\vartheta\sin\Delta\vartheta \quad (6)$$

one obtains $\Delta\vartheta$ expressed by Δx:

$$\Delta\vartheta=-\sin^{-1}\frac{\Delta x}{(1-x^2)^{1/2}}$$
$$=-\left(\frac{\Delta x}{(1-x^2)^{1/2}}+\frac{1}{2}\frac{(\Delta x)^3}{3(1-x^2)^{3/2}}+\frac{1\cdot 3}{2\cdot 4}\frac{(\Delta x)^5}{5(1-x^2)^{5/2}}+\cdots\right)$$

$$\frac{\sin\Delta\vartheta}{\Delta\vartheta}=1-\frac{1}{2}\frac{(\Delta x)^2}{3(1-x^2)}+\cdots$$
$$\frac{1-\cos\Delta\vartheta}{\Delta\vartheta}=-\frac{1}{2}\frac{\Delta x}{(1-x^2)^{1/2}}\left(1+\frac{1}{24}\frac{(\Delta x)^2}{1-x^2}+\cdots\right)$$
$$\frac{(\Delta x)^2}{1-x^2}\le 1 \quad (7)$$

Only even powers of $(1-x^2)^{1/2}$ appear in the series expansions of the terms $(\sin\Delta\vartheta)/\Delta\vartheta$, $(\sin\Delta\vartheta)^2/(\Delta\vartheta)^2$, $(1-\cos\Delta\vartheta)/\Delta\vartheta$, and $(1-\cos\Delta\vartheta)^2/(\Delta\vartheta)^2$.

Equation (5) assumes the following form if the explicit difference quotients are substituted for the symbolic ones:

$$\left\{\left(\frac{x}{\Delta x}\right)^2\left[\left(\frac{\sin\Delta\vartheta}{\Delta\vartheta}\right)^2-\left(\frac{1-\cos\Delta\vartheta}{\Delta\vartheta}\right)^2\right]\right.$$
$$\left.+\frac{x}{\Delta x}\left(\frac{\sin\Delta\theta}{2\Delta\vartheta}+\frac{1-\cos\Delta\vartheta}{(\Delta\vartheta)^2}\right)-\frac{1}{(\Delta x)^2}\left(\frac{\sin\Delta\vartheta}{\Delta\vartheta}\right)^2\right\}\varphi_\mu(x+\Delta x)$$

$$+\left\{2\left(\frac{x}{\Delta x}\right)^2\left[\left(\frac{1-\cos\Delta\vartheta}{\Delta\vartheta}\right)^2-\left(\frac{\sin\Delta\vartheta}{\Delta\vartheta}\right)^2\right]\right.$$
$$\left.+\frac{2}{(\Delta x)^2}\left(\frac{\sin\Delta\vartheta}{\Delta\vartheta}\right)^2-\lambda_\vartheta+\frac{(\nu\Delta\varphi)^2}{1-x^2}\right\}\varphi_\mu(x)$$
$$+\left\{\left(\frac{x}{\Delta x}\right)^2\left[\left(\frac{\sin\Delta\vartheta}{\Delta\vartheta}\right)^2-\left(\frac{1-\cos\Delta\vartheta}{\Delta\vartheta}\right)^2\right]\right.$$
$$-\frac{x}{\Delta x}\left(\frac{\sin\Delta\vartheta}{2\Delta\vartheta}+\frac{1-\cos\Delta\vartheta}{(\Delta\vartheta)^2}\right)$$
$$\left.-\frac{1}{(\Delta x)^2}\left(\frac{\sin\Delta\vartheta}{\Delta\vartheta}\right)^2\right\}\varphi_\mu(x-\Delta x)=0 \quad (8)$$

This difference equation has transcendental coefficients. They can be replaced by polynomials of arbitrarily high degree by means of the series expansions of Eq.(7) multiplied by $(1-x^2)^k$.

$$\left[X^j+X^{j-1}+\left(\frac{7}{12}-\frac{j}{2(\Delta x)^2}\right)X^{j-2}+\cdots\right]\varphi_\mu(X+1)$$
$$+\left[-2X^j+\left(\frac{j}{(\Delta x)^2}+\lambda_\vartheta-\frac{7}{6}\right)X^{j-2}+\cdots\right]\varphi_\mu(X)$$
$$+\left[X^j-X^{j-1}+\left(\frac{7}{12}-\frac{j}{2(\Delta x)^2}\right)X^{j-2}+\cdots\right]\varphi_\mu(X-1)=0$$
$$X=x/\Delta X,\ \varphi_\mu(X)=\varphi_\mu(x),\ j=2k+2 \quad (9)$$

Equations (9) and (6.6-1) become equal if the following values are substituted for the coefficients d_{ij} in Eq.(6.6-2):

$$d_{10}=1 \qquad d_{11}=1 \qquad d_{12}=\frac{7}{12}-\frac{j}{2(\Delta x)^2}$$
$$d_{00}=-2 \qquad d_{01}=0 \qquad d_{02}=-\frac{7}{6}+\lambda_\vartheta+\frac{j}{2(\Delta x)^2}$$
$$d_{-10}=1 \qquad d_{-11}=-1 \quad d_{-12}=\frac{7}{12}-\frac{j}{2(\Delta x)^2} \quad (10)$$

One obtains from Eqs.(6.6-8), (6.6-15), and (6.6-17):

$$s_2=s_3=s_4=1,\quad s_5=\frac{2+j(j-3)}{j(j+1)-2},\quad \lim_{j\to\infty}s_5=1,\quad p(p+1)=-\lambda_\vartheta \quad (11)$$

The solution of Eq.(9) in the point $s_2 = s_3 = 1$ may be written as factorial series:

$$\varphi_\mu(X) = \frac{\Gamma(X)}{\Gamma(X+p+1)} \sum_{\mu=0}^{\infty} (-1)^\mu C_\mu \frac{(p+1)\cdots(p+\mu)}{(X+p+1)\cdots(X+p+\mu)} \tag{12}$$

The differential equation (6.6-6) belonging to the difference equation (9) has the singular points $s = 1$, $s = 0$, and $s = \infty$. A power series in the point $s = 1$ converges inside a circle that passes through $s = 0$. The factorial series of Eq.(12) converges in the half plane $X > X_0$, where X_0 remains to be determined.

Let ϑ run from 0 to π. The variable x runs then from 1 to -1, and X runs from $1/\Delta x$ to $-1/\Delta x$. The difference Δx is arbitrarily small. Hence, only factorial series with an abscissa of convergence $X_0 < -1/\Delta x$ are of interest as solutions. Let us assume Eq.(12) converges for $X > X_0 < -1/\Delta x$. The gamma function $\Gamma(X)$ has poles for negative integer values of X, which must be compensated either by zeros in the factorial series or by poles of $\Gamma(X+p+1)$. The gamma functions $\Gamma(X+p+1)$ and $\Gamma(X)$ have common poles if p is an integer:

$$p = -(l+1), \quad p = +l, \quad l = 0,\ 1,\ 2, \ldots \tag{13}$$

Only nonpositive integer values of p avoid poles for $X = 0, -1/\Delta x, \ldots$. The series of Eq.(12) becomes a polynomial if p is a negative integer, and the abscissa of convergence X_0 becomes $-\infty$.

$$\varphi_\mu(X) = \varphi_\mu(\vartheta) = \sum_{\mu=0}^{l} C_\mu \binom{X-1}{l-\mu}, \quad X = -\frac{\cot\vartheta}{\sin\Delta\vartheta} \tag{14}$$

Equations (13) and (11) yield the eigenvalues:

$$\lambda_\vartheta = -l(l+1) \tag{15}$$

This result is surprising, since it does not depend on $\Delta\vartheta$ or $\Delta\varphi$. It is impossible for $\Delta\vartheta$ and $\Delta\varphi$ to appear, since the only connection between the equations for $\varphi_\nu(\xi)$ and $\varphi_\mu(\eta)$ with that for $u_\mu(\rho)$ in Eqs.(5.4-31), (5.4-37); (5.4-39) is provided by the parameter λ_ϑ.

The condition of Eq.(15) is necessary for the existence of polynomials $\varphi_\mu(X)$. An investigation of the sufficient condition has so far not been made (Harmuth 1989, Sec.12.7).

Let us assume that Eq.(12) does not terminate. The factorial series must then have zeros in the points $X = 0, -\Delta x, \ldots$ and the coefficients C_μ must be functions of the arbitrary differences Δx. Hence, the polynomial solutions are the only ones that have no poles and that become independent of Δx for small values of Δx.

The difference equation (2) is satisfied in first approximation by spherical harmonics. Let us substitute $(\nu\Delta\varphi)^2 = -m^2 = 0$ and replace the difference quotients by means of a Taylor expansion by differential quotients:

$$\frac{\tilde{\Delta}_s}{\tilde{\Delta}\vartheta} = \sum_{\nu=0}^{\infty} \frac{(\Delta\vartheta)^{2\nu}}{(2\nu+1)!} \frac{d^{2\nu+1}}{d\vartheta^{\nu+1}}$$

$$\frac{\tilde{\Delta}^2}{\tilde{\Delta}\vartheta^2} = 2\sum_{\nu=0}^{\infty} \frac{(\Delta\vartheta)^{2\nu}}{(2\nu+1)!} \frac{d^{2\nu+2}}{d\vartheta^{2\nu+2}} \tag{16}$$

One obtains:

$$\left[\frac{d^2}{d\vartheta^2} + 2\sum_{\nu=0}^{\infty} \frac{(\Delta\vartheta)^{2\nu}}{(2\nu+2)!} \frac{d^{2\nu+2}}{d\vartheta^{2\nu+2}} + \frac{\cos\vartheta}{\sin\vartheta}\left(\frac{d}{d\vartheta} + \sum_{\nu=0}^{\infty} \frac{(\Delta\vartheta)^{2\nu}}{(2\nu+1)!} \frac{d^{2\nu+1}}{d\vartheta^{2\nu+1}}\right) - \lambda_\vartheta\right]\varphi_\mu(\vartheta) = 0 \tag{17}$$

The terms not containing $\Delta\vartheta$ are satisfied by spherical harmonics $P_l(\cos\vartheta)$ for $\lambda_\vartheta = -l(l+1)$:

$$P_l(\cos\vartheta) = 2\frac{1\cdot 3\cdot 5\cdots(2l-1)}{2^l\cdot l!}\left(\cos l\vartheta + \frac{1}{1}\frac{l}{2l+1}\cos(l-2)\vartheta + \frac{1\cdot 3}{1\cdot 2}\frac{l(l-1)}{(2l-1)(2l-3)}\cos(l-4)\vartheta + \cdots\right) \tag{18}$$

Differentiation yields:

$$(\Delta\vartheta)^2\frac{d^{2\nu+2}}{d\theta^{2\nu+2}}P_l(\cos\vartheta) = i^{2\nu}(l\Delta\vartheta)^2 2\frac{1\cdot 3\cdot 5\cdots(2l-1)}{2^{l-2}\cdot l!}\left[\cos l\vartheta + \frac{1}{1}\frac{l}{2l-1}\left(\frac{l-2}{l}\right)^{2\nu}\cos(l-2)\vartheta + \cdots\right] \tag{19}$$

The terms in Eq.(17) that contain $\Delta\vartheta$ may be made arbitrarily small by choosing $\Delta\vartheta$ sufficiently small. The Compton wavelength limits only the spatial resolution, not an angular resolution. Hence, $\varphi_\mu(\vartheta)$ has the form $P_l(\cos\vartheta)+O(\Delta\vartheta)$. Taking the derivative of order m of Eq.(17) with respect to $\cos\vartheta$ causes the terms not containing $\Delta\vartheta$ to be satisfied by the associated spherical harmonics $P_l^m(\cos\vartheta)$. The finite number of differentiations does not affect the principle of the vanishing of the terms multiplied by $\Delta\vartheta$. Hence, the difference equation (2) has eigenfunctions of the form $P_l^m(\cos\vartheta) + O(\Delta\vartheta) + O(\Delta\varphi)$.

6.8 Solution of the Differential Equation (5.5-12)

The differential equation (5.5-12) is to be solved with a power series centered at the point s_3. We substitute $s = (s - s_3) + s_3$ and use the relation according to Eq.(5.5-13)

$$s_3^2 + gs_3 + 1 = 0$$

to obtain:

$$\begin{aligned}&[(s-s_3)^3 + (2s_3 - s_2)(s-s_3)^2 + s_3(s_3-s_2)(s-s_3)]w''(s-s_3)\\&\quad + [2(s-s_3)^2 + (4s_3 + g - \tilde{\lambda}\tilde{\delta})(s-s_3) + (2s_3 + g - \tilde{\lambda}\tilde{\delta})s_3]w'(s-s_3)\\&\qquad\qquad - [l(l+1) - \tilde{\gamma}^2]w(s-s_3) = 0 \quad (1)\end{aligned}$$

From Eq.(5.5-15) we get:

$$w(s-s_3) = \sum_{\nu=0}^{N} q_3(\nu)(s-s_3)^{p_3+\nu}, \quad w'(s-s_3) = \sum_{\nu=0}^{N} q_3(\nu)(p_3+\nu)(s-s_3)^{p_3+\nu-1}$$

$$w''(s-s_3) = \sum_{\nu=0}^{N} q_3(\nu)(p_3+\nu)(p_3+\nu-1)(s-s_3)^{p_3+\nu-2} \quad (2)$$

Substitution into Eq.(1) yields for $\nu = 0$ and the lowest power $(s - s_3)^{p_3-1}$ an equation that determines the initial power p_3:

$$[(s_3 - s_2)(p_3 - 1) + 2s_3 + g - \tilde{\lambda}\tilde{\delta}]p_3 s_3 q_3(0)(s-s_3)^{p_3-1} = 0$$

Using the relation $s_2 + s_3 = -g$ of Eq.(5.5-14) we obtain:

$$p_3 = \frac{\tilde{\lambda}\tilde{\delta}}{s_3 - s_2} \quad (3)$$

Equation (1) is rewritten as follows with $s_2 + s_3 = -g$, Eq.(2), and Eq.(3):

$$\begin{aligned}&[(s-s_3)^3 + (2s_3 - s_2)(s-s_3)^2 + s_3(s_3-s_2)(s-s_3)]\\&\qquad\times \sum_{\nu=0}^{N} q_3(\nu)(p_3+\nu)(p_3+\nu-1)(s-s_3)^{p_3+\nu-2}\\&\quad + \{2(s-s_3)^2 + [2s_3 + (1-p_3)(s_3-s_2)](s-s_3) + (1-p_3)(s_3-s_2)s_3\}\\&\qquad\times \sum_{\nu=0}^{N} q_3(\nu)(p_3+\nu)(s-s_3)^{p_3+\nu-1}\\&\qquad - [l(l+1) - \tilde{\gamma}^2]\sum_{\nu=0}^{N} q_3(\nu)(s-s_3)^{p_3+\nu} = 0 \quad (4)\end{aligned}$$

For $\nu = 0$ we obtain from the terms multiplied by $(s - s_3)^{p_3}$ the following relation between $q_3(0)$ and $q_3(1)$:

$$\begin{aligned}\alpha_{3,1}(0)q_3(1) + \alpha_{3,0}(0)q_3(0) &= 0\\ \alpha_{3,1}(0) &= s_3(s_3 - s_2)(p_3 + 1)\\ \alpha_{3,0}(0) &= s_3p_3(p_3 + 1) - l(l+1) + \tilde{\gamma}^2 \end{aligned} \tag{5}$$

Generally we write Eq.(4) for $q_3(\nu - 1)$, $q_3(\nu)$, and $q_3(\nu + 1)$. The terms in boldface will be explained presently:

$$\begin{aligned}
\cdots &+ [\mathbf{(s - s_3)^3} + (2s_3 - s_2)(s - s_3)^2 + s_3(s_3 - s_2)(s - s_3)]\\
&\qquad \times q_3(\nu - 1)(p_3 + \nu - 1)(p_3 + \nu - 2)(s - s_3)^{p_3+\nu-3}\\
&+ \{\mathbf{2(s - s_3)^2} + [2s_3 - (p_3 - 1)(s_3 - s_2)](s - s_3) - (p_3 - 1)(s_3 - s_2)s_3\}\\
&\qquad \times q_3(\nu - 1)(p_3 + \nu - 1)(s - s_3)^{p_3+\nu-2}\\
&\qquad - [l(l+1) - \tilde{\gamma}^2]q_3(\nu - 1)(s - s_3)^{p_3+\nu-1}\\
&+ [(s - s_3)^3 + \mathbf{(2s_3 - s_2)(s - s_3)^2} + s_3(s_3 - s_2)(s - s_3)]\\
&\qquad \times q_3(\nu)(p_3 + \nu)(p_3 + \nu - 1)(s - s_3)^{p_3+\nu-2}\\
&+ \{2(s - s_3)^2 + \mathbf{[2s_3 - (p_3 - 1)(s_3 - s_2)](s - s_3)} - (p_3 - 1)(s_3 - s_2)s_3\}\\
&\qquad \times q_3(\nu)(p_3 + \nu)(s - s_3)^{p_3+\nu-1}\\
&\qquad \mathbf{-[l(l+1) - \tilde{\gamma}^2]}\, q_3(\nu)(s - s_3)^{p_3+\nu}\\
&+ [(s - s_3)^3 + (2s_3 - s_2)(s - s_3)^2 + \mathbf{s_3(s_3 - s_2)(s - s_3)}]\\
&\qquad \times q_3(\nu + 1)(p_3 + \nu + 1)(p_3 + \nu)(s - s_3)^{p_3+\nu-1}\\
&+ \{2(s - s_3)^2 + [2s_3 - (p_3 - 1)(s_3 - s_2)](s - s_3) \mathbf{-(p_3 - 1)(s_3 - s_2)s_3}\}\\
&\qquad \times q_3(\nu + 1)(p_3 + \nu + 1)(s - s_3)^{p_3+\nu}\\
&\qquad - [l(l+1) - \tilde{\gamma}^2]q_3(\nu + 1)(s - s_3)^{p_3+\nu+1} + \ldots
\end{aligned} \tag{6}$$

The terms emphasized by boldface are all multiplied by $(s - s_3)^{p_3+\nu}$. Their sum must be zero:

$$\begin{aligned}
&[(p_3+\nu-1)(p_3+\nu-2)+2(p_3+\nu-1)]q_3(\nu-1)\\
&+\{(2s_3-s_2)(p_3+\nu)(p_3+\nu-1)+[2s_3-(p_3-1)(s_3-s_2)](p_3+\nu)-l[(l+1)-\tilde{\gamma}^2]\}q_3(\nu)\\
&\quad +[s_3(s_3-s_2)(p_3+\nu+1)(p_3+\nu)-(p_3-1)(s_3-s_2)s_3(p_3+\nu+1)]q_3(\nu+1) = 0
\end{aligned} \tag{7}$$

This equation may be simplified:

$$
\begin{aligned}
\alpha_{3,1}(\nu)q_3(\nu+1)+\alpha_{3,0}(\nu)q_3(\nu)+\alpha_{3,-1}(\nu)q_3(\nu-1)&=0\\
\alpha_{3,1}(\nu)&=s_3(s_3-s_2)(p_3+\nu+1)(\nu+1)\\
\alpha_{3,0}(\nu)&=(2s_3-s_2)(p_3+\nu)^2+[s_3-p_3(s_3-s_2)](p_3+\nu)\\
\alpha_{3,-1}(\nu)&=(p_3+\nu)(p_3+\nu-1)
\end{aligned}\tag{8}
$$

6.9 CONFORMAL MAPPING FOR SECTION 5.7

According to Figs.5.7-1 and 5.5-2 we have to resort to conformal mapping if a power series in the point s_3 is to converge as far as $s=0$ for $s_3>1/\sqrt{2}$. We start from Eq.(6.8-1) and replace the terms $(s-s_3)^n$ by $s_3^n(s/s_3-1)^n$. This calls for a change of $w'(s-s_3)$ and $w''(s-s_3)$ in Eq.(6.8-1):

$$
w'(s-s_3)=\frac{dw(s-s_3)}{ds}\to\frac{dw(s/s_3-1)}{d(s/s_3-1)}\frac{d(s/s_3-1)}{ds}=\frac{1}{s_3}w'(s/s_3-1)
$$

$$
w''(s-s_3)\to\frac{1}{s_3^2}w''(s/s_3-1)\tag{1}
$$

Equation (6.8-1) becomes:

$$
\begin{aligned}
&\left[s_3\left(\frac{s}{s_3}-1\right)^3+(2s_3-s_2)\left(\frac{s}{s_3}-1\right)^2+(s_3-s_2)\left(\frac{s}{s_3}-1\right)\right]w''(s/s_3-1)\\
&\quad+\left[2s_3\left(\frac{s}{s_3}-1\right)^2+(4s_3+g-\tilde{\lambda}\tilde{\delta})\left(\frac{s}{s_3}-1\right)+2s_3+g-\tilde{\lambda}\tilde{\delta}\right]w'(s/s_3-1)\\
&\qquad-[l(l+1)-\tilde{\gamma}^2]w(s/s_3-1)=0
\end{aligned}\tag{2}
$$

The substitution $s/s_3=\zeta^{1/P}$ transforms the differential equation from the s-plane to the ζ-plane:

$$
\begin{aligned}
&[s_3(\zeta^{1/P}-1)^3+(2s_3-s_2)(\zeta^{1/P}-1)^2+(s_3-s_2)(\zeta^{1/P}-1)]w''(\zeta^{1/P}-1)\\
&\quad+[2s_3(\zeta^{1/P}-1)^2+(4s_3+g-\tilde{\lambda}\tilde{\delta})(\zeta^{1/P}-1)+2s_3+g-\tilde{\lambda}\tilde{\delta}]w'(\zeta^{1/P}-1)\\
&\qquad-[l(l+1)-\tilde{\gamma}^2]w(\zeta^{1/P}-1)=0
\end{aligned}\tag{3}
$$

In order to obtain a differential equation of the Fuchs-type we must transform the terms $(\zeta^{1/P}-1)^n$ into power series of $\zeta-1$ or $1-\zeta$. This is possible by expanding $\zeta^{1/P}-1=-(1-\zeta^{1/P})$ in a binomial series:

$$1-\zeta^{1/P} = 1-[1-(1-\zeta)]^{1/P} = 1-\left[1-\frac{1}{P}(1-\zeta)+\frac{1}{2!}\frac{1}{P}\left(\frac{1}{P}-1\right)(1-\zeta)^2-\dots\right]$$

$$= \frac{1-\zeta}{P}X$$

$$X=1+\frac{P-1}{2!P}(1-\zeta)+\frac{(P-1)(2P-1)}{3!P^2}(1-\zeta)^2+\frac{(P-1)\dots(3P-1)}{4!P^3}(1-\zeta)^3+\dots$$

$$=1+b_{11}(1-\zeta)+b_{12}(1-\zeta)^2+\dots+b_{1j}(1-\zeta)^j+\dots$$

$$b_{1j} = \frac{(P-1)\dots(jP-1)}{(j+1)!P^j} \tag{4}$$

The series is absolutely convergent in the interval $-1 < 1-\zeta < +1$. We can derive the following powers of $(1-\zeta^{1/P})^n$ from Eq.(4):

$$\left(1-\zeta^{1/P}\right)^n = \left(\frac{1-\zeta}{P}\right)^n X^n, \quad \left(\zeta^{1/P}-1\right)^n = \left(\frac{\zeta-1}{P}\right)^n X^n, \quad n = 1, 2, \dots \tag{5}$$

We obtain for $X^2, X^3, \dots, X^6$:

$$X^2 = 1 + b_{21}(1-\zeta) + b_{22}(1-\zeta)^2 + b_{23}(1-\zeta)^3 + b_{24}(1-\zeta)^4 + \dots$$
$$= 1 + 2b_{11}(1-\zeta) + (2b_{12}+b_{11}^2)(1-\zeta)^2 + 2(b_{13}+b_{11}b_{12})(1-\zeta)^3$$
$$+ (2b_{14}+2b_{11}b_{13}+b_{12}^2)(1-\zeta)^4 + \dots$$
$$b_{21} = 2b_{11},\ b_{22} = 2b_{12}+b_{11}^2,\ b_{23} = 2(b_{13}+b_{11}b_{12}),$$
$$b_{24} = 2b_{14}+2b_{11}b_{13}+b_{12}^2 \tag{6}$$

$$X^3 = 1 + b_{31}(1-\zeta) + b_{32}(1-\zeta)^2 + b_{33}(1-\zeta)^3 + \dots$$
$$b_{31} = 3b_{11},\ b_{32} = 3(b_{11}^2+b_{12}),\ b_{33} = 6b_{11}b_{12}+3b_{13}+b_{11}^3 \tag{7}$$

$$X^4 = 1 + b_{41}(1-\zeta) + b_{42}(1-\zeta)^2 + \dots$$
$$b_{41} = 4b_{11},\ b_{42} = 6b_{11}^2 + 4b_{12} \tag{8}$$

$$X^5 = 1 + b_{51}(1-\zeta) + \dots, \qquad b_{51} = 5b_{11} \tag{9}$$

$$X^6 = 1 + \dots \tag{10}$$

If we substitute $1-\zeta = -(\zeta-1)$ into Eq.(4) for X we obtain alternating series, e.g.,

$$X = 1 - b_{11}(\zeta-1) + b_{12}(\zeta-1)^2 - b_{13}(\zeta-1)^3 + \dots \tag{11}$$

Hence, the substitution of $(\zeta^{1/P}-1)^n$ of Eq.(5) into Eq.(3) brings:

$$
\begin{aligned}
&\{s_3P^{-3}(\zeta-1)^3[1-b_{31}(\zeta-1)+b_{32}(\zeta-2)^2-\ldots]\\
&\qquad +(2s_3-s_2)P^{-2}(\zeta-1)^2[1-b_{21}(\zeta-1)+b_{22}(\zeta-1)^2-\ldots]\\
&\ +(s_3-s_2)P^{-1}(\zeta-1)[1-b_{11}(\zeta-1)+b_{12}(\zeta-1)^2-\ldots]\}w''(\zeta^{1/P}-1)\\
&+\{2s_3P^{-2}(\zeta-1)^2[1-b_{21}(\zeta-1)+b_{22}(\zeta-1)^2-\ldots]\\
&\qquad +(4s_3+g-\tilde{\lambda}\tilde{\delta})P^{-1}(\zeta-1)[1-b_{11}(\zeta-1)+b_{12}(\zeta-1)^2-\ldots]\\
&\qquad\qquad +2s_3+g-\tilde{\lambda}\tilde{\delta}\}w'(\zeta^{1/P}-1)\\
&\qquad\qquad -[l(l+1)-\tilde{\gamma}^2]w(\zeta^{1/P}-1)=0 \qquad (12)
\end{aligned}
$$

This is the differential equation (2) in the ζ-plane. Since it is of the Fuchs-type we write its solution $w(\zeta^{1/P}-1)$ as a power series in $\zeta-1$. The notation $\tilde{q}(\nu)$ and $\tilde{p}$ is used in the ζ-plane instead of the notation $q(\nu)$ and p in the s-plane:

$$
\begin{aligned}
w(\zeta^{1/P}-1)&=\sum_{\nu=0}^{\infty}\tilde{q}(\nu)(\zeta-1)^{\tilde{p}+\nu}\\
w'(\zeta^{1/P}-1)&=\sum_{\nu=0}^{\infty}\tilde{q}(\nu)(\tilde{p}+\nu)(\zeta-1)^{\tilde{p}+\nu-1}\\
w''(\zeta^{1/P}-1)&=\sum_{\nu=0}^{\infty}\tilde{q}(\nu)(\tilde{p}+\nu)(\tilde{p}+\nu-1)(\zeta-1)^{\tilde{p}+\nu-2} \qquad (13)
\end{aligned}
$$

Substitution into Eq.(12) yields:

$$
\begin{aligned}
&\{s_3P^{-3}(\zeta-1)^3[1-b_{31}(\zeta-1)+b_{32}(\zeta-1)^2-\ldots]\\
&\qquad +(2s_3-s_2)P^{-2}(\zeta-1)^2[1-b_{21}(\zeta-1)+b_{22}(\zeta-1)^2-\ldots]\\
&\qquad +(s_3-s_2)P^{-1}(\zeta-1)[1-b_{11}(\zeta-1)+b_{12}(\zeta-1)^2-\ldots]\}\\
&\qquad\qquad \times\sum_{\nu=0}^{\infty}\tilde{q}(\nu)(\tilde{p}+\nu)(\tilde{p}+\nu-1)(\zeta-1)^{\tilde{p}+\nu-2}\\
&+\{2s_3P^{-2}(\zeta-1)^2[1-b_{21}(\zeta-1)+b_{22}(\zeta-1)^2-\ldots]\\
&\qquad +(4s_3+g-\tilde{\lambda}\tilde{\delta})P^{-1}(\zeta-1)[1-b_{11}(\zeta-1)+b_{12}(\zeta-1)^2-\ldots]\\
&\qquad\qquad +2s_3+g-\tilde{\lambda}\tilde{\delta}\}\sum_{\nu=0}^{\infty}\tilde{q}(\nu)(\tilde{p}+\nu)(\zeta-1)^{\tilde{p}+\nu-1}\\
&\qquad\qquad -[l(l+1)-\tilde{\gamma}^2]\sum_{\nu=0}^{\infty}\tilde{q}(\nu)(\zeta-1)^{\tilde{p}+\nu}=0 \qquad (14)
\end{aligned}
$$

The sum of all coefficients of a certain power $(\zeta - 1)^{\tilde{p}+n}$ must be zero. The lowest power in Eq.(14) is $(\zeta - 1)^{\tilde{p}-1}$ for $\nu = 0$. There is only the coefficient $\tilde{q}(0)$ for this power:

$$[(s_3 - s_2)P^{-1}\tilde{p}(\tilde{p} - 1) + (2s_3 + g - \tilde{\lambda}\tilde{\delta})\tilde{p}]\,\tilde{q}(0) = 0 \tag{15}$$

Using the relation $-g = s_2 + s_3$ of Eq.(5.5-14) we obtain the following value for the initial power $\tilde{p}$:

$$\tilde{p} = P\left(\frac{\tilde{\lambda}\tilde{\delta}}{s_3 - s_2} - 1\right) + 1 \tag{16}$$

For $P = 1$ we get the same value for $\tilde{p}$ as for p in Eq.(5.5-16). We derive from Eq.(16) an expression for $\tilde{\lambda}\tilde{\delta}$ that will simplify some of the following equations:

$$\tilde{\lambda}\tilde{\delta} = (s_3 - s_2)[P^{-1}(\tilde{p} - 1) + 1] \tag{17}$$

The second lowest power in Eq.(14) is $(\zeta - 1)^{\tilde{p}}$. It contains the coefficients $\tilde{q}(1)$ and $\tilde{q}(0)$. It yields $\tilde{q}(1)$ if we take $\tilde{q}(0)$ as the choosable constant:

$$\begin{aligned}(2s_3 - s_2)P^{-2}\tilde{q}(0)\,\tilde{p}\,(\tilde{p} - 1) + (s_3 - s_2)P^{-1}[\tilde{q}(1)(\tilde{p} + 1)\tilde{p} - b_{11}\tilde{q}(0)\,\tilde{p}\,(\tilde{p} - 1)]\\ + (4s_3 + g - \tilde{\lambda}\tilde{\delta})P^{-1}\tilde{q}(0) + (2s_3 + g - \tilde{\lambda}\tilde{\delta})\tilde{q}(1)(\tilde{p} + 1)\\ - [l(l + 1) - \tilde{\gamma}^2]\tilde{q}(0) = 0\end{aligned} \tag{18}$$

We may rewrite this equation with the help of Eq.(17) and eventually obtain the following simpler form:

$$\alpha_{\mathrm{P},1}(0)\tilde{q}(1) + \alpha_{\mathrm{P},0}(0)\tilde{q}(0) = 0 \tag{19}$$

$$\alpha_{\mathrm{P},1}(0) = P^{-1}(s_3 - s_2)(\tilde{p} + 1) \tag{20}$$

$$\begin{aligned}\alpha_{\mathrm{P},0}(0) = P^{-1}\tilde{p}\left(s_3P^{-1}(\tilde{p} - 1 + 2P) - \frac{P - 1}{2!P}(s_3 - s_2)(\tilde{p} - 1)\right)\\ - l(l + 1) + \tilde{\gamma}^2\end{aligned} \tag{21}$$

For $P = 1$ and thus $\tilde{p} = p$ we obtain:

$$\alpha_{\mathrm{P},1}(0) = (s_3 - s_2)(p + 1) \tag{22}$$

$$\alpha_{\mathrm{P},0}(0) = s_3p(p + 1) - l(l + 1) + \tilde{\gamma}^2 \tag{23}$$

Equation (23) equals $\alpha_{3,0}(0)$ in Eq.(5.5-18) while Eq.(22) differs by a factor s_3 from $\alpha_{3,1}(0)$. We expect that the inverse transformation from the ζ-plane to the s-plane will provide the missing factor s_3.

The next step would be to determine the coefficient of $(\zeta-1)^{\tilde{p}+1}$ in Eq.(14) that yields in analogy to Eq.(19):

$$\alpha_{P,2}(1)\tilde{q}(2)+\alpha_{P,1}(1)\tilde{q}(1)+\alpha_{P,0}(1)\tilde{q}(0)=0 \tag{24}$$

The elaboration of $\alpha_{P,2}(1)$, $\alpha_{P,1}(1)$, $\alpha_{P,0}(1)$ is a challenge but it is not needed here. As $(\zeta-1)^{\tilde{p}+1}$ increases to $(\zeta-1)^{\tilde{p}+2}$, $(\zeta-1)^{\tilde{p}+3}$, ... the recursion formulas according to Eqs.(19) and (24) increase to 4, 5, ... terms.

Let us turn to the inverse transformation $\zeta=(s/s_3)^P$ from the ζ-plane to the s-plane and see what becomes of $\tilde{p}$ and $\tilde{q}(\nu)$ of Eqs.(16) and (19) for $\nu=1$. We write:

$$\begin{aligned}w(\zeta^{1/P}-1)&=\sum_{\nu=0}^{\infty}\tilde{q}(\nu)(\zeta-1)^{\tilde{p}+\nu}=\sum_{\nu=0}^{\infty}\tilde{q}(\nu)\left[\left(\frac{s}{s_3}\right)^P-1\right]^{\tilde{p}+\nu}\\&=\left(\frac{(s/s_3)^P-1}{s/s_3-1}\right)^{\tilde{p}}\left(\frac{s}{s_3}-1\right)^{\tilde{p}}\sum_{\nu=0}^{\infty}\tilde{q}(\nu)(-1)^{\nu}\left[1-\left(\frac{s}{s_3}\right)^P\right]^{\nu}\end{aligned} \tag{25}$$

The reason for this notation is that we can change the signs of the first factor:

$$\left(\frac{(s/s_3)^P-1}{s/s_3-1}\right)^{\tilde{p}}=\left(\frac{1-(s/s_3)^P}{1-s/s_3}\right)^{\tilde{p}} \tag{26}$$

There is no problem with a sign change for integer powers $\nu=1,2,\ldots$:

$$\left[(s/s_3)^P-1\right]^{\nu}=(-1)^{\nu}\left[1-(s/s_3)^P\right]^{\nu} \tag{27}$$

We use the binomial series to expand $1-(s/s_3)^P$:

$$\begin{aligned}1-\left(\frac{s}{s_3}\right)^P&=1-\left[1-\left(1-\frac{s}{s_3}\right)\right]^P=1-\left[1-P\left(1-\frac{s}{s_3}\right)\right.\\&\quad\left.+\frac{P(P-1)}{2!}\left(1-\frac{s}{s_3}\right)^2-\frac{P(P-1)(P-2)}{3!}\left(1-\frac{s}{s_3}\right)^3+\ldots\right]\\&=P\left(1-\frac{s}{s_3}\right)\left[1-\frac{P-1}{2!}\left(1-\frac{s}{s_3}\right)\right.\\&\quad\left.+\frac{(P-1)(P-2)}{3!}\left(1-\frac{s}{s_3}\right)^2-\ldots\right]=P\left(1-\frac{s}{s_3}\right)Y\end{aligned} \tag{28}$$

The new variable Y is defined as follows:

$$Y = 1 - \frac{P-1}{2!}\left(1 - \frac{s}{s_3}\right) + \frac{(P-1)(P-2)}{3!}\left(1 - \frac{s}{s_3}\right)^2 + \dots$$
$$\dots + (-1)^j \frac{(P-1)\dots(P-j)}{(j+1)!}\left(1 - \frac{s}{s_3}\right)^j + \dots$$
$$= 1 - c_1\left(1 - \frac{s}{s_3}\right) - c_2\left(1 - \frac{s}{s_3}\right)^2 - \dots - c_j\left(1 - \frac{s}{s_3}\right)^j - \dots$$

$$c_1 = \frac{P-1}{2!},\ c_2 = -\frac{(P-1)(P-2)}{3!}, \dots,\ c_j = -(-1)^j \frac{(P-1)\dots(P-j)}{(j+1)!} \tag{29}$$

The choice of the negative sign for the coefficients c_j will become evident presently. Equation (26) may be rewritten:

$$\left(\frac{(s/s_3)^P - 1}{s/s_3 - 1}\right)^{\tilde{p}} = \left(\frac{1 - (s/s_3)^P}{1 - s/s_3}\right)^{\tilde{p}} = P^{\tilde{p}} Y^{\tilde{p}} \tag{30}$$

The series of Eq.(29) for Y terminates for positive integer values of P. Hence, the question of convergence is avoided for $P = 1, 2, \dots$. The expression $Y^{\tilde{p}}$ in Eq.(30) can again be expanded into a binomial series

$$P^{\tilde{p}} Y^{\tilde{p}} = P^{\tilde{p}}[1 - (1 - Y)]^{\tilde{p}} \tag{31}$$

and Eq.(30) becomes:

$$\left(\frac{1 - (s/s_3)^P}{1 - s/s_3}\right)^{\tilde{p}} = P^{\tilde{p}} Y^{\tilde{p}} = P^{\tilde{p}}\left[1 - \tilde{p}(1 - Y) + \frac{\tilde{p}(\tilde{p} - 1)}{2!}(1 - Y)^2 - \dots\right] \tag{32}$$

The positive integer powers of $1 - Y$ will be needed. From Eq.(29) we get:

$$1 - Y = \left(1 - \frac{s}{s_3}\right)\left[c_1 + c_2\left(1 - \frac{s}{s_3}\right) + c_3\left(1 - \frac{s}{s_3}\right)^2 + c_4\left(1 - \frac{s}{s_3}\right)^3 + \dots\right] \tag{33}$$

The reason for the choice of the signs of the coefficients c_j in Eq.(29) becomes clear. For higher powers of $1 - Y$ we obtain:

$$(1 - Y)^2 = \left(1 - \frac{s}{s_3}\right)^2 \left[c_1^2 + 2c_1c_2\left(1 - \frac{s}{s_3}\right) + (2c_1c_3 + c_2^2)\left(1 - \frac{s}{s_3}\right)^2\right.$$
$$\left. + (2c_1c_4 + 2c_2c_3)\left(1 - \frac{s}{s_3}\right)^3 + (2c_1c_5 + 2c_3c_4 + c_3^2)\left(1 - \frac{s}{s_3}\right)^4 + \dots\right] \tag{34}$$

$$(1-Y)^3 = \left(1-\frac{s}{s_3}\right)^3 \left[c_1^3 + 3c_1^2c_2\left(1-\frac{s}{s_3}\right) + (3c_1^2c_3 + 3c_1c_2^2)\left(1-\frac{s}{s_3}\right)^2 + (3c_1^2c_4 + 6c_1c_2c_3 + c_2^3)\left(1-\frac{s}{s_3}\right)^3 + \ldots\right] \quad (35)$$

$$(1-Y)^4 = \left(1-\frac{s}{s_3}\right)^4 \left[c_1^4 + 4c_1^3c_2\left(1-\frac{s}{s_3}\right) + (4c_1^3c_3 + 6c_1^2c_2^2)\left(1-\frac{s}{s_3}\right)^2 + \ldots\right] \quad (36)$$

$$(1-Y)^5 = \left(1-\frac{s}{s_3}\right)^5 \left[c_1^5 + 5c_1^4c_2\left(1-\frac{s}{s_3}\right) + \ldots\right] \quad (37)$$

$$(1-Y)^6 = \left(1-\frac{s}{s_3}\right)^6 \left[c_1^6 + \ldots\right] \quad (38)$$

Substitution of Eqs.(33) to (38) into Eq.(32) produces a very long formula that we bring into a printable form by defining new coefficients d_j:

$$\begin{aligned}\left(\frac{1-(s/s_3)^P}{1-s/s_3}\right)^{\tilde{p}} &= P^{\tilde{p}}\left[1 + d_1\left(1-\frac{s}{s_3}\right) + d_2\left(1-\frac{s}{s_3}\right)^2 + d_3\left(1-\frac{s}{s_3}\right)^3 + \ldots\right] \\ &= P^{\tilde{p}}\left[1 - d_1\left(\frac{s}{s_3}-1\right) + d_2\left(\frac{s}{s_3}-1\right)^2 - d_3\left(\frac{s}{s_3}-1\right)^3 + \ldots\right]\end{aligned}$$

$$\begin{aligned}
d_1 &= \tilde{p}c_1 \\
d_2 &= \tilde{p}c_2 + \frac{1}{2!}\tilde{p}(\tilde{p}-1)c_1^2 \\
d_3 &= \tilde{p}c_3 + \frac{1}{2!}\tilde{p}(\tilde{p}-1)2c_1c_2 + \frac{1}{3!}\tilde{p}(\tilde{p}-1)(\tilde{p}-2)c_1^3 \\
d_4 &= \tilde{p}c_4 + \frac{1}{2!}\tilde{p}(\tilde{p}-1)(2c_1c_3 + c_2^2) + \frac{1}{3!}\tilde{p}(\tilde{p}-1)(\tilde{p}-2)3c_1^2c_2. + \frac{1}{4!}\tilde{p}\ldots(\tilde{p}-3)c_1^4 \\
d_5 &= \tilde{p}c_5 + \frac{1}{2!}\tilde{p}(\tilde{p}-1)(2c_1c_4 + 2c_2c_3) + \frac{1}{3!}\tilde{p}(\tilde{p}-1)(\tilde{p}-2)(3c_1^2c_3 - 3c_1c_2^2) \\
&\quad + \frac{1}{4!}\tilde{p}\ldots(\tilde{p}-3)4c_1^3c_2 + \frac{1}{5!}\tilde{p}\ldots(\tilde{p}-4)c_1^5
\end{aligned}$$

$$d_6 = \tilde{p}c_6 + \frac{1}{2!}\tilde{p}(\tilde{p}-1)(2c_1c_5 + 2c_2c_4 + c_3^2) + \frac{1}{3!}\tilde{p}(\tilde{p}-1)(\tilde{p}-2)(3c_1^2c_4 + 6c_1c_2c_3$$
$$+ c_2^3) + \frac{1}{4!}\tilde{p}\ldots(\tilde{p}-3)(4c_1^3c_2 + 6c_1^2c_2^2 + \frac{1}{5!}\tilde{p}\ldots(\tilde{p}-4)5c_1^4c_2$$
$$+ \frac{1}{6!}\tilde{p}\ldots(\tilde{p}-5)c_1^6 \qquad (39)$$

The terms $[(s/s_3)^P - 1]^\nu$ in Eq.(25) also require a series expansion. Using Eq.(28) we obtain:

$$\left[\left(\frac{s}{s_3}\right)^P - 1\right]^\nu = (-1)^\nu\left[1 - \left(\frac{s}{s_3}\right)^P\right]^\nu = (-1)^\nu P^\nu\left(1 - \frac{s}{s_3}\right)^\nu Y^\nu$$
$$\left(1 - \frac{s}{s_3}\right)^\nu Y^\nu = 1 + e_{\nu 1}\left(1 - \frac{s}{s_3}\right) + e_{\nu 2}\left(1 - \frac{s}{s_3}\right)^2 + \ldots \qquad (40)$$

For $\nu = 1, 2, \ldots, 6$ we obtain with the help of Eq.(29):

$$\left(1 - \frac{s}{s_3}\right)Y = \left(1 - \frac{s}{s_3}\right)\left[1 - c_1\left(1 - \frac{s}{s_3}\right) - c_2\left(1 - \frac{s}{s_3}\right)^2 - \ldots\right]$$
$$e_{11} = -c_1,\ e_{12} = -c_2,\ e_{13} = -c_3, \ldots \qquad (41)$$

$$\left(1 - \frac{s}{s_3}\right)^2 Y^2 = \left(1 - \frac{s}{s_3}\right)^2\left[1 - 2c_1\left(1 - \frac{s}{s_3}\right) + (-2c_2 + c_1)^2\left(1 - \frac{s}{s_3}\right)^2\right.$$
$$\left. + (-2c_3 + 2c_1c_2)\left(1 - \frac{s}{s_3}\right)^2 + (-2c_4 + 2c_1c_3 + c_2^2)\left(1 - \frac{s}{s_3}\right)^4 + \ldots\right]$$
$$e_{21} = -2c_1,\ e_{22} = -2c_2 + c_1^2,\ e_{23} = -2c_3 + 2c_1c_2,$$
$$e_{24} = -2c_4 + 2c_1c_3 + c_2^2 \qquad (42)$$

$$\left(1 - \frac{s}{s_3}\right)^3 Y^3 = \left(1 - \frac{s}{s_3}\right)^3\left[1 - 3c_1\left(1 - \frac{s}{s_3}\right) + (3c_1^2 - 3c_2)\left(1 - \frac{s}{s_3}\right)^2\right.$$
$$\left. + (-c_1^3 + 6c_1c_2 - 3c_3)\left(1 - \frac{s}{s_3}\right)^3 + \ldots\right]$$
$$e_{31} = -3c_1,\ e_{32} = 3c_1^2 - 3c_2,\ e_{33} = -c_1^3 + 6c_1c_2 - 3c_3 \qquad (43)$$

$$\left(1 - \frac{s}{s_3}\right)^4 Y^4 = \left(1 - \frac{s}{s_3}\right)^4\left[1 - 4c_1\left(1 - \frac{s}{s_3}\right) + (6c_1^2 - 4c_2)\left(1 - \frac{s}{s_3}\right)^2 + \ldots\right]$$
$$e_{41} = -4c_1,\ e_{42} = 6c_1^2 - 4c_2 \qquad (44)$$

$$\left(1-\frac{s}{s_3}\right)^5 Y^5 = \left(1-\frac{s}{s_3}\right)^5 \left[1 - 5c_1\left(1-\frac{s}{s_3}\right) + \ldots\right]$$

$$e_{51} = -5c_1 \tag{45}$$

$$\left(1-\frac{s}{s_3}\right)^6 Y^6 = \left(1-\frac{s}{s_3}\right)^6 \left[1 + \ldots\right] \tag{46}$$

We may rewrite the last sum in Eq.(25) as follows:

$$\begin{aligned}
\sum_{\nu=0}^{\infty} \tilde{q}(\nu)(-1)^\nu \left[1-\left(\frac{s}{s_3}\right)^P\right]^\nu &= \sum_{\nu=0}^{\infty} \tilde{q}(\nu)(-1)^\nu P^\nu \left(1-\frac{s}{s_3}\right)^\nu Y^\nu \\
&= \tilde{q}(0)P^0\left(1-\frac{s}{s_3}\right)^0 Y^0 \\
&- \tilde{q}(1)P^1\left(1-\frac{s}{s_3}\right)^1 \left[1 + e_{11}\left(1-\frac{s}{s_3}\right) + \cdots + e_{15}\left(1-\frac{s}{s_3}\right)^5 + \ldots\right] \\
&+ \tilde{q}(2)P^2\left(1-\frac{s}{s_3}\right)^2 \left[1 + e_{21}\left(1-\frac{s}{s_3}\right) + \cdots + e_{24}\left(1-\frac{s}{s_3}\right)^4 + \ldots\right] \\
&- \tilde{q}(3)P^3\left(1-\frac{s}{s_3}\right)^3 \left[1 + e_{31}\left(1-\frac{s}{s_3}\right) + \cdots + e_{33}\left(1-\frac{s}{s_3}\right)^3 + \ldots\right] \\
&+ \tilde{q}(4)P^4\left(1-\frac{s}{s_3}\right)^4 \left[1 + e_{41}\left(1-\frac{s}{s_3}\right) + e_{42}\left(1-\frac{s}{s_3}\right)^2 + \ldots\right] \\
&- \tilde{q}(5)P^5\left(1-\frac{s}{s_3}\right)^5 \left[1 + e_{51}\left(1-\frac{s}{s_3}\right) + \ldots\right] \\
&+ \tilde{q}(6)P^6\left(1-\frac{s}{s_3}\right)^6 \left[1 + \ldots\right] + O(1-s/s_3)^7
\end{aligned} \tag{47}$$

We collect terms multiplied by the same power $(1-s/s_3)^\mu$ with $\mu = 0, 1, 2, \ldots, 6$. The coefficients $e_{\nu\mu}$ used in Eqs.(40) to (45) must be distinguished from the new coefficients e_μ:

$$\begin{aligned}
\sum_{\nu=0}^{\infty} \tilde{q}(\nu)(-1)^\nu \left[1-\left(\frac{s}{s_3}\right)^P\right]^\nu &= \sum_{\nu=0}^{\infty} \tilde{q}(\nu)\left[\left(\frac{s}{s_3}\right)^P - 1\right]^\nu \\
&= e_0 + e_1\left(1-\frac{s}{s_3}\right) + e_2\left(1-\frac{s}{s_3}\right)^2 + e_3\left(1-\frac{s}{s_3}\right)^3 + \ldots \\
&= e_0 - e_1\left(\frac{s}{s_3}-1\right) + e_2\left(\frac{s}{s_3}-1\right)^2 - e_3\left(\frac{s}{s_3}-1\right)^3 + \ldots
\end{aligned}$$

$$
\begin{aligned}
e_0 &= \tilde{q}(0) \\
e_1 &= -\tilde{q}(1)P \\
e_2 &= -\tilde{q}(1)Pe_{11} + \tilde{q}(2)P^2 \\
e_3 &= -\tilde{q}(1)Pe_{12} + \tilde{q}(2)P^2e_{21} - \tilde{q}(3)P^3 \\
e_4 &= -\tilde{q}(1)Pe_{13} + \tilde{q}(2)P^2e_{22} - \tilde{q}(3)P^3e_{31} + \tilde{q}(4)P^4 \\
e_5 &= -\tilde{q}(1)Pe_{14} + \tilde{q}(2)P^2e_{23} - \tilde{q}(3)P^3e_{32} + \tilde{q}(4)P^4e_{41} - \tilde{q}(5)P^5 \\
e_6 &= -\tilde{q}(1)Pe_{15} + \tilde{q}(2)P^2e_{24} - \tilde{q}(3)P^3e_{33} + \tilde{q}(4)P^4e_{42} - \tilde{q}(5)P^5e_{51} + \tilde{q}(6)P^6
\end{aligned}
\tag{48}
$$

Equation (25) may be written as follows with the help of Eqs.(39) and (48):

$$
\begin{aligned}
w_P\left(\frac{s}{s_3}-1\right) = P^{\tilde{p}}\left(\frac{s}{s_3}-1\right)^{\tilde{p}}\left[1-d_1\left(\frac{s}{s_3}-1\right)+d_2\left(\frac{s}{s_3}-1\right)^2-d_3\left(\frac{s}{s_3}-1\right)^3+\dots\right] \\
\times\left[e_0 - e_1\left(\frac{s}{s_3}-1\right) + e_2\left(\frac{s}{s_3}-1\right)^2 - e_3\left(\frac{s}{s_3}-1\right)^3 + \dots\right]
\end{aligned}
$$

$$
\begin{aligned}
&= P^{\tilde{p}}\left(\frac{s}{s_3}-1\right)^{\tilde{p}}\Bigg[e_0 \\
&- (e_1 + e_0d_1)\left(\frac{s}{s_3}-1\right) \\
&+ (e_2 + e_1d_1 + e_0d_2)\left(\frac{s}{s_3}-1\right)^2 \\
&- (e_3 + e_2d_1 + e_1d_2 + e_0d_3)\left(\frac{s}{s_3}-1\right)^3 \\
&+ (e_4 + e_3d_1 + e_2d_2 + e_1d_3 + e_0d_4)\left(\frac{s}{s_3}-1\right)^4 \\
&- (e_5 + e_4d_1 + e_3d_2 + e_2d_3 + e_1d_4 + e_0d_5)\left(\frac{s}{s_3}-1\right)^5 \\
&+ (e_6 + e_5d_1 + e_4d_2 + e_3d_3 + e_2d_4 + e_1d_5 + e_0d_6)\left(\frac{s}{s_3}-1\right)^6\Bigg] \\
&\qquad + O(s/s_3-1)^7
\end{aligned}
\tag{49}
$$

With the final transformation $s/s_3 - 1 = s_3^{-1}(s - s_3)$ and $w(s/s_3 - 1) = w(s - s_3) = w(s)$ we bring Eq.(49) into the form of Eq.(5.5-15):

$$w_{\mathrm{P}}(s-s_3) = w_{\mathrm{P}}(s) = \sum_{\nu=0}^{\infty} \tilde{q}_{\mathrm{P}}(\nu)(s-s_3)^{\tilde{p}+\nu} \doteq \sum_{\nu=0}^{N} \tilde{q}_{\mathrm{P}}(\nu)(s-s_3)^{\tilde{p}+\nu}$$

$$\begin{aligned} \tilde{q}_{\mathrm{P}}(0) &= +P^{\tilde{p}} s_3^{-\tilde{p}} e_0 \\ \tilde{q}_{\mathrm{P}}(1) &= -q_{\mathrm{P}}(0)(e_1 + e_0 d_1)/s_3 \\ \tilde{q}_{\mathrm{P}}(2) &= +q_{\mathrm{P}}(0)(e_2 + e_1 d_1 + e_0 d_1)/s_3^2 \\ &\vdots \end{aligned} \tag{50}$$

The upper limit of the sum was reduced from ∞ to N since an arbitrarily large but finite interval can have only a finite number of arbitrarily small but finite subintervals.

Let us check our result. The constant $q_{\mathrm{P}}(0)$ is of little interest since it can be chosen. For $q_{\mathrm{P}}(1)$ we get from Eqs.(29), (39), and (48) the components

$$e_1 = -\tilde{q}(1)P,\ e_0 = \tilde{q}(0),\ d_1 = \tilde{p}c_1,\ c_1 = \frac{P-1}{2!} \tag{51}$$

Equations (19)–(21) yield:

$$\begin{aligned} \frac{\tilde{q}(1)}{\tilde{q}(0)} &= -\frac{\alpha_{\mathrm{P},0}(0)}{\alpha_{\mathrm{P},1}(0)} \\ &= -\frac{P^{-1}\tilde{p}\left(s_3P^{-1}(\tilde{p}-1+2P)-(s_3-s_2)(\tilde{p}-1)\dfrac{P-1}{2!P}\right)-l(l+1)+\tilde{\gamma}^2}{P^{-1}(s_3-s_2)(\tilde{p}+1)} \end{aligned} \tag{52}$$

From Eqs.(50) and (51) we get:

$$\tilde{q}_{\mathrm{P}}(1) = -\tilde{q}_{\mathrm{P}}(0)\tilde{q}(0)\left(\frac{\tilde{q}(1)}{\tilde{q}(0)}\frac{P}{s_3} + \frac{\tilde{p}}{s_3}\frac{P-1}{2!}\right) \tag{53}$$

We reduce the choosable constant $q_{\mathrm{P}}(0)\tilde{q}(0)$ to $q_{\mathrm{P}}(0)$ by choosing $\tilde{q}(0) = 1$:

$$\frac{\tilde{q}_{\mathrm{P}}(1)}{\tilde{q}_{\mathrm{P}}(0)} = -\left(\frac{\tilde{p}[s_3P^{-1}(\tilde{p}-1+2P)-(s_3-s_2)(\tilde{p}-1)(P-1)/2P]-P[l(l+1)-\tilde{\gamma}^2]}{P^{-1}s_3(s_3-s_2)(\tilde{p}+1)} + \frac{\tilde{p}}{s_3}\frac{P-1}{2!}\right) \tag{54}$$

For $P = 1$, $\tilde{p} = p$ we again obtain Eq.(5.5-18):

$$\frac{\tilde{q}_{\mathrm{P}}(1)}{\tilde{q}_{\mathrm{P}}(0)} = -\frac{s_3p(p+1) - l(l+1) + \tilde{\gamma}^2}{s_3(s_3-s_2)(p+1)} = \frac{q_3(1)}{q_3(0)} \tag{55}$$

6.10 Conformal Mapping for Section 5.10

According to Fig.5.10-1b we must use conformal mapping if a power series in the point s_3 is to converge as far as $s = 0$ for $g > -2\cos\pi/6 = -1.73205$. We start from Eq.(6.8-1) and use the substitutions of Eq.(6.9-1). In principle we can use all the equations of Section 6.9 that do not specify that s_3 is a real number, but we must write $\hat{p}$ and $\hat{q}$ rather than $\tilde{p}$ and $\tilde{q}$. Furthermore we recognize the relation $s_2 = s_3^*$ from Fig.5.10-1a and Eq.(5.10-1). Hence, in Section 6.9 we go directly to Eq.(6.9-16)

$$\begin{aligned}\hat{p} &= P\left(\frac{\tilde{\lambda}\tilde{\delta}}{s_3 - s_3^*} - 1\right) + 1, \quad s_2 = s_3^*,\ s_3 = e^{-i\kappa'\Delta r} \\ &= P\left(i\frac{\tilde{\lambda}\tilde{\delta}}{2\kappa'\Delta r} - 1\right) + 1 \end{aligned} \tag{1}$$

and Eq.(6.9-17):

$$\tilde{\lambda}\tilde{\delta} = -2i\kappa'\Delta r[P^{-1}(\hat{p} - 1) + 1] \tag{2}$$

Equation (6.9-18) assumes the form:

$$\begin{aligned}(2s_3 - s_3^*)P^{-2}\hat{q}(0)\hat{p}(\hat{p} - 1) + (s_3 - s_3^*)P^{-1}[\hat{q}(1)(\hat{p} + 1)\hat{p} - b_{11}\hat{q}(0)\hat{p}(\hat{p} - 1)] \\ + (4s_3 + g - \tilde{\lambda}\tilde{\delta})P^{-1}\hat{q}(0) + (2s_3 + g - \tilde{\lambda}\tilde{\delta})\hat{q}(1)(\hat{p} + 1) \\ - [l(l + 1) - \tilde{\gamma}^2]\hat{q}(0) = 0 \\ b_{11} = (P - 1)/2!P \end{aligned} \tag{3}$$

Using Eq.(2) we may bring this equation into the following form:

$$\alpha_{\mathrm{P},1}(0)\hat{q}(1) + \alpha_{\mathrm{P},0}(0)\hat{q}(0) = 0 \tag{4}$$

$$\alpha_{\mathrm{P},1}(0) = P^{-1}(s_3 - s_3^*)(\hat{p} + 1) \tag{5}$$

$$\begin{aligned}\alpha_{\mathrm{P},0}(0) = P^{-1}\hat{p}\left(s_3 P^{-1}(\hat{p} - 1 + 2P) - \frac{P - 1}{2!P}(s_3 - s_3^*)(\hat{p} - 1)\right) \\ - l(l + 1) + \tilde{\gamma}^2 \end{aligned} \tag{6}$$

As in Section 6.9 we do not attempt to work out the formulas for the higher coefficients $\hat{q}(2)$, $\hat{q}(3)$, ...

The inverse transformation from the ζ-plane to the s-plane follows Eq.(6.9-25). We have only to replace $\tilde{p}$, $\tilde{q}$ by $\hat{p}$, $\hat{q}$:

$$w(\zeta^{1/P}-1)=\left(\frac{(s/s_3)^P-1}{s/s_3-1}\right)^{\hat{p}}\left(\frac{s}{s_3}-1\right)\sum_{\nu=0}^{\infty}\hat{q}(\nu)(-1)^{\nu}\left[1-\left(\frac{s}{s_3}\right)^P\right]^{\nu} \tag{7}$$

For the evaluation of this equation we go from Eq.(6.9-25) to (6.9-39):

$$\begin{aligned}\left(\frac{1-(s/s_3)^P}{1-s/s_3}\right)^{\hat{p}}&=P^{\hat{p}}\left[1-d_1\left(\frac{s}{s_3}-1\right)+d_2\left(\frac{s}{s_3}-1\right)^2-d_3\left(\frac{s}{s_3}-1\right)^3+\dots\right]\\ d_1&=\hat{p}c_1 \qquad\qquad \text{see Eq.(6.9-29) for } c_1,\ c_2,\dots\\ d_2&=\hat{p}c_2+\frac{1}{2!}\hat{p}(\hat{p}-1)c_1^2\\ &\vdots\end{aligned} \tag{8}$$

The sum of Eq.(7) is worked out in Eq.(6.9-48):

$$\begin{aligned}\sum_{\nu=0}^{\infty}\hat{q}(\nu)(-1)^{\nu}\left[1-\left(\frac{s}{s_3}\right)^P\right]^{\nu}&=\sum_{\nu=0}^{\infty}\hat{q}(\nu)\left[\left(\frac{s}{s_3}\right)^P-1\right]^{\nu}\\ &=e_0-e_1\left(\frac{s}{s_3}-1\right)+e_2\left(\frac{s}{s_3}-1\right)^2-e_3\left(\frac{s}{s_3}-1\right)^3+\dots\\ e_0&=\hat{q}(0)\\ e_1&=-\hat{q}(1)P\\ &\vdots\end{aligned} \tag{9}$$

For the whole Eq.(7) we obtain from Eq.(6.9-50):

$$\begin{aligned}w(s-s_3)=w(s)&=\sum_{\nu=0}^{\infty}\hat{q}_{\mathrm{P}}(\nu)(s-s_3)^{\hat{p}+\nu}\doteq\sum_{\nu=0}^{N}\hat{q}_{\mathrm{P}}(\nu)(s-s_3)^{\hat{p}+\nu}\\ \hat{q}_{\mathrm{P}}(0)&=+P^{\hat{p}}s_3^{-\hat{p}}e_0\\ \hat{q}_{\mathrm{P}}(1)&=-q_{\mathrm{P}}(0)(e_1+e_0d_1)/s_3\\ &\vdots\end{aligned} \tag{10}$$

The upper limit of the sum was reduced from ∞ to N since an arbitrarily large but finite interval can have only a finite number of arbitrarily small but finite subintervals.

The constant $q_{\rm P}(0)$ is of little interest since it can be chosen. For $g_{\rm P}(1)$ we get from Eqs.(9), (8), and (6.9-29):

$$e_1 = -\hat{q}(1)P,\ e_0 = \hat{q}(0),\ d_1 = \hat{p}c_1,\ c_1 = \frac{P-1}{2!} \tag{11}$$

Equations (4)–(6) yield:

$$\begin{aligned}\frac{\hat{q}(1)}{\hat{q}(0)} &= \frac{\alpha_{\rm P,0}(0)}{\alpha_{\rm P,1}(0)}\\ &= -\frac{P^{-1}\hat{p}\left(s_3P^{-1}(\hat{p}-1+2P)-(s_3-s_3^*)(\hat{p}-1)\dfrac{P-1}{2!P}\right)-l(l+1)+\tilde{\gamma}^2}{P^{-1}(s_3-s_3^*)(\hat{p}+1)}\end{aligned} \tag{12}$$

From Eqs.(10) and (11) we get:

$$\hat{q}_{\rm P}(1) = -\hat{q}_{\rm P}(0)\hat{q}(0)\left(\frac{\hat{q}(1)}{\hat{q}(0)}\frac{P}{s_3} + \frac{\hat{p}}{s_3}\frac{P-1}{2!}\right) \tag{13}$$

We choose $\hat{q}(0) = 1$ to reduce $q_{\rm P}(0)\hat{q}(0)$ to $q_{\rm P}(0)$:

$$\begin{aligned}\frac{\hat{q}_{\rm P}(1)}{\hat{q}_{\rm P}(0)} = -\Bigg(&\frac{\hat{p}[s_3P^{-1}(\hat{p}-1+2P)-(s_3-s_3^*)(\hat{p}-1)(P-1)/2P]-P[l(l+1)-\tilde{\gamma}^2]}{P^{-1}s_3(s_3-s_3^*)(\hat{p}+1)}\\ &+\frac{\hat{p}}{s_3}\frac{P-1}{2!}\Bigg)\end{aligned} \tag{14}$$

For $P = 1$ we obtain the simpler equation:

$$\begin{aligned}\frac{\hat{q}_{\rm P}(1)}{\hat{q}_{\rm P}(0)} &= -\frac{s_3p'(p'+1)-l(l+1)+\tilde{\gamma}^2}{s_3(s_3-s_3^*)(p'+1)}\\ p' &= \frac{\tilde{\lambda}\tilde{\delta}}{2\kappa'\Delta r},\ s_3-s_3^* = -2i\kappa'\Delta r,\ s_3 \doteq 1-i\kappa'\Delta r,\ P=1\end{aligned} \tag{15}$$

References and Bibliography

Abramowitz, M. and Stegun, I.A., eds. (1964). *Handbook of Mathematical Functions*, National Bureau of Standards, Applied Mathematics Series 55. US Government Printing Office, Washington, DC.

Anastasovski, P.K., Bearden, T.E., Ciubotariu, C., Coffey, W.T., Crowell, L.B., Evans, G.J., Evans, M.W., Flower, R., Jeffers, S., Labounsky, A., Lehnert, B., Mészaŕos, M., Molnár, P.R., Vigier, J.-P., and Roy, S. (2001). Empirical evidence for non-Abelian electrodynamics and theoretical development. *Annales Fondation Louis de Broglie*, vol. 26, no. 4, 653–672.

Apostle, H.G. (1969). *Aristotle's Physics, Translated with Commentaries and Glossary*. Indiana University Press, Bloomington.

Aristotle (1930). *The Works of Aristotle, vol. II, Physics*. R.R. Hardie and R.K. Gaye transl., Clarendon Press, Oxford.

Barrett, T.W. (1993). Electromagnetic phenomena not explained by Maxwell's equations. *Essays on the Formal Aspects of Electromagnetic Theory*, A. Lakhtakia, ed., pp. 6–86. World Scientific Publishing Co., Singapore.

Barrett, T.W. and Grimes, D.M. (Editors). (1996). *Advanced Electromagnetism: Foundations, Theory, and Applications*. World Scientific Publishing Co., Singapore.

Becker, R. (1963). *Theorie der Elektrizität*, vol. 2 (revised by G. Leibfried and W. Breuig), 9th ed. Teubner, Stuttgart.

Becker, R. (1964). *Electromagnetic Fields and Interactions* (transl. by A.W. Knudsen of vol. 1, 16th ed. and by I. de Teissier of vol. 2, 8th ed. of *Theorie der Elektrizität*). Blaisdell, New York. Reprinted 1982 by Dover, New York.

Belfrage, C. (1954). *Seeds of Destruction. The Truth About the US-Occupation of Germany*. Cameron and Kahn, New York. Copyrighted by the Library of Congress but not listed in the online catalog www.loc.gov either under the name Belfrage, Cedric or the book title. Available at Deutsche Staatsbibliothek, Berlin, Signatur 8-29 MA 265 or at Bayrische Staatsbibliothek, München, Germany, Signatur BAY: L 1089.

Berestezki, W.B., Lifschitz, E.M., and Pitajewski, L.P. (1970). *Relativistische Quantentheorie*, vol. IVa of *Lehrbuch der Theoretischen Physik*, L.D. Landau and E.M. Lifschitz eds.; transl. from Russian. Akademie Verlag, Berlin.

Berestetskii, V.B., Lifshitz, E.M., and Pitajevskii, L.P. (1982). *Quantum Electrodynamics*, transl. from Russian. Pergamon Press, New York.

Blum, J.M. (1965). *From the Morgenthau Diaries, vol. 2: Years of Urgency, 1938–1941*. Houghton Mifflin Co., Boston.

Blum, J.M. (1967). *From the Morgenthau Diaries, vol. 3: Years of War˙1941–45*. Houghton Mifflin Co., Boston. German translation by U. Heinemann and I. Goldschmidt: *Deutschland ein Ackerland? (The Morgenthau Diaries)*. Droste Verlag und Druckerei, Düsseldorf 1968.

Bub, J. (1999). *Interpreting the Quantum World*. Cambridge University Press, Cambridge.

Callender, C. and Hugget, N. (2000). *Physics Meets Philosophy at the Planck Scale*. Cambridge University Press, Cambridge.

Cao, T.Y. (1999). *Conceptual Foundations of Quantum Field Theory*. Cambridge University Press, Cambridge.

Chuang, I.L. and Nielsen, M.A. (2000). *Quantum Computation and Information.* Cambridge University Press, Cambridge.

Cole, E.A.B. (1970). Transition from continuous to a discrete space–time scheme. *Nuovo Cimento*, vol. 66A, 645–655.

Cole, E.A.B. (1973a). Perception and operation in the definition of observables. *Int. J. Theor. Phys.*, 8, 155–170.

Das, A. (1966). The quantized complex space-time and quantum theory of free fields. *J. Math. Phys.*, 7 , 52–60.

Dickson, W.M. (1998). *Quantum Chance and Nonlocality.* Cambridge University Press, Cambridge.

Einstein, A. and Infeld, L. (1938). *The Evolution of Physics.* Simon and Schuster, New York.

Elaydi, S.N. (1999). *An Introduction to Difference Equations*, 2nd ed. Springer-Verlag, New York.

Flint, H.T. (1948). The quantization of space and time. *Phys. Rev.*, vol. 74, 209–210.

Frobenius, G. (1873). Über die Integration der linearen Differentialgleichungen durch Reihen. *Journal für die reine und angewandte Mathematik*, vol. LXXVI, 214–235.

Fuchs, L. (1866). Zur Theorie der linearen Differentialgleichungen mit veränderlichen Coefficienten. *Journal für die reine und angewandte Mathematik*, vol. LXVI, 121–160; (1868) vol. LXVIII, 354–385.

Gelfond, A.O. (1958). *Differenzenrechnung* (transl. of Russian original, Moscow 1952). Deutscher Verlag der Wissenschaften, Berlin. French edition 1963.

Ghose, P. (1999). *Testing Quantum Mechanics on New Ground.* Cambridge University Press, Cambridge.

Gradshteyn, I.S. and Ryzhik, I.M. (1980). *Tables of Integrals, Series, and Products.* Academic Press, New York.

Greiner, B. (1995). *Die Morgenthau-Legende.* Hamburger Edition HIS Verlagsgesellschaft, Hamburg. References 133 books and 39 journal articles.

Griffiths, D.J. (1989). *Introduction to Electrodynamics.* Prentice-Hall, Englewood Cliffs, NJ.

Guldberg, A. and Wallenberg, G. (1911). *Theorie der Linearen Differenzengleichungen.* B.G. Teubner, Leipzig

Habermann, R. (1987). *Elementary Applied Partial Differential Equations with Fourier Series and Boundary Value Problems*, 2nd ed., Prentice Hall, Englewood Cliffs, NJ.

Harmuth, H.F. (1986a). *Propagation of Nonsinusoidal Electromagnetic Waves.* Academic Press, New York.

Harmuth, H.F. (1986b,c). Correction of Maxwell's equations for signals I, II. *IEEE Trans. Electromagn. Compat.*, vol. EMC 28, 250–258, 259–266

Harmuth, H.F., (1986d). Propagation velocity of electromagnetic waves. *IEEE Trans. Electromagn. Compat.*, vol. EMC-28, 267–271.

Harmuth, H.F. (1989). *Information Theory Applied to Space-Time Physics* (in Russian). MIR, Moscow. English edition by World Scientific Publishing Co., Singapore 1992.

Harmuth, H.F., Barrett, T.W., and Meffert, B. (2001). *Modified Maxwell Equations in Quantum Electrodynamics.* World Scientific Publishing Co., Singapore. Copies distributed to university libraries in Eastern Europe, Asia, North Africa, and Latin Amrica contain one extra page.

Harmuth, H.F. and Hussain, M.G.M. (1994). *Propagation of Electromagnetic Signals.* World Scientific Publishing Co., Singapore.

Harmuth, H.F. and Meffert, B. (2003). *Calculus of Finite Differences in Quantum Electrodynamics.* Elsevier/Academic Press (Advances in Imaging and Electron Physics, vol. 129), London.

Hasebe, K. (1972). Quantum space–time. *Prog. Theor. Phys.*, vol. 48, 1742–1750.

Hellund, E.J. and Tanaka, K. (1954). Quantized space–time. *Phys. Rev.*, vol. 94, 192–195.

Hill, E.L. (1955). Relativistic theory of discrete momentum space and discrete space–time. *Phys. Rev.*, vol. 100, 1780–1783.

Hillion, P. (1991). Remarks on Harmuth's 'Correction of Maxwell's equations for signals I'. *IEEE Trans. Electromagn. Compat.*, vol. EMC-33, 144.

Hillion, P. (1992a). Response to 'The magnetic conductivity and wave propagation'. *IEEE Trans. Electromagn. Compat.*, vol. EMC-34, 376–377.

Hillion, P. (1992b). A further remark on Harmuth's problem. *IEEE Trans. Electromagn. Compat.*, vol. EMC-34, 377–378.

Hillion, P. (1993). Some comments on electromagnetic signals; in *Essays on the Formal Aspects of Electromagnetic Theory.* A. Lakhtakia ed., 127–137. World Scientific Publishing Co., Singapore.

Hölder, O. (1887). Über die Eigenschaft der Gammafunktion keiner algebraischen Differentialgleichung zu genügen. *Math. Annalen*, vol. 28, 1–13.

Hussain, M.G.M. (2005). Mathematical model for the electromagnetic conductivity of lossy materials. *J. of Electromagn. Waves and Appl.*, vol. 19, no. 2, 271-279; VSP Publishers, Zeist, Netherlands.

Icke, V. (1995). *The force of Symmetry.* Cambrdidge University Press, Cambridge.

Jackson, J.D. (1999). *Classical Electrodynamics*, 3rd ed.. J.Wiley, New York.

King, R.W. (1993). The propagation of a Gaussian pulse in seawater and its application to remote sensing. *IEEE Trans. Geoscience, Remote Sensing*, vol. 31, 595–605.

King, R.W. and Harrison, C.W. (1968). The transmission of electromagnetic waves and pulses into the Earth. *J. Appl. Physics*, vol. 39, 4444–4452.

Landau, L.D. and Lifschitz, E.M. (1966). *Lehrbuch der Theoretischen Physik* (German edition of Russian original by G. Heber). Akademie-Verlag, Berlin.

Leader, E. and Predazzi, E. (1996). *An Introduction to Gauge Theories and Modern Particle Physics.* Cambridge University Press, Cambridge.

Lehner, G. (1990). *Elektromagnetische Feldtheorie.* Springer Verlag, Berlin.

Lehnert, B. (1995). Total reflection process including waves of an extended electromagnetic theory. *Optik*, vol. 99, No. 1, 113–119.

Lehnert, B. (1996). Minimum electric charge of an extended electromagnetic field theory. *Physica Scripta*, vol. 53, 204–211.

Lehnert, B. and Roy, S. (1998). *Extended Electromagnetic Theory.* World Scientific Publishing Co., Singapore.

Levy, H. and Lessman, F. (1961). *Finite Difference Equations.* Macmillan, New York; reprinted Dover, New York.

Marsal, F. (1989). *Finite Differenzen und Elemente: numerische Lösung von Variationsproblemen und Partiellen Differentialgleichungen.* Springer-Verlag, Berlin.

Messiah, A. (1959). *Mécanique Quntique.* Dunod, Paris. English transl by G.M. Temmer, *Quantum Mechanics.* North Holland Publishing Co., Amsterdam 1962. Reprinted Dover, New York.

Milne-Thomson, L.M. (1951). *The Calculus of Finite Differences.* MacMillan, London.

Moon, P. and Spencer, D.E. (1988). *Field Theory Handbook*, 2nd ed., Springer-Verlag, New York.

Morgenthau, H. (1945). *Germany Is Our Problem.* Harper & Brothers, New York and London. See also Belfrage, C., Blum, J.M., Greiner, B. Free download from www.ety.com/berlin/morgenthau.htm .

Morse, P.M and Feshbach, H. (1953). *Methods of Theoretical Physics.* McGraw-Hill, New York.

Newton, I. (1971). *Mathematical Principles.* A. Motte transl., F. Cajori ed. University of California Press, Berkeley.

New York Times (1945a). *Morgenthau Gives Views on Germany.* Oct. 5, p. 7.

New York Times (1945b). *Morgenthau Volume Sent to U.S. Staffs.* Dec. 30, p. 10.

New York Times (1945c). *Nazi Reform Slow, Morgenthau Says*, Nov. 12, p. 3.

New York Times (1945d). *Engineers Offer Plan for Germany.* Sep. 27, p. 14.

Nielsen, N. (1906). *Handbuch der Theorie der Gammafunktion.* Teubner, Leipzig.

Nörlund, N.E. (1910). Fractions continues et différences réciproques. *Acta Math.*, **34**, 1–108.

Nörlund, N.E. (1914). Sur les séries de facultés. *Acta Math.*, **37**, 327–387.

Nörlund, N.E. (1915). Sur les équations linéaires aux différences finis à coefficients rationels. *Acta Math.*, **40**, 191–249.

Nörlund, N.E. (1924). *Vorlesungen über Differenzenrechnung.* Springer-Verlag, Berlin; reprinted Chelsea Publishing Co., New York 1954.

Nörlund, N.E. (1929). *Leçons sur les Équations Linéaires aux Différences Finies*, ed. R. Lagrange. Gauthier-Villars, Paris.

Ostrowski, A. (1919). Neuer Beweis des Hölderschen Satzes, daß die Gammafunktion keiner algebraischen Differentialgleichung genügt. *Math. Annalen*, vol. 79, 286–288.

Ptolemy, C. (1952). *The Almagest*, by Ptolemy, R.C. Taliaferro transl.; *On the Revolution of the Heavenly Spheres*, by N. Copernicus, C.G. Wallis transl.; *The Harmonies of the World*, by J. Kepler, C.G. Wallis transl. Encyclopedia Britannica, Chicago.

Richtmyer, R.D. (1957). *Difference Methods for Initial Value Problems.* Wiley Interscience, New York.

Rohrlich, F. (1990). *Classical Charged Particles.* Addison-Wesley, Reading, MA.

Schiff, L.I. (1949). *Quantum Mechanics.* McGraw-Hill, New York.

Schild, A. (1949). Discrete space–time and integral Lorentz transformations. *Canadian J. Math.*, vol. 1, 29–47.

Schrödinger, E. (1950). *Space-Time Structure.* University Press, Cambridge, Great Britain.

Schrödinger, E. (1956). Causality and wave mechanics, in *The World of Mathematics*, vol. 2, 1056–1068. Simon and Schuster, New York.

Simon, E. (1945). *Mr.Morgenthau's Program for Removing the German Menace.* New York Times Book Review, Oct. 7, p. BR2.

Smirnov, W.I. (1961). *Lehrgang der höheren Mathematik*, 4th ed. (transl. of the 12th Russian edition). Deutscher Verlag der Wissenschaften, Berlin. English edition: *A Course in Higher Mathematics.* Pergamon Press, Oxford 1964.

Smith, G.D. (1982). *Theory and Problems of Calculus of Finite Differences and Difference Equations.* McGraw-Hill, New York.

Snyder, H.S. (1947a). Quantized space–time. *Phys. Rev.*, vol. 71, 38–41.

Snyder, H.S. (1947b). The electromagnetic field in quantized space-time. *Phys. Rev.*, vol. 72, 68–71.

Spiegel, M.R. (1994). *Theory and Problems of Calculus of Finite Differences and Difference Equations*; reprint of edition of 1971. McGraw-Hill (Schaum's Outline Series), New York. *Endliche Differenzen und Differenzengleichungen*; transl., McGraw-Hill, New York 1982. A basic introduction containing useful formulas.

Stratton, J.A. (1941). *Electromagnetic Theory.* McGraw-Hill, New York.

van der Waerden, B.L. (1966). *Algebra.* Springer-Verlag, Berlin and New York.

Wei, Y. and Zhang, Q. (2000). *Common Waveform Analysis.* Kluwer Academic Publishers, Boston.

Weinberg, S. (1995). *The Quantum Theory of Fields: I. Introduction.* Cambridge University Press, Cambridge.

Weinberg, S. (1996). *The Quantum Theory of Fields: II. Modern Applications.* Cambridge University Press, Cambridge.

Weinberg, S. (2000). *The Quantum Theory of Fields: III. Supersymmetry.* Cambridge University Press, Cambridge.

Weyl, H. (1921). Das Raumproblem. *Jahresberichte der Deutschen Mathematikervereinigung*, vol. 30, 92–93.

Weyl, H. (1968). *Gesammelte Abhandlungen.* Springer-Verlag, Berlin.

Welch, L.C. (1976). Quantum mechanics in a discrete space–time. *Nuovo Cimento*, vol. 31B, 279–288.

Yukawa, H. (1966). Atomistics and the divisibility of space and time. *Suppl. Prog. Theor. Phys.*, vol. 37–38, 512–523.

Index